Werner Kuich   Grzegorz Rozenberg
Arto Salomaa (Eds.)

Developments in Language Theory

5th International Conference, DLT 2001
Wien, Austria, July 16-21, 2001
Revised Papers

Springer

Series Editors

Gerhard Goos, Karlsruhe University, Germany
Juris Hartmanis, Cornell University, NY, USA
Jan van Leeuwen, Utrecht University, The Netherlands

Volume Editors

Werner Kuich
Technische Universität Wien, Institut für Algebra und Computermathematik
Abt. für Theoretische Informatik
Wiedner Hauptstraße 8-10, 1040 Wien, Austria
E-mail: kuich@tuwien.ac.at

Grzegorz Rozenberg
Leiden University, Department of Computer Science
P.O. Box 9512, 2300 RA Leiden, The Netherlands
rozenber@wi.leidenuniv.nl

Arto Salomaa
Turku Centre for Computer Science
Lemminkaisenkatu 14A, 20520 Turku, Finland
asalomaa@utu.fi

Cataloging-in-Publication Data applied for

Die Deutsche Bibliothek - CIP-Einheitsaufnahme

Developments in language theory : 5th international conference ; revised
papers / DLT 2001, Wien, Austria, July 16 - 21, 2001. Werner Kuich ...
(ed.). - Berlin ; Heidelberg ; New York ; Barcelona ; Hong Kong ; London ;
Milan ; Paris ; Tokyo : Springer, 2002
  (Lecture notes in computer science ; Vol. 2295)
  ISBN 3-540-43453-4

CR Subject Classification (1998): F.4.3, F.4.2, F.4, F.3, F.1, G.2

ISSN 0302-9743
ISBN 3-540-43453-4 Springer-Verlag Berlin Heidelberg New York

This work is subject to copyright. All rights are reserved, whether the whole or part of the material is
concerned, specifically the rights of translation, reprinting, re-use of illustrations, recitation, broadcasting,
reproduction on microfilms or in any other way, and storage in data banks. Duplication of this publication
or parts thereof is permitted only under the provisions of the German Copyright Law of September 9, 1965,
in its current version, and permission for use must always be obtained from Springer-Verlag. Violations are
liable for prosecution under the German Copyright Law.

Springer-Verlag Berlin Heidelberg New York
a member of BertelsmannSpringer Science+Business Media GmbH

http://www.springer.de

© Springer-Verlag Berlin Heidelberg 2002
Printed in Germany

Typesetting: Camera-ready by author, data conversion by PTP-Berlin, Stefan Sossna e.K.
Printed on acid-free paper   SPIN: 10846416   06/3142   5 4 3 2 1 0

QA 66.5 LEC/2295

Lecture Notes in Computer Science
Edited by G. Goos, J. Hartmanis, and J.

QM Library

23 1204664 4

WITHDRAWN FROM STOCK QMUL LIBRARY

DATE DUE FOR RETURN

NEW ACCESSIONS

Springer
*Berlin
Heidelberg
New York
Barcelona
Hong Kong
London
Milan
Paris
Tokyo*

Preface

DLT 2001 was the fifth Conference on Developments in Language Theory. It was a broadly based conference covering all aspects of Language Theory: grammars and acceptors for strings, graphs, arrays, etc.; efficient algorithms for languages; combinatorial and algebraic properties of languages; decision problems; relations to complexity theory; logic; picture description and analysis; DNA computing; cryptography; concurrency.

DLT 2001 was held at Technische Universität Wien from July 16 to July 21, 2001. The Organizing Committee consisted of Rudolf Freund, Werner Kuich (chairman), Christiane Nikoll, Margarethe Soukup, Friedrich Urbanek.

Previous DLTs were held in Turku (1993), Magdeburg (1995), Thessalonike (1997), Aachen (1999).

The Program Committee of DLT 2001 consisted of Christian Choffrut (Paris), Jürgen Dassow (Magdeburg), Masami Ito (Kyoto), Werner Kuich (Wien, chairman), Giancarlo Mauri (Milano), Gheorghe Păun (Bucureşti), Grzegorz Rozenberg (Leiden), Arto Salomaa (Turku), Wolfgang Thomas (Aachen). It selected 24 papers from 64 papers submitted in response to the call for papers. These papers came from the following countries: Australia, Austria, Belgium, Brazil, Bulgaria, Canada, Czech Republic, Estonia, Finland, France, Germany, Hungary, India, Italy, Japan, Moldova, The Netherlands, Phillipines, Poland, Romania, Russia, Slovakia, Spain, United Kingdom, USA. Each submitted paper was evaluated by at least four members of the Program Committee. Together with 10 invited presentations all 24 selected papers are contained in this volume. The papers in this volume are printed according to the order of presentation at DLT 2001 and thus grouped into sessions, most of which are thematic.

It is a pleasure for the editors to thank the members of the Program Committee for the evaluation of the papers, and the many referees who assisted in this process. We are grateful to the contributors of DLT 2001, in particular to the invited speakers, for their willingness to present interesting new developments in a condensed and accessible form.

Finally, we would like to express our thanks to those whose work behind the scenes made this volume possible: Rudolf Freund for his technical support during the submission period, Herbert Pötzl and Tatjana Svizensky for implementing the electronic submission tool, Margarethe Soukup for designing the websites, Alois Panholzer for compiling the proceedings, and Christiane Nikoll for the secretarial work connected with this conference.

January 2002

Werner Kuich
Grzegorz Rozenberg
Arto Salomaa

Contents

Invited Presentations

Automata: From Uncertainty to Quantum 1
  C.S. Calude, E. Calude (Auckland, New Zealand)

Elementary Theory of Ordinals with Addition and Left Translation by ω 15
  C. Choffrut (Paris, France)

The Equational Theory of Fixed Points with Applications to
Generalized Language Theory 21
  Z. Ésik (Szeged, Hungary)

Second-Order Logic over Strings: Regular and Non-regular Fragments 37
  T. Eiter, G. Gottlob (Wien, Austria), T. Schwentick (Jena, Germany)

Decision Questions on Integer Matrices 57
  T. Harju (Turku, Finland)

Some Petri Net Languages and Codes 69
  M. Ito (Kyoto, Japan), Y. Kunimoch (Fukuroi, Japan)

Words, Permutations, and Representations of Numbers 81
  H. Prodinger (Johannesburg, South Africa)

Proof Complexity of Pigeonhole Principles 100
  A.A. Razborov (Moscow, Russia)

Words and Patterns 117
  A. Restivo, S. Salemi (Palermo, Italy)

A Short Introduction to Infinite Automata 130
  W. Thomas (Aachen, Germany)

Contributions

The Power of One-Letter Rational Languages 145
  T. Cachat (Aachen, Germany)

The Entropy of Lukasiewicz-Languages 155
  L. Staiger (Halle, Germany)

Collapsing Words vs. Synchronizing Words 166
  D.S. Ananichev, M.V. Volkov (Ekaterinburg, Russia)

A Note on Synchronized Automata and Road Coloring Problem 175
  K. Culik (Columbia, USA), J. Karhumäki (Turku, Finland), J. Kari
  (Iowa City, USA)

Shuffle Quotient and Decompositions 186
  C. Câmpeanu, K. Salomaa (Kingston, Canada), S. Vágvölgyi
  (Szeged, Hungary)

The Growing Context-Sensitive Languages Are the
Acyclic Context-Sensitive Languages 197
  G. Niemann (Kassel, Germany), J.R. Woinowski (Darmstadt, Germany)

Recognizable Sets of N-Free Pomsets Are Monadically Axiomatizable 206
  D. Kuske (Leicester, United Kingdom)

Automata on Series-Parallel Biposets 217
  Z. Ésik, Z.L. Németh (Szeged, Hungary)

Hierarchies of String Languages Generated by Deterministic Tree Transducers . . . 228
  J. Engelfriet, S. Maneth (Leiden, The Netherlands)

Partially-Ordered Two-Way Automata: A New Characterization of DA 239
  T. Schwentick (Marburg, Germany), D. Thérien (Montréal, Canada),
  H. Vollmer (Würzburg, Germany)

Level 5/2 of the Straubing-Thérien Hierarchy for Two-Letter Alphabets 251
  C. Glaßer (Würzburg, Germany), H. Schmitz (München, Germany)

On the Power of Randomized Pushdown Automata 262
  J. Hromkovič (Aachen, Germany), G. Schnitger
  (Frankfurt am Main, Germany)

The Root of a Language and Its Complexity 272
  G. Lischke (Jena, Germany)

Valuated and Valence Grammars: An Algebraic View 281
  H. Fernau (Callaghan, Australia), R. Stiebe (Magdeburg, Germany)

Context-Free Valence Grammars – Revisited 293
  H.J. Hoogeboom (Leiden, The Netherlands)

An Undecidability Result Concerning Periodic Morphisms 304
  V. Halava, T. Harju (Turku, Finland)

A Universal Turing Machine with 3 States and 9 Symbols 311
  M. Kudlek (Hamburg, Germany), Y. Rogozhin (Chişinău, Moldova)

Minimal Covers of Formal Languages 319
  M. Domaratzki, J. Shallit (Waterloo, Canada), S. Yu (London, Canada)

Some Regular Languages That Are Church-Rosser Congruential 330
    G. Niemann (Kassel, Germany), J. Waldmann (Leipzig, Germany)

On the Relationship between the McNaughton Families of Languages and the
Chomsky Hierarchy 340
    M. Beaudry (Sherbrooke, Canada), M. Holzer (München, Germany),
    G. Niemann, F. Otto (Kassel, Germany)

Forbidden Factors and Fragment Assembly 349
    F. Mignosi, A. Restivo, M. Sciortino (Palermo, Italy)

Parallel Communicating Grammar Systems with
Incomplete Information Communication 359
    E. Csuhaj-Varjú, G. Vaszil (Budapest, Hungary)

Eliminating Communication by Parallel Rewriting 369
    B. Rovan, M. Slašt'an (Bratislava, Slovakia)

String Rewriting Sequential P-Systems and Regulated Rewriting 379
    P. Sosík (Opava, Czech Republic), R. Freund (Wien, Austria)

Author Index 389

To Professor G. Rozenberg

for His 60th Birthday

Automata: From Uncertainty to Quantum

Cristian S. Calude and Elena Calude

Department of Computer Science, University of Auckland, New Zealand
Institute of Information Sciences, Massey University at Albany, New Zealand
cristian@cs.auckland.ac.nz, e.calude@massey.ac.nz

Abstract. Automata are simple mathematical objects with unexpected computational, mathematical, modelling and explanatory capabilities. This paper examines some relations between automata and physics. Automata will be used to model quantum uncertainty and quantum computation. Finally, mathematical proofs will be discussed from the perspective of quantum automata.

1 Modelling Quantum Uncertainty with Automata

1.1 Moore Automata

All automata we are going to consider are *finite* in the sense that they have a finite number of states, a finite number of input symbols, and a finite number of output symbols. The deterministic or non-deterministic behaviour of such a machine will be contextually clear.

First we will look at *deterministic automata* each of which consists of a finite set S_A of states, an input alphabet Σ, and a transition function $\delta_A : S_A \times \Sigma \to S_A$. Sometimes a fixed state, say 1, is considered to be the *initial state*, and a subset of S_A denotes the *final states*. A *Moore automaton* is a deterministic automaton having an *output function* $F_A : S_A \to O$, where O is a finite set of output symbols. At each time the automaton is in a given state q and is continuously emitting the output $F_A(q)$. The automaton remains in state q until it receives an input signal σ, when it assumes the state $\delta(q, \sigma)$ and starts emitting $F_A(\delta_A(q, \sigma))$. In what follows $\Sigma = \{0, 1\}$ having $O = \Sigma$, so, from now on, a Moore automaton will be just a triple $A = (S_A, \delta_A, F_A)$.

Let Σ^* be the set of all finite sequences (words) over the alphabet Σ, including the empty word e. The transition function δ can be extended to a function $\overline{\delta}_A : S_A \times \Sigma^* \to S_A$, as follows: $\overline{\delta}_A(q, e) = q$, for all $q \in S_A$, $\overline{\delta}_A(q, \sigma w) = \overline{\delta}_A(\delta_A(q, \sigma), w)$, for all $q \in S_A, \sigma \in \Sigma, w \in \Sigma^*$.

The output produced by an experiment started in state q with input $w \in \Sigma^*$ is described by the *total response* of the automaton A, given by the function $R_A : S_A \times \Sigma^* \to \Sigma^*$ defined by $R_A(q, e) = f(q)$, $R_A(q, \sigma w) = f(q) R_A(\delta(q, \sigma), w), q \in S_A, \sigma \in \Sigma, w \in \Sigma^*$, and the output function f.

1.2 Moore's Uncertainty Revisited

Moore [38] has studied some experiments on deterministic automata trying to understand what kind of conclusions about the internal conditions of a machine it is possible to draw from input-output experiments. To emphasize the conceptual nature of his experiments, Moore has borrowed from physics the word "Gedanken".

A (simple) Moore experiment can be described as follows: a copy of a deterministic machine will be experimentally observed, i.e. the experimenter will input a finite sequence of input symbols to the machine and will observe the sequence of output symbols. The correspondence between input and output symbols depends on the particular chosen machine and on its initial state. The experimenter will study sequences of input and output symbols and will try to conclude that "the machine being experimented on was in state q at the beginning of the experiment".[1] Moore's experiments have been studied from a mathematical point of view by various researchers, notably by Ginsburg [27], Gill [26], Chaitin [17], Conway [20], Brauer [6], Salomaa [42].

Following Moore [38] we shall say that a state q is "indistinguishable" from a state q' (with respect to Moore's automaton $A = (S_A, \delta_A, F_A)$) if every experiment performed on A starting in state q produces the same outcome as it would starting in state q'. Formally, $R_A(q, x) = R_A(q', x)$, for all words $x \in \Sigma^+$. An equivalent way to express the indistinguishability of the states q and q' is to require, following Conway [20], that for all $w \in \Sigma^*$, $F_A(\overline{\delta}_A(q, w)) = F_A(\overline{\delta}_A(q', w))$.

A pair of states will be said to be "distinguishable" if they are not "indistinguishable", i.e. if there exists a string $x \in \Sigma^+$, such that $R_A(q, x) \neq R_A(q', x)$.

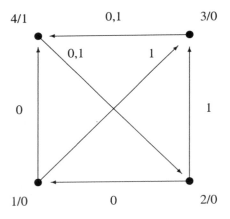

Fig. 1.

Moore [38] has proven the following important theorem: *There exists a Moore automaton A such that any pair of its distinct states are distinguishable, but there is no*

---
[1] This is often referred to as a *state identification experiment*.

experiment which can determine what state the machine was in at the beginning of the experiment. He used the automaton displayed in Figure 1 and the argument is simple. Indeed, each pair of distinct states can be distinguished by an experiment; however, there is no (unique) experiment capable to distinguish between every pair of arbitrary distinct states. If the *experiment starts with* 1, then x cannot distinguish between the states 1, 2 and if the *experiment starts with* 0, then x cannot distinguish between the states 1, 3.

Moore's theorem can be thought of as being a *discrete analogue* of the Heisenberg uncertainty principle. The state of an electron E is considered specified if both its velocity and its position are known. Experiments can be performed with the aim of answering either of the following:

1. What was the position of E at the beginning of the experiment?
2. What was the velocity of E at the beginning of the experiment?

For a Moore automaton, experiments can be performed with the aim of answering either of the following:

1. Was the automaton in state 1 at the beginning of the experiment?
2. Was the automaton in state 2 at the beginning of the experiment?

In either case, performing the experiment to answer question 1 changes the state of the system, so that the answer to question 2 cannot be obtained. This means that it is only possible to gain partial information about the previous history of the system, since performing experiments causes the system to "forget" about its past.

An exact quantum mechanical analogue has been given by Foulis and Randall [24, Example III]: Consider a device which, from time to time, emits a particle and projects it along a linear scale. We perform two experiments. In experiment α, the observer determines if there is a particle present. If there is not, the observer records the outcome of α as the outcome $\{4\}$. If there is, the observer measures its position coordinate x. If $x \geq 1$, the observer records the outcome $\{2\}$, otherwise $\{3\}$. A similar procedure applies for experiment β: If there is no particle, the observer records the outcome of β as $\{4\}$. If there is, the observer measures the x-component p_x of the particle's momentum. If $p_x \geq 1$, the observer records the outcome $\{1, 2\}$, otherwise the outcome $\{1, 3\}$. Still another quantum mechanical analogue has been proposed by Giuntini [28]. A pseudo-classical analogue has been proposed by Cohen [19] and by Wright [44].

Moore's automaton is a simple model featuring an "uncertainty principle" (cf. Conway [20]), later termed "computational complementarity" by Finkelstein and Finkelstein [23].

It would be misleading to assume that any automaton state corresponds to a *bona fide* element of physical reality (though, perhaps, hidden). Because, whereas in models of automaton complementarity it might still be possible to pretend that initially the automaton *actually is in a single automaton state*, which we just do not know (such a state can be seen if the automaton is "screwed open"), quantum mechanically this assumption leads to a Kochen-Specker contradiction [32,43].

Two non-equivalent concepts of computational complementarity based on automata have been proposed and studied in Calude, Calude, Svozil and Yu [15]. Informally, they can be expressed as follows. Consider the class of all elements of reality (or "properties", and "observables") and consider the following properties.

A Any two distinct elements of reality can be mutually distinguished by a suitably chosen measurement procedure, Bridgman [7].
B For any element of reality, there exists a measurement which distinguishes between this element and all the others. That is, a distinction between any one of them and all the others is operational.
C There exists a measurement which distinguishes between any two elements of reality. That is, a single pre-defined experiment operationally exists to distinguish between an arbitrary pair of elements of reality. (Classical case.)

It is easy to see that there exist automata with property **C**. More interestingly, there exist automata which have *CI* that is **A** but not **B** (and therefore not **C**) as well as automata with *CII*, i.e. **B** but not **C**. Properties *CI*, *CII* are called *complementarity principles*. Moore's automaton in Figure 1 has indeed *CI* . To get *CII* we can use again Moore's automaton but with different output functions, for example those in Figure 2:

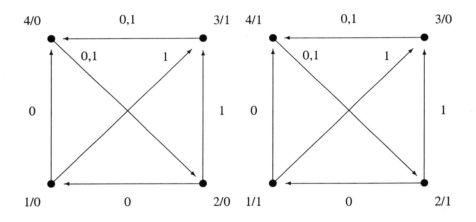

Fig. 2.

According to the philosophical view called realism, *reality* exists and has definite properties irrespective whether they are observed by some agent. Motivated by this view point, Einstein, Podolsky and Rosen [22] suggested a classical argument showing that quantum mechanics is incomplete. EPR assumed a) the non-existence of action-at-a-distance, b) that some of the statistical predictions of quantum mechanics are correct, and c) a reasonable criterion defining the existence of an element of physical reality. They considered a system of two spatially separated but quantum mechanically correlated particles. A "mysterious" feature appears: By counterfactual reasoning, quantum mechanical experiments yield outcomes which cannot be predicted by quantum theory; hence the quantum mechanical description of the system is incomplete!

One possibility to complete the quantum mechanical description is to postulate additional "hidden-variables" in the hope that completeness, determinism and causality will

q' upon reading symbol σ), the state q_0 is the initial configuration of the system, and F is the set of accepting states. For all states $q_1, q_2 \in Q$ and symbol $\sigma \in \Sigma$ the function δ must be unitary, thus satisfying the condition:

$$\sum_{q' \in S} \overline{\delta(q_1, \sigma, q')} \delta(q_2, \sigma, q') = \begin{cases} 1, & \text{if } q_1 = q_2, \\ 0, & \text{otherwise.} \end{cases}$$

The end-marker $ is assumed to be the last symbol of each input and is the last symbol read before the computation terminates. At the end of a computation M measures its configuration; if it is in an accepting state then it accepts the input, otherwise it rejects. The configuration of M is a superposition of states and it is represented by an n-dimensional complex unit vector, where n is the number of states. This vector is denoted by $|\psi\rangle = \Sigma_{i=1}^n \alpha_i |q_i\rangle$, where $\{|q_i\rangle\}$ is an orthonormal basis corresponding to the states of M. The coefficient α_i is the probability density amplitude of M being in state q_i. Since $|\psi\rangle$ is a unit vector, it follows that $\Sigma_{i=1}^n |\alpha_i|^2 = 1$. The transition function δ is represented by a set of unitary matrices $U_\sigma, \sigma \in \Sigma$, where U_σ represents the unitary transitions of M upon reading σ. If M is in configuration $|\psi\rangle$ and reads symbol σ, then the new configuration of M is

$$|\psi'\rangle = U_\sigma |\psi\rangle = \sum_{q_i, q_j \in S} \alpha_i \delta(q_i, \sigma, q_j) |q_j\rangle.$$

A measurement is represented by a diagonal zero-one projection matrix P where p_{ii} is 1 or 0 depending whether $q_i \in F$. The probability of M accepting string x is

$$p_M(x) = \langle \psi_x | P | \psi_x \rangle = ||P|\psi_x\rangle||^2,$$

where $|\psi_x\rangle = U(x)|q_0\rangle = U_{x_n} U_{x_{n-1}} \ldots U_{x_1} |q_0\rangle$.

Physically, this can be interpreted as follows. We have a quantum system prepared in a superposition of initial states. We expose it over time to a sequence of input symbols, one time-step per symbol. At the end of this process, we perform a measurement on the system and $p_M(x)$ is the probability of this measurement having an accepting outcome. Note that p_M is the probability of a particular event, not a general measure (on a space coded by strings).

The power of $MO - QFA$ depends on the type of acceptance, i.e. accept with bounded/unbounded-error probability. A language L is accepted with bounded-error probability by an $MO - QFA$ if there exists an $\varepsilon > 0$ such that every string in L is accepted with probability at least $\frac{1}{2} + \varepsilon$ and every string not in L is rejected with probability at least $\frac{1}{2} + \varepsilon$. The language L is accepted with unbounded-error probability by an $MO - QFA$ if every string in L is accepted with probability at least $\frac{1}{2}$ and every string not in L is rejected with probability at least $\frac{1}{2}$.

The main results are due to Brodsky and Pippenger [8]:

1. *The class of languages accepted by $MO - QFA$ with bounded-error probability coincides with the class of group languages, a proper subset of regular languages.*[3]

---

[3] A group automaton (GFA) is a DFA such that for every state q and input symbol σ, there exists exactly one state q' such that $\delta(q', \sigma) = q$. Equivalently, a DFA is reversible if for every $\sigma \in \Sigma$ there exists a string $x \in \Sigma^*$ such that for every state q, $\delta(q, \sigma x) = q$; see Bavel and Muller [3].

2. Any language accepted by an $MO - QFA$ with bounded-error probability can also be accepted by a deterministic probabilistic automaton with bounded-error probability.
3. Some $MO-QFA$ with unbounded-error probability accept non-regular languages, for example, the language $\{x \in \{0,1\}^* \mid x$ has an equal number of 0's and 1's$\}$.

3.2 Measure-Many Quantum Automata

For bounded-error acceptance, the power of $MO - QFA$ is too limited. One way of adding power to QFA is by introducing intermediate measurements. However, doing a measurement that causes the superposition to collapse to a single state would turn the QFA into a probabilistic automaton. A possible solution is to partition the set of states in three subsets–the accepting, rejecting and non-halting states–and use the spans of these sets as observables. A measurement is performed after every transition.

Inspired by the classical one-tape deterministic automata two types of $MM-QFA$, namely 1-way QFA (1QFA) and 2-way QFA (2QFA), have been introduced by Kondacs and Watrous [33].

An one-way measure-many quantum automaton (1QFA) is a 6 tuple $M = (S, \Sigma, q_0, S_a, S_r, \delta)$, where Σ is the finite alphabet with two end-marker symbols #, \$, S is the finite set of states, q_0 is the initial state, $S_a \subseteq S$ is the set of accepting states, $S_r \subseteq S$ is the set of rejecting states, $S_a \cap S_r = \emptyset$. The transition function δ is given by: $\delta : S \times \Sigma \times S \to \mathcal{C}$.

The computation of M is performed in the inner-product space $l_2(S)$, i.e. with the basis $\{|q\rangle \mid q \in S\}$, using the unary linear operators $V_\sigma, \sigma \in \Sigma$, defined by $V_\sigma(|q\rangle) = \sum_{q' \in S} \delta(q, \sigma, q')|q'\rangle$.

1. Any language recognized by an 1QFA with bounded-error probability is regular, cf. Kondacs and Watrous [33].
2. All group languages (i.e., languages recognized by group automata) are recognized by 1QFA (cf. Brodsky and Pippenger [8]), *but not all regular languages are recognized by* 1QFA; *for example, the language* $\{ab\}^*a$ *cannot be recognized by* 1QFA *with bounded-error probability*, cf. Kondacs and Watrous [33].
3. An 1QFA can accept a language with probability higher than $\frac{7}{9}$ iff the language is accepted by a deterministic reversible automata,[4]

The definition of two-way measure-many quantum automata (2QFA) is more complex, because of the effort to make their evolution unitary. The price paid is in the "size of quantum memory" which can grow proportional to the size of the input. 2QFA accept with bounded-error probability all regular languages in linear time. Their capability goes beyond regular languages; for example, the non-context-free language $\{a^n b^n c^n \mid n \geq 0\}$ is recognized by a 2QFA, cf. Kondacs and Watrous [33].

---
[4] According to Brodsky and Pippenger [8], a DFA is reversible if for every state q and input symbol σ, there exists at most one state q' such that $\delta(q', \sigma) = q$, and if there exist distinct states q_1, q_2 such that $\delta(q_1, \sigma) = q = \delta(q_2, \sigma)$, then $\delta(q, \Sigma) = \{q\}$. This notion of reversibility is equivalent to the one used by Ambainis and Freivalds [1]. Group automata are reversible in the above sense, but the converse is false.

Ambainis and Watrous [2] have introduced a model of two-way measure-many quantum automata in which both quantum and classical states are used, but the tape head position is classical. These automata have better computational capabilities than $2QFA$. For example, *the language $\{a^n b^n | n \geq 0\}$ can be recognized by a two-way automata with quantum and classical states in polynomial time* (a classical probabilistic automaton recognizes it in exponential time).

3.3 Ancilla QFA

To avoid the restriction to unitary transitions (which is quite strong) ancilla qubits have been added: with them, each transition can be unitary. Formally, this is done by adding an output tape to the QFA, cf. Paschen [39].

An ancilla QFA is a 6-tuple $M = (S, \Sigma, \Omega, \delta, q_0, F)$, where S, Σ, q_0 and F are as for $MO-QFA$, Ω is the output alphabet and the transition function $\delta : S \times \Sigma \times S \times \Omega \to \mathcal{C}$ verifies the following condition: for all states $q_1, q_2 \in S$ and $\sigma \in \Sigma$

$$\sum_{q \in S, \omega \in \Omega} \overline{\delta(q_1, \sigma, q, \omega)}, \delta(q_2, \sigma, q, \omega) = \begin{cases} 1, & \text{if } q_1 = q_2, \\ 0, & \text{otherwise.} \end{cases}$$

The main result in Paschen [39] is: *For every regular language L, there is a non-negative integer k such that an ancilla QFA using k ancilla qubits can recognize L exactly*. These quantum automata can recognize with one-sided unbounded error some non-regular languages.[5]

3.4 More Comments

Several types of quantum automata (QFA) have been proposed in the literature (see more in Gruska [30]). Some of them are more powerful than their classical counterpart. Others, as we have seen, *are less powerful*. This is the first problem: in principle, any quantum computational system is a generalization of a classical counterpart, so its computational power should not be *less than that of the classical system*. What is the explanation of this anomalie?

According to Moore [36], "The only case in which quantum automata are weaker than classical ones is when they are required to be unitary throughout their evolution, i.e. when measurements are only allowed at the end. This imposes a strict kind of reversibility, and (for instance) prevents languages like $\{w \in (a+b)^* \mid w$ contains no $aa\}$ from being recognized by a finite-state quantum machine. If you allow measurements during the computation as well as at the end (which seems reasonable) they include all classical automata."

Ciamarra [18] suggests a different reason, namely the lack of reversibility. A quantum computation is performed through a unitary operator, which is *reversible*, so the computation performed by a quantum automaton should be reversible till measurement. However, no model of quantum automata is reversible as from the final state $U_w |q_0\rangle$ one cannot retrace the computation because w is unknown; one can compute backward

---

[5] An automaton M accepts a language L with one-sided unbounded error if M accepts all strings of L with certainty and rejects strings not in L with some positive probability (or vice-versa).

from the operator U_w, but this information *is not* encoded in the final state. In spite of this, the class of languages recognized accepted by $MO - QFA$ with bounded-error probability coincides with the class of group languages, that is languages recognized by reversible classical automata! Classically, reversibility can be guaranteed by introducing the so-called *garbage* which can be recycled so that it grows linearly with the input size. Quantum mechanically, recycling is forbidden as the *garbage* might be entangled with the computational system. Ciamarra [18] suggests a model of quantum automaton which is strictly reversible (modeling classical reversible automata) and has at least the power of classical automata.

Reversibility is a very important notion in both classical and quantum computing (see for example, Frank, Knight, Margolus [25]). It seems that to date we don't have a satisfactory formal definition, which may be the cause of various anomalies in quantum computing, as the one discussed above.

4 Proofs and "Quantum" Proofs

Classically, there are two equivalent ways to look at the mathematical notion of proof: a) as a finite sequence of sentences strictly obeying some axioms and inference rules, b) as a specific type of computation. Indeed, from a proof given as a sequence of sentences one can easily construct a machine producing that sequence as the result of some finite computation and, conversely, giving a machine computing a proof we can just print all sentences produced during the computation and arrange them in a sequence. This gives mathematics an immense advantage over any science: any proof is an explicit sequence of reasoning steps that can be inspected at *leisure*; *in theory*, if followed with care, such a sequence either reveals a gap or mistake, or can convince a skeptic of its conclusion, in which case the theorem *is considered proven*. We said, *in theory*, because the game of mathematical proofs is ultimately a social experience, so it is contaminated to some degree by all "social maladies".

This equivalence has stimulated the construction of programs which perform like *artificial mathematicians*.[6] From proving simple theorems of Euclidean geometry to the proof of the four-color theorem, these "theorem provers" have been very successful. Of course, this was a good reason for sparking lots of controversies (see [9]).

Artificial mathematicians are far less ingenious and subtle than human mathematicians, but they surpass their human counterparts by being infinitely more patient and diligent. What about making errors? Are human mathematicians less prone to errors? This is a difficult question which requires more attention.

If a conventional proof is replaced by a "quantum computational proof" (or a proof produced as a result of a molecular experiment), then the conversion from a computation to a sequence of sentences may be impossible, e.g., due to the size of the computation. For example, a quantum automaton could be used to create some proof that relied on quantum interference among all the computations going on in superposition. The quantum automaton would say "your conjecture is true", but there will be no way to exhibit all trajectories followed by the quantum automaton in reaching that conclusion.

---

[6] Other types of "reasoning" such as medical diagnosis or legal inference have been successfully modeled and implemented; see, for example, the British National Act which has been encoded in first-order logic and a machine has been used to uncover its potential logical inconsistencies.

In other words, the quantum automaton has the ability to check a proof, but it may fail to reveal a "trace" of the proof for the human being operating the quantum automaton. Even worse, any attempt to *watch* the inner working of the quantum automaton (e.g. by "looking" at any information concerning the state of the on going proof) may compromise for ever the proof itself!

These facts may not affect the essence of mathematical objects and constructions (which have an autonomous reality quite independent of the physical reality), but they seem to have an impact of how we learn/understand mathematics (which is thorough the physical world). Indeed, our glimpses of mathematics are revealed only through physical objects, human brains, silicon computers, quantum automata, etc., hence, according to Deutsch [21], they have to obey not only the axioms and the inference rules of the theory, but the *laws of physics* as well.

References

1. AMBAINIS, A., FREIVALDS, R. 1-way quantum finite automata: strengths, weaknesses and generalizations, *Proceedings of 39th IEEE FOCS* (1998), 332-341.
2. AMBAINIS, A., WATROUS, J. Two-way finite automata with quantum and classical states, *Technical Report*, CC/9911009, 1999.
3. BAVEL, Z., AND MULLER, D. E. Connectivity and reversibility in automata, *J. Assoc. Comput. Mach.* 17 (1970), 231–240.
4. BELL, J. S. On the Einstein Podolsky Rosen paradox, *Physics*, 1 (1964), 195–200. Reprinted in [5] pp. 14–21.
5. BELL, J. S. *Speakable and Unspeakable in Quantum Mechanics*, Cambridge University Press, Cambridge, 1987.
6. BRAUER, W. *Automatentheorie*, Teubner, Stuttgart, 1984.
7. BRIDGMAN, P. W. A physicists second reaction to Mengenlehre, *Scripta Mathematica 2* (1934), 101–117, 224–234.
8. BRODSKY, A., PIPPENGER, N. Characterisation of 1-way quantum finite automata, quant-ph/9903014, 1999.
9. CALUDE, A. S. The journey of the four colour theorem through time, *The New Zealand Mathematics Magazine* 38, 3 (2001), 1-10.
10. CALUDE, C. S., CALUDE, E. Bisimulations and behaviour of nondeterministic automata, in G. Rozenberg, W. Thomas (eds.) *Developments in Language Theory. Foundations, Applications, and Perspectives*, World Scientific, Singapore, 2000, 60-70.
11. CALUDE, C. S., CALUDE, E., CHIU, T., DUMITRESCU, M., AND NICOLESCU, R. Testing computational complementarity for Mermin automata, *J. Multi Valued Logic*, 6 (2001), 47-65.
12. CALUDE, C. S., CALUDE, E., KHOUSSAINOV, B. Deterministic automata: Simulation, universality and minimality, *Annals of Applied and Pure Logic* 90, 1-3 (1997), 263-276.
13. CALUDE, C. S., CALUDE, E., KHOUSSAINOV, B. Finite nondeterministic automata: Simulation and minimality, *Theoret. Comput. Sci.* 242, 1-2 (2000), 219–235.
14. CALUDE, C. S., CALUDE, E., SVOZIL, K. Computational complementarity for probabilistic automata, in C. Martin-Vide, V. Mitrana (eds.). *Where Mathematics, Computer Science, Linguistics and Biology Meet*, Kluwer, Amsterdam 2000, 99-113.
15. CALUDE, C. S., CALUDE, E., SVOZIL, K., AND YU, S. Physical versus computational complementarity I, *International Journal of Theoretical Physics* 36 (1997), 1495–1523.
16. CALUDE, C. S., CALUDE, E., ŞTEFĂNESCU, C. Computational complementarity for Mealy automata, *EATCS Bull.* 66 (1998), 139–149.

17. CHAITIN, G. J. ewblock An improvement on a theorem by E. F. Moore. *IEEE Transactions on Electronic Computers EC-14* (1965), 466–467.
18. CIAMARRA, M. P. Quantum reversibility and a new model of quantum automaton, in R. Freivalds (ed.). *The 13th International Symposium on Foundations of Computation Theory (FCT'2001)*, Riga, Latvia, Springer-Verlag, Lect. Notes Comput. Sci. 2138, 2001, 376-379.
19. COHEN, D. W. *An Introduction to Hilbert Space and Quantum Logic*, Springer, New York, 1989.
20. CONWAY, J. H. *Regular Algebra and Finite Machines*, Chapman and Hall Ltd., London, 1971.
21. DEUTSCH, D. Quantum theory, the Church-Turing principle and the universal quantum computer, *Proceedings of the Royal Society London*, A 400 (1985), 97–119.
22. EINSTEIN, A., PODOLSKY, B., AND ROSEN, N. Can quantum-mechanical description of physical reality be considered complete? *Physical Review 47* (1935), 777–780.
23. FINKELSTEIN, D., AND FINKELSTEIN, S. R. Computational complementarity, *International Journal of Theoretical Physics 22*, 8 (1983), 753–779.
24. FOULIS, D. J., AND RANDALL, C. Operational statistics. i. Basic concepts, *Journal of Mathematical Physics 13* (1972), 1667–1675.
25. FRANK, M., KNIGHT, T., MARGOLUS, N. Reversibility in optimally scalable computer architectures, In *Unconventional Models of Computation*, C. S. Calude, J. Casti, M. Dinneen, Eds., Springer-Verlag, 1998, 165–182.
26. GILL, A. State-identification experiments in finite automata, *Information and Control 4* (1961), 132–154.
27. GINSBURG, S. On the length of the smallest uniform experiment which distinguishes the terminal states of the machine, *J. Assoc. Comput. Mach. 5* (1958), 266–280.
28. GIUNTINI, R. *Quantum Logic and Hidden Variables*. BI Wissenschaftsverlag, Mannheim, 1991.
29. GREENBERGER, D. B., HORNE, M., AND ZEILINGER, A. Multiparticle interferometry and the superposition principle, *Physics Today 46* (August 1993), 22–29.
30. GRUSKA, J. *Quantum Computing*, McGraw-Hill, London, 1999.
31. HOPCROFT, J. E., AND ULLMAN, J. D. *Introduction to Automata Theory, Languages, and Computation*, Addison-Wesley, Reading, MA, 1979.
32. KOCHEN, S., AND SPECKER, E. P. The problem of hidden variables in quantum mechanics, *Journal of Mathematics and Mechanics 17*, 1 (1967), 59–87.
33. KONDACS, A., WATROUS, J. On the power of quantum finite state automata, *Proceedings of 38th IEEE FOCS*, 1997, 66-75.
34. KOZEN, D. *Automata and Computability*, Springer-Verlag, New York, 1997.
35. MOORE, C. Dynamical recognizers: real-time language recognition by analogue computers, *Theoret. Comput. Sci.* 201 (1998), 99–136.
36. MOORE, C. Email to C. S. Calude, 10 May 2001.
37. MOORE, C. CRUTCHFIELD, J. P. Quantum automata and quantum grammars, *Theoret. Comput. Sci.* 237 (2000), 275–306.
38. MOORE, E. F. Gedanken-experiments on sequential machines, In *Automata Studies*, C. E. Shannon and J. McCarthy, Eds., Princeton University Press, Princeton, 1956, 129–153.
39. PASCHEN, K. Quantum finite automata using ancilla qubits, manuscript, May 2001.
40. PENROSE, R. *Shadows of the Minds, A Search for the Missing Science of Consciousness*, Oxford University Press, Oxford, 1994.
41. RABIN, M.O. Probabilistic automata, *Information and Control 6* (1963), 230–244.
42. SALOMAA, A. *Computation and Automata*, Cambridge University Press, Cambridge, 1985.
43. SVOZIL, K. *Randomness & Undecidability in Physics*, World Scientific, Singapore, 1993.
44. WRIGHT, R. Generalized urn models, *Foundations of Physics 20* (1990), 881–903.

Elementary Theory of Ordinals with Addition and Left Translation by ω

Christian Choffrut

LIAFA, Université Paris 7, 2, Pl. Jussieu
75 251 Paris Cedex 05, France
cc@liafa.jussieu.fr

1 Background

After Büchi it has become very natural to interprete formulae of certain logical theories as finite automata, i.e., as recognizing devices. This recognition aspect though, was neglected by the inventor of the concept and the study of the families of linear structures that could be accepted in the language theory sense of the term, was carried out by other authors. The most popular field of application of Büchi type automata is nowadays connected with model checking by considering a process as a possibly infinite sequence of events. For over a decade, the original model has been enriched by adding a time parameter in order to model reactive systems and their properties. Originally Büchi was interested in the monadic second order theory with the successor over ω but he later considered the theory of countable ordinals for which he was led to propose new notions of finite automata. Again these constructs can be viewed as recognizing devices of words over a finite alphabet whose length are countable ordinals. They were studied by other authors, mainly Choueka and Wojciechowski to who we owe two Theorems "à la Kleene" asserting the equivalence between expressions using suitable rational operators and subsets (languages) of transfinite words, [6] and [13].

Lately, there has been a renewed interest for strings of transfinite lengths as such by shifting the emphasis from logic to language theory. This is testified by various attempts to extend results from finite to transfinite strings. E.g., Eilenberg's famous varieties theorem suggested by Schützenberger's characterization of star-free languages, asserts that there exists a "natural" bijection between certain families of languages (i.e., subsets of finite length words) and certain families of finite semigroups. In both cases these families are defined by simple closure properties, [8]. In order to extend this result to words of transfinite length, Bedon and Carton extended Wilke's ω-semigroups to so-called ω_1-semigroups and were able that Eilenberg's theorem extends naturally, [1]. Also, the theory of rational relations which studies the rational subsets of pairs of words was extended to pairs of transfinite words in [4] where it is shown that the two traditional notions of rational relations still coincide when properly re-defined. Finally, we mention the beginning of a systematic study of the combinatorial properties of words of transfinte length by showing, for example, how equations in two unknowns can be "solved", [5].

2 Ordinals with Addition and Multiplication by Constants

The first order theory of ordinal addition is long known to be decidable (Ehrenfeucht, 1957 [7], Büchi, 1964 [3]). Furthermore, its complexity is a linear tower of exponentials (Maurin, 1996 [10,11]). There are two methods to obtain these results: one uses Fraïsse-Ehrenfeucht games ([7,11]), the other relies on Büchi's theory [3] of sets of transfinite strings recognized by finite automata.

Here, we are concerned with the theory of ordinal addition enriched with the multiplication by a constant. Left and right ordinal multiplications have different properties. E.g., multiplication distributes over addition with a left multiplicand but not with a right one: $(\omega+1)\omega = \omega^2 \neq \omega^2 + \omega$. We shall state and briefly sketch the proofs of two results concerning the theory of the ordinals with the usual addition and the left multiplication by ω. Actullay, contrary to theory of ordinal addition, the Π_2^0-fragment of the first-order theory of $\langle \omega^\omega; =, +, x \mapsto \omega x \rangle$ is already undecidable while the existential fragment is decidable. Observe that enriching the language with the right multiplication by ω does not increase the power of expression.

Let us mention a recent paper though not directly connected to ours, which tackles the decidability issue of the theory of ordinal multiplication over the ordinal α and shows that it is decidable if and only if α is less than ω^ω, [2].

We refer the reader to the numerous standard handbooks such as [12] or [9] for a comprehensive exposition of the theory on ordinals. We recall that every ordinal α has a unique representation, known as Cantor's normal form, as a finite sum of non-increasing prime components. By grouping the equal prime components, all non-zero ordinals α can thus be written as

$$\alpha = \omega^{\lambda_n} a_n + \omega^{\lambda_{n-1}} a_{n-1} + \ldots + \omega^{\lambda_1} a_1 + \omega^{\lambda_0} a_0 \tag{1}$$

where $n \geq 0, 0 < a_n, \ldots, a_1, a_0 < \omega$ and $\lambda_n > \lambda_{n-1} > \ldots > \lambda_0 \geq 0$. The ordinal λ_n is the *degree* of α.

We consider the theory of the ordinals less than ω^ω with equality as unique predicate, the ordinal addition and the left ordinal multiplication by ω: $x \to \omega x$ as operations, in other words we consider the theory ThL = Th$\langle \omega^\omega; +, x \to \omega x \rangle$ and we prove that it is undecidable by reducing the halting problem of Turing machines to it. We make the usual assumption that the formulas are in the prenex normal form

$$Q_1 y_1 \ldots Q_p y_p : \phi(x_1, \ldots, x_n) \tag{2}$$

where each Q_i is a universal or an existential quantifier, each x_i and each y_j is a variable and ϕ is a Boolean combination of formulas of the form

$$L(x_1, \ldots, x_n) = R(x_1, \ldots, x_n) \text{ or } L(x_1, \ldots, x_n) \neq R(x_1, \ldots, x_n) \tag{3}$$

with L and R linear combinations of terms such as ax and a where x is a variable and a is a constant.

It is not difficult to see that the case of roght multiplication is trivial. Indeed, for every $\xi < \omega^{\omega^\omega}$ the relations $y = \xi$ and $y = x\xi$ can be defined in every $\langle \lambda; =, + \rangle$ for $\lambda > 0$ and thus, in particular, the theory $\langle \lambda; +, x \to x\omega \rangle$ is decidable for all ordinals λ, [3].

2.1 Undecidability

Presburger arithmetics of the integers is cloesely related to the n-ary rational relations over ω and is therefore decidable. This connection no longer holds under our assumptions and ThL can be shown to be undecidable by reducing the halting problem of Turing machines to it.

Theorem 1. *The elementary theory* $\mathrm{ThL} = \mathrm{Th}\langle \omega^\omega; +, x \to \omega x \rangle$ *is undecidable.*

Sketch of the proof. A computation of a Turing machine is a sequence of configurations $(c_i)_{0 \le i \le n}$ such that c_0 is the initial configuration, c_n is a final configuration and c_{i+1} is a next configuration of c_i for all $0 \le i \le n-1$. We view the set $\Sigma \cup Q$ as digits. To the sequence $(c_i)_{0 \le i \le n}$ we assign the ordinal

$$\sum_{0 \le i \le n} \omega^i ||c_i||$$

where $||c_i||$ is the integer whose representation is c_i in the appropriate base. The problem reduces to expressing the fact that there exists an ordinal α which encodes a valid computation by using no other operations than those of the logic. The first task is to decompose α into its Cantor normal form $\sum_{0 \le i \le n} \omega^i a_i$ and to verify that the sequence of its coefficients a_i, once interpreted as strings c_i over $\Sigma \cup Q$, defines a computation. More precisely, α encodes a valid computation if a_0 can be interpreted as the initial configuration (with the input word on the tape and the initial state as current state), a_n as a final configuration and if for all $0 \le i \le n-1$, a_i and a_{i+1} can be interpreted as two successive configurations.

Now we explain why this cannot work in such a simple way. Indeed, a configuration is traditionnaly a word of the form $w_1 q w_2$ where $w_1 w_2 \in \Sigma^*$ is the content of the minimal initial segment of the tape comprising all cells containing a non-blank symbol along with the cell where the head is positionned, q is the current state and the reading head is at position $|w_1|$ from the left border (starting from position 0). Interpret a_i and a_{i+1} as the two successive configurations $w_1 q w_2$ and $y_1 p y_2$. Since a_i and a_{i+1} are ordinary integers, the only operations at hand is the addition. It seems intuitively impossible to "extract" the values of the states q and p by a mere use of this operation. As a consequence, instead of encoding the configuration $w_1 q w_2$ with an integer, we will encode it with the ordinal $\omega^2 a_2 + \omega a_1 + a_0$ where $a_2 = ||w_1||$, $a_1 = ||q||$ and $a_0 = ||w_2||$.

2.2 Existential Fragment of ThL

We mentioned earlier the connection between rational relations and Presburger formulae. For ThL this connection no longer holds even in the case of the existential fragment.

In order to convince the reader we introduce some definitions inspired by the rational subsets of the free commutative monoids.

Define the ω-rational relations over ω^ω of arity n as the least family of relations containing the single n-tuples of ordinals and closed by set union, componentwise addition, Kleene and ω-closures wher by the Kleene closure of a subset X of ω we mean all possible finite (possibly empty) sums of elements in X and by the ω-closure of X we mean all ω-sums of elements in X.

A subset of $(\omega^\omega)^k$ is *linear* if it is of the form

$$\{\beta_0 + \alpha_1 x_1 + \beta_1 + \alpha_2 x_2 + \ldots + \alpha_r x_r + \beta_r \mid x_1, \ldots, x_r < \omega\}$$

where $\beta_0, \alpha_1, \beta_1, \ldots, \alpha_r, \beta_{r+1}$ are elements in $(\omega^\omega)^k$. It is *semilinear* if it is a finite union of linear subsets.

Proposition 1 *For an arbitrary subset $X \subseteq \omega^\omega \times \omega^\omega$ the following properties are equivalent*

1) *X is ω-rational*

2) *X is semilinear*

3) *X is a finite union of subsets of the form*

$$\alpha_1 R_1 + \alpha_2 R_2 + \ldots + \alpha_k R_k \qquad (4)$$

where the α_i's are arbitrary k-tuples of ordinals in ω^ω and the R_i's are rational relations of $\mathbb{N}^k$.

Sketch of the proof. In view of the previous discussion it suffices to prove that **3)** implies **1)**. But this is trivial since every pair of ordinals is a rational relation of $\omega^\omega \times \omega^\omega$ reduced to an element. ∎

Now we return to the theory ThR and observe that the set of values of the free variables satisfying a given the formula is an ω-rational relation. This no longer holds for ThL as shown by the following formula

$$\phi \equiv (\omega^2 x + x = y) \wedge (z = x + z = \omega^2)$$

where the set of pairs of ordinals satisfying ϕ is equal to $\{(\omega^3 n + \omega^2 p + \omega n + p, \omega n + p) \mid n, p < \omega\}$.

We prove that the existential fragment of ThL is decidable. After possible introduction of new individual variables and equalities for transforming all disequalities, the general form of such an existential formula is a disjunction of conjunctions of equalities

$$L(x_1, \ldots, x_n) = R(x_1, \ldots, x_n) \qquad (5)$$

prefixed by a collection of existential quantifiers. Actually each handside is a linear combination

$$\alpha_1 y_1 + \beta_1 + \alpha_2 y_2 + \beta_2 + \ldots \alpha_p y_p + \beta_p$$

where the α_i's and the β_i's are ordinals and the y_i's are (possibly repeated) unknowns. We consider a system of equations

$$L_j(x_1, x_2 \ldots, x_n) = R_j(x_1, x_2 \ldots, x_n) \; j = 1, \ldots t \qquad (6)$$

where each left- and right-handside is a linear combination. A *monomial* of the above system is an expression of the form $\alpha_i y_i$ which occurs in a left- or right-handside of some equation of the system.

Theorem 2. *The existential fragment of* ThL *is decidable*

Sketch of the proof. It suffices to show that the existence of a solution for the system (6) is decidable. Consider Cantor's normal form of the value of an arbitrary unkown

$$x = \omega^m a_m + \omega^{m-1} a_{m-1} + \ldots + \omega a_1 + a_0 \qquad (7)$$

If we can effectively bound the degree m of each of these variables then we can also effectively bound the degrees of each handside of each equation the system say these degrees are less than N. By equating the coefficients of degree less than N, each equation of the system splits into up to $N+1$ linear equations where the unknowns are the coefficients a_i of the variables as in (7). Denote by δ the maximum degree of the coefficients α_i in (6) and by K the least common multiple of all integers less than or equal to δ. Let $d_1 \leq d_2 \leq \ldots \ldots d_p$ be the sequence, in non-decreasing order, of the degrees of the monomials occurring in an equation of the system. We can show that if the sytem has a solution, then it has one that satisfies for all $1 \leq i \leq p$

$$\delta \leq d_i < d_{i+1} \Rightarrow d_{i+1} \leq d_i + 3K$$

which proves the claim. ∎

References

1. N. Bedon and O. Carton. An Eilenberg theorem for words on countable ordinals. In *Proceedings of Latin'98 Theoretical Informatics*, number 1380 in LNCS, pages 53–64. Springer-Verlag, 1998.
2. Alexis Bès. Decision problems related to the elementary theory of oordinal multiplication. *Fund. Math.*, xxx(xxx):xxx. to appear.
3. J. Büchi. Transfinite automata recursions and weak second ordre theory of ordinals. In *International Congress in Logic, Methodology and Philosophy of Sciences*, pages 3–23. North-Holland, 1964.
4. C. Choffrut and S. Grigorieff. Uniformization of rational relations. In G. Paun J. Karhumäki, H. Maurer and G. Rozenberg, editors, *Jewels are Forever*, pages 59–71. Springer Verlag, 1999.
5. C. Choffrut and S. Horváth. Transfinite euqations in transfinite strings. *Internat. J. Algebra Comput.*, 10(5):625–649, 2000.
6. S. C. Choueka. Finite automata, definable sets and regular expressions over ω^n-tapes. *J. Comput. System Sci.*, 17:81–97, 1978.

7. A. Ehrenfeucht. Applications of games to some problems of mathematical logic. *Bull. Académie Polonaise des Sciences*, 5:35–37, 1957.
8. S. Eilenberg. *Automata, Languages and Machines*, volume B. Academic Press, 1976.
9. D. Klaua. *Allgemeine Mengenlehre*. Akademie Verlag, 1969.
10. F. Maurin. Exact complexity bounds for ordinal addition. *Theoret. Comput. Sci.*, 165(2):247–273, 1996.
11. F. Maurin. Ehrenfeucht games and ordinal addition. *Annals of Pure and Applied Logic*, 89(1):53–73, 1997.
12. W. Sierpiński. *Cardinal and Ordinal Numbers*. Warsaw: PWN, 1958.
13. J. Wojciechowski. Finite automata on transfinite sequences and regular expressions. *Fundamenta Informaticae*, 8(3–4):379–396, 1985.

The Equational Theory of Fixed Points with Applications to Generalized Language Theory
Dedicated to Prof. Werner Kuich on the Occasion of His 60th Birthday

Z. Ésik[*]

Dept. of Computer Science, University of Szeged, P.O.B. 652, 6720 Szeged, Hungary
esik@inf.u-szeged.hu

Abstract. We review the rudiments of the equational logic of (least) fixed points and provide some of its applications for axiomatization problems with respect to regular languages, tree languages, and synchronization trees.

1 Introduction

A classic result of the theory of context-free languages is Parikh's theorem [32] that asserts that the letter occurrence vectors (Parikh vectors) corresponding to the words of a context-free language on a k-letter alphabet form a semilinear subset of $\mathcal{N}^k$, the free commutative monoid of k-dimensional vectors over the naturals. The theorem is usually proved by combinatorial arguments on the derivation trees of the context-free grammar. However, as Pilling [35] observed, Parikh's theorem may be formulated as an assertion about "rational functions" on the (free) commutative idempotent continuous semiring of all subsets of $\mathcal{N}^k$. Subsequently, Kuich [28] generalized Parikh's result to all commutative idempotent continuous semirings (l-semirings). (See also [27] for a related treatment.) In fact, by introducing rational terms that denote rational functions, or more generally, recursion terms or μ-terms denoting functions that arise as least solutions of systems of polynomial fixed point equations, Parikh's theorem can be translated into a statement about the equational theory of commutative idempotent continuous semirings: For every μ-term t there exists a rational term r such that the equation $t = r$ holds in all commutative idempotent continuous semirings. Alternatively, one may just consider rational terms and prove that for each rational term $t(x, y_1, \ldots, y_n)$ in the variables $x, y_1, \ldots, y_n$ there is a rational term $r(y_1, \ldots, y_n)$ containing no occurrence of x that provides least solution to the fixed point equation $x = t(x, y_1, \ldots, y_n)$ over *all* commutative idempotent continuous semirings. This approach has been pursued by Hopkins and Kozen in [23], in their argument lifting Parikh's theorem to all commutative idempotent semirings with enough least fixed points to provide solutions to recursion equations. By proving this more general result, Hopkins and Kozen have shown how to replace the analytic arguments of Pilling and Kuich by arguments based only on the the least (pre-)fixed point rule (also known as the Park induction rule [33]), the fixed point

---

[*] Partially supported by grants T30511 and T35163 from the National Foundation of Hungary for Scientific Research and the Austrian-Hungarian Action Foundation.

equation, and the algebraic laws of the sum and product operations. But since Parikh's theorem is a claim about equational theories, one would eventually like to have a purely equational proof of it. This question has been addressed recently in [1]. In this paper, Parikh's theorem is derived from a small set of purely equational axioms involving fixed points.

Parikh's theorem is not the only result of automata and language theory that can be derived by simple *equational* reasoning from the algebraic properties of fixed points. Other applications of the equational logic of fixed points include Kleene's theorem and its generalizations [7], see also [28,29,9], where the presentation is not fully based on equational reasoning, and Greibach's theorem [19]. The methods employed in the papers [26,17] even indicate that one can embed the Krohn-Rhodes decomposition theorem [21] for finite automata and semigroups within equational logic. Further applications of fixed point theory include an algebraic proof of the soundness and relative completeness of Hoare's logic [5,6]. See also [25].

The aim of this paper is to provide an introduction to the basics of the equational theory of fixed points and to show some of its applications in the solution of axiomatization problems for "generalized languages". In his book [10], Conway formulated several conjectures regarding the equational theory of the regular sets. Some of his conjectures have since been confirmed (e.g., the completeness of the group-identities, settled by Krob in [26]), but some are still open. In particular, Conway's "letter reduction" conjecture is still open. In this paper, we generalize both Conway's group-identities and his letter reduction conjecture to continuous algebras over any signature. Just as in the classical setting of regular languages, the group-identities are complete, in conjunction with the classical identities. The generalized letter reduction conjecture remains open.

2 Preiteration Algebras

Terms, or μ-*terms, over the signature* Σ are defined by the syntax

$$T ::= x \mid \sigma(\overbrace{T,\ldots,T}^{n-\text{times}}) \mid \mu x.T,$$

where x ranges over a countably infinite set V of variables, and for each $n \geq 0$, σ ranges over Σ_n, the set of n-ary *function symbols* in Σ. *Free* and *bound* occurrences of variables in a term are defined as usual. We identify any two μ-terms that differ only in the bound variables. Moreover, for any μ-terms $t, t_1, \ldots, t_n$ and distinct variables $x_1, \ldots, x_n$, we write

$$t[t_1/x_1, \ldots, t_n/x_n] \quad \text{or} \quad t[(t_1, \ldots, t_n)/(x_1, \ldots, x_n)]$$

for the term obtained by simultaneously substituting t_i for x_i, for each $i \in [n] = \{1, \ldots, n\}$. Since we may assume that the bound variables in t are different from the variables that have a free occurrence in the terms t_i, no free variable in any t_i may become bound as the result of the substitution. Below, we will write $t(x_1, \ldots, x_n)$ or $t(\boldsymbol{x})$, where $\boldsymbol{x} = (x_1, \ldots, x_n)$ to denote that t is a term with free variables in the set $\{x_1, \ldots, x_n\}$. When writing μ-terms, we assume that the scope of a prefix μx extends to the right as far as possible.

The Equational Theory of Fixed Points 23

We will be interested in interpretations where $\mu x.t$ provides a "canonical solution" to the fixed point equation $x = t$.

A *preiteration Σ-algebra* is a nonempty set A together with an interpretation of the terms t as functions $t_A : A^V \to A$ such that the following hold:

1. When t is a variable $x \in V$, then t_A is the corresponding projection $A^V \to A$, i.e., $t_A(\rho) = \rho(x)$, for all $\rho : V \to A$.
2. For any terms $t, t_1, \ldots, t_n$ and different variables $x_1, \ldots, x_n$,

$$(t[(t_1, \ldots, t_n)/(x_1, \ldots, x_n)])_A(\rho) = t_A(\rho[x_i \mapsto (t_i)_A(\rho) : i \in [n]]),$$

for all $\rho : V \to A$, where for any $(a_1, \ldots, a_n) \in A^n$, the function $\rho[x_i \mapsto a_i : i \in [n]]$ is the same as ρ except that it maps each x_i to a_i.
3. If t, t' are terms with $t_A = t'_A$, then for all variables x, also $(\mu x.t)_A = (\mu x.t')_A$.
4. For any t and variable x which has no free occurrence in t, the function t_A does not depend on its argument corresponding to x, i.e., $(\mu x.t)_A(\rho) = (\mu x.t)_A(\rho[x \mapsto b])$, for all $\rho : X \to A$ and $b \in A$.

A *strong preiteration algebra* is a preiteration algebra that satisfies the following strengthened version of the third condition above: For all terms t, t' and for all $\rho : V \to A$ and $x \in V$, if $t_A(\rho[x \mapsto a]) = t'_A(\rho[x \mapsto a])$, for all $a \in A$, then $(\mu x.t)_A(\rho) = (\mu x.t')_A(\rho)$.

We will also consider *ordered preiteration Σ-algebras* which are preiteration Σ-algebras A equipped with a partial order $\leq$ such that each term function t_A is monotonic with respect to the pointwise order on A^V, and such that the following stronger version of the third condition above is satisfied: If t, t' are terms over Σ with $t_A \leq t'_A$ in the pointwise order, then for all variables x, also $(\mu x.t)_A \leq (\mu x.t')_A$. The "ordered notion" corresponding to strong preiteration Σ-algebras is the notion of *strong ordered preiteration Σ-algebra* which is defined in the obvious way.

Below, if $t = t(x_1, \ldots, x_n)$ and if A is an (ordered) preiteration Σ-algebra with $a_1, \ldots, a_n \in A$, we write $t_A(a_1, \ldots, a_n)$ for $t_A(\rho)$, where $\rho : V \to A$ maps each x_i to a_i, $i \in [n]$.

A *morphism* of (strong) preiteration Σ-algebras A and B is a function $h : A \to B$ such that

$$h(t_A(a_1, \ldots, a_n)) = t_B((h(a_1), \ldots, h(a_n)),$$

for all terms $t(x_1, \ldots, x_n)$, and for all $(a_1, \ldots, a_n) \in A^n$. A morphism of (strong) ordered preiteration Σ-algebras also preserves the partial order.

Note that any preiteration Σ-algebra A determines a Σ-algebra: For each $\sigma \in \Sigma_n$ and $(a_1, \ldots, a_n)$, we define $\sigma_A(a_1, \ldots, a_n) = t_A(a_1, \ldots, a_n)$, where t is the term $\sigma(x_1, \ldots, x_n)$ for some sequence of different variables $x_1, \ldots, x_n$. Also, any preiteration algebra morphism is a Σ-algebra homomorphism.

First-order formulas involving μ-terms over Σ are constructed form atomic formulas $t = t'$, where t and t' are μ-terms, in the expected way. In the ordered setting, expressions $t \leq t'$ are also atomic formulas. Free and bound occurrences of variables in a formula and substitution are defined as usual. A formula with no free variables is called a *sentence*. The semantic notion of satisfaction is defined in the usual Tarskian style. Suppose that

A is an (ordered) preiteration algebra, ρ is a function $V \to A$ and φ is a formula. We write $(A, \rho) \models \varphi$ to denote that A satisfies φ under the given evaluation of the variables. When ϕ is a sentence, we say that A *satisfies* φ, or that φ *holds in* A, notation $A \models \varphi$, if $(A, \rho) \models \varphi$ for all, or for some ρ. (Note that a preiteration algebra is not empty.)

Most sentences that we will encounter in this paper fall into three categories. First of all, we will have *equations* and *inequations* that are sentences of the form $\forall x_1 \ldots \forall x_n (t = t')$ and $\forall x_1 \ldots \forall x_n (t \leq t')$, respectively, also denoted as $\forall \boldsymbol{x}(t = t')$ and $\forall \boldsymbol{x}(t \leq t')$. An equation is also called an *identity*. Second, we will consider *implications* of the form $\forall \boldsymbol{x}(t_1 = t'_1 \wedge \ldots \wedge t_k = t'_k) \Rightarrow t = t'$) and $\forall \boldsymbol{x}(t_1 \leq t'_1 \wedge \ldots \wedge t_k \leq t'_k) \Rightarrow t \leq t')$. Finally, we will also have *implications between equations (or inequations)* that are of the form $\forall \boldsymbol{x}(t_1 = t'_1 \wedge \ldots \wedge t_k = t'_k) \Rightarrow \forall \boldsymbol{y}(s = s')$. As usual, we will abbreviate an equation as $t = t'$, an inequation as $t \leq t'$, and an implication as $t_1 = t'_1 \wedge \ldots \wedge t_k = t'_k \Rightarrow t = t'$, etc.

Example 1. Every preiteration Σ-algebra satisfies the implication between equations

$$\forall x \forall y (t = t') \Rightarrow \forall y (\mu x.t = \mu x.t'),$$

for all terms $t(x, y), t'(x, y)$ over Σ. A preiteration algebra is strong iff it satisfies all sentences

$$\forall y (\forall x (t = t') \Rightarrow \mu x.t = \mu x.t').$$

3 Continuous Algebras

Recall that a *cpo* is a poset $(A, \leq)$ which has a least element, denoted $\bot_A$, and such that each directed set[1] $D \subseteq A$ has a supremum $\bigvee D$. Note that when A is a cpo, so is any direct power of A equipped with the pointwise ordering, as is the direct product of any number of cpo's.

Suppose that A and B are cpo's and f is a function $A \to B$. We call f *monotonic* if $f(a) \leq f(b)$ whenever $a \leq b$ in A. Moreover, we call f *continuous* if f is monotonic and $f(\bigvee D) = \bigvee f(D)$ holds for all directed sets $D \subseteq A$. Finally, we call f *strict* if $f(\bot_A) = \bot_B$.

Below we will make use of the following well-known properties of continuous functions.

Proposition 1. *Each projection function $\prod_{i \in I} A_i \to A_j$ from a direct product of cpo's A_i, $i \in I$ to a cpo A_j is continuous. Moreover, if $f : A_1 \times \ldots \times A_n \to A$ and $g_1 : B \to A_1, \ldots, g_n : B \to A_n$ are continuous, where $A, B, A_1, \ldots, A_n$ are all cpo's, then so is the function $h : B \to A$ defined by $h(b) = f(g_1(b), \ldots, g_n(b))$, for all $b \in B$.*

When $f : A \times B \to A$ and b is a fixed element of B, let f_b denote the function $A \to A$ defined by $f_b(a) = f(a, b)$.

---
[1] A set $D \subseteq A$ is called directed if it is not empty and each pair of elements in D has an upper bound in D.

Proposition 2. *Suppose that A and B are cpo's and f is a continuous function $A \times B \to A$. Then for each $b \in B$ there is a least $a = f^\dagger(b)$ which is a pre-fixed point of f_b, i.e., such that $f(a, b) \leq a$. Moreover, $f^\dagger : B \to A$, as a function of b, is continuous.*

The least pre-fixed point a is in fact a *fixed point* of f_b, i.e., $f(a, b) = a$. Indeed, since $f(a, b) \leq a$ and f is monotonic, also $f(f(a, b), b) \leq f(a, b)$, showing that $f(a, b)$ is a pre-fixed point. But since a is the least pre-fixed point, we have that $a \leq f(a, b)$, which, together with $f(a, b) \leq a$ gives $f(a, b) = a$.

A *continuous Σ-algebra* consists of a cpo $(A, \leq)$ and a Σ-algebra (A, Σ) such that each operation $\sigma_A : A^n \to A$, $\sigma \in \Sigma_n$ is continuous. A morphism of continuous Σ-algebras is a strict continuous Σ-algebra homomorphism.

Each continuous Σ-algebra A gives rise to a strong ordered preiteration Σ-algebra. We define t_A by induction on the structure of the term t. Suppose that $\rho : V \to A$.

1. When t is the variable x, we define $t_A(\rho) = \rho(x)$.
2. When t is a term of the form $\sigma(t_1, \ldots, t_n)$, we let
   $t_A(\rho) = \sigma_A((t_1)_A(\rho), \ldots, (t_n)_A(\rho))$.
3. When t is of the form $\mu x.t'$, then we define $t_A(\rho)$ to be the least $a \in A$ with $t_A(\rho[x \mapsto a]) \leq a$, in fact $t_A(\rho[x \mapsto a]) = a$.

The fact that t_A is a well-defined continuous function $A^V \to A$ follows from Propositions 1 and 2. Since strict continuous functions preserve least pre-fixed points, it follows that each strict continuous morphism $A \to B$, where A and B are continuous Σ-algebras, is a (strong) preiteration algebra morphism.

Since in continuous algebras, μ-terms are interpreted by least pre-fixed points, we have:

Proposition 3. *Each continuous Σ-algebra satisfies the* fixed point equation

$$\mu x.t = t[\mu x.t/x] \tag{1}$$

and the implication

$$t[y/x] \leq y \Rightarrow \mu x.t \leq y, \tag{2}$$

for all terms t over Σ, and for all variables x, y.

Note that the fixed point equation (1) is not a single equation, but in fact an equation scheme. Nevertheless, following standard practice, we will call such schemes just equations.

The above implication (2) is sometimes referred to as the *Park induction rule* [33], or the *least pre-fixed point rule*. It is an instance of a more general induction principle attributed to Scott. See also [2]. The Park induction rule has a weak version, which is an implication between inequations: For all terms $t(x, \boldsymbol{y})$ and $t'(\boldsymbol{y})$

$$\forall \boldsymbol{y}(t[t'/x] \leq t') \Rightarrow \forall \boldsymbol{y}(\mu x.t \leq t').$$

4 Conway Algebras

A (strong) *Conway Σ-algebra* [7] is a (strong) preiteration algebra satisfying the following *diagonal* (3) and *rolling equations* (4) for all terms t, t' over Σ and for all variables x, y:

$$\mu x.t[x/y] = \mu x.\mu y.t \tag{3}$$

$$\mu x.t[t'/x] = t[\mu x.t'[t/x]/x]. \tag{4}$$

The above equations are by now classic, see [22,2,31], to mention a few early references. A morphism of Conway Σ-algebras is a preiteration Σ-algebra morphism. Note that when t' is the variable x, (4) reduces to the fixed point equation defined above. Thus, in Conway algebras, $\mu x.t$ provides a *canonical* solution to the fixed point equation $x = t$. A strong Conway algebra is also a strong preiteration algebra. Strong Conway Σ-algebras satisfy the same equations as Conway Σ-algebras.[2] Morphisms of (strong) Conway algebras are preiteration algebra morphisms.

It turns out that in Conway algebras it is also possible to solve *systems* of fixed point equations. Below we will often consider term vectors $\boldsymbol{t} = (t_1, \ldots, t_n)$, where n is any positive integer. We say that a variable has a free occurrence in $\boldsymbol{t}$ if it has a free occurrence in one of the t_i. Bound occurrences are defined in the same way. Substitution into a term vector is defined component-wise. When $\boldsymbol{t} = (t_1, \ldots, t_n)$ and A is a preiteration algebra, $\boldsymbol{t}_A : A^V \to A^n$ is the function $\rho \mapsto ((t_1)^A(\rho), \ldots, (t_n)_A(\rho))$. We identify any vector of dimension 1 with its unique component. A formula $\boldsymbol{t} = \boldsymbol{s}$, where $\boldsymbol{t} = (t_1, \ldots, t_n)$ and $\boldsymbol{s} = (s_1, \ldots, s_n)$, is viewed as an abbreviation for the formula $t_1 = s_1 \wedge \ldots \wedge t_n = s_n$. Formulas $\boldsymbol{t} \leq \boldsymbol{s}$ are abbreviations in the same way.

Suppose that $\boldsymbol{t} = (t_1, \ldots, t_n)$ is a vector of terms and $\boldsymbol{x} = (x_1, \ldots, x_n)$ is a vector of different variables of the same dimension $n \geq 1$. We define the term vector $\mu \boldsymbol{x}.\boldsymbol{t}$ by induction on n. When $n = 1$, we define $\mu \boldsymbol{x}.\boldsymbol{t} = (\mu x_1.t_1)$. When $n > 1$, we let

$$\mu \boldsymbol{x}.\boldsymbol{t} = (\mu(x_1, \ldots, x_{n-1}).(t_1, \ldots, t_{n-1})[\mu x_n.t_n/x_n],$$
$$\mu x_n.t_n[\mu(x_1, \ldots, x_{n-1}).(t_1, \ldots, t_{n-1})/(x_1, \ldots, x_{n-1})]).$$

The above definition is motivated by the Bekić–De Bakker–Scott rule [3,2]. See also Pilling [35].

Proposition 4. *Suppose that $t(x, y)$ and $s(x, y)$ are n-dimensional term vectors where x and y are vectors of distinct variables such that the dimension of x is n.*

1. *If A is a preiteration algebra then*

$$A \models \forall \boldsymbol{x}, \boldsymbol{y}(\boldsymbol{t} = \boldsymbol{s}) \Rightarrow \forall \boldsymbol{y}(\mu \boldsymbol{x}.\boldsymbol{t} = \mu \boldsymbol{x}.\boldsymbol{s}).$$

2. *If A is a strong preiteration algebra then*

$$A \models \forall \boldsymbol{y}(\forall \boldsymbol{x}(\boldsymbol{t} = \boldsymbol{s}) \Rightarrow \forall \boldsymbol{y}(\mu \boldsymbol{x}.\boldsymbol{t} = \mu \boldsymbol{x}.\boldsymbol{s})).$$

---

[2] An equation holds in a "variety" of preiteration algebras iff it holds in the strong preiteration algebras belonging to that variety. See [7].

Theorem 1. [7] *If A is a Conway algebra, then the "vector forms" of (3) and (4) hold in A: For all term vectors $t(x, y, z)$, where t, x and y have the same dimension,*

$$A \models \mu x.t[x/y] = \mu x.\mu y.t. \tag{5}$$

Moreover, for all term vectors $t(y, z)$ and $s(x, z)$, where the dimension of t agrees with that of x and the dimension of s with that of y,

$$A \models \mu x.t[s/y] = t[\mu x.s[t/x]/y]. \tag{6}$$

Corollary 1. *For each term vector t and vector of distinct variables x of the same dimension, the equation*

$$\mu x.t = t[\mu x.t/x] \tag{7}$$

holds in all Conway algebras.

Equation (7) is the vector form of the fixed point equation (1). Since by Theorem 1, the vector forms of (3) and (4) hold in any Conway Σ-algebra, so does the vector form of any other equation that holds in these algebras.

Corollary 2. *If an equation holds in all Conway Σ-algebras, then so does its vector form.*

For a full characterization of the equations of Conway Σ-algebras we refer to [4]. It is shown in op. cit. that when Σ contains a symbol of rank > 1 then it is PSPACE-complete to decide whether an equation holds in all Conway Σ-algebras, whereas the problem easily lies in P if each symbol in Σ has rank at most 1.

We now give a characterization of Conway algebras based on the vector form of the fixed point identity.

Theorem 2. *The following three conditions are equivalent for a preiteration Σ-algebra A.*

1. *A is a Conway Σ-algebra.*
2. *The vector form of the fixed point equation holds in A.*
3. *A satisfies the fixed point equation for binary vectors.*

Below we will abbreviate the term $\mu x.x$ as $\bot$. Suppose that X and Y are finite disjoint sets of variables. We call a term t over Σ *primitive* with respect to (X, Y) if it is either $\bot$, or a variable in Y, or a term $\sigma(x_1, \ldots, x_k)$, where $\sigma \in \Sigma_k$ and the not necessarily different variables $x_1, \ldots, x_k$ are all in X. The following fact is a version of Salomaa's equational characterization of regular languages [36], see also [12,30]. In [14,7], the result is derived only from the equational axioms of Conway algebras.

Theorem 3. *Normal forms [14,7] For each term $t(y)$ in the free variables $y = (y_1, \ldots, y_m)$ there exists an integer $n \geq 1$, a vector $x = (x_1, \ldots, x_n)$ of fresh variables* and terms $s_1, \ldots, s_n$, all primitive with respect to (X, Y), where $X = \{x_1, \ldots, x_n\}$ and $Y = \{y_1, \ldots, y_m\}$, such that equation

$$t = (\mu(x_1, \ldots, x_n).(s_1, \ldots, s_n))_1$$

holds in all Conway algebras, where the right side of the equation is the first component of the term vector $\mu(x_1, \ldots, x_n).(s_1, \ldots, s_n)$.

The following result is essentially due to Bekić and De Bakker and Scott.

Theorem 4. [2,3] *Suppose that A is an ordered preiteration Σ-algebra satisfying the fixed point equation (1) and the Park induction principle (2). Then A is a strong preiteration algebra. Moreover, the vector form of the fixed point equation (7) and the vector form of the Park induction rule (8) hold in A:*

$$t[\boldsymbol{y}/\boldsymbol{x}] \leq \boldsymbol{y} \Rightarrow \mu\boldsymbol{x}.t \leq \boldsymbol{y}, \tag{8}$$

for all term vectors t over Σ of dimension n, and all vectors of distinct variables $\boldsymbol{x}, \boldsymbol{y}$ of dimension n.

We call such algebras *Park Σ-algebras*. Morphisms of Park Σ-algebras are order preserving preiteration algebra morphisms. Any such morphism is strict.

Remark 1. Each ordered preiteration Σ-algebra satisfying the fixed point equation and the weak version of the Park induction rule satisfies the vector forms of these axioms.

5 Iteration Algebras

In his book [10], John H. Conway associated an equation of regular sets with every finite group and conjectured that a finite set of classical identities together with the equations associated with the finite (simple) groups form a complete set of equations for the regular sets. Conway's equations can be generalized.

Suppose that G is a finite group with elements $\{g_1, \ldots, g_n\}$, and let $t(\boldsymbol{x}, \boldsymbol{y})$ denote a term over some signature Σ, where $\boldsymbol{x} = (x_1, \ldots, x_n)$ and $\boldsymbol{y} = (y_1, \ldots, y_m)$, so that the dimension of $\boldsymbol{x}$ agrees with the order of G. For each $i \in [n]$, let $\pi_i : [n] \to [n]$ denote the function $j \mapsto k$ iff $g_i \cdot g_j = g_k$. Define

$$s_i = t[(x_{\pi_i(1)}, \ldots, x_{\pi_i(n)})/(x_1, \ldots, x_n)], \quad i \in [n].$$

Let x denote a fresh variable. The *group-equation* or *group-identity* [17] associated with G is:

$$(\mu(x_1, \ldots, x_n).(s_1, \ldots, s_n))_1 = \mu x.t[(x, \ldots, x)/(x_1, \ldots, x_n)]. \tag{9}$$

(The definition of the equation associated with G also depends on the ordering of the group elements $g_1, \ldots, g_n$. However, with respect to the Conway identities, different orderings result in equivalent equations.) A *(strong) iteration Σ-algebra* is a *(strong) Conway Σ-algebra* satisfying the group-equations associated with the finite groups. Strong iteration Σ-algebras and iteration Σ-algebras satisfy the same equations. A morphism of (strong) iteration algebras is a preiteration algebra morphism.

Theorem 5. [17] *If an equation holds in iteration Σ-algebras, then so does its vector form.*

Theorem 6. [17] *An equation holds in all iteration Σ-algebras iff it holds in all continuous Σ-algebras.*

For the axiomatization of iteration algebras based on variants of the Conway identities and the *commutative identity*, the above result was established in [14]. See [7] for a thorough treatment of the earlier results.

Suppose that A is a set disjoint from Σ and does not contain the special symbol $\bot$. A *partial (Σ, A)-tree* [22,12] is an at most countable, ordered rooted tree whose nodes are labeled by the elements of $\Sigma \cup A \cup \{\bot\}$ such that nodes labeled in Σ_n have n descendants and all the nodes labeled in $A \cup \{\bot\}$ are leaves. Say that $T \leq T'$, for trees T and T', if T' can be constructed from T by attaching non-uniformly (Σ, A)-trees to the leaves of T labeled $\bot$. Equipped with this partial order, the set $(\Sigma, A)\mathbf{T}$ of (Σ, A)-trees is a cpo whose bottom element is the one-node tree labeled $\bot$. Moreover, equipped with the usual Σ-operations, $(\Sigma, A)\mathbf{T}$ is a continuous Σ-algebra, in fact, the free continuous Σ-algebra on A.

Theorem 7. [22] *For each set A, the algebra $(\Sigma, A)\mathbf{T}$ is freely generated by A in the class of all continuous Σ-algebras.*

Corollary 3. *An equation holds in all iteration Σ-algebras iff it holds in continuous Σ-algebras $(\Sigma, A)\mathbf{T}$.*

Call a tree (Σ, A) *total* if it has no leaves labeled $\bot$. Moreover, call a tree (Σ, A)-tree *regular* if it has a finite number of (nonisomorphic) subtrees. Note that every finite tree is regular. It turns out that the free iteration Σ-algebras may also be represented by trees.

Theorem 8. [14] *The free iteration Σ-algebra on a set A can be represented as the algebra $(\Sigma, A)\mathbf{R}$ of regular (Σ, A)-trees.*

Remark 2. The algebra of regular (Σ, A)-trees is also free in the class of *regular Σ-algebras* [22,37], and the algebra of total regular trees is free in the class of *iterative Σ-algebras* [12].

Corollary 4. [11] *There is a polynomial time algorithm to decide for an equation between terms over Σ whether it holds in all iteration Σ-algebras.*

For later use we recall:

Theorem 9. [15] *Every Park Σ-algebra, or ordered iteration algebra satisfying the fixed point equation and the weak version of the Park induction rule, is an iteration algebra. An equation between terms over Σ holds in all Park Σ-algebras iff it holds in all iteration Σ-algebras.*

Corollary 5. *For each set A, the algebra of regular trees $(\Sigma, A)\mathbf{R}$, equipped with the partial order inherited from $(\Sigma, A)\mathbf{R}$, is freely generated by A in the class of all Park Σ-algebras.*

Theorem 9 is a hidden completeness result. It follows that an equation between μ-terms over Σ holds in all continuous Σ-algebras iff it can be derived from (instances of) the fixed point equation using the usual rules of (in)equational logic and a non-standard rule corresponding to the weak version of the Park induction principle. This logic was proposed in [31].

5.1 A Conjecture

In addition to the completeness of the group-identities and the classical identities for the equational theory of the regular sets, Conway [10] conjectured that the system consisting of the classical identities and an equation derived for each $n \geq 3$ from the n-state automaton with an input letter inducing a transposition and a letter inducing a cyclic permutation is also complete. As a consequence of the conjecture, it would follow that the regular identities in at most three variables form a complete system, whereas no upper bound on the number of variables is known to date.

In this section, we formulate a related conjecture for continuous Σ-algebras (or equivalently, by Theorem 6, for iteration Σ-algebras).

For each $n \geq 3$ and term t over Σ, consider the equation

$$\mu z.t[(z,z)/(x,y)] = \mu x.t[(t[t'/y], t[t'/y])/(x,y)], \tag{10}$$

where t' is the term $(\mu y.t)^{n-2}$ obtained by substituting $(n-3)$-times the term $\mu y.t$ for x in $\mu y.t$. (Thus, e.g., $(\mu y.t)^2 = (\mu y.t)[\mu y.t/x]$.)

Conjecture 1. A preiteration Σ-algebra is an iteration Σ-algebra iff it is a Conway Σ-algebra and satisfies the equation (10), for each $n \geq 3$.

If this conjecture holds, then so does Conway's.

6 Algebras with a Semilattice Structure

We will consider preiteration Σ-algebras equipped with a commutative idempotent additive structure. These are in fact preiteration Δ-algebras for the signature $\Delta = \Sigma_+$ that results by adding the binary symbol $+$ to Δ. Such a preiteration algebra is called a *semilattice preiteration Σ-algebra* if it satisfies the equations:

$$x + (y + z) = (x + y) + z \tag{11}$$
$$x + y = y + x \tag{12}$$
$$x + x = x \tag{13}$$
$$x + \bot = x. \tag{14}$$

Thus, semilattice preiteration Σ-algebras have the structure of a commutative idempotent monoid with neutral element the constant denoted by the term $\bot$, i.e., $\mu x.x$. Each such algebra A comes with the *semilattice order* defined by $a \leq b$ iff $a + b = b$. Note that $\bot_A$ is least with respect to this order.

A semilattice Σ-algebra is called continuous if, equipped with the semilattice order, it is a continuous Σ-algebra. Note that if A is a continuous semilattice algebra then A is in fact a *complete semilattice*, i.e., it has all suprema. Moreover, since the + operation is automatically continuous, A is a continuous Δ-algebra for the enlarged signature $\Delta = \Sigma_+$. Hence, by Theorem 6, any continuous semilattice Σ-algebra is a strong iteration Δ-algebra and satisfies the Park induction rule, i.e., it is a Park Δ-algebra. We call such algebras *semilattice Park Σ-algebras*. In a similar fashion, we define a semilattice Conway Σ-algebra (semilattice iteration Σ-algebra, respectively) to be a semilattice preiteration Σ-algebra which is a Conway Δ-algebra (iteration Δ-algebra, respectively). Morphisms of continuous semilattice Σ-algebras are continuous Δ-algebra morphisms. Morphisms of semilattice preiteration Σ-algebras, semilattice Conway algebras and semilattice iteration algebras and semilattice Park Σ-algebras are just preiteration Δ-algebra morphisms. Note that morphisms are automatically monotonic.

We end this section with a normal form theorem that applies to all semilattice Conway Σ-algebras, and thus to continuous semilattice Σ-algebras, semilattice Park Σ-algebras and semilattice iteration Σ-algebras. We will return to semilattice iteration Σ-algebras, and in particular to continuous semilattice Σ-algebras in Section 8.

A *simple term* over Σ_+ is a term that is the finite sum of different primitive terms over Σ excluding $\bot$. More precisely, given a pair (X, Y) of disjoint sets of variables, a simple term over (X, Y) is a finite sum of pairwise different terms of two types: Terms $\sigma(x_1, \ldots, x_k)$, where σ is in Σ and each x_i is in X, and terms y, for y a variable in Y. By assumption, the empty sum is $\bot$, so that the term $\bot$ itself is simple. The next result is a version of Milner's equational characterization of regular processes, cf. [30].

Theorem 10. [18] *For each term t over Σ_+ with free variables in $Y = \{y_1, \ldots, y_p\}$ there exist a set $X = \{x_1, \ldots, x_n\}$ and simple terms $t_1, \ldots, t_n$ over (X, Y) such that*

$$t = (\mu(x_1, \ldots, x_n).(t_1, \ldots, t_n))_1$$

holds in all additive Conway Σ-algebras satisfying

$$\mu x.x + y = y. \tag{15}$$

Remark 3. In Conway Σ_+-algebras, (15) is a strengthened form of idempotence. In fact, (13) follows from the fixed point equation and (15).

7 Regular Languages and Tree Languages

Suppose that A is a Σ-algebra. Then we may turn $\mathbf{P}(A)$, the power set of A into a Δ-algebra, where $\Delta = \Sigma_+$. For each $\sigma \in \Sigma_n$ and $B_1, \ldots, B_n \in P(A)$, we define

$$\sigma(B_1, \ldots, B_n) = \{\sigma(b_1, \ldots, b_n) : b_i \in B_i, \ i \in [n]\}.$$

Moreover, we define $B_1 + B_2 = B_1 \cup B_2$, for all $B_1, B_2 \in \mathbf{P}(A)$. Equipped with these operations, $\mathbf{P}(A)$ is a continuous semilattice Σ-algebra, hence a (strong) semilattice iteration Σ-algebra and a semilattice Park Σ-algebra. Note that when A is the free Σ-algebra $(\Sigma, B)\mathbf{FT}$ of finite (complete) (Σ, B)-trees, then $\mathbf{P}(A)$ is the algebra of all (finite complete) (Σ, B)-tree languages that we denote by $(\Sigma, B)\mathbf{TL}$.

Theorem 11. [13] *For each set A, $(\Sigma, A)\mathbf{TL}$ is freely generated by A in the class of all continuous semilattice Σ-algebras satisfying the following equations for all $\sigma \in \Sigma_n$ and $i \in [n]$, $n > 0$:*

$$\sigma(x_1 + y_1, \ldots, x_n + y_n) = \sum_{z_i \in \{x_i, y_i\}} \sigma(z_1, \ldots, z_n) \qquad (16)$$

$$\sigma(x_1, \ldots, \bot, \ldots, x_n) = \bot. \qquad (17)$$

On the left-hand side of (17), the symbol $\bot$ is on the ith position.

Suppose now that Σ is a unary signature, i.e., each symbol in Σ has rank one. Then a finite total tree in $(\Sigma, A)\mathbf{FT}$ may be represented as a word in $\Sigma^* A$. Hence $(\Sigma, A)\mathbf{TL}$ is just the algebra of languages in $\Sigma^* A$ equipped with the prefixing operations $L \mapsto \sigma L$, $\sigma \in \Sigma$, and the operation of set union as its sum operation. We let $(\Sigma^* A)\mathbf{L}$ denote this continuous semilattice Σ-algebra. By our general results, $(\Sigma^* A)\mathbf{L}$ is a (strong) semilattice iteration Σ-algebra and a semilattice Park Σ-algebra. The regular sets in $(\Sigma^* A)\mathbf{L}$ determine a subalgebra, denoted $(\Sigma^* A)\mathbf{RL}$, which is also a strong semilattice iteration Σ-algebra and a semilattice Park Σ-algebra. The following result is a version of Krob's theorem [26] that confirms a conjecture of Conway [10]. In [17], Krob's result is derived from Theorem 6.

Theorem 12. [26] *For each set A and unary signature Σ, the algebra $(\Sigma^* A)\mathbf{RL}$ is freely generated by A in the class of all (strong) semilattice iteration Σ-algebras satisfying (15) and*

$$t[y + z/x] = t[y/x] + t[z/x] \qquad (18)$$

$$t[0/x] = 0, \qquad (19)$$

for all terms $t = t[x]$ over Σ_+ containing at most the free variable x.

Since each semilattice Park Σ-algebra is an iteration Σ-algebra (Theorem 9), and since $(\Sigma^* A)\mathbf{RL}$ is a semilattice Park Σ-algebra, by Krob's theorem we have

Corollary 6. *For each set A and each unary signature Σ, the algebra $(\Sigma^* A)\mathbf{RL}$ is freely generated by A in the class of all semilattice Park Σ-algebras satisfying (18) and (19).*

This corollary may be viewed as a strengthened version of Kozen's axiomatization [24] of the equational theory of the regular sets.

Corollary 7. *The following conditions are equivalent for an equation $t = t'$ between μ-terms over Σ_+, where Σ is a unary signature.*

1. *The equation holds in all continuous semilattice Σ-algebras equipped with operations satisfying (16) and (17).*

2. The equation holds in all algebras $(\Sigma^*A)\mathbf{L}$ of languages, or $(\Sigma^*A)\mathbf{RL}$ of regular languages.
3. The equation holds in all semilattice Park Σ-algebras satisfying (16) and (17).
4. The equation holds in all (strong) semilattice iteration Σ-algebras satisfying (18), (19) and (15).

We do not know how Theorem 12 carries over to arbitrary signatures. Nevertheless the following result holds for all signatures containing symbols of arbitrary rank. For *regular tree languages* we refer to [20].

Theorem 13. [16] *For each signature Σ and set A, the algebra $(\Sigma, A)\mathbf{RL}$ of regular tree languages is freely generated by the set A in the class of all semilattice Park Σ-algebras satisfying (16) and (17).*

Note that since distributivity is required here only for basic symbols, this result is stronger for unary signatures than Corollary 6.

Corollary 8. *The following conditions are equivalent for an equation $t = t'$ between μ-terms over Σ_+.*

1. *The equation $t = t'$ holds in all continuous semilattice Σ-algebras equipped with operations satisfying (16) and (17).*
2. *The equation $t = t'$ holds in all complex algebras $\mathbf{P}(A)$ derived from Σ-algebras.*
3. *The equation $t = t'$ holds in all algebras $(\Sigma, A)\mathbf{TL}$ of languages, or $(\Sigma, A)\mathbf{RL}$ of regular tree languages.*
4. *The equation $t = t'$ holds in all semilattice Park Σ-algebras satisfying (16) and (17).*

8 Synchronization Trees

In this section, we consider the class of all continuous semilattice Σ-algebras, where Σ is any signature. The basic question we seek answer to is to provide a basis of identities for these structures. We refer to [30] and in particular [18] for the definition of (Σ, A)-*labeled synchronization trees* and the definition of *simulation*, originally introduced in [34].

Theorem 14. [18] *For each signature Σ and set A, the algebra $(\Sigma, A)\mathbf{RST}$ of regular synchronization trees over A modulo simulation is freely generated by A in the class of all semilattice iteration Σ_+-algebras satisfying (15) and*

$$t \leq t[x+y] \qquad (20)$$
$$\mu x.t \leq \mu x.t + t' \qquad (21)$$

for all terms t over Σ_+ and variables x, y.

Note that the meaning of (20) is that each function induced by any term t is monotonic, while (21) can be rephrased as an implication between equations:

$$\forall x, y(t \leq t') \Rightarrow \forall y(t \leq t'),$$

where $t(x, y)$ and $t'(x, y)$ are any terms over Σ_+.

Since $(\Sigma, A)\mathbf{RST}$ can be embedded in a continuous semilattice Σ-algebra, as shown in [18], we have:

Theorem 15. [18] *The following conditions are equivalent for an equation $t = t'$ between terms over Σ_+:*

1. $t = t'$ *holds in all continuous semilattice Σ-algebras.*
2. $t = t'$ *holds in all semilattice Park Σ-algebras.*
3. $t = t'$ *holds in all iteration semilattice Σ-algebras satisfying (15), (20) and (21).*
4. $t = t'$ *holds in all algebras of regular synchronization trees modulo simulation.*

It is natural to ask what happens if we drop (20) and (21) in Theorem 14. The answer to this question was given in [8]: The free algebras can be described as *bisimulation* equivalence classes of regular synchronization trees.

9 Conclusion

Several models studied in computer science have some facility of recursion, usually modeled by fixed points. Continuous algebras are those cpo's equipped with a Σ-algebra structure whose operations are continuous, giving rise to recursive definitions by least fixed points. Unfortunately, for μ-terms, the equational theory of the class of models studied is not always recursive, or recursively enumerable. For example, the equational theory of languages equipped with sum (set union) and concatenation as basic operations is not r.e., due to the fact that the equivalence of context-free grammars is not semidecidable. However, the theory of all continuous algebras is decidable, and in fact lies in P, and several equational theories are finitely axiomatizable over it. Moreover, the equations of fixed points in continuous algebras are quite general, and can thus be taken as a basis for the study of the respective equational theories. The relative axiomatization results presented here and elsewhere also provide a classification of the models by their logical properties. We have seen such examples in Sections 7 and 8. It would be of great importance to explore the boundary between axiomatizability and nonaxiomatizability, and decidability and undecidability.

Acknowledgment. This paper was prepared during the author's visit at BRICS (Basic Research in Computer Science, Centre of the Danish Research Foundation) and the Department of Computer Science of the University of Aalborg. He would like to thank the members of the Department, and in particular Anna Ingólfsdóttir, Luca Aceto and Kim G. Larsen, for their hospitality.

References

1. L. Aceto, Z. Ésik and A. Ingólfsdóttir. A fully equational proof of Parikh's theorem, *BRICS Report Series*, RS-01-28, 2001.
2. J. W. De Bakker and D. Scott. *A theory of programs*, IBM Seminar, Vienna, 1969.
3. H. Bekić. *Definable operations in general algebra. Technical Report, IBM Laboratory*, Vienna, 1969.
4. L. Bernátsky and Z. Ésik, Semantics of flowchart programs and the free Conway theories. *RAIRO Inform. Théor. Appl.*, 32(1998), 35–78.
5. S. L. Bloom and Z. Ésik. Floyd-Hoare logic in iteration theories. *J. Assoc. Comput. Mach.*, 38(1991), 887–934.
6. S. L. Bloom and Z. Ésik. Program correctness and matricial iteration theories. In: *Proc. Mathematical Foundations of Programming Semantics '91*, LNCS 598, Springer-Verlag, 1992, 457–475.
7. S. L. Bloom and Z. Ésik. *Iteration Theories.* Springer-Verlag, 1993.
8. S. L. Bloom, Z. Ésik and D. Taubner. Iteration theories of synchronization trees. *Inform. and Comput.*, 102(1993), 1–55.
9. S. Bozapalidis. Equational elements in additive algebras. *Theory Comput. Syst.*, 32(1999), 1–33.
10. J. H. Conway. *Regular Algebra and Finite Machines*, Chapman and Hall, 1971.
11. B. Courcelle, G. Kahn and J. Vuillemin. Algorithmes d'équivalence et de réduction á des expressions minimales dans une classe d'équations récursives simples. In *Proc. ICALP 74, Saarbrücken*, LNCS 14, Springer, 200–213.
12. C. C. Elgot, S. L. Bloom and R. Tindell. On the algebraic structure of rooted trees. *J. Comput. System Sci.*, 16(1978), 362–399.
13. J. Engelfriet and E. M. Schmidt. IO and OI. I. *J. Comput. System Sci.*, 15(1977), 328–353.
14. Z. Ésik. Identities in iterative and rational algebraic theories. *Comput. Linguist. Comput. Lang.*, 14(1980), 183–207.
15. Z. Ésik. Completeness of Park induction. *Theoret. Comput. Sci.*, 177(1997), 217–283.
16. Z. Ésik Axiomatizing the equational theory of regular tree languages. In: *Proc. STACS 98, Paris*, LNCS 1373, Springer, 1998, 455–466.
17. Z. Ésik. Group axioms for iteration. *Inform. and Comput.*, 148(1999), 131–180.
18. Z. Ésik. Axiomatizing the least fixed point operation and binary supremum. In: *Proc. CSL 2000*, LNCS 1862, Springer-Verlag, 302–316.
19. Z. Ésik and H. Leiß. In preparation.
20. F. Gécseg and M. Steinby. *Tree Automata.* Akadémiai Kiadó. Budapest, 1984.
21. A. Ginzburg. Algebraic theory of automata. *Academic Press*, New York–London, 1968.
22. J. A. Goguen, J. W. Thatcher, E. G. Wagner and J. B. Wright. Initial algebra semantics and continuous algebras. *J. Assoc. Comput. Mach.*, 24(1977), 68–95.
23. M. W. Hopkins and D. Kozen. Parikh's theorem in commutative Kleene algebra. In: *Proc. IEEE Conf. Logic in Computer Science (LICS'99)*, IEEE Press, July 1999, 394-401.
24. D. Kozen. A completeness theorem for Kleene algebras and the algebra of regular events. In: *Proc. 1991 IEEE Symposium on Logic in Computer Science (Amsterdam, 1991)*, and *Inform. and Comput.*, 110(1994), 366–390.
25. D. Kozen. On Hoare logic and Kleene algebra with tests. In: *Proc. IEEE Conf. Logic in Computer Science (LICS'99), IEEE, July 1999, 167–172*, and *ACM Trans. Computational Logic*, 1(2000), 60–76.
26. D. Krob. Complete systems of B-rational identities. *Theoret. Comput. Sci.*, 89(1991), 207–343.

27. I. Takanami and N. Honda. A characterization of Parikh's theorem and semilinear sets by commutative semigroups with length. *Electronics and Communications in Japan*, 52(1969), 179–184.
28. W. Kuich. The Kleene and the Parikh theorem in complete semirings. In: *Proc. ICALP '97*, LNCS 267, Springer-Verlag, 212–225.
29. W. Kuich. Gaussian elimination and a characterization of algebraic power series. In: *Proc. Mathematical Foundations of Computer Science, 1998*, LNCS 1450, Springer, Berlin, 1998, 512–521.
30. R. Milner. A complete inference system for a class of regular behaviors. *J. Comput. System Sci.*, 28(1984), 439–466.
31. D. Niwinski. Equational μ-calculus. In: *Computation Theory (Zaborów, 1984)*, LNCS 208, Springer, 1985, 169–176.
32. R. J. Parikh. {esik, zlnemeth}@inf.u-szeged.hu On context-free languages. *J. Assoc. Comput. Mach.*, 4(1996), 570–581.
33. D. Park. Fixpoint induction and proofs of program properties. In: *Machine Intelligence*, 5, American Elsevier, New York, 1970, 59–78.
34. D. Park. Concurrency and automata on infinite sequences. In: *Proc. GI Conf.*, LNCS 104, Springer, 1981, 167–183.
35. D. L. Pilling. Commutative regular equations and Parikh's theorem. *J. London Math. Soc.*, 6(1973), 663–666.
36. A. Salomaa. Two complete axiom systems for the algebra of regular events. *J. Assoc. Comput. Mach.*, 13(1966), 158–169.
37. J. Tiuryn. Fixed-points and algebras with infinitely long expressions. I. Regular algebras. *Fund. Inform.*, 2(1978/79), 103–127.

Second-Order Logic over Strings: Regular and Non-regular Fragments

Thomas Eiter[1], Georg Gottlob[1], and Thomas Schwentick[2]

[1] Institut für Informationssysteme, Technische Universität Wien
Favoritenstraße 9-11, A-1040 Wien, Austria
eiter@kr.tuwien.ac.at, gottlob@dbai.tuwien.ac.at
[2] Institut für Informatik, Friedrich-Schiller-Universität Jena
Ernst-Abbe-Platz 1-4, D-07743 Jena, Germany
tick@minet.uni-jena.de

Abstract. By a well-known result due to Büchi and Trakhtenbrot, all monadic second-order sentences over words describe regular languages. In this paper, we investigate prefix classes of general second-order logic. Such a prefix class is called *regular*, if each of its sentences describes a regular language, and *nonregular* otherwise. Recently, the regular and nonregular prefix classes of existential second-order logic (Σ_1^1) were exhaustively determined. We briefly recall these results and continue this line of research by systematically investigating the syntactically more complex prefix classes $\Sigma_k^1(\mathcal{Q})$ of second-order logic for each integer $k > 1$ and for each first-order quantifier prefix $\mathcal{Q}$. We give an exhaustive classification of the regular and nonregular prefix classes of this form, and derive of complexity results for the corresponding model checking problems. We also give a brief survey of recent results on the complexity of evaluating existential second-order logic over graphs, and a list of interesting open problems.

1 Introduction

Ever since the embryogenesis of computer science, the following scientific method has been successfully used: Represent computational problems by means of logical formulas, and derive insights into the structure (and thus the solvability and complexity) of a problem from the analysis of the structure of the corresponding formula. In particular, logicians and computer scientists have been studying for a long time the relationship between fragments of predicate logic and the solvability and complexity of decision problems that can be expressed within such fragments. Among the studied fragments, quantifier prefix classes play a predominant role. This can be explained by the syntactical simplicity of such prefix classes and by the fact that they form a natural hierarchy of increasingly complex fragments of logic that appears to be deeply related to core issues of decidability and complexity. In fact, one of the most fruitful research programs that kept logicians and computer scientists busy for decades was the exhaustive solution of Hilbert's classical decision problem (cf. [2]), i.e., of the problem of determining those prefix classes of first-order logic for which formula-satisfiability (resp. finite satisfiability) of formulas) is decidable.

Quantifier prefixes emerged not only in the context of decidability theory (a common branch of recursion theory and theoretical computer science), but also in core areas of computer science such as formal language and automata theory, and later in complexity theory. In automata theory, Büchi [4,3] and Trakhtenbrot [20] independently proved that a language is regular iff it can be described by a sentence of monadic second-order logic, in particular, by a sentence of monadic existential second-order logic. In complexity theory, Fagin [9] showed that a problem on finite structures is in NP iff it can be described by a sentence of existential second-order logic (ESO). These fundamental results have engendered a large number of further investigations and results on characterizing language and complexity classes by fragments of logic (see, e.g. the monographs [17,15,6, 12]).

What has not been achieved so far is the exhaustive classification of the complexity of second-order prefix classes over finite structures. By complexity of a prefix class C we mean the complexity of the following model checking problem: Given a fixed sentence Φ in C, decide for variable finite structures A whether A is a model of Φ, which we denote by $A \models \Phi$. Determining the complexity of all prefix classes is an ambitious research programme, in particular the analysis of various types of finite structures such as *strings*, i.e., finite word structures with successor, *trees*, or *graphs*. Over strings and trees, one of the main goals of this classification is to determine the *regular* prefix classes, i.e., those whose formulas express regular languages only; note that by Büchi's theorem, regular fragments over strings are (semantically) included in monadic second-order logic.

In the context of this research programme, two systematic studies were carried out recently, that shed light on the prefix classes of the existential fragment ESO (also denoted by Σ_1^1) of second-order logic:

- In [8], the ESO prefix-classes over strings are exhaustively classified. In particular, the precise frontier between regular and nonregular classes is traced, and it is shown that every class that expresses some nonregular language also expresses some NP-complete language. There is thus a huge complexity gap in ESO: some prefix classes can express only regular languages (which are well-known to have extremely low complexity), while all others are intractable. The results of [8] are briefly reviewed in Section 3.
- In [10], the complexity of all ESO prefix-classes over graphs and arbitrary relational structures is analyzed, and the tractability/intractability frontier is completely delineated. Unsurprisingly, several classes that are regular over strings become NP-hard over graphs. Interestingly, the analysis shows that one of the NP-hard classes becomes polynomial for the restriction to undirected graphs without self-loops. A brief account of these results is given in Section 5.

In the present paper, we continue the research on strings initiated in [8]. But rather than considering fragments of ESO = Σ_1^1, we will venture into spheres of the higher prefix classes Σ_k^1, where $k > 1$. For a first-order quantifier pattern $\mathcal{Q}$, we denote by $\Sigma_k^1(\mathcal{Q})$ the class of all second-order formulas over strings of the form

$$\exists \mathbf{R}_1 \forall \mathbf{R}_2 \cdots Q_k \mathbf{R}_k \; \mathcal{Q}[\mathbf{x}] \; \varphi(\mathbf{R}_1, \ldots, \mathbf{R}_k, \mathbf{x}),$$

where, for $1 \leq i \leq k$, $\mathbf{R}_i$ is a list of relations, and $\mathcal{Q}([\mathbf{x}])$ denotes a quantifier prefix according to pattern $\mathcal{Q}$ quantifying over first-order variables from the list $\mathbf{x}$.

We give an exhaustive classification of the regular and nonregular second-order prefix classes described by the notation $\Sigma_k^1(\mathcal{Q})$, for all integers $k > 1$ and all first-order quantification patterns $\mathcal{Q}$. We also present complexity results for the corresponding model checking problems. However, at this point, we do not completely describe the tractability frontier but leave a few complexity problems for further work (see Section 6).

The rest of this paper is organized as follows. In Section 2, we state the relevant definitions and recall some classical results. In Section 3, we briefly survey the results of [8] on the classification of ESO prefix-classes over strings. Section 4, in which we present our new results, is the core of this paper. In Section 5, we give a brief account on the classification of ESO over graphs from [10]. Finally, we conclude the paper with a list of interesting open problems in Section 6.

2 Preliminaries and Classical Results

We consider second-order logic with equality (unless explicitly stated otherwise) and without function symbols of positive arity. Predicates are denoted by capitals and individual variables by lower case letters; a bold face version of a letter denotes a tuple of corresponding symbols.

A *prefix* is any string over the alphabet $\{\exists, \forall\}$, and a *prefix set* is any language $\mathcal{Q} \subseteq \{\exists, \forall\}^*$ of prefixes. A prefix set $\mathcal{Q}$ is *trivial*, if $\mathcal{Q} = \emptyset$ or $\mathcal{Q} = \{\lambda\}$, i.e., it consists of the empty prefix. In the rest of this paper, we focus on nontrivial prefix sets. We often view a prefix Q as the prefix class $\{Q\}$. A *generalized prefix* is any string over the extended prefix alphabet $\{\exists, \forall, \exists^*, \forall^*\}$. A prefix set $\mathcal{Q}$ is *standard*, if either $\mathcal{Q} = \{\exists, \forall\}^*$ or $\mathcal{Q}$ can be given by some generalized prefix.

For any prefix Q, the class $\Sigma_0^1(Q)$ is the set of all prenex first-order formulas (which may contain free variables and constants) with prefix Q, and for every $k \geq 0$, $\Sigma_{k+1}^1(Q)$ (resp., Π_{k+1}^1) is the set of all formulas $\exists \mathbf{R} \Phi$ (resp., $\forall \mathbf{R} \Phi$) where Φ is from Π_k^1 (resp., Σ_k^1). For any prefix set $\mathcal{Q}$, the class $\Sigma_k^1(\mathcal{Q})$ is the union $\Sigma_k^1(\mathcal{Q}) = \bigcup_{Q \in \mathcal{Q}} \Sigma_k^1(Q)$. We write also ESO for Σ_1^1. For example, ESO($\exists^*\forall\exists^*$) is the class of all formulas $\exists \mathbf{R} \exists \mathbf{y} \forall x \exists \mathbf{z} \varphi$, where φ is quantifier-free; this is the class of ESO-prefix formulas, whose first-order part is in the well-known Ackermann class with equality.

Let $A = \{a_1, \ldots, a_m\}$ be a finite alphabet. A *string* over A is a finite first-order structure $W = \langle U, C_{a_1}^W, \ldots, C_{a_m}^W, Succ^W, min^W, max^W \rangle$, for the vocabulary $\sigma_A = \{C_{a_1}, \ldots, C_{a_m}, Succ, min, max\}$, where

- U is a nonempty finite initial segment $\{1, 2, \ldots, n\}$ of the positive integers;
- each $C_{a_i}^W$ is a unary relation over U (i.e., a subset of U) for the unary predicate C_{a_i}, for $i = 1, \ldots, m$, such that the $C_{a_i}^W$ are pairwise disjoint and $\bigcup_i C_{a_i}^W = U$.
- $Succ^W$ is the usual successor relation on U and min^W and max^W are the first and the last element in U, respectively.

Observe that this representation of a string is a *successor structure* as discussed e.g. in [7]. An alternative representation uses a standard linear order $<$ on U instead of the successor $Succ$. In full ESO or second-order logic, $<$ is tantamount to $Succ$ since either predicate can be defined in terms of the other.

The strings W for A correspond to the nonempty finite words over A in the obvious way; in abuse of notation, we often use W in place of the corresponding word from A^* and vice versa.

A *SO sentence* Φ over the vocabulary σ_A is a second-order formula whose only free variables are the predicate variables of the signature σ_A, and in which no constant symbols except min and max occur. Such a sentence defines a language over A, denoted $\mathcal{L}(\Phi)$, given by $\mathcal{L}(\Phi) = \{W \in A^* \mid W \models \Phi\}$. We say that a language $L \subseteq A^*$ is *expressed* by Φ, if $\mathcal{L}(\Phi) = L \cap A^+$ (thus, for technical reasons, without loss of generality we disregard the empty string); L is *expressed by a set* S of sentences, if L is expressed by some $\Phi \in S$. We say that S *captures* a class C of languages, if S expresses all and only the languages in C.

Example 1. Let us consider some languages over the alphabet $A = \{a, b\}$, and how they can be expressed using logical sentences.

- $L_1 = \{a,b\}^*b\{a,b\}^*$: This language is expressed by the simple sentence

$$\exists x. C_b(x).$$

- $L_2 = a^*b$: This language is expressed by the sentence

$$C_b(max) \wedge \forall x \neq max. C_a(x).$$

- $L_3 = (ab)^*$: Using the successor predicate, we can express this language by

$$C_a(min) \wedge C_b(max) \wedge \forall x, y. Succ(x,y) \to (C_a(x) \leftrightarrow \neg C_a(y)).$$

- $L_4 = \{w \in \{a,b\}^* \mid |w| = 2n, n \geq 1\}$: We express this language by the sentence

$$\exists E \, \forall x, y. \neg E(min) \wedge E(max) \wedge Succ(x,y) \to (E(x) \leftrightarrow \neg E(y)).$$

Note that this a monadic ESO sentence. It postulates the existence of a monadic predicate E, i.e., a "coloring" of the string such that neighbored positions have different color, and the first and last position are uncolored and colored, respectively.
- $L_5 = \{a^n b^n \mid n \geq 1\}$: Expressing this language is more involved:

$$\exists R \, \forall x, x^+, y, y^-. R(min, max) \wedge [R(x,y) \to (C_a(x) \wedge C_b(y))] \wedge \\ [(Succ(x, x^+) \wedge Succ(y^-, y) \wedge R(x,y)) \to R(x^+, y^-)].$$

Observe that this sentence is not monadic. Informally, it postulates the existence of an arc from the first to the last position of the string W, which must be an a and a b, respectively, and recursively arcs from the i-th to the $(|W| - i + 1)$-th position.

Let A be a finite alphabet. A sentence Φ over σ_A is called *regular*, if $\mathcal{L}(\Phi)$ is a regular language. A set of sentences S (in particular, any ESO-prefix class) is *regular*, if for every finite alphabet A, all sentences $\Phi \in S$ over σ_A are regular.

Büchi [4] has shown the following fundamental theorem, which was independently found by Trakhtenbrot [20]. Denote by MSO the fragment of second-order logic in which all predicate variables have arity at most one,[1] and let REG denote the class of regular languages.

---
[1] Observe that we assume MSO allows one to use nullary predicate variables (i.e., propositional variables) along with unary predicate variables. Obviously, Büchi's Theorem survives.

Proposition 1 (Büchi's Theorem). MSO *captures* REG.

That MSO can express all regular languages is easy to see, since it is straightforward to describe runs of a finite state automaton by an existential MSO sentence. In fact, this is easily possible in monadic $ESO(\forall\exists)$ as well as in monadic $ESO(\forall\forall)$. Thus, we have the following lower expressiveness bound on ESO-prefix classes over strings.

Proposition 2. *Let $\mathcal{Q}$ be any prefix set. If $\mathcal{Q} \cap \{\exists,\forall\}^*\forall\{\exists,\forall\}^+ \neq \emptyset$, then $ESO(\mathcal{Q})$ expresses all languages in* REG.

On the other hand, with non-monadic predicates allowed ESO has much higher expressivity. In particular, by Fagin's result [9], we have the following.

Proposition 3 (Fagin's Theorem). ESO *captures* NP.

This theorem can be sharpened to various fragments of ESO. In particular, by Leivant's results [14,7], in the presence of a successor and constants min and max, the fragment $ESO(\forall^*)$ captures NP; thus, $ESO(\forall^*)$ expresses all languages in NP.

3 Recent Results on ESO over Strings

Combining and extending the results of Büchi and Fagin, it is natural to ask: What about (nonmonadic) prefix classes $ESO(\mathcal{Q})$ over finite strings? We know by Fagin's theorem that all these classes describe languages in NP. But there is a large spectrum of languages contained in NP ranging from regular languages (at the bottom) to NP-hard languages at the top. What can be said about the languages expressed by a given prefix class $ESO(\mathcal{Q})$? Can the expressive power of these fragments be characterized? In order to clarify these issues, the following particular problems where investigated in [8]:

- Which classes $ESO(\mathcal{Q})$ express only regular languages? In other terms, for which fragments $ESO(\mathcal{Q})$ is it true that for any sentence $\Phi \in ESO(\mathcal{Q})$ the set $Mod(\Phi) = \{W \in A^* \mid W \models \Phi\}$ of all finite strings (over a given finite alphabet A) satisfying Φ constitutes a regular language? By Büchi's Theorem, this question is identical to the following: Which prefix classes of ESO are (semantically) included in MSO?

Note that by Gurevich's classifiability theorem (cf. [2]) and by elementary closure properties of regular languages, it follows that there is *a finite number of maximal regular prefix classes* $ESO(\mathcal{Q})$, and similarly, of minimal nonregular prefix classes; the latter are, moreover, standard prefix classes (cf. Section 2). It was the aim of [8] to determine the maximal regular prefix classes and the minimal nonregular prefix classes.
- What is the complexity of model checking (over strings) for the *nonregular* classes $ESO(\mathcal{Q})$, i.e., deciding whether $W \models \Phi$ for a given W (where Φ is fixed)?

Model checking for regular classes $ESO(\mathcal{Q})$ is easy: it is feasible by a finite state automaton. We also know (e.g. by Fagin's Theorem) that some classes $ESO(\mathcal{Q})$ allow us to express NP-complete languages. It is therefore important to know (i) which classes $ESO(\mathcal{Q})$ can express NP-complete languages, and (ii) whether there are prefix classes $ESO(\mathcal{Q})$ of intermediate complexity between regular and NP-complete classes.

- Which classes $ESO(\mathcal{Q})$ *capture* the class REG? By Büchi's Theorem, this question is equivalent to the question of which classes $ESO(\mathcal{Q})$ have exactly the expressive power of MSO over strings.
- For which classes $ESO(\mathcal{Q})$ is finite satisfiability decidable, i.e., given a formula $\Phi \in ESO(\mathcal{Q})$, decide whether Φ is true on some finite string ?

Reference [8] answers all the above questions exhaustively. Some of the results are rather unexpected. In particular, a surprising dichotomy theorem is proven, which sharply classifies all $ESO(\mathcal{Q})$ classes as either regular or intractable. Among the main results of [8] are the following findings.

(1) The class $ESO(\exists^*\forall\exists^*)$ is regular. This theorem is the technically most involved result of [8]. Since this class is nonmonadic, it was not possible to exploit any of the ideas underlying Büchi's proof for proving it regular. The main difficulty consists in the fact that relations of higher arity may connect elements of a string that are very distant from one another; it was not *a priori* clear how a finite state automaton could guess such connections and check their global consistency. To solve this problem, new combinatorial methods (related to hypergraph transversals) were developed.

Interestingly, model checking for the fragment $ESO(\exists^*\forall\exists^*)$ is NP-complete over *graphs*. For example, the well-known set-splitting problem can be expressed in it. Thus the fact that our input structures are monadic *strings* is essential (just as for MSO).

(2) The class $ESO(\exists^*\forall\forall)$ is regular. The regularity proof for this fragment is easier but also required the development of new techniques (more of logical than of combinatorial nature). Note that model checking for this class, too, is NP-complete over graphs.

(3) Any class $ESO(\mathcal{Q})$ not contained in the union of $ESO(\exists^*\forall\exists^*)$ and $ESO(\exists^*\forall\forall)$ is not regular.

Thus $ESO(\exists^*\forall\exists^*)$ and $ESO(\exists^*\forall\forall)$ are the *maximal regular standard prefix classes*. The unique maximal (general) regular ESO-prefix class is the union of these two classes, i.e, $ESO(\exists^*\forall\exists^*) \cup ESO(\exists^*\forall\forall) = ESO(\exists^*\forall(\forall \cup \exists^*))$.

As shown in [8], it turns out that there are three minimal nonregular ESO-prefix classes, namely the standard prefix classes $ESO(\forall\forall\forall)$, $ESO(\forall\forall\exists)$, and $ESO(\forall\exists\forall)$. All these classes express nonregular languages by sentences whose list of second-order variables consists of a single binary predicate variable.

Thus, 1.-3. give a *complete characterization* of the regular $ESO(\mathcal{Q})$ classes.

(4) The following dichotomy theorem is derived: Let $ESO(\mathcal{Q})$ be any prefix class. Then, either $ESO(\mathcal{Q})$ is regular, or $ESO(\mathcal{Q})$ expresses some NP-complete language. This means that model checking for $ESO(\mathcal{Q})$ is either possible by a deterministic finite automaton (and thus in constant space and linear time) or it is already NP-complete. Moreover, for all NP-complete classes $ESO(\mathcal{Q})$, NP-hardness holds already for sentences whose list of second-order variables consists of a single binary predicate variable. There are no fragments of intermediate difficulty between REG and NP.

(5) The above dichotomy theorem is paralleled by the solvability of the finite satisfiability problem for ESO (and thus FO) over strings. As shown in [8], over finite strings satisfiability of a given $ESO(\mathcal{Q})$ sentence is decidable iff $ESO(\mathcal{Q})$ is regular.

(6) In [8], a precise characterization is given of those prefix classes of ESO which are equivalent to MSO over strings, i.e. of those prefix fragments that *capture* the class REG

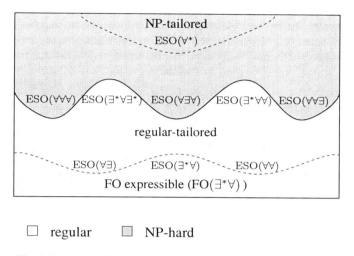

Fig. 1. Complete picture of the ESO-prefix classes on finite strings

of regular languages. This provides new logical characterizations of REG. Moreover, in [8] it is established that any regular ESO-prefix class is over strings either equivalent to full MSO, or is contained in first-order logic, in fact, in FO(∃*∀).

It is further shown that ESO(∀*) is the unique minimal ESO prefix class which captures NP. The proof uses results in [14,7] and well-known hierarchy theorems.

The main results of [8] are summarized in Figure 1. In this figure, the ESO-prefix classes are divided into four regions. The upper two contain all classes that express nonregular languages, and thus also NP-complete languages. The uppermost region contains those classes which capture NP; these classes are called NP-*tailored*. The region next below, separated by a dashed line, contains those classes which can express some NP-hard languages, but not all languages in NP. Its bottom is constituted by the minimal nonregular classes, ESO(∀∀∀), ESO(∀∃∀), and ESO(∀∀∃). The lower two regions contain all regular classes. The maximal regular standard prefix classes are ESO(∃*∀∃*) and ESO(∃*∀∀). The dashed line separates the classes which capture REG(called *regular-tailored*), from those which do not; the expressive capability of the latter classes is restricted to first-order logic (in fact, to FO(∃*∀)) [8]. The minimal classes which capture REG are ESO(∀∃) and ESO(∀∀).

As an example for an NP-complete language, we consider a natural string encoding of the classical satisfiability problem (SAT) and show how it can be expressed by a sentence from the minimal non-regular class ESO(∀∀∀).

Example 2 ([8]). For expressing SAT, we encode instances $\mathcal{C} = \{C^1, \ldots C^m\}$, where the C^i are clauses on propositional variables $p_1, \ldots, p_n$, to strings $enc(\mathcal{C})$ over the alphabet $A = \{0, 1, +, -, [,], (,)\}$ as follows. The variables p_i, $1 \leq i \leq n$ are encoded by binary strings of length $\lceil \log n \rceil$. Each string encoding p_i is enclosed by parentheses

'(',')'. The polarity of a literal $p_i/\neg p_i$ is represented by the letter '+' or '−', respectively, which immediately follows the closing parenthesis ')' of the encoding of p_i. A clause is encoded as a sequence of literals which is enclosed in square brackets '[',']'. Without loss of generality, we assume that $\mathcal{C} \neq \emptyset$ and that each clause $C^i \in \mathcal{C}$ contains at least one literal.

For example, $\mathcal{C} = \{\,\{p,q,\neg r\},\ \{\neg p,\neg q,r\}\,\}$ is encoded by the following string:

$$enc(\mathcal{C}) = \texttt{[(00)+(01)+(10)-][(00)-(01)-(10)+]}\,.$$

Here, p, q, r are encoded by the binary strings 00, 01, 10, respectively. Clearly, $enc(\mathcal{C})$ is obtainable from any standard representation of $\mathcal{C}$ in logspace.

In what follows, we will use the formulas

$$eqcol(x,y) = \bigvee_{\ell \in A} (C_\ell(x) \wedge C_\ell(y)), \tag{1}$$

$$varenc(x) = C_{(}(x) \vee C_0(x) \vee C_1(x) \vee C_{)}(x) \tag{2}$$

which state that the string has at positions x and y the same letter from A and that x is a letter of a variable encoding, respectively.

Then, satisfiability of $\mathcal{C}$ is expressed through the following $\Sigma_1^1(\forall\forall\forall)$ sentence:

$$\Phi = \exists V \exists G \exists R \exists R' \forall x \forall y \forall z.\, \varphi(x,y,z),$$

where G and V are unary, R' and R are binary, and $\varphi(x,y,z)$ is the conjunction of the following quantifier-free formulas φ_G, φ_V, φ_R, and $\varphi_{R'}$:

$$\varphi_G = \varphi_{G,1} \wedge \varphi_{G,2} \wedge \varphi_{G,3},$$

where

$$\varphi_{G,1} = \Big(C_{[}(x) \to \neg G(x)\Big) \wedge \Big(C_{]}(x) \to G(x)\Big),$$
$$\varphi_{G,2} = \Big(Succ(x,y) \wedge \neg C_{[}(y) \wedge \neg C_{)}(y)\Big) \to \Big(G(y) \leftrightarrow G(x)\Big),$$
$$\varphi_{G,3} = \Big(C_{)}(y) \wedge Succ(x,y) \wedge Succ(y,z)\Big) \to$$
$$\Big(G(y) \leftrightarrow \big[G(x) \vee (V(y) \wedge C_+(z)) \vee (\neg V(y) \wedge C_-(z))\big]\Big);$$

next,

$$\varphi_V = \Big(C_{)}(x) \wedge C_{)}(y) \wedge R(x,y)\Big) \to \Big(V(x) \leftrightarrow V(y)\Big),$$
$$\varphi_R = \Big[R(x,y) \to (eqcol(x,y) \wedge varenc(x))\Big] \wedge$$
$$\Big[(C_{(}(x) \wedge C_{(}(y)) \to R(x,y)\Big] \wedge$$
$$\Big[\big(\neg C_{(}(x) \wedge Succ(z,x)\big) \to \Big(R(x,y) \leftrightarrow (R'(z,y) \wedge eqcol(x,y))\Big)\Big], \tag{3}$$

and

$$\varphi_{R'} = Succ(z,y) \to \Big[R'(x,y) \leftrightarrow \Big(R(x,z) \wedge \neg C_{)}(z)\Big)\Big]$$

Informally, the predicate V assigns a truth value to each occurrence of a variable p_i in $\mathcal{C}$, which is given by the value of V at the closing parenthesis of this occurrence. $\mathcal{C}$ is satisfiable, precisely if there exists such a V which assigns all occurrences of p_i the same truth value and satisfies every clause. This is checked using G, R, and R'.

The predicate G is used for checking whether each clause $C \in \mathcal{C}$ is satisfied by the assignment V. To this end, G is set to false at the '[' marking the beginning of C, and set to true at the ']' marking the end of C by formula $\varphi_{G,1}$; the formulas $\varphi_{G,2}$ and $\varphi_{G,3}$ propagate the value of G from a position x in the clause representation to the successor position y, where the value switches from false to true if y marks the sign of a literal which is satisfied by V; the conjunct $\neg C_{[}$ in φ_G, prohibits the transfer of G from the end of C to the beginning of the next clause, for which G must be initialized to false.

The predicate R is used to identify the closing parentheses ')' of the representations of occurrences of the same variables. For positions x and y at which the string W has letter ')', the predicate $R(x, y)$ is true precisely if x and y mark the end of the same variable name. This is used in the formula φ_V, which then simply states that V assigns every occurrence of a variable p in $\mathcal{C}$ the same truth value.

The purpose of φ_R and $\varphi_{R'}$ is to ensure that $R(x, y)$ has for positions x and y which mark the ends of occurrences of the same variable the desired meaning. This is accomplished inductively. Intuitively, $R(x, y)$ expresses that x and y have the same color, must be part of the encodings $o_x = (\cdots)$ and $o_y = (\cdots)$ of variable occurrences in $\mathcal{C}$, and are at the same distance from the beginnings of these encodings. By reference to the predicate R', $R(x, y)$ furthermore expresses that this also holds for predecessors x^- and y^- of x in o_x and y in o_y, respectively. The predicate R' is needed since we may use only three first-order variables, while the natural statement of the inductive property of R uses four first-order variables.

As shown in [8], $enc(\mathcal{C}) \models \Phi$ iff $\mathcal{C}$ is satisfiable. Furthermore, using packing techniques, Φ can be replaced by an equivalent ESO($\forall\forall\forall$) sentence of the form $\Phi' = \exists R \forall x \forall y \forall z \varphi(x, y, z)$, where R is binary.

4 SO over Strings

In this section, we investigate an extension of the results in the previous section from ESO to full second-order logic over strings. Our focus of attention will be the SO prefix classes which are regular vs. those which are not. In particular, we are interested to know how adding second-order variables affects regular fragments.

Generalizing Fagin's theorem, Stockmeyer [16] has shown that full SO captures the polynomial hierarchy (PH). Second-order variables turn out to be quite powerful. In fact, already two first-order variables, a single binary predicate variable, and further monadic predicate variables are sufficient to express languages that are complete for the levels of PH.

4.1 Regular Fragments

Let us first consider possible generalizations of regular ESO prefix classes. It is easy to see that every ESO sentence in which only a single first-order variable occurs is equivalent to a monadic second-order sentence. This can be shown by eliminating predicate variables of arity > 1 through introducing new monadic predicate variables. The result generalizes to sentences in full SO with the same first-order part, which can be shown using the same technique. Thus, the class $\Sigma_k^1(\{\exists, \forall\})$ is regular for every $k \geq 0$.

Similarly, it can be shown that every ESO sentence in which only two first-order variables occur is equivalent to a MSO sentence [11]. However, as follows from the results of the next section, this does not generalize beyond ESO. Nonetheless, there are still higher-rank Σ_k^1 fragments with two first-order variables, in particular Σ_2^1 prefix classes, which are regular. Here, similar elimination techniques as in the case of a single first-order variable are applicable. The following lemmas are useful for this purpose.

Lemma 1 ([13]). *Every formula $\exists \mathbf{R} \exists y_1 \ldots \exists y_k.\varphi$, where φ is quantifier-free, is equivalent to a formula $\exists y_1 \ldots \exists y_k.\psi$, where ψ is quantifier-free.*

Proposition 4. *Every formula in $\Sigma_k^1(\exists^j)$, where $k \geq 1$ is odd, is equivalent to some formula in $\Sigma_{k-1}^1(\exists^j)$, and every formula in $\Sigma_k^1(\forall^j)$, where $k \geq 2$ is even, is equivalent to some formula in $\Sigma_{k-1}^1(\forall^j)$.*

A generalization of the proof of Theorem 9.1 in [8] yields the following.

Theorem 1. *Over strings, $\Sigma_2^1(\forall\forall) = $ MSO.*

Proof. By the previous proposition, every formula $\exists \mathbf{S} \forall \mathbf{R} \forall x \forall y \varphi$ in $\Sigma_2^1(\forall\forall)$ is equivalent to some formula in $\Sigma_1^1(\forall\forall)$ (rewrite $\exists \mathbf{S} \forall \mathbf{R} \forall x \forall y \varphi$ to $\exists \mathbf{S} \neg (\exists \mathbf{R} \exists x \exists y \neg \varphi)$). Thus, $\Sigma_2^1(\forall\forall) \subseteq$ MSO. On the other hand, over strings MSO $= \Sigma_1^1(\forall\forall)$, which is trivially included in $\Sigma_2^1(\forall\forall)$. □

A similar elimination of second-order variables is possible in a slightly more general case.

Lemma 2 ([8]). *Every formula $\Phi = \exists \mathbf{R} \exists y_1 \ldots \exists y_k \forall x.\varphi$, where φ is quantifier-free, is equivalent to a finite disjunction of first-order formulas $\exists y_1 \ldots \exists y_k \forall x.\psi$, where ψ is quantifier-free.*

The lemma was proven in [8] for strings, but the proof easily generalizes to arbitrary structures.

Let $\Sigma_k^1(\bigvee \mathcal{Q})$ (resp., $\Sigma_k^1(\bigwedge \mathcal{Q})$) denote the class of Σ_k^1 sentences where the first-order part is a finite disjunction (resp., conjunction) of prefix formulas with quantifier in $\mathcal{Q}$. Then we have the following.

Proposition 5. *Every formula in $\Sigma_k^1(\exists^j \forall)$, where $k \geq 1$ is odd and $j \geq 0$, is equivalent to some formula in $\Sigma_{k-1}^1(\bigvee \exists^j \forall)$, and every formula in $\Sigma_k^1(\forall^j \exists)$, where $k \geq 2$ is even and $j \geq 1$, is equivalent to some formula in $\Sigma_{k-1}^1(\bigwedge \forall^j \exists)$.*

Proof. The first part follows immediately from Lemma 2, by which also the second is easily shown. Let $\Phi = \exists \mathbf{S} \cdots \forall \mathbf{R} \forall \mathbf{y} \exists x.\varphi(\mathbf{y}, x)$ be a sentence in $\Sigma_k^1(\forall^j \exists)$ for even $k \geq 2$. Then, Φ is equivalent to

$$\exists \mathbf{S} \cdots \neg (\exists \mathbf{R} \exists \mathbf{x} \forall x. \neg \varphi(\mathbf{y}, x))$$

By Lemma 2, this formula is equivalent to

$$\exists \mathbf{S} \cdots \neg (\bigvee_{i=1}^{n} \exists \mathbf{x} \forall x . \psi_i(\mathbf{y}, x))$$

where each $\psi_i(\mathbf{y}, x)$ is quantifier-free. By moving negation inside, we obtain the result. □

Corollary 1. *Over strings, $\Sigma_2^1(\forall\exists) =$ MSO.*

Proof. By the previous proposition, every sentence Φ in $\Sigma_2^1(\forall\exists)$ is equivalent to some sentence Φ' in $\Sigma_1^1(\bigwedge \forall\exists)$. We may rewrite Φ' as a sentence Φ'' of form $\exists \mathbf{R} \forall x \exists \mathbf{y} \varphi$, where φ is quantifier-free. Recall that $\mathbf{y}$ denotes a tuples of variables. By Corollary 8.4 in [8], Φ'' is in MSO. Thus $\Sigma_2^1(\forall\exists) \subseteq$ MSO. The reverse inclusion holds since, as well-known, over strings MSO $= \Sigma_1^1(\forall\exists)$. □

4.2 Non-regular Fragments

$\Sigma_2^1(Q)$ where $|Q| \leq 2$. While for the first-order prefixes $Q = \forall\exists$ and $Q = \forall\forall$, regularity of ESO(Q) generalizes to Σ_2^1, this is not the case for the other two-variables prefixes $\exists\forall$ and $\exists\exists$.

Theorem 2. *$\Sigma_2^1(\exists\forall)$ is nonregular.*

For a proof of this result, we consider the following example.

Example 3. Let $A = \{a, b\}$ and consider the following sentence:

$$\Phi = \exists R \forall X \exists x \forall y . [C_a(y) \vee (C_a(x) \wedge (X(y) \leftrightarrow R(x, y))]$$

Informally, this sentence is true for a string W, just if the number of b's in W (denoted $\#b(W)$) is at most logarithmic in the number of a's in W (denoted $\#a(W)$). More formally, $L(\Phi) = \{w \in \{a, b\}^* \mid \#b(W) \leq \log \#a(W)\}$; by well-known results, this language is not regular.

Indeed, for any assignment to the monadic variable X, there must be a position i colored with C_a in the string such that, for every position j colored with C_b, an arc goes from i to j just if X is true for j. For assignments to X which are different on the positions colored with C_b, different positions i must be chosen. If $\#b(W) \leq \log \#a(W)$, this choice can be satisfied; otherwise, there must exist two different assignments to X which require that, for some position i colored with C_a and some position j colored with C_b, both $R(i, j)$ and $\neg R(i, j)$ hold, which is not possible.

Lemma 3. *Every first-order formula $\forall x \forall y \varphi(x, y)$, where φ is quantifier-free, is equivalent to some first-order formula $\forall X \exists x \forall y \psi(x, y)$ where X is a fresh monadic variable and φ is quantifier-free.*

Proof. (Sketch) Define
$$\psi(x,y) = (X(x) \land \varphi(x,y)) \lor (\neg X(x) \land \neg X(y))$$
Then, for every structure and assignment to the free variables, $\forall x \forall y \varphi(x,y)$ (resp., $\forall x \exists y \varphi(x,y)$) is equivalent to $\forall X \exists x \forall y \psi(x,y)$. □

Using the lemma, we establish the following result about the expressiveness of $\Sigma_2^1(\exists\forall)$.

Theorem 3. *Every $\Phi \in \Sigma_2^1(\bigwedge\{\exists\forall, \forall\forall\})$ is equivalent to some $\Sigma_2^1(\exists\forall)$ formula.*

Proof. (Sketch) Let $\Phi = \exists \mathbf{R} \forall \mathbf{T} \varphi$, where $\varphi = \bigwedge_{i=1}^n \varphi_i$ is a conjunction of prefix first-order formulas φ_i with prefix from $\{\exists\forall, \forall\exists, \forall\forall\}$. By Lemma 3, each φ_i of the form $\forall x \forall y \varphi'_i(x,y)$ (resp. $\forall x \exists y \varphi(x,y)$), can be equivalently replaced by a formula $\exists x \forall y \psi_i(x,y,X_i)$ and adding $\forall X_i$ to the universal second-order part of Φ, where X_i is a fresh monadic variable. Then, rewrite the resulting conjunction $\bigwedge_{i=1}^n \exists x \forall y \alpha_i(x,y)$ in the first-order part to $\forall Z_1 \cdots \forall Z_n \exists x \forall y (\beta \to (\bigwedge_{i=1}^n Z_i \to \alpha(x,y)))$, where the Z_i are fresh Boolean variables, and β states that exactly one out of $Z_1, \ldots, Z_n$ is true. The resulting formula Φ', which is in $\Sigma_2^1(\exists\forall)$, is equivalent to Φ. □

For the class $\Sigma_2^1(\exists\exists)$, our proof of non-regularity is more involved. The proof of the following lemma shows how to emulate universal first-order quantifiers using universal second-order quantifiers and existential quantifiers over strings.

Lemma 4. *Over strings, every universal first-order formula $\forall \mathbf{x} \varphi(\mathbf{x})$ which contains no predicates of arity > 2 is equivalent to some $\Pi_1^1(\exists\exists)$ formula.*

Proof. (Sketch) The idea is to emulate the universal quantifier $\forall x_i$ for every x_i from $\mathbf{x} = x_1, \ldots, x_k$ using a universally quantified variable S_i which ranges over singletons, and express "x_i" by "$\exists x_i.S_i(X_i)$." Then, $\forall \mathbf{x} \varphi(\mathbf{x})$ is equivalent to $\forall \mathbf{S} \exists \mathbf{x} \bigwedge_{i=1}^k S_i(x_k) \land \varphi(\mathbf{x})$.

We can eliminate all existential variables $\mathbf{x}$ but two in this formula as follows. Rewrite the quantifier-free part in CNF $\bigwedge_{i=1}^\ell \delta_i(\mathbf{x})$, where each $\delta_i(\mathbf{x})$ is a disjunction of literals. Denote by $\delta_i^{j,j'}(x_j, x_{j'})$ the clause obtained from $\delta_i(\mathbf{x})$ by removing every literal which contains some variable from $\mathbf{x}$ different from x_j and $x_{j'}$. Since no predicate in φ has arity > 2, formula $\forall \mathbf{x} \varphi(\mathbf{x})$ is equivalent to the formula

$$\forall \mathbf{S} \exists \mathbf{x} \bigwedge_{i=1}^\ell \left(\bigvee_{j \neq j'} \exists x \exists y \delta_i^{j,j'}(x,y) \right).$$

The conjunction $\bigwedge_{i=1}^\ell$ can be simulated similarly as in the proof of Theorem 3, by using universally quantified Boolean variables $Z_1, \ldots, Z_\ell$ and a control formula β which states that exactly one out of $Z_1, \ldots, Z_n$ is true. By pulling existential quantifiers, we thus obtain

$$\forall \mathbf{S} \exists x \exists y \gamma,$$

where

$$\gamma = \beta \to \bigwedge_{i=1}^\ell \left(Z_i \to \bigvee_{j \neq j'} \delta_i^{j,j'}(x,y) \right).$$

Thus, it remains to express the variables S_i ranging over singletons. For this, we use a technique to express S_i as the difference $X_{i,1} \setminus X_{i,2}$ of two monadic predicates $X_{i,1}$ and $X_{i,2}$ which describe initial segments of the string. Fortunately, the fact that $X_{i,1}$ and $X_{i,2}$ are *not* initial segments or their difference is *not* a singleton can be expressed by a first-order formula $\exists x \exists y \psi_i(x, y)$, where ψ_i is quantifier-free, by using the successor predicate. Thus, we obtain

$$\forall \mathbf{X}_1 \mathbf{X}_2 \left(\bigvee_{i=1}^{k} \exists x \exists y \psi_i(x, y) \right) \vee \exists x \exists y \gamma^*,$$

where γ^* results from γ by replacing each S_i with $X_{i,1}$ and $X_{i,2}$. By pulling existential quantifiers, we obtain a $\Sigma_2^1(\exists\exists)$ formula, as desired. □

For example, the lemma is applicable to the sentence Φ in Example 2. Thus,

Theorem 4. $\Sigma_2^1(\exists\exists)$ *is nonregular.*

We remark that a similar argument as in the proof of Lemma 4 can be used to show that over strings, every universal first-order formula $\forall \mathbf{x} \varphi(\mathbf{x})$ which contains no predicates of arity > 2 is equivalent to some $\Pi_1^1(\bigvee \exists\forall)$ formula. Consequently, the SAT encoding in Example 2 can be expressed in $\Sigma_2^1(\bigvee \exists\forall)$.

$\Sigma_2^1(Q)$ where $|Q| > 2$. By the result of the previous subsection, and the results in Section 3, we can derive that there is no $\Sigma_2^1(Q)$ prefix class where Q contains more than two variables which is regular.

Indeed, Theorem 2 implies this for every prefix Q which contains $\exists$ followed by $\forall$, and Theorem 4 implies this for every prefix Q which contains at least two existential quantifiers. For the remaining minimal prefixes $Q \in \{\forall\forall\forall, \forall\forall\exists\}$, non-regularity of $\Sigma_2^1(Q)$ follows from the results summarized in Figure 1. Thus,

Theorem 5. $\Sigma_2^1(Q)$ *is nonregular for any prefix Q such that $|Q| > 2$.*

$\Sigma_k^1(Q)$ where $k > 2$. Let us now consider the higher fragments of SO over strings. The question is whether any of the regular two-variable prefixes $Q \in \{\forall\forall, \forall\exists\}$ for $\Sigma_2^1(Q)$ survives. However, as we shall see this is not the case.

Since $\Pi_2^1(\forall\forall)$ is contained in $\Sigma_3^1(\forall\forall)$, it follows from Theorem 4 that $\Sigma_3^1(\forall\forall)$ is nonregular. For the remaining class $\Sigma_3^1(\forall\exists)$, we use the following result, which simulates an existential first-order quantifier using an existential second-order quantifier and a first-order universal quantifier.

Lemma 5. *Suppose a signature contains constants. Then every formula $\exists x \exists y \varphi$ in which some constant c occurs is equivalent to some formula $\exists X \forall x \exists y \psi$, where X is a fresh monadic variable.*

Proof. Write ψ as $(x = c \rightarrow \varphi(x, y) \vee X(y)) \wedge (X(x) \rightarrow \varphi(x, y))$. □

We thus easily obtain the following result.

Theorem 6. *Over strings, $\Sigma_k^1(\exists\exists) \subseteq \Sigma_k^1(\forall\exists)$ for every odd k, and $\Sigma_k^1(\forall\forall) \subseteq \Sigma_k^1(\exists\forall)$ for every even $k \geq 2$.*

Thus, combined with Theorem 4, we obtain that $\Sigma_3^1(\forall\exists)$ is nonregular.

We note that under very general conditions a universal first-order quantifier can be replaced by a universal second-order quantifier and an existential first-order quantifier.

Lemma 6. *Let φ be a prenex first-order formula (possibly with free first and second order variables) of the form $\varphi = \forall y\, Q \forall z \alpha(y, \ldots, z)$, where Q is a (possible empty) block of first-order quantifiers. Then, φ is equivalent to*

$$\psi = \forall Y \exists y\, Q \forall z [(Y(y) \wedge \alpha) \vee (\neg Y(y) \wedge \neg Y(z))],$$

where Y is a fresh monadic SO variable.

Using the lemma, we obtain the following result.

Theorem 7. *Let $P_1 \in \{\forall\}^*$ and $P_2 \in \{\exists, \forall\}^* \forall \{\exists, \forall\}^*$ be first-order prefixes. Then,*

- *for any odd $k \geq 1$, $\Sigma_k^1(P_1 \forall P_2) \subseteq \Sigma_{k+1}^1(P_1 \exists P_2)$ and $\Pi_k^1(P_1 \forall P_2) \subseteq \Pi_k^1(P_1 \exists P_2)$,*
- *for any even $k \geq 2$, $\Sigma_k^1(P_1 \forall P_2) \subseteq \Sigma_k^1(P_1 \exists P_2)$, $\Pi_k^1(P_1 \forall P_2) \subseteq \Pi_{k+1}^1(P_1 \exists P_2)$.*

Thus, for example we obtain $\Sigma_2^1(\forall\forall\forall) \subseteq \Sigma_2^1(\exists\forall\forall)$, and by repeated application $\Sigma_2^1(\forall\forall\forall) \subseteq \Sigma_2^1(\exists\exists\forall)$. Since by the results in Figure 1, $\Sigma_2^1(\forall\forall\forall)$ is intractable, these fragments inherit intractability.

4.3 Summary of New Results

The maximal standard Σ_k^1 prefix classes which are regular and the minimal standard Σ_k^1 prefix classes which are non-regular are summarized in Figure 2.

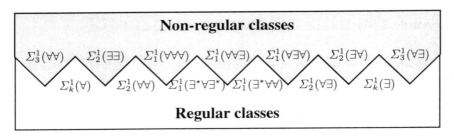

Fig. 2. Maximal regular and minimal non-regular SO prefix classes on strings.

Thus, no $\Sigma_k^1(Q)$ fragment, where $k \geq 3$ and Q contains at least two variables, is regular, and for $k = 2$, only two such fragments ($Q = \exists\forall$ or $Q = \forall\exists$) are regular. Note that Grädel and Rosen have shown [11] that $\Sigma_1^1(\mathrm{FO}^2)$, i.e., existential second order logic with two first-order variables, is over strings regular. By our results, $\Sigma_k^1(\mathrm{FO}^2)$, for $k \geq 2$, is intractable.

$$\begin{array}{ccc} \Sigma_2^1(\exists\forall) & & \Sigma_2^1(\exists\exists) \\ \supseteq & & \subseteq \\ \Sigma_2^1(\forall\forall) & = & \Sigma_2^1(\forall\exists) \end{array}$$

Fig. 3. Semantic inclusion relations between $\Sigma_2^1(Q)$ classes over strings, $|Q| = 2$

Figure 3 shows inclusion relationship between the classes $\Sigma_2^1(Q)$ where Q contains 2 quantifiers. Similar relationships hold for $\Sigma_k^1(Q)$ classes. Furthermore, by Proposition 5 we have that $\Sigma_2^1(\forall\exists) = \Sigma_1^1(\bigwedge \forall\exists)$ and $\Sigma_3^1(\exists\forall) = \Sigma_2^1(\bigvee \exists\forall)$.

As for the complexity of model checking, the results from [8] and above imply that deciding whether $W \models \Phi$ for a fixed formula Φ and a given string W is intractable for all prefix classes $\Sigma_k^1(Q)$ which are (syntactically) not included in the maximal regular prefix classes shown in Figure 2, with the exception $\Sigma_2^1(\exists\forall)$, for which the tractability is currently open.

In more detail, the complexity of SO over strings increases with the number of SO quantifier alternations. In particular, $\Sigma_2^1(\exists\exists)$ can express Σ_2^p-complete languages, where $\Sigma_2^p = \mathrm{NP}^{\mathrm{NP}}$ is from the second level of the polynomial hierarchy. This can be shown by encoding, generalizing the SAT encoding in Example 2, quantified boolean formulas (QBFs) of the form $\exists p_1 \cdots \exists p_n \forall q_1 \cdots \forall q_m \neg \varphi$, where φ is a propositional CNF over the atom $p_1, \ldots, p_n, q_1, \ldots, q_m$, to model checking for $\Sigma_2^1(\exists\exists)$. Roughly, we can mark in the string the occurrences of the variables q_i with an additional predicate, and represent truth assignments to them by a monadic variable V'. The sentence Φ in Example 2 is rewritten to

$$\Psi = \exists R \exists R' \exists V \forall V'[\alpha_1 \wedge (\alpha_2 \vee \alpha_3)]$$

where α_1 is a universal first-order formula which defines proper R, R' and V using φ_R, $\varphi_{R'}$, and φ_V; α_2 is a $\exists\exists$-prenex first-order formula which states that V' assigns two different occurrences of some universally quantified atom q_i different truth values; and α_3 states that the assignment to $p_1, \ldots, p_n, q_1, \ldots, q_m$ given by V and V' violates φ. The latter can be easily checked by a finite state automaton, and thus is expressible as a monadic $\Pi_1^1(\exists\exists)$ sentence. Since Ψ contains no predicate of arity > 2, by applying the techniques in the previous section we can rewrite it to an equivalent $\Sigma_2^1(\exists\exists)$ sentence.

Thus, model checking for the fragment $\Sigma_2^1(\exists\exists)$ is Σ_2^p-complete. Other fragments of Σ_2^1 have lower complexity; for example, Proposition 5 and Figure 1 imply that $\Sigma_2^1(\forall^*\exists)$ is NP-complete. By adding further second-order variables, we can encode evaluating Σ_k^p-complete QBFs into $\Sigma_k^1(Q)$, where $Q = \exists\exists$ if $k > 2$ is even and $Q = \forall\forall$ if $k > 1$ is odd. Note that, after eliminating R' from the sentence Φ in Example 2 as described in [8], the sentence contains only a single binary predicate variable.

5 ESO over Graphs: A Short Account

In this section, we briefly describe the main results of [10], where the computational complexity of ESO-prefix classes of is investigated and completely characterized in three contexts: over (1) directed graphs, (2) undirected graphs with self-loops, and (3) undirected graphs without self-loops. A main theorem of [10] is that a *dichotomy* holds in these contexts, that is to say, each prefix class of ESO either contains sentences that can express NP-complete problems or each of its sentences expresses a polynomial-time solvable problem. Although the boundary of the dichotomy coincides for 1. and 2. (which we refer ot as *general graphs* from now on), it changes if one moves to 3. The key difference is that a certain prefix class, based on the well-known *Ackermann class*, contains sentences that can express NP-complete problems over general graphs, but becomes tractable over undirected graphs without self-loops. Moreover, establishing the dichotomy in case 3. turned out to be technically challenging, and required the use of sophisticated machinery from graph theory and combinatorics, including results about graphs of bounded tree-width and Ramsey's theorem.

In [10], a special notation for ESO-prefix classes was used in order to describe the results with the tightest possible precision involving both the number of SO quantifiers and their arities.[2] Expressions in this notation are built according to the following rules:

- E (resp., E_i) denotes the existential quantification over a single predicate of arbitrary arity (arity $\leq i$).

- a (resp., e) denotes the universal (existential) quantification of a single first-order variable.

- If η is a quantification pattern, then η^* denotes all patterns obtained by repeating η zero or more times.

An expression $\mathcal{E}$ in the special notation consists of a string of ESO quantification patterns (E-patterns) followed by a string of first-order quantification patterns (a or e patterns); such an expression represents the class of all prenex ESO-formulas whose quantifier prefix corresponds to a (not-necessarily contiguous) substring of $\mathcal{E}$.

For example, $E_1^* eaa$ denotes the class of formulas $\exists P_1 \cdots \exists P_r \exists x \forall y \forall z \varphi$, where each P_i is monadic, x, y, and z are first-order variables, and φ is quantifier-free.

A prefix class C is NP-*hard* on a class $\mathcal{K}$ of relational structures, if some sentence in C expresses an NP-hard property on $\mathcal{K}$, and C is *polynomial-time* (PTIME) on $\mathcal{K}$, if for each sentence $\Phi \in C$, model checking is polynomial. Furthermore, C is called *first-order* (FO), if every $\Phi \in C$ is equivalent to a first-order formula.

The first result of [10] completely characterizes the computational complexity of ESO-prefix classes on general graphs. In fact, the same characterization holds on the collection of all finite structures over any relational vocabulary that contains a relation symbol of arity ≥ 2. This characterization is obtained by showing (assuming P $\neq$ NP) that there are four *minimal* NP-hard and three *maximal* PTIME prefix classes, and that these seven classes combine to give complete information about all other prefix classes. This means that every other prefix either contains one of the minimal NP-hard prefix classes as a substring (and, hence, is NP-hard) or is a substring of a maximal PTIME prefix class (and, hence, is in PTIME). Figure 4 depicts the characterization of the NP-hard and PTIME prefix classes of ESO on general graphs.

As seen in Figure 4, the four minimal NP-hard classes are $E_2 eaa$, $E_1 ae$, $E_1 aaa$, and $E_1 E_1 aa$, while the three maximal PTIME classes are $E^* e^* a$, $E_1 e^* aa$, and Eaa. The NP-hardness results are established by showing that each of the four minimal prefix

---

[2] For ESO over strings [8], the same level of precision was reached with simpler notation.

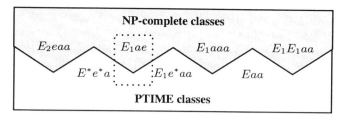

Fig. 4. ESO on arbitrary structures, directed graphs and undirected graphs with self-loops.

classes contains ESO-sentences expressing NP-complete problems. For example, a SAT encoding on general graphs can be expressed by an $E_1 ae$ sentence. Note that the first-order prefix class ae played a key role in the study of the classical decision problem for fragments of first-order logic (see [2]). As regards the maximal PTIME classes, $E^* e^* a$ is actually FO, while the model checking problem for fixed sentences in $E_1 e^* aa$ and Eaa is reducible to 2SAT and, thus, is in PTIME (in fact, in NL).

The second result of [10] completely characterizes the computational complexity of prefix classes of ESO on undirected graphs without self-loops. As mentioned earlier, it was shown that a dichotomy still holds, but its boundary changes. The key difference is that $E^* ae$ turns out to be PTIME on undirected graphs without self-loops, while its subclass $E_1 ae$ is NP-hard on general graphs. It can be seen that interesting properties of graphs are expressible by $E^* ae$-sentences. Specifically, for each integer $m > 0$, there is a $E^* ae$-sentence expressing that a connected graph contains a cycle whose length is divisible by m. This was shown to be decidable in polynomial time by Thomassen [19]. $E^* ae$ constitutes a maximal PTIME class, because all four extensions of $E_1 ae$ by any single first-order quantifier are NP-hard on undirected graphs without self-loops [10]. The other minimal NP-hard prefixes on general graphs remain NP-hard also on undirected graphs without self-loops. Consequently, over such graphs, there are seven minimal NP-hard and four maximal PTIME prefix classes that determine the computational complexity of all other ESO-prefix classes (see Figure 5).

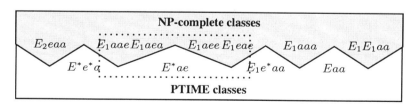

Fig. 5. ESO on undirected graphs without self-loops. The dotted boxes in Figures 4 and 5 indicate the difference between the two cases.

Technically, the most difficult result of [10] is the proof that $E^* ae$ is PTIME on undirected graphs without self-loops. First, using syntactic methods, it is shown that each $E^* ae$-sentence is equivalent to some $E_1^* ae$-sentence. After this, it is shown that for each $E_1^* ae$-sentence the model checking problem over undirected graphs without self-loops is equivalent to a natural coloring problem called the *saturation problem*. This problem asks whether there is a particular mapping from a given undirected graph without self-loops to a fixed, directed *pattern graph* P which is extracted from the E_1^*ae-formula under consideration. Depending on the labelings of cycles in P, two cases of the saturation problem are distinguished, namely *pure pattern graphs* and *mixed pattern graphs*. For each case, a polynomial-time algorithm is designed. In simplified terms and focussed on the case of connected graphs, the one for pure pattern graphs has three main ingredients. First, adapting results by Thomassen [19] and using a new graph coloring method, it is shown that if a E_1^*ae-sentence Φ gives rise to a pure pattern graph, then a fixed integer k can be found such that every undirected graph without self-loops and having tree-width bigger than k satisfies Φ. Second, Courcelle's theorem [5] is used by which model checking for MSO sentences is polynomial on graphs of bounded tree-width. Third, Bodlaender's result [1] is used that, for each fixed k, there is a polynomial-time algorithm to check if a given graph has tree-width at most k.

The polynomial-time algorithm for mixed pattern graphs has a similar architecture, but requires the development of substantial additional technical machinery, including a generalization of the concept of graphs of bounded tree-width. The results of [10] can be summarized in the following theorem.

Theorem 8. *Figures 4 and 5 provide a complete classification of the complexity of all ESO prefix classes on graphs.*

6 Conclusion and Open Problems

The aim of this paper was twofold, namely, to give a status report on the ambitious research programme of determining the complexity of prefix classes over finite structures, and to present new results consisting of a complete characterization of those prefix classes $\Sigma_k^1(\mathcal{Q})$ which over strings express regular languages only.

As for the first goal, we hope to have convinced the reader that this is an important and interesting research programme. In particular, let us stress that this programme is anything but boring. Many of the prefix classes we have analyzed so far represented mathematical challenges and required novel solution methods. Some of them could be solved with automata-theoretic techniques, others with techniques of purely combinatorial nature, and yet others required graph-theoretic arguments.

While we have traced the exact "regularity frontier" for the $\Sigma_k^1(\mathcal{Q})$ classes (see Fig. 2), we have not yet completely determined their tractability frontier with respect to model checking. A single result is missing, namely, the complexity of the nonregular class $\Sigma_2^1(\exists\forall)$. If model checking turns out to be NP-hard for it, then the tractability frontier coincides with the regularity frontier (just as for ESO, cf. [8]). If, on the other hand, model checking for $\Sigma_2^1(\exists\forall)$ turns out to be tractable, then the picture is slightly different. We plan to clarify this issue in the future. Moreover, it would be interesting to sharpen our analysis of the Σ_k^1 fragments over strings by studying the second-order quantification patterns at a finer level, taking into account the number of the second-order variables and their arities, as done for ESO over graphs in [10] (cf. Section 5).

Let us conclude this paper by pointing out a few interesting (and in our opinion important) issues that should eventually be settled.

- While the work on word structures concentrated so far on strings with a successor relation $Succ$, one should also consider the case where in addition a predefined linear order $<$ is available on the word structures, and the case where the successor relation $Succ$ is replaced by such a linear order. While for full ESO or SO, $Succ$ and $<$ are freely interchangeable, because either predicate can be defined in terms of the other, this is not so for many of the limited ESO-prefix classes. Preliminary results suggest that most of the results in this paper carry over to the $<$ case.

- Delineate the tractability/intractability frontier for all SO prefix classes over graphs.

- Study SO prefix classes over trees and other interesting classes of structures (e.g. planar graphs).

- The scope of [8] and of Section 3 in this paper are *finite* strings. However, infinite strings or ω-words are another important area of research. In particular, Büchi has shown that an analogue of his theorem (Proposition 1) also holds for ω-words [3]. For an overview of this and many other important results on ω-words, we refer the reader to the excellent survey paper [18]. In this context, it would be interesting to see which of the results established so far survive for ω-words. For some results, such as the regularity of $\mathrm{ESO}(\exists^*\forall\forall)$ this is obviously the case since no finiteness assumption on the input word structures was made in the proof. For determining the regularity or nonregularity of some other clases such as $\mathrm{ESO}(\exists^*\forall\exists^*)$, further research is needed.

Acknowledgments. This work was supported by the Austrian Science Fund (FWF) Project Z29-INF.

References

1. H. L. Bodlaender. A linear-time algorithm for finding tree-decompositions of small treewidth. *SIAM Journal on Computing*, 25(6):1305–1317, 1996.
2. E. Börger, E. Grädel, and Y. Gurevich. *The Classical Decision Problem*. Springer, 1997.
3. J. R. Büchi. On a decision method in restricted second-order arithmetic. In E. Nagel et al., editor, *Proc. International Congress on Logic, Methodology and Philosophy of Science*, pp. 1–11, Stanford, CA, 1960. Stanford University Press.
4. J. R. Büchi. Weak second-order arithmetic and finite automata. *Zeitschrift für mathematische Logik und Grundlagen der Mathematik*, 6:66–92, 1960.
5. B. Courcelle. The monadic second-order logic of graphs I: recognizable sets of finite graphs. *Information and Computation*, 85:12–75, 1990.
6. H.-D. Ebbinghaus and J. Flum. *Finite Model Theory*. Springer, 1995.
7. T. Eiter, G. Gottlob, and Y. Gurevich. Normal forms for second-order logic over finite structures, and classification of NP optimization problems. *Annals of Pure and Applied Logic*, 78:111–125, 1996.
8. T. Eiter, G. Gottlob, and Y. Gurevich. Existential second-order logic over strings. *Journal of the ACM*, 47(1):77–131, 2000.
9. R. Fagin. Generalized first-order spectra and polynomial-time recognizable sets. In R. M. Karp, editor, *Complexity of Computation*, pp. 43–74. AMS, 1974.
10. G. Gottlob, P. Kolaitis, and T. Schwentick. Existential second-order logic over graphs: charting the tractability frontier. In *Proc. 41st Annual Symposium on Foundations of Computer Science (FOCS 2000)*, pp. 664–674. IEEE Computer Society Press, 2000.

11. E. Grädel and E. Rosen. Two-variable descriptions of regularity. In *Proc. 14th Annual Symposium on Logic in Computer Science (LICS-99)*, pp. 14–23. IEEE CS Press, 1999.
12. N. Immerman. *Descriptive Complexity*. Springer, 1997.
13. P. Kolaitis and C. Papadimitriou. Some computational aspects of circumscription. *Journal of the ACM*, 37(1):1–15, 1990.
14. D. Leivant. Descriptive characterizations of computational complexity. *Journal of Computer and System Sciences*, 39:51–83, 1989.
15. J.-E. Pin. Logic on words. *Bulletin of the EATCS*, 54:145–165, 1994.
16. L. J. Stockmeyer. The polynomial-time hierarchy. *Theoretical Comp. Sc.*, 3:1–22, 1977.
17. H. Straubing. *Finite Automata, Formal Logic, and Circuit Complexity*. Birkhäuser, 1994.
18. W. Thomas. Automata on infinite objects. In J. van Leeuwen, editor, *Handbook of Theoretical Computer Science*, volume B, chapter 4. Elsevier Science Pub., 1990.
19. C. Thomassen. On the presence of disjoint subgraphs of a specified type. *Journal of Graph Theory*, 12:1, 101-111, 1988.
20. B. Trakhtenbrot. Finite automata and the logic of monadic predicates. *Dokl. Akad. Nauk SSSR*, 140:326–329, 1961.

Decision Questions on Integer Matrices

Tero Harju

Department of Mathematics
University of Turku
FIN-20014 Turku, Finland
harju@utu.fi

Abstract. We give a survey of simple undecidability results and open problems concerning matrices of low order with integer entries. Connections to the theory of finite automata (with multiplicities) are also provided.

1 Introduction

Representations of words by integer matrices is a powerful tool for proving results in combinatorics of words. In this respect we mention the proof of the Ehrenfeucht's Conjecture by Guba [10] (independently solved by Albert and Lawrence [1]). By this result, known also as the Compactness Theorem for equations, any system of equations on words with finitely many variables has a finite equivalent subsystem, or equivalently, each subset $L \subseteq \Sigma^*$ has a finite set $F \subseteq \Sigma^*$ such that for any two morphisms $h, g \colon \Sigma^* \to \Sigma^*$, $h(u) = g(u)$ for all $u \in F$ implies $h(u) = g(u)$ for all $u \in L$. The problem is reduced to the Hilbert's Basis Theorem by using matrix representation provided by suitable embeddings of words into $\mathbb{Z}[X]^{2 \times 2}$. For the proof and several corollaries of the result, see also [15] and [16].

Let $\mathbb{Z}^{n \times n}$ be the semigroup of all $n \times n$ integer matrices with respect to matrix multiplication. The corresponding matrix semigroups with nonnegative (resp., rational, complex) entries are denoted by $\mathbb{N}^{n \times n}$ (resp., $\mathbb{Q}^{n \times n}$, $\mathbb{C}^{n \times n}$).

We shall consider finitely generated matrix semigroups with integer entries. Such semigroups can be presented by finite automata with multiplicities. Indeed, let $M_1, M_2, \ldots, M_k \in \mathbb{Z}^{n \times n}$ be a finite set of matrices. We define an n-state $\mathbb{Z}$-*automaton* $\mathcal{A}$ (a nondeterministic automaton with integer multiplicities) as follows. Let $\mathcal{A} = (\Sigma, Q, \delta, J, F)$, where

- $\Sigma = \{a_1, a_2, \ldots, a_k\}$ is the input alphabet, where the letter a_i corresponds to the matrix M_i;
- $Q = \{1, 2, \ldots, n\}$ is the state set (i corresponds to the ith row and column of the matrices);
- J is the set of the initial states, and $F \subseteq Q$ is the set of final states;
- δ is the set of transitions that provide the rules:

$$r \xrightarrow{\binom{a_i}{m}} s,$$

where $a_i \in \Sigma$, and $m = (M_i)_{rs}$ is the multiplicity of the rule.

The multiplicity of a path (*i.e.*, a computation)

$$\pi = s_1 \xrightarrow{\binom{b_1}{m_1}} s_2 \xrightarrow{\binom{b_2}{m_2}} s_3 \longrightarrow \cdots \longrightarrow s_t \xrightarrow{\binom{b_t}{m_t}} s_{t+1}$$

(where the consumed word is $w = b_1 b_2 \ldots b_t \in \Sigma^*$) in $\mathcal{A}$ is equal to $\|\pi\| = m_1 m_2 \ldots m_t \in \mathbb{Z}$. For a word $w \in \Sigma^*$, denote by $\Pi_{rs}(w)$ the set of the paths in $\mathcal{A}$ from the state r to the state s that read w. The multiplicity of $w = a_{i_1} a_{i_2} \ldots a_{i_t} \in \Sigma^*$ from r to s is then the sum

$$\mathcal{A}_{rs}(w) = \sum_{\pi \in \Pi_{rs}(w)} \|\pi\| = (M_{i_1} M_{i_2} \ldots M_{i_t})_{rs}.$$

The *multiplicity* of w in $\mathcal{A}$ is obtained from the accepting paths:

$$\mathcal{A}(w) = \sum_{r \in J, s \in F} \mathcal{A}_{rs}(w) = \sum_{r \in J, s \in F} (M_{i_1} M_{i_2} \ldots M_{i_t})_{rs}.$$

Problems concerning finite automata with multiplicities tend to be difficult. For example, the following was shown in Eilenberg [9].

Theorem 1. *It is undecidable for two $\mathbb{N}$-automata $\mathcal{A}$ and $\mathcal{B}$ whether or not $\mathcal{A}(w) \leq \mathcal{B}(w)$ for all words w.*

However, the problem whether two $\mathbb{Z}$-automata $\mathcal{A}$ and $\mathcal{B}$ accept every word with the same multiplicity, is decidable, see Eilenberg [9], or Harju and Karhumäki [14] for a more general result.

Theorem 2. *It is decidable for two $\mathbb{Z}$-automata $\mathcal{A}$ and $\mathcal{B}$ whether or not $\mathcal{A}(w) = \mathcal{B}(w)$ for all words w.*

The letters Σ and Δ are reserved for finite alphabets. The empty word is denoted by ε. Let

$$\Sigma_\kappa = \{a_0, a_1, \ldots, a_{\kappa-1}\}.$$

The undecidability results of this paper are all reduced from the *Post Correspondence Problem*, or the *PCP*:

Theorem 3. *It is undecidable for morphisms $g, h \colon \Sigma^* \to \Delta^*$ whether there exists a nonempty word w such that $g(w) = h(w)$.*

If $\Delta = \{b_1, b_2, \ldots, b_n\}$, then the morphism $\mu \colon \Delta^* \to \Sigma_2^*$ defined by $\mu(b_i) = a_0 a_1^i$ is injective, and hence Theorem 3 remains undecidable, when the image alphabet Δ is restricted to a binary alphabet (of two letters).

2 Embeddings

2.1 Embedding Free Monoids to a Matrix Monoid

Consider first the order 2 for nonnegative integer matrices, that is, the multiplicative semigroup $\mathbb{N}^{2 \times 2}$. We shall start with a proof for the following embedding property.

Theorem 4. *Every free monoid Σ^* embeds into the matrix semigroup $\mathbb{N}^{2\times 2}$.*

There are many different embeddings (injective morphisms) that prove the above result. Let $\mathbf{SL}(2,\mathbb{N})$ denote the *special linear monoid*, a multiplicative submonoid of $\mathbb{N}^{2\times 2}$ consisting of the unimodular matrices M, *i.e.*, the matrices with determinants $\det(M) = 1$. Notice that $\mathbf{SL}(2,\mathbb{N})$ is a submonoid of the special linear group $\mathbf{SL}(2,\mathbb{Z})$ consisting of unimodular integer matrices.

The following result was proved by Nielsen [28] in 1924.

Theorem 5. $\mathbf{SL}(2,\mathbb{N})$ *is a free monoid generated by the matrices*

$$A = \begin{pmatrix} 1 & 1 \\ 0 & 1 \end{pmatrix} \quad \text{and} \quad B = \begin{pmatrix} 1 & 0 \\ 1 & 1 \end{pmatrix}.$$

Proof. The matrices A and B have the inverses

$$A^{-1} = \begin{pmatrix} 1 & -1 \\ 0 & 1 \end{pmatrix} \quad \text{and} \quad B^{-1} = \begin{pmatrix} 1 & 0 \\ -1 & 1 \end{pmatrix}$$

in the group $\mathbf{SL}(2,\mathbb{Z})$. Let

$$M = \begin{pmatrix} m_{11} & m_{12} \\ m_{21} & m_{22} \end{pmatrix} \in \mathbf{SL}(2,\mathbb{N}).$$

We obtain

$$MA^{-1} = \begin{pmatrix} m_{11} & m_{12} - m_{11} \\ m_{21} & m_{22} - m_{21} \end{pmatrix} \quad \text{and} \quad MB^{-1} = \begin{pmatrix} m_{11} - m_{12} & m_{12} \\ m_{21} - m_{22} & m_{22} \end{pmatrix}.$$

Here both MA^{-1} and MB^{-1} have strictly smaller entry sums than M. If $m_{11} \leq m_{12}$, then $m_{21} \leq m_{22}$, since $\det(M) = 1$. Hence $MA^{-1} \in \mathbf{SL}(2,\mathbb{N})$, but $MB^{-1} \notin \mathbf{SL}(2,\mathbb{N})$. Similarly for $m_{11} > m_{12}$, $MB^{-1} \in \mathbf{SL}(2,\mathbb{N})$, but $MA^{-1} \notin \mathbf{SL}(2,\mathbb{N})$. Therefore there is a unique sequence $A_1, A_2, \ldots, A_k \in \{A, B\}$ such that $MA_1^{-1} A_2^{-1} \ldots A_k^{-1} = I$, where I is the identity matrix. It follows that M can be factored uniquely as $M = A_k A_{k-1} \ldots A_1$, which shows that $\mathbf{SL}(2,\mathbb{N})$ is freely generated by A and B. □

Theorem 4 follows from Theorem 5, since the matrices $AB, AB^2, \ldots$ generate a free submonoid of $\mathbf{SL}(2,\mathbb{N})$. Here

$$AB^n = \begin{pmatrix} n+1 & 1 \\ n & 1 \end{pmatrix}.$$

As another example, consider the matrices

$$F_1 = \begin{pmatrix} 1 & 1 \\ 1 & 0 \end{pmatrix}, \quad F_2 = \begin{pmatrix} 0 & 1 \\ 1 & 1 \end{pmatrix}, \quad E = \begin{pmatrix} 0 & 1 \\ 1 & 0 \end{pmatrix}.$$

We have $F_1 = EB = AE$ and $F_2 = EA = BE$, and for $n \geq 3$,

$$F_1^n = \begin{pmatrix} f_n & f_{n-1} \\ f_{n-1} & f_{n-2} \end{pmatrix} \quad \text{and} \quad F_2^n = \begin{pmatrix} f_{n-2} & f_{n-1} \\ f_{n-1} & f_n \end{pmatrix},$$

where $f(n)$ is the nth Fibonacci number: $f_1 = 1$, $f_2 = 2$, $f_n = f_{n-1} + f_{n-2}$. Every 2-element subset of $\{A, B, F_1, F_2\}$ generates a free semigroup. This is easily seen from the above partial commutations, $EB = AE$ and $EA = BE$, and the fact that $E^2 = I$. It is also clear that these are all the cases for two free generators where the entries take their values from $\{0, 1\}$, see Lemma 1.

One can construct new free matrix semigroups from the known ones by using similarity transformations. Indeed, if $P \in \mathbb{Q}^{n \times n}$ is a nonsingular matrix (i.e., $\det(P) \neq 0$), then the transformation $X \mapsto PXP^{-1}$ is an injective morphism on $\mathbb{Q}^{n \times n}$. Therefore a set $M_1, M_2, \ldots, M_m \in \mathbb{Q}^{n \times n}$ of matrices generates a free semigroup if and only if the matrices PM_iP^{-1} do so for $i = 1, 2, \ldots, m$.

Lemma 1. *If the matrices $M_1, M_2, \ldots, M_m \in \mathbb{Q}^{2 \times 2}$ with $m \geq 2$ generate a free semigroup S, then they are nonsingular. In particular, every matrix in S is nonsingular.*

Proof. Suppose that one of the matrices, say M_1, is singular. Then $M_1^2 = tM_1$ for the trace $t = m_{11} + m_{22}$ of M_1, and hence $M_1^2 M_2 M_1 = M_1 M_2 M_1^2$, which means that the generators of the semigroup S satisfy a nontrivial identity, and, hence S is not freely generated by M_i, $i = 1, 2, \ldots, m$. □

The following open problem is likely to be very difficult, see Problem 5.

Problem 1. *Characterize those matrices $M_1, M_2 \in \mathbb{Z}^{2 \times 2}$ that generate a free semigroup.*

For the rest of the paper we shall choose an embedding using triangular matrices, because these allow an embedding of $\Sigma^* \times \Sigma^*$ into $\mathbb{Z}^{3 \times 3}$.

Let $\kappa \geq 2$ be a constant integer, and let $\mathbb{M}_{2 \times 2}(\kappa)$ be the subset of $\mathbb{N}^{2 \times 2}$ consisting of the upper triangular matrices

$$M = \begin{pmatrix} \kappa^m & i \\ 0 & 1 \end{pmatrix}, \quad 0 \leq m, \ 0 \leq i < \kappa^m. \tag{1}$$

Note that $I \in \mathbb{M}_{2 \times 2}(\kappa)$, and for $i < \kappa^n$ and $j < \kappa^m$,

$$\begin{pmatrix} \kappa^n & i \\ 0 & 1 \end{pmatrix} \begin{pmatrix} \kappa^m & j \\ 0 & 1 \end{pmatrix} = \begin{pmatrix} \kappa^{m+n} & j\kappa^n + i \\ 0 & 1 \end{pmatrix} \in \mathbb{M}_{2 \times 2}(\kappa).$$

Indeed, $j\kappa^n + i \leq (\kappa^m - 1)\kappa^n + \kappa^n - 1 = \kappa^{n+m} - 1$. Therefore $\mathbb{M}_{2 \times 2}(\kappa)$ is a submonoid of $\mathbb{N}^{2 \times 2}$.

Lemma 2. *The monoid $\mathbb{M}_{2 \times 2}(\kappa)$ is free. To be more precise, define a morphism $\beta_\kappa \colon \Sigma_\kappa^* \to \mathbb{N}^{2 \times 2}$ by*

$$\beta_\kappa(a_i) = \begin{pmatrix} \kappa & i \\ 0 & 1 \end{pmatrix}.$$

Then β_κ is injective and onto $\mathbb{M}_{2 \times 2}(\kappa)$.

Proof. Denote $M_i = \beta_\kappa(a_i)$. Then for a matrix M as in (1),

$$M_j^{-1}M = \begin{pmatrix} (1/\kappa) & -(i/\kappa) \\ 0 & 1 \end{pmatrix} \begin{pmatrix} \kappa^m & i \\ 0 & 1 \end{pmatrix} = \begin{pmatrix} \kappa^{m-1} & (1/\kappa)(i-j) \\ 0 & 1 \end{pmatrix}.$$

Now $M_j^{-1}M \in \mathbb{M}_{2\times 2}(\kappa)$ for a unique $0 \leq j < \kappa$, namely, $j \equiv i \pmod{\kappa}$. Consequently, each matrix $M \in \mathbb{M}_{2\times 2}(\kappa)$ has a unique factorization $M = M_{i_1}M_{i_2}\ldots M_{i_r}$ in terms of the matrices M_i. This shows that $\mathbb{M}_{2\times 2}(\kappa)$ is freely generated by the matrices M_i, and the claim follows from this. □

Define $\sigma_\kappa \colon \Sigma_\kappa^* \to \mathbb{N}$ by $\sigma_\kappa(\varepsilon) = 0$ and

$$\sigma_\kappa(a_{i_1}a_{i_2}\ldots a_{i_n}) = \sum_{j=1}^n i_j \kappa^j.$$

The function $w \mapsto (|w|, \sigma_\kappa(w))$ is injective on Σ_κ^*, and when σ_κ is restricted to $(\Sigma_\kappa \setminus \{a_0\})^*$, it is injective. We have,

$$\sigma_\kappa(uv) = \sigma_\kappa(u) + \kappa^{|u|}\sigma_\kappa(v), \qquad (2)$$

$$\beta_\kappa(w) = \begin{pmatrix} \kappa^{|w|} & \sigma_\kappa(w) \\ 0 & 1 \end{pmatrix}. \qquad (3)$$

2.2 Embedding Direct Products

The Post Correspondence Problem involves pairs of words $(h(a), g(a))$ $(a \in \Sigma)$, and in order to exploit the undecidability of the PCP, we need to embed the direct products $\Sigma^* \times \Sigma^*$ into $\mathbb{Z}^{n\times n}$. We note first that according to Cassaigne, Harju and Karhumäki [6], one has to increase the order n above two:

Theorem 6. *The direct product $\Sigma^* \times \Sigma^*$ of two free monoids with $|\Sigma| \geq 2$ does not embed into the matrix semigroup $\mathbb{C}^{2\times 2}$.*

However, $\Sigma^* \times \Sigma^*$ can be embedded into $\mathbb{N}^{4\times 4}$ simply by placing two representations of Σ^* on the diagonal. For $\Sigma = \Sigma_\kappa$, we obtain

$$(u, v) \mapsto \begin{pmatrix} \beta_\kappa(u) & 0 \\ 0 & \beta_\kappa(v) \end{pmatrix},$$

where 0 is the zero 2×2-matrix.

Using the upper triangular matrices, we can do better:

Theorem 7. *The direct product $\Sigma^* \times \Sigma^*$ embeds into $\mathbb{N}^{3\times 3}$.*

Proof. Let $\Sigma = \Sigma_\kappa$. We place two copies of β_κ simultaneously in 3×3-matrices to obtain the following injective monoid morphism $\gamma_\kappa \colon \Sigma_\kappa^* \times \Sigma_\kappa^* \to \mathbb{N}^{3\times 3}$:

$$\gamma_\kappa(u, v) = \begin{pmatrix} \kappa^{|u|} & 0 & \sigma_\kappa(u) \\ 0 & \kappa^{|v|} & \sigma_\kappa(v) \\ 0 & 0 & 1 \end{pmatrix}.$$

The injectivity of γ_κ is obvious, and that it is a morphism is easily checked. □

Note that a morphism $h \colon \Sigma_\kappa^* \to \Sigma_\kappa^*$ can be presented as a set of pairs of words, namely by $\{(a, h(a)) \mid a \in \Sigma_\kappa\}$. Therefore h has a presentation as a matrix semigroup.

3 Undecidability Results

The undecidability results are mostly derived from the proof of the following lemma. For an instance (h, g) of the PCP, where $h, g \colon \Sigma^* \to \Gamma^*$ with $\Gamma = \Sigma_\kappa \setminus \{a_0\} = \{a_1, \ldots, a_{\kappa-1}\}$ for $\kappa \geq 3$, define

$$M_a(h, g) = \gamma_\kappa(h(a), g(a)) \quad \text{for } a \in \Sigma. \tag{4}$$

Lemma 3. *It is undecidable for finitely generated matrix semigroups S whether S contains a matrix M such that $M_{13} = M_{23}$.*

Proof. Let $M_a(h, g) = M_a$ for an instance (h, g) of the PCP as in the above. Then $M_{13} = M_{23}$ for a matrix $M = M_{b_1} M_{b_2} \ldots M_{b_m}$ if and only if for $w = b_1 b_2 \ldots b_m$, $\sigma_\kappa(h(w)) = \sigma_\kappa(g(w))$, i.e., if and only if $h(w) = g(w)$, since σ_κ is injective on Γ. Hence the claim follows from the undecidability of the PCP. □

3.1 Zeros in the Corners

One can use various similarity transformations of matrices to improve Lemma 3. The following result was attributed to R.W. Floyd in [22], see also [7]. It states that it is undecidable whether a finitely generated matrix semigroup has a matrix with zero in its upper right corner. A similar result for the lower left corner follows by transposing the matrices.

Theorem 8. *It is undecidable for finitely generated subsemigroups S of $\mathbb{Z}^{3\times 3}$ whether S contains a matrix M with $M_{13} = 0$.*

Proof. Using a similarity transformation $X \mapsto PXP^{-1}$, for $P = \begin{pmatrix} 1 & -1 & 0 \\ 0 & 1 & 0 \\ 0 & 0 & 1 \end{pmatrix}$, we obtain

$$P\gamma_\kappa(u, v)P^{-1} = \begin{pmatrix} \kappa^{|u|} & \kappa^{|u|} - \kappa^{|v|} & \sigma_\kappa(u) - \sigma_\kappa(v) \\ 0 & \kappa^{|v|} & \sigma_\kappa(v) \\ 0 & 0 & 1 \end{pmatrix}$$

and the claim follows as in the proof of Lemma 3. □

The previous result has a nice interpretation for finite automata. Indeed, by taking in the automata the special case, where $n = 3$, $J = \{1\}$ and $F = \{3\}$, we obtain

Corollary 1. *It is undecidable whether a 3-state $\mathbb{Z}$-automaton $\mathcal{A}$ accepts a word w with multiplicity zero, $\mathcal{A}(w) = 0$.*

The automata needed in Corollary 1 are rather simple. They have only three states, and they do not have cycles visiting more than one state, that is, the states of the automata are linearly ordered: a transition $p \xrightarrow{\binom{a}{m}} q$ implies that $p \leq q$.

Corollary 1 was obtained by using the injective morphism γ_κ into $\mathbb{Z}^{3\times 3}$. We can also use the function β_κ into $\mathbb{N}^{2\times 2}$ and Lemma 3 to prove the following result.

Corollary 2. *It is undecidable for two 2-state $\mathbb{N}$-automata $\mathcal{A}$ and $\mathcal{B}$ whether they accept a word w with the same multiplicity, $\mathcal{A}(w) = \mathcal{B}(w)$.*

The occurrence of zero in the diagonal corner needs a bit more work. Following an idea of Paterson [29], the next result was proved in [12].

Theorem 9. *It is undecidable for finitely generated subsemigroups S of $\mathbb{Z}^{3 \times 3}$ whether S contains a matrix M with $M_{11} = 0$.*

We can then derive

Corollary 3. *Let $1 \leq i, j \leq 3$. It is undecidable for finitely generated subsemigroups S of $\mathbb{Z}^{3 \times 3}$ whether S contains a matrix M with $M_{ij} = 0$.*

The above *proofs* do need the order at least 3. The decidability status of the case for the order $n = 2$ is still an open problem.

Problem 2. *Is it decidable for finitely generated semigroups $S \subseteq \mathbb{Z}^{2 \times 2}$ if S contains a matrix M such that (a) $M_{11} = 0$, (b) $M_{12} = 0$?*

The following open problem is the famous *Skolem's Problem*.

Problem 3. *Is it decidable for matrices $M \in \mathbb{Z}^{n \times n}$ whether there exists an integer m such that $(M^m)_{1n} = 0$?*

From the point of view of automata theory, the problem can be restated as follows: given a $\mathbb{Z}$-automaton $\mathcal{A}$ with *a single input letter a*, does there exist a word $w \in a^*$ accepted by $\mathcal{A}$ with multiplicity zero.

As an example, consider the matrix

$$M = \begin{pmatrix} 0 & 1 & -3 \\ 0 & 1 & 1 \\ 1 & 0 & 1 \end{pmatrix}.$$

The least power n for which the right upper corner of M^n equals 0 is 39,

$$M^{39} = \begin{pmatrix} -3456106496 & 301989888 & 0 \\ 905969664 & -436207616 & 1207959552 \\ 301989888 & 905969664 & -3154116608 \end{pmatrix}.$$

Consider a linear recurrence equation $\sum_{i=0}^{m-1} a_i x_{n+i} = 0$, where the coefficients a_i are fixed integers. By the Skolem-Mahler-Lech theorem, for any $k \in \mathbb{Z}$, the solutions $x_i = k$ lie either in a finite set or in a finite union of arithmetic progressions, see Hansel [13] for an elementary proof of this result. The Skolem-Mahler-Lech theorem was used in proving the following result of Berstel and Mignotte [2], see also Berstel and Reutenauer [3].

Theorem 10. *It is decidable for $M \in \mathbb{Z}^{n \times n}$ whether there exist infinitely many powers M^i such that $(M^i)_{1n} = 0$,*

Now, if $(M^i)_{12} = 0$, then also $(M^{ir})_{12} = 0$ for all $r \geq 1$, and therefore there are infinitely many powers M^j, where the upper corner equals zero. Hence

Corollary 4. *Skolem's problem is decidable in $\mathbb{Z}^{2 \times 2}$.*

For two matrices (of higher order), the upper corner problem remains undecidable. This result is obtained by encoding several matrices into the diagonal of one matrix. This encoding was first described by Turakainen [33], and then the undecidability was shown by Claus [8] for two matrices in $\mathbb{Z}^{29 \times 29}$, see also Cassaigne and Karhumäki [5]. The present state of the Post Correspondence Problem, see [25], gives the following result by [8].

Theorem 11. *It is undecidable for subsemigroups S of $\mathbb{Z}^{23 \times 23}$ generated by two matrices whether or not S has a matrix M with a zero in the upper right corner.*

3.2 Mortality of Matrices

Paterson [29] proved in 1970 a beautiful undecidability result for 3×3-integer matrices, the mortality problem for finitely generated matrix semigroups, see also [7], [18] and [19]. In the *mortality problem* we ask whether the zero matrix belongs to a given finitely generated semigroup S of $n \times n$-matrices from $\mathbb{Z}^{n \times n}$, that is, whether there is a sequence of matrices $M_1, M_2, \ldots, M_k$ of S such that $M_1 M_2 \ldots M_k = 0$.

We reduce the mortality problem to Theorem 9.

Theorem 12. *The mortality problem is undecidable in $\mathbb{Z}^{3 \times 3}$.*

Proof. Consider the idempotent matrix $U = \begin{pmatrix} 1 & 0 & 0 \\ 0 & 0 & 0 \\ 0 & 0 & 0 \end{pmatrix}$, and let R be any finitely generated semigroup of $\mathbb{Z}^{3 \times 3}$, and let S be generated by the generators of R and U. For any $M \in S$, $(UMU)_{ij} = 0$ for $i, j \neq 1$ and $(UMU)_{11} = M_{11}$. Hence, if there exists an element $M \in R$ with $M_{11} = 0$, then $0 \in S$. Assume then that $0 \in S$, say, $0 = UM_1UM_2 \ldots M_nU$, for some $n \geq 1$ and $M_i \in R$. Now,

$$0 = UM_1UM_2 \ldots M_tU = UM_1U \cdot UM_2U \ldots UM_nU$$
$$\iff 0 = (M_1)_{11}(M_2)_{11} \ldots (M_n)_{11},$$

and therefore $(M_i)_{11} = 0$ for some i. Consequently, $0 \in S$ if and only if $M_{11} = 0$ for some $M \in R$, and the claim follows by Theorem 9. □

For one matrix M, we can decide whether $M^m = 0$ for some power m, that is, whether M is nilpotent. (Consider the characteristic polynomial of M.) This follows also from a general result of Jacob [17] and Mandel and Simon [21]:

Theorem 13. *It is decidable whether or not a finitely generated submonoid of $\mathbb{Z}^{n \times n}$ is finite.*

The following problem was raised by Schultz [31].

Problem 4. *Is the mortality problem decidable for 2×2 integer matrices?*

For upper triangular $n \times n$ integer matrices, the mortality problem is decidable. Indeed, a finite set D of upper triangular generates 0 if and only if for each i, there exists a matrix $M_i \in D$ with $(M_i)_{ii} = 0$. In this case $M_{ii} = 0$, where $M = M_1 M_2 \ldots M_n$, and $M^n = 0$.

The mortality problem is undecidable already for semigroups generated by two matrices. The order was bounded by 45 by Cassaigne and Karhumäki [5] and Blondel and Tsitsiklis [4]. It was noticed by Halava and Harju [12] that the proof can be modified to give the bound 24 for the order of the matrices.

3.3 The Problem of Freeness

Using the matrix representation γ_κ we will obtain a result of Klarner, Birget and Satterfield [18] according to which the freeness problem of 3×3 nonnegative integer matrices is undecidable. An improvement for the upper triangular matrices was proven by Cassaigne, Harju and Karhumäki [6], the proof of which is bases on a 'mixed' modification of the PCP.

Theorem 14. *It is undecidable for morphisms* $g, h \colon \Sigma^* \to \Sigma^*$ *whether there exists a word* $w = a_1 a_2 \ldots a_n$ *such that*

$$g_1(a_1) g_2(a_2) \ldots g_n(a_n) = h_1(a_1) h_2(a_2) \ldots h_n(a_n),$$

where $g_i, h_i \in \{g, h\}$ *and* $g_j \neq h_j$ *for some* j.

Theorem 15. *It is undecidable whether a finitely generated subsemigroup of upper triangular matrices of* $\mathbb{N}^{3 \times 3}$ *is free.*

Proof. Let $h, g \colon \Sigma^* \to \Sigma^*$ be two morphisms, where $\Sigma = \Sigma_\kappa$ for $\kappa \geq 3$. Let S be the semigroup generated by the matrices $H_a = \gamma_\kappa(a, h(a))$ and $G_a = \gamma_\kappa(a, g(a))$ for $a \in \Sigma$. Assume that $H'_{a_1} H'_{a_2} \ldots H'_{a_k} = G'_{b_1} G'_{b_2} \ldots G'_{b_t}$, where $H'_{a_i} = \gamma_\kappa(a_i, h_i(a_i))$ and $G'_{b_i} = \gamma_\kappa(b_i, g_i(b_i))$ for some $h_i, g_i \in \{h, g\}$. It follows that $a_1 a_2 \ldots a_k = b_1 b_2 \ldots b_t$ and $h_1(a_1) \ldots h_t(a_k) = g_1(a_1) \ldots g_t(a_k)$, and conversely. Therefore the claim follows from Theorem 14. □

Problem 1 is open even for two upper triangular matrices:

Problem 5. *Is it decidable for two upper triangular matrices* $M_1, M_2 \in \mathbb{N}^{2 \times 2}$ *whether they generate a free semigroup?*

Problem 5 was considered in more details in [6]. In particular, the following reduction result was shown there.

Theorem 16. *Problem 5 is decidable if and only if it is decidable for rational matrices of the form*

$$M_1 = \begin{pmatrix} a & 0 \\ 0 & 1 \end{pmatrix} \text{ and } M_2 = \begin{pmatrix} b & 1 \\ 0 & 1 \end{pmatrix},$$

where $a, b \in \mathbb{Q} \setminus \{-1, 0, 1\}$, $|a| + |b| > 1$, *and* $\frac{1}{|a|} + \frac{1}{|b|} > 1$.

The freeness problem was left open in [6] for the concrete instance, where $a = \frac{2}{3}$ and $b = \frac{3}{5}$.

For an $\mathbb{N}$-automaton $\mathcal{A}$, define two different words u and v to be *indistinguishable*, if they give the same multiplicities between every pair of states, that is, if $\mathcal{A}_{rs}(u) = \mathcal{A}_{rs}(v)$ for all states r and s. Theorem 15 then has the following immediate corollary.

Corollary 5. *It is undecidable for 3-state $\mathbb{N}$-automata $\mathcal{A}$ whether $\mathcal{A}$ has indistinguishable words u and v.*

3.4 Generalized Word Problem

In the *generalized word problem* for a semigroup S we are given a finitely generated subsemigroup S_1 and an element $s \in S$, and we ask whether $s \in S_1$.

Since the mortality problem is undecidable for matrices in $\mathbb{Z}^{3\times 3}$, we have the following corollary.

Corollary 6. *The generalized word problem is undecidable for $\mathbb{Z}^{3\times 3}$.*

The following result was proven by Markov [24] in 1958, see also Miller [27].

Theorem 17. *The generalized word problem is undecidable for subsemigroups of $\mathbf{SL}(4, \mathbb{Z})$ of unimodular matrices.*

Proof. Let F_2 be a free group generated by two elements. By Miller [26] and Schupp [32], it is undecidable for finite subsets $D \subseteq F_2 \times F_2$ whether D generates the whole direct product $F_2 \times F_2$. Clearly, D generates $F_2 \times F_2$ if and only if it generates all the pairs (ε, a) and (a, ε) for $a \in F_2$. Therefore the problem for a single a whether (ε, a) is generated by D is undecidable.

The matrices $\begin{pmatrix} 1 & 2 \\ 0 & 1 \end{pmatrix}$ and $\begin{pmatrix} 1 & 0 \\ 2 & 1 \end{pmatrix}$ generate a free group (isomorphic to F_2) with integer entries, and therefore $F_2 \times F_2$ can be embedded into $\mathbb{Z}^{4\times 4}$. The claim then follows from the above. □

The next problem is mentioned in [20].

Problem 6. *Is the generalized word problem decidable for $\mathbf{SL}(3, \mathbb{Z})$?*

Of course, for matrix *monoids* the question whether $I \in S$ is trivial, but semigroups do not only have a mortality problem, they do have an identity problem as well:

Problem 7. *Is it decidable for finitely generated subsemigroups S of matrices from $\mathbb{Z}^{n\times n}$ whether or not $I \in S$?*

The above problem is equivalent to the following one: given a finitely generated semigroup $S \subseteq \mathbb{Z}^{n\times n}$, is it decidable whether a subset of the generators of S generates a group? Indeed, if S contains a group, then $I \in S$, and if $M_1 M_2 \ldots M_k = I$ for some generators M_i of S, then $M_{i+1} \ldots M_k M_1 \ldots M_{i-1} = M_i^{-1}$ for all i, and therefore $M_1, \ldots, M_k$ generate a group.

References

1. Albert, M. H., Lawrence, J.: A proof of Ehrenfeucht's Conjecture. *Theoret. Comput. Sci.* **41** (1985) 121 – 123.
2. Berstel, J., Mignotte, M.: Deux propriétés décidables des suites récurrentes linéaires. *Bull. Soc. Math. France* **104** (1976) 175 – 184.
3. Berstel, J., Reutenauer, C.: "Rational Series and Their Languages", Springer-Verlag, 1988.
4. Blondel, V. D., Tsitsiklis, J. N.: When is a pair of matrices mortal. *Inf. Proc. Letter* **63** (1997) 283 – 286.
5. Cassaigne, J., Karhumäki, J.: Examples of undecidable problems for 2-generator matrix semigroups. *Theoret. Comput. Sci.* **204** (1998) 29 – 34.
6. Cassaigne, J., Harju, T., Karhumäki, J.: On the undecidability of freeness of matrix semigroups. *Int. J. Algebra Comput.* **9** (1999) 295 – 305.
7. Claus, V.: Some remarks on PCP(k) and related problems. *Bull. EATCS* **12** (1980) 54 – 61.
8. Claus, V.: The (n,k)-bounded emptiness-problem for probabilistic acceptors and related problems. *Acta Inform.* **16** (1981) 139 – 160.
9. Eilenberg, S.: "Automata, Languages, and Machines", Vol A, Academic Press, New York, 1974.
10. Guba, V. S.: The equivalence of infinite systems of equations in free groups and semigroups with finite subsystems. *Mat. Zametki* **40** (1986) 321 – 324 (in Russian).
11. Halava, V.: "Decidable and Undecidable Problems in Matrix Theory", Master's Thesis, University of Turku, 1997.
12. Halava, V., Harju, T.: Mortality in matrix semigroups. *Amer. Math. Monthly* **108** (2001) 649 – 653.
13. Hansel, G.: Une démonstration simple du théorème de Skolem-Mahler-Lech. *Theoret. Comput. Sci.* **43** (1986) 1 – 10.
14. Harju, T., Karhumäki, J.: The equivalence problem of multitape automata. *Theoret. Comput. Sci.* **78** (1991) 347 – 355.
15. Harju, T., Karhumäki, J.: Morphisms. In Handbook of Formal Languages, Vol. 1, (A. Salomaa and G. Rozenberg, eds), Springer-Verlag, 1997, pp. 439 – 510.
16. Harju, T., Karhumäki, J., Plandowski, W.: Independent system of equations, in M. Lothaire, *Algebraic Combinatorics on Words*, Cambridge University Press, to appear.
17. Jacob, G.: La finitude des reprèsentations linéaires de semi-groupes est décidable. *J. Algebra* **52** (1978) 437 – 459.
18. Klarner, D. A., Birget, J.-C., Satterfield, W.: On the undecidability of the freeness of integer matrix semigroups. *Int. J. Algebra Comp.* **1** (1991) 223 – 226.
19. Krom, M.: An unsolvable problem with products of matrices. *Math. System. Theory* **14** (1981) 335 – 337.
20. Open problems in group theory: http://zebra.sci.ccny.cuny.edu/cgi-bin/LINK.CGI?/www/web/problems/oproblems.html .
21. Mandel, A., Simon, I.: On finite semigroups of matrices. *Theoret. Comput. Sci.* **5** (1977) 101 – 111.
22. Manna, Z.: "Mathematical Theory of Computations", McGraw-Hill, 1974.
23. Markov, A. A.: On certain insoluble problems concerning matrices. *Doklady Akad. Nauk SSSR (N.S.)* **57** (1947) 539 – 542 (in Russian).
24. Markov, A. A.: On the problem of representability of matrices. *Z. Math. Logik Grundlagen Math.* **4** (1958) 157 – 168 (in Russian).
25. Matiyasevich, Y., Sénizergues, G.: Decision problems for semi-Thue systems with a few rules. In Proceedings of the 11th IEEE Symposium on Logic in Computer Science, pages 523–531, 1996.

26. Miller III, C. F.: "On Group Theoretic Decision Problems and Their Classification", Annals of Math. Study **68**, Princeton Univ. Press 1971.
27. Miller III, C. F.: Decision problems for groups – Survey and reflections, in "Algorithms and Classification in Combinatorial Group Theory" (G. Baumslag and C. F. Miller III, eds.), Springer-Verlag, 1992, 1 – 59.
28. Nielsen, J.: Die Gruppe der dreidimensionalen Gittertransformationen. *Danske Vid. Selsk. Math.-Fys. Medd. V* **12** (1924) 1 – 29.
29. Paterson, M. S.: Unsolvability in 3×3-matrices. *Studies in Appl. Math.* **49** (1970) 105 – 107.
30. Post, E.: A variant of a recursively unsolvable problem. *Bulletin of Amer. Math. Soc.* **52** (1946) 264 – 268.
31. Schultz, P.: Mortality of 2×2-matrices. *Amer. Math. Monthly* **84** (1977) 463 – 464.
32. Schupp, P. E.: Embeddings into simple groups. *J. London Math. Soc.* **13** (1976) 90 – 94.
33. Turakainen, P.: On multistochastic automata. *Inform. Control* **23** (1973) 183 – 203.

Some Petri Net Languages and Codes*

Masami Ito[1] and Yoshiyuki Kunimoch[2]

[1] Department of Mathematics, Kyoto Sangyo University, Kyoto 603-8555, Japan
[2] Department of Computer Science, Shizuoka Institute of Science and Technology, Fukuroi 437-8555, Japan

Abstract. In this paper, we consider the language over an alphabet T generated by a given Petri net with a positive initial marking, called a *CPN language*. This language becomes a prefix code over T. We are interested in CPN languages which are maximal prefix codes, called *mCPN languages* over T. We will investigate various properties of $mCPN$ languages. Moreover, we will prove that a CPN language is a context-sensitive language in two different ways.

1 $mCPN$ Languages of the Form C^n

Let $D = (P, T, \delta, \mu_0)$ be a Petri net with an initial marking μ_0 where P is the set of places, T is the set of transitions, δ is the transition function and $\mu_0 \in N_+^P$ is a positive marking, i.e. $\pi_p(\mu_0) > 0$ for any $p \in P$. Notice that $\pi_p(\mu_0)$ is meant the number of tokens at p of the marking μ_0. A language C is called a *CPN language* over T generated by D and denoted by $C = \mathcal{L}(D)$ if $C = \{u \in T^+ \mid \exists p \in P, \pi_p(\delta(\mu_0, u)) = 0, \forall q \in P, \pi_q(\delta(\mu_0, u)) \geq 0, \text{ and } \forall q' \in P, \pi_{q'}(\delta(\mu_0, u')) > 0 \text{ for } u' \in P_r(u) \setminus \{u\}$ where $P_r(u)$ is the set of all prefixes of $u\}$. Then it is obvious that $C = \mathcal{L}(D)$ is a prefix code over T if $C = \mathcal{L}(D) \neq \emptyset$. Notice that CPN languages were introduced in [6]. If C is a maximal prefix code, then C is called an *mCPN langauge* over T. Now let $u = a_1 a_2 \ldots a_r \in T^*$ where $a_i \in T$. Then, for any $p \in P$, by $p(u)$ we denote $(\#(p, O(a_1)) - \#(p, I(a_1))) + (\#(p, O(a_2)) - \#(p, I(a_2))) + \cdots + (\#(p, O(a_r)) - \#(p, I(a_r)))$. Regarding notations and definitions which are not explained in this paper, refer to [1,3,4].

Lemma 1.1 *Let $C = \mathcal{L}(D)$ be a finite mCPN language where $D = (P, T, \delta, \mu_0)$. By t_p we denote $\pi_p(\mu_0)$ for any $p \in P$. For any $u, v \in C$, if there exists a $p \in P$ such that $t_p = p(u) = p(v)$, then C is a full uniform code over T, i.e. $C = T^n$ for some n, $n \geq 1$.*

Proof. Let $a_1 a_2 \ldots a_{n-1} a_n \in C$ be a word with the longest length in C. Since C is a finite maximal prefix code over T, $a_1 a_2 \ldots a_{n-1} T \subseteq C$. Assume $a_1 a_2 \ldots a_{i-1} a_i T^{n-i} \subseteq C$. Suppose $a_1 a_2 \ldots a_{i-1} b\alpha \in C$ where $a_i \neq b \in T$ and $\alpha \in T^*$ with $|\alpha| < n - i$. Then there exists $\beta \in T^*$ such that $a_1 a_2 \ldots a_{i-1} a_i \alpha \beta a_i \in C$. Notice that $|\alpha \beta| = n - i - 1$. Let $u = a_1 a_2 \ldots a_{i-1} a_i \alpha \beta a_i$ and $v = a_1 a_2 \ldots a_{i-1} b\alpha$. By assumption, there exists $p \in P$ such that $t_p = p(u) = p(v)$. Consider $w = a_1 a_2 \ldots a_{i-1} a_i b \alpha \beta \in C$. Since $p(a_1 a_2 \ldots a_{i-1} a_i \alpha \beta a_i) = t_p, p(a_1 a_2 \ldots a_{i-1} a_i) > p(a_1 a_2 \ldots a_{i-1})$. On the other

---

* This research was supported by Grant-in-Aid for Science Research 10440034, Japan Society for the Promotion of Science

hand, notice that $t_p = p(a_1a_2\ldots a_{i-1}b\alpha)$. This means that $\delta(\delta(\mu_0, a_1a_2\ldots a_{i-1}), b\alpha))$ can be computed only when at least $t_p - p(a_1a_2\ldots a_{i-1})$ tokens are placed at $p \in P$. Consequently, this contradicts the fact $p(w) = p(a_1a_2\ldots a_{i-1}a_ib\alpha\beta) \leq t_p$. Hence $a_1a_2\ldots a_{i-1}bT^{n-i} \subseteq C$. This yealds $a_1a_2\ldots a_{i-1}T^{n-i+1} \subseteq C$. By induction, we have $T^n \subseteq C$ and $T^n = C$. □

Lemma 1.2 *Let A, B be finite maximal prefix codes over T. If AB is an $mCPN$ language over T, then B is an $mCPN$ language over T.*

Proof. Let $AB = \mathcal{L}(D)$ where $D = (P, T, \delta, \mu_0)$. Let $u \in A$ and let $D' = (P, T, \delta, \delta(\mu_0, u))$. Then obviously $\delta(\mu_0, u) \in N_+^P$ and $B \subseteq \mathcal{L}(D')$. Since B is a maximal prefix code over T, $B = \mathcal{L}(D')$, i.e. B is an $mCPN$ language over T. □

Corollary 1.1 *Let C^n be a finite $mCPN$ language over T for some n, $n \geq 2$. Then C^k is an $mCPN$ language over T for any k, $1 \leq k \leq n$.*

Now we provide a fundamental result.

Proposition 1.1 *Let C^n be a finite $mCPN$ language over T for some $n, n \geq 2$. Then C is a full uniform code over T.*

Proof. By Corollary 1.1, we can assume that $n = 2$. Let $D = (P, T, \delta, \mu_0)$ be a Petri net such that $C^2 = \mathcal{L}(D)$. Let $t_p = \pi_p(\mu_0)$ for any $p \in P$. Since $u^2 \in C^2$, there exists $p \in P$ such that $t_p = p(u^2) = 2p(u)$. Let $s_p = t_p/2$ for such a $p \in P$. Then s_p can be well-defined. For remaining $q \in P$, we define $s_q = t_q$. Let define μ_0' as $\pi_p(\mu_0') = s_p$ for any $p \in P$. First we check whether $\delta(\mu_0', u)$ is computable if $p(u) = s_p = t_p/2$. The problem is whether the number of tokens at each place $p \in P$ is enough to make fire the sequence of transitions u. By the proof of Lemma 1.2, we have at most $|C|$ Petri nets recognizing C. Let $C = \{u_i \mid i = 1, 2, \ldots, r\}$ and let $D_i = (P, T, \delta, \mu_{0,i})$ where $\mu_{0,i} = \delta(\mu_0, u_i), i = 1, 2, \ldots, r$. Notice that $\mathcal{L}(D_i) = C$ for any $i, i = 1, 2, \ldots, r$. Since $u_i^2 \in C^2$ for any $i, i = 1, 2, \ldots, r$, there exists $p \in P$ such that $p(u_i^2) = t_p$, i.e. $p(u_i) = t_p/2$. Moreover, notice that, for another $u_j \in C$, $p(u_j) \leq t_p/2$ because $p(u_j^2) \leq t_p$. Hence $min\{\pi_p(\mu_{0,i})\} = s_p$ if $s_p = t_p/2$. Since $\mathcal{L}(D_i) = C$ for any $i, i = 1, 2, \ldots, r$, it follows that the marking μ_0' with $\pi_p(\mu_0') = s_p, p \in P$ has enough tokens at each place to make fire the sequence of transitions $u \in C$. Let $D' = (P, T, \delta, \mu_0')$. Then $C \subseteq \mathcal{L}(D')$. On the other hand, since C is a maximal prefix code over T, $C = \mathcal{L}(D')$. Let $u, v \in C$. Since $uv \in C^2$, there exists $p \in P$ such that $t_p = p(uv)$. Notice that $p(uv) = p(u) + p(v)$. If $p(u) \neq p(v)$, then $p(uv) < max\{p(uu), p(vv)\}$. However, since $uu, vv \in C^2$, $max\{p(uu), p(vv)\} \leq t_p$, a contradiction. Hence $s_p = p(u) = p(v)$. By Lemma 1.1, C is a full uniform code over T. □

Corollary 1.2 *The property being an $mCPN$ language over T is not preserved under concatenation.*

Proof. Let $C \subseteq T^+$ be an $mCPN$ language over T which is not a full uniform code over T. Suppose C^2 is an $mCPN$ language over T. Then, by Proposition 1.1, C is a full uniform code over T, a contradiction. □

Remark 1.1 We can prove the above corollary in a different way. Let $T = \{a, b\}$, let $A = \{a, ba, bb\}$ and let $B = \{b, ab, aa\}$. Then $ab, aaa, bbb \in AB$ and $|aaa|, |bbb| > |ab|$. By the following lemma, AB is not an $mCPN$ language over $\{a, b\}$.

Lemma 1.3 *Let $C \subseteq T^+$ be an mCPN language over T. Then there exists $a \in T$ such that $a^{min\{|u| \mid u \in C\}} \in C$.*

Proof. Let $D = (P, T, \delta, \mu_0)$ be a Petri net with $C = \mathcal{L}(D)$. Moreover, let $v \in C$ such that $|v| = min\{|u| \mid u \in C\}$. Let $v = v'a^i$ where $i \geq 1, v' \in T^*, a \in T$ and $v' \notin T^*a$. If $v' = \lambda$, then we have done. Let $v' = v''b$, i.e. $v = v''ba^i$ where $b \in T$ and $b \neq a$. Since $v \in C$, there exists $p \in P$ such that $p(v) = t_p = \pi_p(\mu_0)$. If $p(a) \geq p(b)$, then we consider $p(v''aa^i)$. If $p(a) > p(b)$, then $p(v''aa^i) > t_p$ and some proper prefix of $v''aa^i$ must be an element of C. However, this contradicts the assumption that v has the minimum length as a word in C. Therefore, $p(a) = p(b)$ and $p(v''aa^i) = t_p$. This implies that $v''aa^i \in C$. Now let $p(b) > p(a)$. In this case, we consider $v''bb^i \in T^+$. It is obvious that $p(v''bb^i) > p(v) = t_p$. This contradicts again that the assumption that v has the minimum length as a words in C. Hence $v''a^{i+1} \in C$ and $|v''a^{i+1}| = |v|$. Contuining the same procedure, we have $a^{|v|} \in C$. This completes the proof of the lemma. □

Remark 1.2 If C is an infinite $mCPN$ language over T, then Proposition 1.1 does not hold true. For instance, let $T = \{a, b\}$ and let $C = b^*a$. Then both C and $C^2 = b^*ab^*a$ are $mCPN$ languages over T.

Remark 1.3 In the following section, we will generalize Proposition 1.1.

2 $mCPN$ Languages of the Form AB

In fact, we can generalize Proposition 1.1 as follows:

Proposition 2.1 *Let A, B be finite maximal prefix codes over T. If AB is an mCPN language over T, then A and B are full uniform codes over T.*

Proof. (i) Let $a_1a_2 \ldots a_{n-1}a_n$ be one of the longest words in A. Since A is a finite maximal prefix code over T, $a_1a_2 \ldots a_{n-1}T \subseteq A$. Assume that $a_1a_2 \ldots a_{i-1}a_iT^{n-i} \subseteq A$ where $1 \leq i < n$ and $a_1a_2 \ldots a_{i-1}T^{n-i+1} \setminus A \neq \emptyset$. Then there exist $b \in T$ and $\alpha \in T^*$ such that $b \neq a_i, |\alpha| < n - i$ and $a_1a_2 \ldots a_{i-1}b\alpha \in A$. Consider the word $a_1a_2 \ldots a_{i-1}a_ib\alpha a_i^t \in A$ where $t \geq 0$ and $|\alpha| + t = n - i - 1$. Since B is a finite maximal prefix code over T, there exists $h, h \geq 1$ such that $a_i^h \in B$. Hence $a_1a_2 \ldots a_{i-1}b\alpha a_i^h \in AB$ and $a_1a_2 \ldots a_{i-1}a_ib\alpha a_i^t a_i^h \in AB$. Then there exists $p \in P$ such that $p(a_1a_2 \ldots a_{i-1}b\alpha a_i^h) = t_p$ and $p(a_i) > 0$. It is obvious that $p(a_1a_2 \ldots a_{i-1}a_ib\alpha\ a_i^t a_i^h) = (t+1)p(a_i) + t_p > t_p$, a contradiction. Therefore, $a_1a_2 \ldots a_{i-1}T^{n-i+1} \subseteq A$. Continuing this procedure, we have $T^n \subseteq A$ and hence $T^n = A$.

(ii) Let $b_1b_2 \ldots b_{m-1}b_m$ be one of the longest words in B. Assume that $b_1b_2 \ldots b_{j-1}b_jT^{m-j} \subseteq B$ where $1 \leq j < m$ and $b_1b_2 \ldots b_{j-1}T^{m-j+1} \setminus B \neq \emptyset$. Then there exist $c \in T$ and $\beta \in T^*$ such that $c \neq b_j$, $|\beta| < m - j$ and $b_1 b_2 \ldots b_{j-1} c\beta \in B$. Let $c\beta \in T^* d$ where $d \in T$. By the above assumption, $b_1 b_2 \ldots b_{j-1} b_j c\beta d^s \in B$ where $s \geq 0$. By (i), $A = T^n$ for some $n, n \geq 1$. Hence $(d^{n-1} b_j) b_1 b_2 \ldots b_{j-1} c\beta \in AB$ and $d^n b_1 b_2 \ldots b_{j-1} b_j c\beta d^s \in AB$. Notice that there exists $q \in P$ such that $t_q = q((d^{n-1} b_j) b_1 b_2 \ldots b_{j-1} c\beta)$ and $q(d) > 0$. Then $q(d^n b_1 b_2 \ldots b_j c\beta d^s) = t_q + (s+1) q(d) > t_q$, a contradiction. Hence $b_1 b_2 \ldots b_{j-1} T^{m-j+1} \subseteq B$. Continuing this procedure, we have $T^m \subseteq B$ and hence $T^m = B$. This completes the proof of the proposition. □

3 Constructions of $mCPN$ Languages

In this section, we provide two construction methods of $mCPN$ languages.

Definition 3.1 Let $A, B \subseteq T^+$. Then by $A \oplus B$ we denote the language $(\cup_{b \in T} \{(P_r(A) \setminus A) \diamond B b^{-1}\} b) \cup (\cup_{a \in T} \{(P_r(B) \setminus B) \diamond A a^{-1}\} a)$ where $\diamond$ is meant the shuffle operation and $Ca^{-1} = \{u \in T^+ | ua \in C\}$ for $C \subseteq T^+$ and $a \in T$.

Proposition 3.1 Let $T = Y \cup Z$ where $Y, Z \neq \emptyset, Y \cap Z = \emptyset$. If $A \subseteq Y^+$ is an $mCPN$ language over Y and $B \subseteq Z^+$ is an $mCPN$ language over Z, then $A \oplus B$ is an $mCPN$ language over T.

Proof. Let $D_1 = (P_1, Y, \delta_1, \mu_1)$ and $D_2 = (P_2, Z, \delta_2, \mu_2)$ be Petri nets such that $\mathcal{L}(D_1) = A$ and $\mathcal{L}(D_2) = B$, respectively, and $P_1 \cap P_2 = \emptyset$. Now we define the following Petri net D: $D = (P_1 \cup P_2, T, \delta, \mu_0)$ where $\pi_p(\mu_0) = \pi_p(\mu_1)$ for $p \in P_1$ and $\pi_q(\mu_0) = \pi_q(\mu_2)$ for $q \in P_2$, and $\delta(\mu, a)|_{P_1} = \delta_1(\mu|_{P_1}, a)$ for $a \in Y$ and $\delta(\mu, b)|_{P_2} = \delta_2(\mu|_{P_2}, b)$ for $b \in Z$. Here $\mu|_{P_i}$ is the restriction of μ to $P_i, i = 1, 2$. Then $A \oplus B = \mathcal{L}(D)$. □

Example 3.1 Let $T = \{a, b\}$. Consider $A = \{a\}$ and $B = \{bb\}$. Then both A and B are $mCPN$ languages over $\{a\}$ and $\{b\}$, respectively. Hence $A \oplus B = \{a, ba, bb\}$ is an $mCPN$ language over T.

Proposition 3.2 Let $A, B \subseteq T^+$ be finite $mCPN$ languages over T. Then $A \oplus B$ is an $mCPN$ language over T if and only if $A = B = T$.

Proof. ($\Leftarrow$) Since $A \oplus B = T \oplus T = T$, $A \oplus B$ is an $mCPN$ language over T.
($\Rightarrow$) Assume $A \neq T$. Then there exists $a \in T$ such that $a^i \in A, i \geq 2$. Moreover, since B is an $mCPN$ language over T, there exists some $j, j \geq 1$ such that $a^j \in B$. Notice that $\lambda, a \in P_r(a^i) \setminus \{a^i\}$. Hence $a^j, a^{j+1} \in A \oplus B$. This implies that $A \oplus B$ is not a prefix code. Consequently, $A \oplus B$ is not even a CPN language. The proof can be done in the same way for the case $B \neq T$. □

Remark 3.1 For the class of infinite $mCPN$ languages over T, the situation is different. For instance, let $T = \{a, b\}$ and let $A = B = b^* a$. Then $A \oplus B = b^* a$ and A, B and $A \oplus B$ are $mCPN$ languages over T.

Proposition 3.3 Let $A, B \subseteq T^+$ be $mCPN$ languages over T. Then there exist an alphabet Y, $D \subseteq Y^+$: an $mPCN$ language over Y and a λ-free homomorphism h of Y^* onto T^* such that $A \oplus B = h(D)$.

Proof. Let $T' = \{a' \mid a \in T\}$. For $u = a_1 a_2 \ldots a_n \in T^n$, we denote $u' = a'_1 a'_2 \ldots a'_n \in T'^n$. Let $Y = T \cup T'$ and let h be the homomorphism of Y^* onto T^* defined as: $h(a) = a, h(a') = a$ for any $a \in T$. Let $B' = \{u' \mid u \in B\} \subseteq T'^+$. Then B' is an $mCPN$ language over T'. By Proposition 3.1, $D = A \oplus B'$ is an $mCPN$ language over Y and $A \oplus B = h(D)$. □

Corollary 3.1 *The property being an $mCPN$ language over T is not preserved under λ-free homomorphism.*

Lemma 3.2 *Let $C \subseteq T^+$ be an mCPN language over T and let $a, b \in T$. If $bbaa \in C$, then $baba \in C$.*

Proof. Let $D = (P, T, \delta, \mu_0)$ be a Petri net with $C = \mathcal{L}(D)$. Since $bbaa \in C$, there exists $p \in P$ such that $p(bbaa) = t_p = \pi_p(\mu_0)$. It follows from the assumption that C is an $mCPN$ language over T that $p(bbaa) = p(baba) = t_p$. Since $p(b), p(bb), p(bba) \leq t_p$, we have $p(b), p(ba) \leq t_p$ and $p(bab) \leq t_p$. Moreover, since $q(b), q(bb), q(bba), q(bbaa) \leq t_q$ for any $q \in P$, we have $q(b), q(ba), q(bab) \leq t_q$ and $q(baba) \leq t_q$. Hence $baba \in C$. □

Remark 3.2 By the above lemma, a maximal prefix code over T having the property in Lemma 3.2 cannot be necessarily realized by a Petri net. For instance, let $T = \{a, b\}$ and let $C = \{a, ba, bbaa, bbab, bbb\}$. Then C is a maximal prefix code over T. However, by Lemma 3.2, it is not an $mCPN$ language over T.

Now we introduce another method to construct $mCPN$ languages.

Definition 3.2 Let $A \subseteq T^+$. By $m(A)$, we denote the language $\{v \in A \mid \forall u, v \in A, \forall x \in T^*, v = ux \Rightarrow x = 1\}$. Obviously, $m(A)$ is a prefix code over T. Let $A, B \subseteq T^+$. By $A \otimes B$, we denote the language $m(A \cup B)$.

Proposition 3.4 *Let A, B be mCPN languages over T. Then, $A \otimes B$ is an mCPN language over T.*

Proof. Let $D = (P, T, \delta, \mu_0)$ and $E = (Q, T, \theta, \gamma_0)$ be Petri nets with a positive initial markings such that $A = \mathcal{L}(D)$, $B = \mathcal{L}(E)$ and $P \cap Q = \emptyset$. Let $D \otimes E = (P \cup Q, T, \delta \times \theta, \mu_0 \times \gamma_0)$ where $\delta \times \theta(\mu \times \gamma, a) = \delta(\mu, a) \times \theta(\gamma, a)$ for a marking $\mu \times \gamma$ and $a \in T$. Notice that $\mu \times \gamma$ means that $\pi_p(\mu \times \gamma) = \pi_p(\mu)$ for $p \in P$ and $\pi_q(\mu \times \gamma) = \pi_q(\gamma)$ for $q \in Q$. We will show that $\mathcal{L}(D \otimes E) = A \otimes B$. First, we show that $A \otimes B$ is a maximal prefix codes over T. Let $u \in T^*$. Since A is a maximal prefix code over T, $u \in AT^*$ or $uT^* \cap A \neq \emptyset$. If $u \in AT^*$, then $m(A \cup B)T^* = (A \otimes B)T^*$. Asssume $uT^* \cap A \neq \emptyset$. Since B is a maximal prefix code over T, $uT^* \cap B \neq \emptyset$ or $u \in BT^*$. Then either $uT^* \cap (A \otimes B) \neq \emptyset$ or $u \in (A \otimes B)T^*$ hold. Hence $A \otimes B$ is a maximal prefix code over T. Now we show that $\mathcal{L}(D \otimes E) = A \otimes B$. Let $u \in A \otimes B$. Then $u \in A \cup B$. Assume $u \in A$. Then there exists $p \in P$ such that $\pi_p(\delta(\mu_0, u)) = 0$. Moreover, $\pi_{p'}(\delta(\mu_0, u)) \geq 0$ and $\pi_{p'}(\delta(\mu_0, u')) > 0$ for any $p' \in P$ and $u' \in P_r(u) \setminus \{u\}$. On the other hand, since $u \in A \otimes B$ and $u \in A$, there exists $x \in T^*$ such that $ux \in B$. Hence, for any $q \in Q$, $\pi_q(\theta(\gamma_0, u)) \geq 0$ and $\pi_q(\theta(\gamma_0, u')) > 0$ for any

$q \in Q$ and $u' \in P_r(u) \setminus \{u\}$. This means that $\pi_p(\delta(\mu_0 \times \gamma_0, u)) = 0$ for $p \in P \cup Q$ and $\pi_r(\delta(\mu_0 \times \gamma_0, u)) \geq 0, \pi_r(\delta(\mu_0 \times \gamma_0, u')) > 0$ for any $r \in P \cup Q$ and $u' \in P_r(u) \setminus \{u\}$. Hence $u \in \mathcal{L}(D \otimes E)$. By the same way, we can prove that $u \in \mathcal{L}(D \otimes E)$ for $u \in B$. Therefore, $A \otimes B \subseteq \mathcal{L}(D \otimes E)$. By the maximality of $A \otimes B$, we have $A \otimes B = \mathcal{L}(D \otimes E)$. □

Example 3.2 It is obvious that a^*b and $(a \cup b)^3$ are $mCPN$ languages over $\{a, b\}$. Hence $a^*b \otimes (a \cup b)^3 = \{b, ab, aaa, aab\}$ is an $mCPN$ language over $\{a, b\}$.

Remark 3.3 Proposition 3.4 does not hold for the classe of CPN languages over T. The reason is the following: Suppose that $A \otimes B$ is a CPN language over T for any two CPN languages A and B over T. Then we can show that, for a given finite CPN language A over T, there exists a finite $mCPN$ language B over T such that $A \subseteq B$ as follows. Let $A \subseteq T^+$ be a finite CPN language over T which is not an $mCPN$ language. Let $n = max\{|u| \mid u \in A\}$. Consider T^n which is an $mCPN$ language over T. By assumption, $A \otimes T^n$ becomes a CPN language (in fact, an $mCPN$ language) over T. By the definition of the operation $\otimes$, it can be also proved that $A \subseteq A \otimes T^n$. However, as the following example shows, there exists a finite CPN language A over T such that there exists no $mCPN$ language B over T with $A \subseteq B$. Hence, Proposition 3.4 does not hold for the class of all CPN language over T.

Example 3.3 Consider the language $A = \{ab, aaba, aaa\} \subseteq \{a, b\}^+$. Then this language becomes a CPN language over $\{a, b\}$ (see Fig. 3.1). Moreover, it can be proved that there is no $mCPN$ language B over $\{a, b\}$ with $A \subseteq B$ as follows: Suppose $B \subseteq T^+$ is an $mCPN$ language with $A \subseteq B$ over T. By Lemma 1.3, $b \in B$ or $b^2 \in B$. Let $b^i \in B$ where $i = 1$ or 2. Let $t_p = p(ab)$ where $p \in P$ and P is the set of places of the Petri net which recognizes B. If $p(a) < 0$. Then $p(b) > t_p$ and hence $p(b^i) > t_p$. This contradicts the fact $b^i \in B$. If $p(a) > 0$, then $p(aaba) = p(ab) + 2p(a) > t_p$. This contradicts the fact $aaba \in B$ as well. Hence $p(a) = 0$ and $p(aab) = t_p$. However, since aab is a prefix of $aaba \in B$, $p(aab) < t_p$. This yealds a contradiction again. Therefore, there is no $mCPN$ language $B \subseteq T^+$ with $A \subseteq B$.

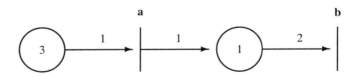

Fig. 3.1 Petri net D with $\mathcal{L}(D) = \{ab, aaba, aaa\}$

Remark 3.4 The set of all $mCPN$ languages over T forms a semigroup under $\otimes$. Moreover, the operation $\otimes$ has the following properties:
(1) $A \otimes B = B \otimes A$, (2) $A \otimes A = A$, (3) $A \otimes T = T$.
Consequently, the set of all $mCPN$ languages over T forms a commutative band with zero under $\otimes$ (for bands, see [2]).

4 Rank of CPN Languages

In this section, we will consider the rank and related decomposition of CPN languages.

Definition 4.1 Let $A \subseteq T^+$ be a CPN language over T. By $r(A)$ we denote the value $min\{|P| \, | \, D = (P, T, \delta, \mu_0), \mathcal{L}(D) = A\}$.

Remark 4.1 Let $A \subseteq T^+$ be a finite $mCPN$ language over T. Then $r(A) \leq |A|$. The proof can be done as follows: Let $D = (P, T, \delta, \mu_0)$ be a Petri net with a positive initial marking μ_0 such that $\mathcal{L}(D) = A$. Let $P' = \{p_u \in P \, | \, u \in A, p_u(u) = \delta(\mu_0, u)\} \subseteq P$. The transition function δ' can be defined as $\delta'(\mu|_{P'}, a) = \delta(\mu, a)|_{P'}$ where $a \in T$. Then $A = \mathcal{L}(D')$ and it is obvious that $r(A) \leq |A|$. However, in general this inequality does not hold for a CPN language over T as the following example shows. In Fig. 4.1, $\mathcal{L}(D) = \{aba\}$ but $r(\{aba\}) \neq 1$ because $aba \in A$ if and only if $baa \in A$ for any CPN language with $r(A) = 1$.

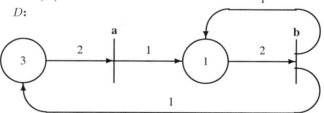

Fig. 4.1   Petri net D with $r(\mathcal{L}(D)) > |\mathcal{L}(D)|$

Now let $A, B \subseteq T^+$ be $mCPN$ languages over T. Then it is easy to see that $|A \otimes B| \leq max\{|A|, |B|\}$. Moreover, if A and B are finite, then $r(A \otimes B) \leq r(A) + r(B)$.

We define three language classes as follows: $\mathcal{L}_{CPN} = \{A \subseteq T^+ \, | \, A \text{ is a } CPN \text{ language over } T\}$, $\mathcal{L}_{mCPN} = \{A \subseteq T^+ \, | \, A \text{ is an } mCPN \text{ language over } T\}$, $\mathcal{L}_{NmCPN} = \{A \subseteq T^+ \, | \, A: \text{ an } mCPN \text{ language over } T, \exists D = (P, T, \delta, \mu_0), \forall p \in P, \forall a \in T, \#(p, I(a)) \leq 1, \mathcal{L}(D) = A\}$. Then it is obvious that we have the following inclusion relations: $\mathcal{L}_{NmCPN} \subseteq \mathcal{L}_{mCPN} \subseteq \mathcal{L}_{CPN}$. It is also obvious that $\mathcal{L}_{mCPN} \neq \mathcal{L}_{CPN}$.

Problem 4.1 Does $\mathcal{L}_{mCPN} = \mathcal{L}_{NmCPN}$ hold?

Proposition 4.1 Let $A \in \mathcal{L}_{mCPN}$. Then there exist a positive integer $k \geq 1$ and $A_1, A_2, \ldots, A_k \in \mathcal{L}_{CPN}$ such that $r(A_i) = 1, i = 1, 2, \ldots, k$ and $A = A_1 \otimes A_2 \otimes \cdots \otimes A_k$. Moreover, in the above, if $A \in \mathcal{L}_{NmCPN}$, then $A_1, A_2, \ldots, A_k$ are in $\mathcal{L}_{NmCPN}$ and context-free.

Proof. Let $k = r(A)$. Then there exists a Petri net $D = (P, T, \delta, \mu_0)$ such that $\mathcal{L}(D) = A$ and $|P| = k$. For any $p \in P$, let $D_p = (\{p\}, T, \delta_p, \{\pi_p(\mu_0)\})$ where δ_p represents $\#(p, I(a)) = \#(p, I(a))$ in D and $\#(p, O(a)) = \#(p, O(a))$ in D for any $a \in T$. Let $A_p = \mathcal{L}(D_p), p \in P$. We prove that $A = \otimes_{p \in P} A_p$. Assume $u \in A$. Then there exists $p \in P$ such that $\pi_p(\mu_0) = p(u), \pi_q(\mu_0) \geq q(u)$ and $\pi_q(\mu_0) > q(u')$ for any $q \in P, u' \in P_r(u) \setminus \{u\}$. Therefore, $u \in A_p$ and $(P_r(u) \setminus \{u\}) \cap A_q = \emptyset$ for any $q \in P$. Therefore, $u \in \otimes_{p \in P} A_p$, i.e. $A \subseteq \otimes_{p \in P} A_p$. Now assume $u \in \otimes_{p \in P} A_p$. Then there exists $p \in P$ such that $u \in A_p$ and $(P_r(u) \setminus \{u\}) \cap A_q = \emptyset$ for any $q \in P$. Hence $\pi_p(\mu_0) = p(u), \pi_q(\mu_0) \geq q(u)$ and $\pi_q(\mu_0) > q(u')$ for any $q \in P, u' \in P_r(u) \setminus \{u\}$. Consequently, $u \in A$, i.e. $\otimes_{p \in P} A_p \subseteq A$. Hence $\otimes_{p \in P} A_p = A$. For the proof of the latter half, notice that there exists a Petri net $D = (P, T, \delta, \mu_0)$ such that $\#(p, I(a)) \leq 1$ for all $(p, a) \in P \times T$ and $\mathcal{L}(D) = A$. Let $k = |P|$ where $k \geq r(A)$. Then it is obvious that $A_1, A_2, \ldots, A_k$ are in $\mathcal{L}_{NmCPN}$. Hence it is enough to show that $\mathcal{L}(D)$ is context-free if $D = (\{p\}, T, \delta, \{n\})$ where $\#(p, I(a)) \leq 1$ for any $a \in T$. Now construct the following context-free grammar G: $G = (\{S, P\}, T, R, S)$ where $R = \{S \to P^n\} \cup \{P \to a \mid a \in T, \#(p, I(a)) = 1, \#(p, O(a)) = 0\} \cup \{p \to p^{\#(p,O(a)) - \#(p,I(a))} \mid a \in T, \#(p, O(a)) - \#(p, I(a)) > 0\}$. Then it can easily be verified that $\mathcal{L}(D) = \mathcal{L}(G)$. $\square$

Problem 4.2 *In the above proposition, can we take $r(A)$ as k if $A \in \mathcal{L}_{NmCPN}$?*

Proposition 4.2 Let $A \subseteq T^+$ be a finite $mCPN$ language with $r(A) = 1$ over T. Then A is a full uniform code over T.

Proof. Let $D = (\{p\}, T, \delta, \{n\})$ be a Petri net with $\mathcal{L}(D) = A$ and let $u, v \in A$. Then $p(u) = p(v) = n$. By Lemma 1.1, A is a full uniform code over T. $\square$

Proposition 4.3 Let $A \subseteq T^+$ be an $mCPN$ language with $r(A) = 1$ over T and let k be a positive integer. Then A^k is an $mCPN$ language with $r(A^k) = 1$ over T.

Proof. Let $D = (\{p\}, T, \delta, \{n\})$ be a Petri net with $\mathcal{L}(D) = A$. Now let $D_k = (\{p\}, T, \delta, \{kn\})$. Then it can easily be seen that A^k is a maximal prefix code over T and $\mathcal{L}(D_k) = A^k$. Hence A^k is an $mCPN$ language over T with $r(A^k) = 1$. $\square$

Proposition 4.4 Let $A \in \mathcal{L}_{NmCPN}$. Then, by Proposition 4.1, there exist $A_1, A_2, \ldots, A_k \in \mathcal{L}_{NmCPN}$ such that $r(A_i) = 1, i = 1, 2, \ldots, k$ and $A = A_1 \otimes A_2 \otimes \cdots \otimes A_k$. Let $n_1, n_2, \ldots, n_k$ be positive integers. Then $A_1^{n_1} \otimes A_2^{n_2} \otimes \ldots, \otimes A_k^{n_k} \in \mathcal{L}_{NmCPN}$.

Proof. Obvious from the above results. $\square$

5 Context-Sensitiveness of CPN Languages

Consider the Petri net $D = (S, T, \delta, \mu_0)$ depicted below. Then $\mathcal{L}(D) \cap a^+b^+c^+ = \cup_{n \geq 1}\{a^n b^i c^{n+i+1} | 1 \leq i \leq n\}$ is not context-free. Hence $\mathcal{L}(D)$ is not context-free. Therefore, the class of all CPN languages over an alphabet T is not necessary included in the class of all context-free languages over T. However, in this section, we will prove the context-sensitiveness of CPN languages in two different ways, i.e. the first one is an indirect proof and the second one is a direct proof.

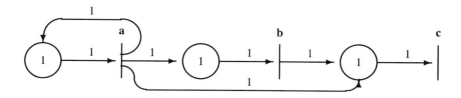

Fig. 5.1 Petri net which generates a non-context-free language

Proposition 5.1 *Let $C \subseteq T^+$ be a CPN language over T. Then C is a context-sensitive language over T.*

For the first proof, we will use the following results: 1) The complement of a context-sensitive language is context-sensitive ([5]). 2) The concatenation of two context-sensitive languages is context-sensitive. 3) A Petri net language of type G is context-sensitive ([4]).

First proof. Let μ_1, μ_2 be markings of a Petri net $D = (P, T, \delta)$. Then, $\mu_1 \leq \mu_2$ means that $\pi_p(\mu_1) \leq \pi_p(\mu_2)$ for any $p \in P$. Now let $D = (P, T, \delta, \mu_0)$ be a Petri net with an initial marking $\mu_0 \in N_+^P$ such that $C = \mathcal{L}(D)$. Let $F = \{\mu \,|\, \Sigma_{p \in P} \pi_p(\mu) < |P|\}$. Moreover, Let $L_1 = \{w \in T^* \,|\, \exists \mu \in F, \delta(\mu_0, w) \geq \mu\}$, i.e. L_1 is a Petri net language of type G. Therefore, L_1 is context-sensitive. Now let $L_2 = \{w \in T^* \,|\, \forall p \in P, \pi_p(\delta(\mu_0, w)) \geq 1\}$. Then L_2 is also context-sensitive. It is obvious that $L_1 \setminus L_2 = \{w \in T^* \,|\, \exists p \in P, \pi_p(\delta(\mu_0, w)) = 0\}$ and this language is context-sensitive. Consider $(L_1 \setminus L_2) \setminus (L_1 \setminus L_2)T^+$. Then $(L_1 \setminus L_2) \setminus (L_1 \setminus L_2)T^+ = \{w \in T^* \,|\, \exists p \in P, \pi_p(\delta(\mu_0, w)) = 0 \text{ and } \forall p \in P, \pi_p(\delta(\mu_0, w')) > 0 \text{ for } w' \in P_r(w) \setminus \{w\}\} = \mathcal{L}(D)$. Hence C is context-sensitive. □

Before giving the second proof, we provide a few notations. Let $\mu_1, \mu_2, \ldots, \mu_r$ and μ be markings of a Petri net. Then $\mu = \mu_1 + \mu_2 + \cdots + \mu_r$ if $\pi_p(\mu) = \pi_p(\mu_1) + \pi_p(\mu_2) + \cdots + \pi_p(\mu_r)$ for any $p \in P$. Now let $D = (P, T, \delta, \mu_0)$ be a Petri net with a positive initial marking μ_0. Let $N_D = max\{\#(p, I(a)), \#(q, O(b)) \,|\, a, b \in T, p, q \in P\}$ and let $M_D = max\{\pi_p(\mu_0) \,|\, p \in P\}$. By Ω_D we denote the set of markings $\{\mu \,|\, \forall p \in P, \pi_p(\mu) \leq M_D + 3N_D\}$. Notice that Ω_D is a finite set.

Second proof. Let $D = (P, T, \delta, \mu_0)$ be a Petri net with a positive marking μ_0. We construct the following context-sensitive grammar $G = (V, T, R, S)$ where V is the set of variables, T is an alphabet, R is a set of productions (rewriting rules) and S is a start symbol, as follows: $V = \{S, [\delta]\} \cup \{[w] \mid w \in T^2 \cup T^3\} \cup \{[\mu] \mid \mu \in \Omega_D\} \cup \{[\pi_p] \mid p \in P\}$ and $R = R_1 \cup R_2 \cup R_3 \cup R_4 \cup R_5 \cup R_6 \cup R_7 \cup R_8$, where

$R_1 = \{S \to w \mid w \in (T \cup T^2 \cup T^3) \cap \mathcal{L}(D)\},$
$R_2 = \{S \to [\delta][\mu_0]\},$
$R_3 = \{[\delta][\mu] \to [w][\delta][\nu][\nu'] \mid \mu \in N_+^P \cap \Omega_D, w \in T^2 \cup T^3,$
$\quad \nu + \nu' = \delta(\mu, w), \nu + \nu' = \delta(\mu, w), \nu, \nu' \in \Omega_D,$
$\quad \forall w' \in P_r(w), \delta(\mu, w') \in N_+^P\},$
$R_4 = \{[\mu][\nu] \to [\mu'][\nu'] \mid \mu + \nu = \mu' + \nu', \mu, \nu, \mu', \nu' \in \Omega_D\},$
$R_5 = \{[\delta][\mu] \to [w][\pi_p] \mid p \in P, \mu \in N_+^P \cap \Omega_D, w \in T^2 \cup T^3,$
$\quad \forall w' \in P_r(w) \setminus \{w\}, \delta(\mu, w') \in N_+^P, \pi_p(\delta(\mu, w)) = 0\},$
$R_6 = \{[\pi_p][\mu] \to [\pi_p][\pi_p] \mid p \in P, \mu \in \Omega_D, \pi_p(\mu) = 0\},$
$R_7 = \{[w][\pi_p] \to [\pi_p][w] \mid p \in P, w \in T^2 \cup T^3\},$
$R_8 = \{[w][\pi_p] \to w \mid p \in P, w \in T^2 \cup T^3\}$

We provide the following lemma.

Lemma 5.1 *Let k be a positive integer. Then $S \Rightarrow^* [w_1][w_2]\ldots[w_k][\delta][\mu_1][\mu_2]\ldots[\mu_k][\mu_{k+1}]$ if and only if $\delta(\mu_0, w_1 w_2 \ldots w_k) = \mu_1 + \mu_2 + \cdots + \mu_k + \mu_{k+1}$ where $\mu_i \in \Omega_D, i = 1, 2, \ldots k+1, w_j \in T^2 \cup T^3, j = 1, 2, \ldots, k$ and $\delta(\mu_0, w) \in N_+^P$ for any $w \in P_r(w_1 w_2 \ldots w_k)$.*

Proof of Lemma 5.1 Consider the case $k = 1$. Then $S \Rightarrow^* [w_1][\delta][\mu_1][\mu_2]$ implies that $S \Rightarrow [\delta][\mu_0] \Rightarrow [w_1][\delta][\mu_1][\mu_2]$ where $\delta(\mu_0, w_1) = \mu_1 + \mu_2$. Moreover, we have $\delta(\mu_0, w) \in N_+^P$ for any $w \in P_r(w_1)$. Now assume that $\delta(\mu_0, w_1) = \mu_1 + \mu_2$ and $\delta(\mu_0, w) \in N_+^P$ for any $w \in P_r(w_1)$. Then, by R_2 and R_3, we have $S \Rightarrow^* [w_1][\delta][\mu_1][\mu_2]$. Thus the lemma holds for $k = 1$. Assume that the lemma holds for $k = 1, 2, \ldots, n$. Let $S \Rightarrow^* [w_1][w_2]\ldots[w_n][w_{n+1}][\delta][\mu_1][\mu_2]\ldots[\mu_{n+1}][\mu_{n+2}]$. Then we have the following derivation:

$S \Rightarrow^* [w_1][w_2]\ldots[w_n][\delta][\nu_1][\nu_2]\ldots[\nu_n][\nu_{n+1}] \Rightarrow [w_1][w_2]\ldots[w_n][w_{n+1}]$
$[\delta][\nu_1'][\nu_2''][\nu_2]\ldots[\nu_n][\nu_{n+1}] \Rightarrow^* [w_1][w_2]\ldots[w_n][w_{n+1}][\delta][\mu_1][\mu_2]\ldots[\mu_{n+1}][\mu_{n+2}]$

In the above, $\delta(\nu_1, w_{n+1}) = \nu_1' + \nu_1''$ and $\delta(\nu_1, w) \in N_+^P$ for any $w \in P_r(w_{n+1})$. Moreover, $\nu_1' + \nu_1'' + \nu_2 + \cdots + \nu_{n+1} = \mu_1 + \mu_2 + \cdots + \mu_{n+1} + \mu_{n+2}$. On the other hand, by assumption, $\delta(\mu_0, w_1 w_2 \ldots w_n) = \nu_1 + \nu_2 + \cdots + \nu_n + \nu_{n+1}$ and $\delta(\mu_0, w) \in N_+^P$ for any $w \in P_r(w_1 w_2 \ldots w_n)$. Since $\delta(\mu_0, w_1 w_2 \ldots w_n) = \nu_1 + \nu_2 + \cdots + \nu_n + \nu_{n+1}$ and $\delta(\nu_1, w_{n+1}) = \nu_1' + \nu_2'', \delta(\mu_0, w_1 w_2 \ldots w_n w_{n+1}) = \delta(\nu_1, w_{n+1}) + \nu_2 + \cdots + \nu_n + \nu_{n+1} = \nu_1' + \nu_2'' + \nu_2 + \cdots + \nu_n + \nu_{n+1} = \mu_1 + \mu_2 + \cdots + \mu_{n+1} + \mu_{n+2}$. Moreover, since $[\delta][\nu_1] \to [w_{n+1}][\delta][\nu_1'][\nu_2'']$ is applied, $\delta(\nu_1, w') \in N_+^P$ for any $w' \in P_r(w_{n+1}) \setminus \{w_{n+1}\}$. Remark that $\delta(\mu_0, w_1 w_2 \ldots w_n) = \nu_1 + \nu_2 + \cdots + \nu_n + \nu_{n+1}$. Therefore, $\delta(\mu_0, w_1 w_2 \ldots w_n w') = \delta(\nu_1, w') + \nu_2 + \cdots + \nu_n + \nu_{n+1} \in N_+^P$ for any $w_1 w_2 \ldots w_n w' \in P_r(w_1 w_2 \ldots w_n w_{n+1})$. Together with the previous assumption, we have $\delta(\mu_0, w) \in N_+^P$ for any $w \in P_r(w_1 w_2 \ldots w_n w_{n+1})$. Now assume that $\delta(\mu_0, w_1 w_2 \ldots w_n w_{n+1}) = \mu_1 + \mu_2 + \cdots + \mu_{n+1} + \mu_{n+2}$ where $\mu_i \in \Omega_D, i = 1, 2, \ldots, n+2$ and $\delta(\mu_0, w) \in N_+^P$ for any $w \in P_r(w_1 w_2 \ldots w_n w_{n+1})$.

Consider $\delta(\mu_0, w_1w_2\ldots w_n) \in N_+^P$. Since $\pi_p(\delta(\mu_0, w_1w_2\ldots w_n)) \leq N_D + 3nM_D \leq n(N_D + 3M_D)$, $\delta(\mu_0, w_1w_2\ldots w_n)$ can be represented as $\nu_1 + \nu_2 + \cdots + \nu_n + \nu_{n+1}$ where $\nu_1, \nu_2, \ldots, \nu_n, \nu_{n+1} \in \Omega_D$. On the other hand, $\delta(\mu_0, w) \in N_+^P$ for any $w \in P_r(w_1w_2\ldots w_n)$. By induction hypothesis, $S \Rightarrow^* [w_1][w_2]\ldots[w_n][\delta][\nu_1][\nu_2]\ldots[\nu_{n+1}]$ where $\delta(\mu_0, w_1w_2\ldots w_nw_{n+1})$ can be computed and $\delta(\mu_0, w_1w_2\ldots w_nw') \in N_+^P$ for any $w' \in P_r(w_{n+1})$. Moreover, since $max\{|\pi_p(\delta(\mu_0, w_1w_2\ldots w_nu))| - |\pi_p(\delta(\mu_0, w_1w_2\ldots w_n))| \,|\, p \in P, |u| \leq 3\} \leq 3M_D$, There exist $\gamma_1, \gamma_2, \ldots, \gamma_{n+1} \in \Omega_D$ such that $\nu_1+\nu_2+\cdots+\nu_{n+1} = \gamma_1 + \gamma_2 + \cdots + \gamma_{n+1}$ and $\delta(\gamma_1, w') \in N_+^*$ for any $w' \in P_r(w_n)$. Therefore, we have $S \Rightarrow^* [w_1][w_2]\ldots[w_n][\delta][\nu_1][\nu_2]\ldots[\nu_{n+1}] \rightarrow [w_1][w_2]\ldots[w_n][\delta][\gamma_1][\gamma_2]\ldots[\gamma_{n+1}]$. Notice that the rule $[\delta][\gamma_1] \rightarrow [w_1][\delta][\gamma_1'][\gamma_1'']$ can be applied. Hence we have $S \Rightarrow^* [w_1][w_2]\ldots[w_n][w_{n+1}][\delta][\gamma_1'][\gamma_1''][\gamma_2]\ldots[\gamma_{n+1}]$. Since $\delta(\mu_0, w_1w_2\ldots w_nw_{n+1}) = \mu_1 + \mu_2 + \cdots + \mu_{n+1} + \mu_{n+2} = \gamma_1' + \gamma_1'' + \gamma_2 + \cdots + \gamma_{n+1}$, $S \Rightarrow^* [w_1][w_2]\ldots[w_n][w_{n+1}][\delta][\gamma_1'][\gamma_1''][\gamma_2]\ldots[\gamma_{n+1}] \Rightarrow^* [w_1][w_2]\ldots[w_n][w_{n+1}][\delta][\mu_1][\mu_2]\ldots[\mu_{n+1}][\mu_{n+2}]$. This completes the proof of the lemma. □

Now we return to the second proof. Let $u \in \mathcal{L}(D)$. If $|u| \leq 3$, then $S \Rightarrow u$ and $u \in \mathcal{L}(G)$. Assume $|u| > 3$. Then $u = w_1w_2\ldots w_n$ for some $w_i \in T^2 \cup T^3$, $i = 1, 2, \ldots, n$. Notice that $|p(w_i)| \leq 3M_D$ for any $p \in P, w_i, i = 1, 2, \ldots, n-1$. Hence there exist $\mu_1, \mu_2, \ldots, \mu_n \in \Omega_D$ such that $\delta(\mu_0, w_1w_2\ldots w_{n-1}) = \mu_1 + \mu_2 + \cdots + \mu_n$. Moreover, it is obvious that $\delta(\mu_0, w) \in N_+^P$ for any $w \in P_r(w_1w_2\ldots w_{n-1})$. By Lemma 5.1, $S \Rightarrow^* [w_1][w_2]\ldots[w_{n-1}][\delta][\mu_1][\mu_2]\ldots[\mu_n]$. Let $\nu_1, \nu_2, \ldots, \nu_n \in \Omega_D$ such that $\pi_p(\nu_1) = min\{\pi_p(\mu_1 + \mu_2 + \cdots + \mu_n), N_D + 3M_D\}$ for any $p \in P$ and $\nu_1 + \nu_2 + \cdots + \nu_n = \mu_1 + \mu_2 + \cdots + \mu_n$. Then $S \Rightarrow^* [w_1][w_2]\ldots[w_{n-1}][\delta][\mu_1][\mu_2]\ldots[\mu_n] \Rightarrow [w_1][w_2]\ldots[w_{n-1}][\delta][\nu_1][\nu_2]\ldots[\nu_n]$. By the definition of ν_1 and $u = w_1w_2\ldots w_{n-1}w_n \in \mathcal{L}(D)$, $\delta(\mu_0, w_1w_2\ldots w_{n-1}w) \in N_+^P$ for any $p \in P$ and $w \in P_r(w) \setminus \{w\}$, and $\pi_q(\delta(\mu_0, w_1w_2\ldots w_{n-1}w_n)) = 0$ for some $q \in P$. Thus the rule $[\delta][\nu_1] \rightarrow [w_n][\pi_q]$ can be applied and we have $S \Rightarrow^* [w_1][w_2]\ldots[w_{n-1}][\pi_q][\nu_2]\ldots[\nu_n]$. Since $\pi_q(\nu_2+\nu_3+\cdots+\nu_n) = 0$, $\pi_q(\nu_i) = 0$ for any $i, i = 2, 3, \ldots, n$. Thus we have $S \Rightarrow^* [w_1][w_2]\ldots[w_{n-1}][w_n][\pi_q]^n \Rightarrow^* [w_1][\pi_q][w_2][\pi_q]\ldots[w_{n-1}][\pi_q][w_n][\pi_q] \Rightarrow^* w_1w_2\ldots w_n$. Consequently, $u = w_1w_2\ldots w_n \in \mathcal{L}(G)$ and $\mathcal{L}(D) \subseteq \mathcal{L}(G)$.

Let $u = w_1w_2\ldots w_{n-1}w_n \in \mathcal{L}(G)$. Then we have the following derivation:

$S \Rightarrow^* [w_1][w_2]\ldots[w_{n-1}][\delta][\mu_1][\mu_2]\ldots[\mu_{n-1}][\mu_n] \Rightarrow [w_1][w_2]\ldots[w_{n-1}][w_n][\pi_p][\mu_2]\ldots[\mu_{n-1}][\mu_n] \Rightarrow^* [w_1][w_2]\ldots[w_{n-1}][w_n][\pi_p]^n \Rightarrow^* [w_1][\pi_p][w_2][\pi_p]\ldots[w_{n-1}][\pi_p][w_n][\pi_p] \Rightarrow^* w_1w_2\ldots w_{n-1}w_n$.

By Lemma 5.1, $\delta(\mu_0, w) \in N_+^P$ for any $w \in P_r(w_1w_2\ldots w_{n-1})$. Since the rules $[\delta][\mu_1] \rightarrow [w_n][\pi_p], [\pi_p][\mu_i] \rightarrow [\pi_p][\pi_p], i = 2, 3, \ldots, n$ are applied, $\delta(\mu_0, w) \in N_+^P$ for any $w \in P_r(w_1w_2\ldots w_{n-1}w_n) \setminus \{w_1w_2\ldots w_{n-1}w_n\}$ and $\pi_p(\delta(\mu_0, w_1w_2\ldots w_{n-1}w_n)) = 0$. This means that $u = w_1w_2\ldots w_{n-1}w_n \in \mathcal{L}(D)$, i.e. $\mathcal{L}(G) \subseteq \mathcal{L}(D)$. Consequently, $\mathcal{L}(G) = \mathcal{L}(D)$ □

References

1. J. Berstel and D. Perrin, *Theory of Codes*, Academic Press, London-New York, 1985.
2. A.H. Clifford and G.B. Preston, *The Algebraic Theory of Semigroups* Vol. 1, American Mathematical Society, Providence R.I.,1961.
3. J.E. Hopcroft and J.D. Ullman, *Introduction to Automata Theory, Languages and Computation*, Addison-Wesley, Reading MA,1979.
4. J.L. Peterson, *Petri Net Theory and the Modeling of Systems*, Printice-Hall, New Jersey,1981.
5. G. Rozenberg and A. Salomaa Eds., *Handbook of Formal Languages* Vol. 1, *Word, Language and Grammar*, Springer-Verlag, Berlin-Heidelberg,1997.
6. G. Tanaka, Prefix codes determined by Petri nets, *Algebra Colloquium* **5** (1998), 255-264.

Words, Permutations, and Representations of Numbers

Helmut Prodinger

The John Knopfmacher Centre for Applicable Analysis and Number Theory
School of Mathematics
University of the Witwatersrand, P. O. Wits
2050 Johannesburg, South Africa
helmut@gauss.cam.wits.ac.za,
http://www.wits.ac.za/helmut/index.htm

Dedicated to Werner Kuich on the Occasion of His Sixtieth Birthday

Abstract. In this survey paper we consider words, where the letters are interpreted to be numbers or digits. In the first part, natural numbers are weighted with probabilities (from the geometric distribution). Several properties and parameters of sets of such words are analyzed probabilistically; the case of permutations is a limiting case. In the second part, the representation of Gaussian integers to the base $-2 + i$ is considered, as well as redundant representations to the base q, where the digits can be arbitrary integers.

1 Introduction

In this survey paper, we consider words $w = a_1 a_2 \ldots a_n$, where the letters a_i are taken from the set $\mathbb{Z}$ or a finite subset thereof. In the first part, we assume that the letters are obtained from the geometric distribution, i. e., $\mathbb{P}\{a = k\} = pq^{k-1}$ with $p + q = 1$. In this way, a probability (weight) is attached to each word w, namely $(p/q)^n q^{a_1 + \cdots + a_n}$ (we assume that the letters are independent from each other). Of course, a probability is then also attached to each set $L \cap \mathbb{N}^n$, where L is an arbitrary language. Most of the time, we will consider probability generating functions $\sum_{n \geq 0} \mathbb{P}\{L \cap \mathbb{N}^n\} z^n$. A typical example is the set of all *up–down* words $L = \{a_1 a_2 \ldots a_n \mid n \geq 0, a_1 \leq a_2 \geq a_3 \leq \ldots\}$. The probability of $L \cap \mathbb{N}^n$ is then the probability that a random word of length n is an up–down word. The interest in the geometric distribution comes from computer science applications, namely a data structure called *skip lists* [11], and *permutation counting*. A permutation $\sigma_1 \sigma_2 \ldots \sigma_n$ does not enjoy the independence property of letters in a word; a letter σ_i can only occur if it was not already used in $\sigma_1 \sigma_2 \ldots \sigma_{i-1}$. This is often cumbersome to model. However, with the present approach, we can (often) consider the limit $q \to 1$. Then, the probability that a letter appears more than once, goes to 0, and each relative ordering of the letters is equiprobable. Hence parameters of permutations which only depend on the "order statistics" will carry over. For example, the parameter "position of the largest element" translates accordingly, but not "value of the largest element." We report about some recent results in the sequel.

The second part of this survey deals with the alphabet $\{0, 1, \ldots, q - 1\}$ and similar ones; we think of the letters as digits, and consider the words as representations of numbers. For example, a word $a_n \ldots a_1 a_0$ represents the number $\sum_{i=0}^{n} a_i q^i$ in the q-ary representation. The kth digit of the q-ary representation of n is given by

$$a_k = \left\lfloor \frac{n}{q^k} \right\rfloor - q \left\lfloor \frac{n}{q^{k+1}} \right\rfloor.$$

We will deal with more exotic number systems: On the one hand, with the basis $-2 + i$ (with the complex unit $i = \sqrt{-1}$), see [6], and on the other hand with redundant representations to the base q, where the digits are allowed to be integers (the alphabet is then $\mathbb{Z}$), see [22], [9].

2 Words, Geometric Probabilities, and Permutations

2.1 Monotone Words

One of the simplest examples deals with the probability that a word is monotone, i. e., $a_1 < a_2 < \ldots$ (or $a_1 \leq a_2 \leq \ldots$). This establishes a nice combinatorial interpretation of *Euler's partition identities* (see [17]). The identities in question are

$$\prod_{n \geq 0} (1 + tq^n) = \sum_{n \geq 0} \frac{t^n q^{\binom{n}{2}}}{(q;q)_n} \quad \text{and} \quad \prod_{n \geq 0} \frac{1}{1 - tq^n} = \sum_{n \geq 0} \frac{t^n}{(q;q)_n},$$

where $(x;q)_n := (1-x)(1-qx)\ldots(1-q^{n-1}x)$. Now the language of monotone words is $\mathcal{M}_< = (\varepsilon + 1)(\varepsilon + 2) \ldots$ resp. $\mathcal{M}_\leq = 1^* \cdot 2^* \ldots$, so that the associated generating functions are given as

$$M_<(z) = \prod_{k \geq 0} (1 + pq^k z) = \sum_{n \geq 0} \frac{p^n z^n q^{\binom{n}{2}}}{(q;q)_n}$$

and

$$M_\leq(z) = \prod_{k \geq 0} \frac{1}{1 - pq^k z} = \sum_{n \geq 0} \frac{p^n z^n}{(q;q)_n}.$$

Therefore the probability that a word of length n is strictly monotone, is $p^n q^{\binom{n}{2}}/(q;q)_n$ and that it is weakly monotone is $p^n/(q;q)_n$. Both quantities tend for $q \to 1$ to $1/n!$, which confirms to the fact that just one permutation of n elements is monotone.

2.2 Alternating Words

Let us now consider up–down words or down–up words; for permutations, this does not make a difference, but for words it does (see [21] for details). Naturally, one can combine the weak and strict forms for the inequalities. Also, one must distinguish the case of odd resp. even lengths of words. The case of odd lengths is more interesting. Let $f^{\leq >}(z)$ be the generating functions where the coefficient of z^{2n+1} is the probability that a word

$a_1 \ldots a_{2n+1}$ satisfies $a_1 \le a_2 > a_3 \le \ldots$, etc. The results can all be expressed with the following "q–tangent function"

$$\tan_q^{[A,B,C,D]}(z) := \sum_{n \ge 0} \frac{(-1)^n z^{2n+1}}{[2n+1]_q!} q^{An^2+Bn} \Big/ \sum_{n \ge 0} \frac{(-1)^n z^{2n}}{[2n]_q!} q^{Cn^2+Dn};$$

here we used the notation of q–factorials: $[n]_q! = (q;q)_n/(1-q)^n$, which converges for $q \to 1$ to $n!$.

$$f^{\ge \le}(z) = \tan_q^{[1,1,1,-1]}(z), \qquad f^{<>}(z) = \tan_q^{[1,1,1,0]}(z),$$
$$f^{><}(z) = \tan_q^{[1,0,1,0]}(z), \qquad f^{\le \ge}(z) = \tan_q^{[1,0,1,-1]}(z),$$
$$f^{\ge <}(z) = f^{\le >}(z) = f^{> \le}(z) = f^{< \ge}(z) = \tan_q^{[0,0,0,0]}(z).$$

Generating functions for patterns like $(\le \le \ge)^*$ lead to *Olivier functions*, see [2]; q–versions of that appear in a joint paper with A. Tsifhumulo [24].

2.3 The q–Path Length of a Binary Search Tree

The path length $\rho(t)$ of a binary search tree t is the defined to be the sum over all distances from the root to any node in the tree (measured in terms of nodes, not edges) and satisfies the recursion $\rho(t) = \rho(t_L) + \rho(t_R) + |t_L| + |t_R|$ where t_L and t_R are the left resp. right subtree of the root. ($|t|$ denotes the size of the tree t, i. e. the number of nodes.) Our aim is to rewrite the definition of the path length in terms of permutations. For a permutation $\pi = \pi_1 \ldots \pi_n$ we define $\rho(\pi)$ by

$$\rho(\pi) = \big|\{(j,k) \mid 1 \le j < k \le n,\ \pi_j = \min\{\pi_j,\ldots,\pi_k\} \text{ or } \pi_k = \min\{\pi_j,\ldots,\pi_k\}\}\big|.$$

Then $\rho(\Box) = 0$ and, if $\pi = \sigma 1 \tau$, then $\rho(\pi) = \rho(\sigma) + \rho(\tau) + |\sigma| + |\tau|$, as pairs with the left coordinate in σ and the right coordinate in τ are definitely not counted. (The correspondence from permutations to trees is as follows: the smallest number forms the root, everything to the left (right) of it (recursively) the left (right) subtree.) This definition gives the path length of the binary search tree associated to the permutation $\pi = \pi_1 \ldots \pi_n$, provided we interpret π_i as the arrival position of key i. But this definition of π can be taken as it is, where $\pi_1 \ldots \pi_n$ now denotes a *word* over the alphabet $\{1, 2, \ldots\}$. We compute the expectation of the parameter ρ, for random words of length n, and define for this purpose random variables

$$L_{jk} = \begin{cases} 1 & \text{if } \pi_j = \min\{\pi_j, \ldots, \pi_k\} \\ 0 & \text{otherwise,} \end{cases}, \quad R_{jk} = \begin{cases} 1 & \text{if } \pi_k = \min\{\pi_j, \ldots, \pi_k\} \\ 0 & \text{otherwise,} \end{cases},$$

$B_{jk} = L_{jk} \cdot R_{jk}$, $N_{jk} = (1-L_{jk}) \cdot (1-R_{jk})$. (The letters L, R, B, N are chosen to indicate *left, right, both, not*.) Then the parameter ρ may be described as

$$\rho = \sum_{1 \le j < k \le n} \big[L_{jk} + R_{jk} - B_{jk}\big].$$

From this one can compute the expected value (q–analogue of the path length in a binary search tree of size n):

$$\mathbf{E}_n = (1-q^2) \sum_{1 \leq i \leq n} \frac{n+1-i}{1-q^i} - n(1+q),$$

which is in the limit $q \to 1$: $\mathbf{E}_n = 2(n+1)H_n - 4n$, as is of course well known. (The notations $H_n = \sum_{1 \leq k \leq n} \frac{1}{k}$ refer to harmonic numbers, see e. g. [7].) The variance can also be computed, but this is much harder.

A similar but simpler parameter appears in [19], namely the number of inversions (for words):

$$\mathbf{E}_n = \frac{n(n-1)}{2} \frac{q}{1+q},$$

$$\mathbf{V}_n = \frac{n(n-1)}{6} \frac{q}{(1+q)^2(1+q+q^2)} \Big(2(1-q+q^2)n - q^2 + 7q - 1\Big).$$

The same paper has also a parameter a of Knuth, which is important in the analysis of *permutation in situ*:

$$a = \big|\{(i,j) \mid 1 \leq i < j \leq n,\ x_i = \min\{x_i, x_{i+1}, \ldots, x_j\}\}\big|.$$

Its expected value can be computed as follows:

$$\mathbf{E}_n = \left(\frac{p}{q}\right)^n \sum_{i_1, \ldots, i_n \geq 1} q^{i_1 + \cdots + i_n} \sum_{1 \leq j < k \leq n} [\![i_j = \min\{i_j, \ldots, i_k\}]\!]$$

$$= \sum_{1 \leq j < k \leq n} \left(\frac{p}{q}\right)^{k+1-j} \sum_{i_j = \min\{i_j, \ldots, i_k\}} q^{i_j + \cdots + i_k}$$

$$= \sum_{1 \leq j < k \leq n} \left(\frac{p}{q}\right)^{k+1-j} \sum_{i \geq 1} q^{i(k+1-j)} \frac{1}{p^{k-j}}$$

$$= p \sum_{1 \leq j < k \leq n} \left(\frac{1}{q}\right)^{k+1-j} \sum_{i \geq 1} q^{i(k+1-j)}$$

$$= p \sum_{1 \leq j < k \leq n} \frac{1}{1 - q^{k+1-j}}$$

$$= p \sum_{2 \leq h \leq n} (n+1-h) \frac{1}{1-q^h}.$$

(We used the Iverson convention: $[\![C]\!] = 1$ if the condition C is true, and $[\![C]\!] = 0$ otherwise.)

2.4 Lengths of Ascending Runs

Knuth in [14] has considered the average length L_k of the kth ascending run in random permutations of n elements (for simplicity, mostly the instance $n \to \infty$ was discussed). We study the concept of ascending runs in our usual setting (see [20], [15]),

i. e., we consider infinite words, with letters $1, 2, \cdots$, and they appear with probabilities $p, pq, pq^2, \cdots$. If we decompose a word into ascending runs

$$a_1 < \cdots < a_r \geq b_1 < \cdots < b_s \geq c_1 < \cdots < c_t \geq \cdots ,$$

then r is the length of the first, s of the second, t of the third run, and so on. We are interested in the averages of these parameters. The generating function

$$\prod_{i \geq 1}(1 + pq^{i-1}z) = (-pz; q)_\infty$$

plays a rôle here, where $(a; q)_\infty = (1-a)(1-aq)(1-aq^2)\cdots$:

$$\Lambda_m(z) = \sum_{n \geq 0} \bigl[\text{Probability that a word of length } n \text{ has (exactly) } m \text{ ascending runs}\bigr] z^n,$$

then $\Lambda_0(z) = 1$ and $\Lambda_1(z) = (-pz; q)_\infty - 1$ and we have the recurrence

$$\Lambda_m(z) = \bigl((-pz; q)_\infty\bigr)^m - \sum_{k=1}^{m} \frac{1}{k!} D^k \bigl(z^k \Lambda_{m-k}(z)\bigr)$$

(D means differentiation w. r. t. z.) In the limiting case $q \to 1$ we can specify these quantities explicitly:

$$\Lambda_m(z) := e^{mz} - \sum_{k=1}^{m} \frac{1}{k!} D^k \bigl(z^k \Lambda_{m-k}(z)\bigr), \qquad \Lambda_0(z) := 1,$$

and for $m \geq 1$ (e. g. via induction)

$$\Lambda_m(z) = \sum_{j=0}^{m} e^{jz} \frac{z^{m-j-1}(-1)^{m-j} j^{m-j-1}(j(z-1)+m)}{(m-j)!}.$$

Knuth has shown that the average length of the mth run is given by

$$L_m = \Lambda_m(1) = m \sum_{j=0}^{m} \frac{(-1)^{m-j} j^{m-j-1}}{(m-j)!} e^j,$$

and that $L_m \to 2$. In the general case, we are able to prove that $L_m \to 1 + q$. To do that we study the double generating function

$$R(w, z) = \sum_{m \geq 0} \Lambda_m(z) w^m = \frac{1-w}{1 - w \prod_{n \geq 0}(1 + pz(1-w)q^n)}.$$

In the limit $q \to 1$ we get $R(w, z) = (1-w)/(1 - we^{z(1-w)})$, which is the generating function for Eulerian numbers $A(n, k)$. Note that $A(n, k)$ is the number of permutations of n elements with exactly k runs. (Sometimes, as in [7], the focus is on the number of descents, which is one less than the number of runs.) Hence $R(w, z)$ provides q–Eulerian numbers which differ from those investigated by Carlitz [3].

2.5 Left–to–Right Maxima

As is easy to guess by the name, this parameter counts how often we meet a number that is larger than all the elements to the left. The results for random permutations are: The harmonic number H_n for the expectation and $H_n - H_n^{(2)}$ for the variance; here, $H_n^{(2)} = 1 + \frac{1}{4} + \cdots + \frac{1}{n^2}$ denotes the nth harmonic number of second order.

For recent research on that in the context of random words, see [18], [12], [23]. We will set up a certain "language" $\mathcal{L}$ and translate it to enumerating generating functions. We decompose all sequences $x_1 x_2 \ldots$ in a canonical way as follows: Each left–to–right maximum k will be combined with the following (smaller or equal) elements. A typical part is described by $\mathcal{A}_k := k\{1, \ldots, k\}^*$. Such a group may be present or not. This observation gives the desired "language," where ε denotes the "empty word:"

$$\mathcal{L} := (\mathcal{A}_1 + \varepsilon) \cdot (\mathcal{A}_2 + \varepsilon) \cdot (\mathcal{A}_3 + \varepsilon) \cdot \ldots .$$

Now we want to mark each letter by a "z" and each left–to–right maximum by a "u". The probability pq^{k-1} for a letter k should of course not being forgotten. The set $\{1, \ldots, k\}$ maps into $z(1 - q^k)$ and its star $\{1, \ldots, k\}^*$ into $1/(1 - z(1 - q^k))$. So we obtain the generating function $F(z, u)$ as an infinite product:

$$F(z, u) = \prod_{k \geq 1} \left(1 + \frac{zupq^{k-1}}{1 - z(1 - q^k)} \right).$$

To be explicit, the coefficient of $z^n u^k$ in $F(z, u)$ is the probability that n random variables have k left–to–right maxima. Let $f(z) = \frac{\partial F(z,y)}{\partial y}\big|_{y=1}$. It is the generating function for the expected values $\mathbf{E}_n$. We obtain

$$f(z) = \frac{pz}{1 - z} \sum_{k \geq 0} \frac{q^k}{1 - z(1 - q^k)},$$

which is also, by partial fraction decomposition,

$$f(z) = p \sum_{k \geq 0} \left[\frac{1}{1 - z} - \frac{1}{1 - z(1 - q^k)} \right].$$

From this the coefficients $\mathbf{E}_n$ are easy to get:

$$\mathbf{E}_n = [z^n] f(z) = p \sum_{k \geq 0} \left[1 - (1 - q^k)^n \right].$$

The asymptotic evaluation of this is well known: The average number $\mathbf{E}_n$ of left–to–right maxima in the context of n independently distributed geometric random variables has the asymptotic expansion

$$\mathbf{E}_n = p \left[\log_Q n + \frac{\gamma}{L} + \frac{1}{2} - \delta\left(\log_Q n\right) \right] + \mathcal{O}\!\left(\frac{1}{n}\right)$$

with a (small) periodic function $\delta(x)$. (Here and later we will use the handy abbreviations $Q = 1/q$ and $L = \log Q$.)

Now we turn to the average *value* and *position* of the rth left–to–right maximum, for fixed r and $n \to \infty$, something that was studied by Wilf in the instance of permutations [27]. Summarizing our results, we obtain the asymptotic formulæ $\frac{r}{p}$ and $\frac{1}{(r-1)!}\left(\frac{p}{q}\log_Q n\right)^{r-1}$. The generating function related to the *value* of interest is

$$\mathsf{Value}(z, u) := \frac{1}{1-z} \prod_{i=1}^{h-1} \left\{ 1 + \frac{pq^{i-1}zu}{1 - (1-q^i)z} \right\} zpq^{h-1},$$

which originates from the (unique) decomposition of a string as $a_1 w_1 \ldots a_{r-1} w_{r-1} a_r w$ where $a_1, \ldots, a_r$ are the left–to–right maxima, the w_i are the strings between them, and w can be anything. Note that if $a_k = l$, then this corresponds to a term $pq^{l-1}zu$. Any letter in w_k can come with the probability $1 - q^k$, and thus w_k corresponds to $\sum_{j\geq 0}(1-q^k)^j z^j = 1/(1-(1-q^k)z)$. A particular value i does not necessarily occur as a left–to–right maximum; that is reflected by the $1 + \ldots$ in the product. However, when we look for the coefficient of u^{r-1}, we have seen $r - 1$ left–to–right maxima, and the rth has value h. What comes after that is irrelevant and covered by the factor $1/(1-z)$. We find it useful to use the abbreviation $[\![i]\!] := 1 - (1-q^i)z$. The coefficients of $z^n u^{r-1}$ in $\mathsf{Value}(z, u)$, call them $\pi_{n,h}^{(r)}$, are not probabilities, because there exist words of length n with fewer than r left–to–right maxima, but the quantities $\pi_{n,h}^{(r)}/\pi_n^{(r)}$ are, where $\pi_n^{(r)}$ is the probability that a string of length n has r left–to–right maxima. We compute $\pi_n^{(r)}$ from $\mathsf{Value}(z, u)$ by summing over all values h;

$$\pi_n^{(r)} := [z^n u^{r-1}] \frac{1}{1-z} \sum_{h\geq 1} \prod_{i=1}^{h-1} \left\{ 1 + \frac{pq^{i-1}zu}{[\![i]\!]} \right\} zpq^{h-1} = \sum_{h\geq 1} \pi_{n,h}^{(r)}.$$

Now we turn to the *position*. Set

$$\sigma_{n,j}^{(r)} := [z^n u^{r-1} v^j] \frac{1}{1-z} \sum_{h\geq 1} \prod_{k=1}^{h-1} \left\{ 1 + \frac{pq^{k-1}zvu}{1 - (1-q^k)zv} \right\} zvpq^{h-1},$$

then $\sigma_{n,j}^{(r)}/\pi_n^{(r)}$ is the probability that a random string of length n has the rth left–to–right maximum in position j. It is the same decomposition as before, however, we are not interested in the value h, so we sum over it. On the other hand, we label the position with the variable v, so we must make sure that every z that does not appear in the factor $1/(1-z)$ must be multiplied by a v. Computationally, we find it easier to work with the parameter "position $-r$". This leads us to the generating function

$$\mathsf{Position}(z, u, v) := \frac{1}{1-z} \sum_{h\geq 1} \prod_{k=1}^{h-1} \left\{ 1 + \frac{pq^{k-1}zu}{1 - (1-q^k)zv} \right\} zpq^{h-1}.$$

There is no space to develop this further; for a full account see [12].

Now we consider the instance of *large* left–to–right maxima (see [23]). We say that the last left–to–right maximum is the instance $r = 1$, the previous left–to–right maximum is the instance $r = 2$, and so on.

The largest element in a random permutation of n elements (which is n) is expected to be in the middle. The second largest is then expected to be in the middle of the first half, and so on. In general, the average position should be about $n/2^r$. The following expression for the average value of the rth left–to–right maximum from the right is obtained as before; just observe that we assume that it is h, and that $r-1$ more left–to–right maxima follow to the right, whereas to the left everything is smaller than h:

$$\mathbf{E}_n^{(r)} = (-1)^r \left(\frac{p}{q}\right)^r \sum_{k=r}^n \binom{n}{k}(-1)^k \times$$

$$\times [w^k] \sum_{1 \le h < i_1 < \cdots < i_{r-1}} \frac{hwq^h}{(1-wq^{h-1})(1-wq^h)} \frac{q^{i_1}w \cdots q^{i_{r-1}}w}{(1-q^{i_1}w)\cdots(1-q^{i_{r-1}}w)}.$$

A lengthy study leads to

$$\mathbf{E}_n^{(r)} = \log_Q n + \sigma_r + \delta_r(\log_Q n) + o(1),$$

where $\delta_r(x)$ is a small periodic function of $\log_Q n$; the constant σ_r is given by

$$\sigma_r = \frac{\gamma}{L} + \frac{1}{2} - p^{r-1} \sum_{j \ge 1} \frac{q^{(j-1)(r-1)}}{(1-q^j)^r} - p \sum_{j \ge 1} \frac{q^j}{(1-q^j)^2}$$

$$+ p^{r-1} \sum_{j \ge 1} \frac{q^{j(r-1)}}{(1-q^j)^2 (1-q^{j+1})^{r-2}}.$$

Turning to the average position of the rth left–to–right maximum from the right we find

$$\mathbf{E}_n^{(r)} = (-1)^{r-1} \left(\frac{p}{q}\right)^r \sum_{k=r+1}^n \binom{n}{k}(-1)^k \times$$

$$\times [w^k] \sum_{1 \le h < i_1 < \cdots < i_{r-1}} \frac{wq^h w(1-q^{h-1})}{(1-wq^{h-1})^2(1-wq^h)} \frac{wq^{i_1}}{1-wq^{i_1}} \cdots \frac{wq^{i_{r-1}}}{1-wq^{i_{r-1}}}$$

and

$$\mathbf{E}_n^{(r)} = (-1)^r \left(\frac{p}{q}\right)^{r-1} \psi_r(1)\, n + \varpi_r(\log_Q n) + o(n),$$

with some complicated constants and periodic functions. In the limit $q \to 1$ (permutations), this simplifies to

$$\mathbf{E}_n^{(r)} = \frac{n}{2^r} + o(n).$$

2.6 Skip Lists

We start from the interpretation of a skip list ρ of size n as a word $\rho = a_1 \ldots a_n$, $n \geq 0$, $a_i \in \{1, 2, 3, \ldots\}$. The (horizontal) path length X_n is a random variable defined on the set of words of length n. For an individual word, the path length $b(\rho)$ is defined as the sum of all numbers of (weak) right–to–left maxima in the subwords (prefixes) $a_1, \ldots, a_i$.

For example, consider 21413. The right–to–left maxima when considering prefixes are 2, 2—1, 4, 4—1, 4—3, giving a total of $1 + 2 + 1 + 2 + 2 = 8$.

In order to obtain an expression for the corresponding probability generating function it is convenient to start from the following combinatorial decomposition:

Let m be the maximal element occurring in ρ. Then we can write ρ in a unique way as

$$\rho = \sigma m \tau, \quad \text{where} \quad \sigma \in \{1, \ldots, m-1\}^*, \; \tau \in \{1, \ldots, m\}^*$$

(fixing the leftmost occurence of the maximum m). From this decomposition it follows that

$$b(\rho) = b(\sigma m \tau) = b(\sigma) + b(\tau) + |\tau| + 1, \quad b(\varepsilon) = 0, \tag{1}$$

since the contribution of the leftmost maximum is 1 plus the number of the succeeding elements. We introduce the bivariate generating functions $P^{=m}(z, u)$, where the coefficient of $z^n u^j$ is the probability that a skip list ρ with n elements has maximum element m and path length j. The generating functions $P^{\leq m}(z, u)$ and $P^{<m}(z, u)$ are defined analogously. Then relation (1) translates into the functional equation

$$P^{=m}(z, u) = pq^{m-1} z u P^{<m}(z, u) P^{\leq m}(zu, u), \quad P^{=0}(z, u) = 1. \tag{2}$$

Although no simple closed formula for the solution of (2) is available, it allows to extract sufficient information on the moments of the path length. If we set $F^*(z) = \frac{\partial P^*}{\partial u}(z, 1)$, where $*$ stands for $= m$, $\leq m$ or $< m$, the generating function $F(z) = \sum_{n \geq 0} \mathbf{E}(X_n) z^n$ of the *expectations* is obtained as

$$F(z) = \lim_{m \to \infty} F^{\leq m}(z).$$

Performing some algebra one derives

$$F^{\leq m}(z) = \frac{p}{q} \frac{z}{[\![m]\!]^2} \sum_{i=1}^{m} \frac{q^i}{[\![i]\!]}.$$

Now we let m tend to infinity and obtain

$$F(z) = \frac{p}{q} \frac{z}{(1-z)^2} \sum_{i \geq 1} \frac{q^i}{[\![i]\!]}.$$

The generating function $H(z)$ of the second factorial moments can be obtained in the same way, although the computations are much more involved, see [11]. It is, however, not too complicated to find the coefficients in $F(z)$:

$$\mathbf{E}(X_n) = [z^n]F(z) = \frac{p}{q}\sum_{i\geq 1}\left[-Q^i + n + 1 + Q^i\left(1-q^i\right)^{n+1}\right]$$

$$= \frac{p}{q}\sum_{k=2}^{n+1}\binom{n+1}{k}(-1)^k\frac{1}{Q^{k-1}-1}$$

$$= (Q-1)n\left(\log_Q n + \frac{\gamma-1}{L} - \frac{1}{2} + \frac{1}{L}\delta_1(\log_Q n)\right) + O(\log n)$$

where γ is Euler's constant and $\delta_1(x) = \sum_{k\neq 0}\Gamma(-1-\frac{2k\pi i}{L})e^{2k\pi i x}$ is a continuous periodic function of period 1 and mean zero.

3 Representations of Numbers

3.1 Complex Bases

There are representations of numbers that are perhaps not very well known, see e. g. [6]. For easier readability, we describe only a special case. Every Gaussian integer $z \in \mathbb{Z}[i]$ (of the form $z = x + iy$, with $x, y \in \mathbb{Z}$) can be uniquely written as

$$\sum_{k=0}^{n}\varepsilon_k(-2+i)^k, \quad \text{with} \quad \varepsilon_k \in \{0,1,2,3,4\}.$$

The minimal polynomial of $-2+i$ is x^2+4x+5. As a consequence, $145 = 0$, and there is the carry rule $5 = 1310$. So if one wants to add 2041 and 4432, one takes 6473, uses the carry rule to get 19523 and the reduction rule to get 5023, and eventually 1310023. Or, one applies the reduction rule earlier, to get 5023 from 6473.

The digits can be computed as follows: Define

$$s_{k+1} = -4\left\lfloor\frac{s_k}{5}\right\rfloor - \left\lfloor\frac{s_{k-1}}{5}\right\rfloor, \quad k \geq 0, \quad s_{-1} = -5y, \quad s_0 = x + 2y,$$

then the kth digit is given by $s_k \bmod 5$.

Addition of 1 (or any fixed number) can be described by a *transducer*; digits are read from right to left, and eventually one of the two states marked by • are reached, where the convention is that all digits not read so far are simply copied. The states represent the possible carries.

The paper [6] deals also with the sum–of–digits function (defined in an obvious way), and *fundamental regions*. Each complex number has a representation in this system, of the form $\sum_{k=-\infty}^{n}\varepsilon_k(-2+i)^k$, which is in general unique; only "a few" numbers have 2 or even 3 representations. (This is akin to the base 10 representation, where $1 = 0.9999\ldots$.)

3.2 Redundant Representations

The paper [22] is based on the Paul system, which is to the base 2, with digits in $\{-1,0,1\}$, and no two adjacent nonzero digits. This leads to unique representations,

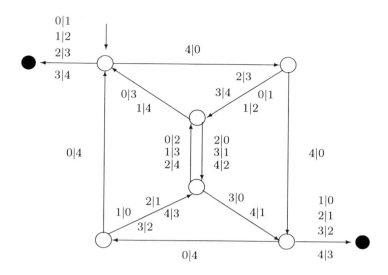

see [8]. However, it was later found out that this system was rediscovered many times. Probably the first appearance is in [25].[1]

Let us write $\bar{1} = -1$ for convenience. It was found out that the kth digit is given by

$$a_k(n) = 1 \iff \left\lfloor \frac{n}{2^{k+2}} + \frac{5}{6} \right\rfloor - \left\lfloor \frac{n}{2^{k+2}} + \frac{4}{6} \right\rfloor = 1$$

and

$$a_k(n) = \bar{1} \iff \left\lfloor \frac{n}{2^{k+2}} + \frac{2}{6} \right\rfloor - \left\lfloor \frac{n}{2^{k+2}} + \frac{1}{6} \right\rfloor = 1.$$

If the expressions on the right hand sides are not 1, they must be 0. Consequently we have the formula

$$n = \sum_{k \geq 0} \left(\left\lfloor \frac{n}{2^{k+2}} + \frac{5}{6} \right\rfloor - \left\lfloor \frac{n}{2^{k+2}} + \frac{4}{6} \right\rfloor - \left\lfloor \frac{n}{2^{k+2}} + \frac{2}{6} \right\rfloor + \left\lfloor \frac{n}{2^{k+2}} + \frac{1}{6} \right\rfloor \right) 2^k.$$

The key formula (see [13, 1.2.4, Ex. 38]) to manipulate such floor functions is

$$\lfloor x \rfloor + \lfloor x + \tfrac{1}{q} \rfloor + \cdots + \lfloor x + \tfrac{q-1}{q} \rfloor = \lfloor qx \rfloor, \tag{3}$$

where q is an integer ≥ 1 and x a real number. We can now consider the binary representation of λn, for a natural number λ: The kth digit is given by

---

[1] That is what I learnt from Donald Knuth who told me that Reitwiesner was earlier than Avizienis [1] (private communication). Christiane Frougny mentions even Cauchy (!), but without giving a reference (private communication).

$$a_k(\lambda n) = \sum_{i=0}^{2\lambda-1} \left\lfloor \frac{n}{2^{k+1}} + \frac{i}{2\lambda} \right\rfloor (-1)^{i-1};$$

this follows from

$$\lambda n = \sum_{k \geq 0} 2^k \left(\left\lfloor \frac{\lambda n}{2^k} \right\rfloor - 2 \left\lfloor \frac{\lambda n}{2^{k+1}} \right\rfloor \right)$$

and the formula (3). From this it follows that the representation investigated by Reitwiesner, Paul and others is simply obtained as follows: One computes the ordinary representations of $3n/2$ and of $n/2$ and takes the bitwise difference! For example, $25 = (11001)_2$, and $\frac{3}{2} \cdot 25 = 37.5 = (100101.1)_2$, $\frac{1}{2} \cdot 25 = 12.5 = (1100.1)_2$, and the bitwise difference is $10\bar{1}001$, which is the representation of 25. Clearly, the result is a representation of n in base 2 using digits $-1, 0, 1$. An obvious generalization is to consider $n = (\alpha+1)n - \alpha n$ and take the bitwise differences of the two representations. We analyzed only the case that α is either an integer or an integer divided by a power of 2 (a dyadic rational number). Here is one example:

$$a_k(4n) - a_k(3n) =$$
$$= \left\lfloor \frac{n}{2^{k+1}} + \frac{7}{8} \right\rfloor - \left\lfloor \frac{n}{2^{k+1}} + \frac{5}{6} \right\rfloor - \left\lfloor \frac{n}{2^{k+1}} + \frac{6}{8} \right\rfloor + \left\lfloor \frac{n}{2^{k+1}} + \frac{4}{6} \right\rfloor$$
$$+ \left\lfloor \frac{n}{2^{k+1}} + \frac{5}{8} \right\rfloor - \left\lfloor \frac{n}{2^{k+1}} + \frac{4}{8} \right\rfloor - \left\lfloor \frac{n}{2^{k+1}} + \frac{3}{6} \right\rfloor + \left\lfloor \frac{n}{2^{k+1}} + \frac{3}{8} \right\rfloor$$
$$+ \left\lfloor \frac{n}{2^{k+1}} + \frac{2}{6} \right\rfloor - \left\lfloor \frac{n}{2^{k+1}} + \frac{2}{8} \right\rfloor - \left\lfloor \frac{n}{2^{k+1}} + \frac{1}{6} \right\rfloor + \left\lfloor \frac{n}{2^{k+1}} + \frac{1}{8} \right\rfloor.$$

We discuss now a few syntactical properties. In [8] the rewriting rules

$$1\bar{1} \longrightarrow 01 \qquad \bar{1}1 \longrightarrow 0\bar{1} \qquad 011 \longrightarrow 10\bar{1} \qquad 0\bar{1}\bar{1} \longrightarrow \bar{1}01$$

are presented, that can be (repeatedly) applied in any order to transform the standard binary representation of $n \in \mathbb{N}$ into the "$3n/2 - n/2$" representation. We say "repeatedly" because of a problem with carries; the following example explains what happens:

$$01111 \longrightarrow 10\bar{1}11 \longrightarrow 100\bar{1}1 \longrightarrow 1000\bar{1}.$$

A more "algorithmic" version is by the following *transducer*. It reads the binary representation of n from right to left (leading zeros must be added if needed). Of course, the output is also produced from right to left. The basic idea is to transform 01^k into $10^{k-1}\bar{1}$ for $k \geq 2$. However, if the next group (to the left) starts immediately with 1, then we can't output the leading 1; it is then a carry, belonging to the next group. To make sure that the transduction ends in the starting state (as it should), we can add two leading zeros (see Figure 1).

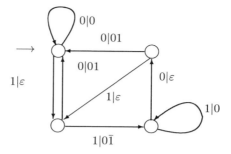

Fig. 1. The binary → "$3n/2 - n/2$" transformer

When dealing with number systems, it is usually very instructive to see how the number 1 can be added. This can normally be done by an automaton. In this paper we want to keep things simple; Christiane Frougny [5] has quite general results about addition performed by automata. The word is processed from right to left. If no continuation is defined, the rest of the word is left unchanged. Again, possibly leading zeros are needed to lead to one of the accepting states to the right. The output is also recorded from right to left, usually replacing only a suffix of the word (representation of a number), when reaching one of the two states to the right.

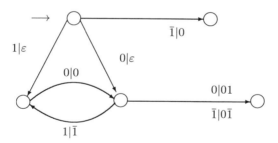

Fig. 2. Adding 1 in the "$3n/2 - n/2$" representation

Now let us consider a few properties of the system induced by writing $n = 5n/4 - n/4$: First, we describe the syntactically correct words. If $B = 1 + \bar{1} + 11(\bar{1}1)^*(\varepsilon + \bar{1}) + \bar{1}\bar{1}(1\bar{1})^*(\varepsilon + 1)$, then the possible representations are given by $\varepsilon + (B000^*)^*B0^*$. It is also possible to give a complete set of rewriting rules, but, again, an automaton (transducer) is more instructive: Input are natural numbers, written in ordinary binary notation. The computation ends in the starting state.

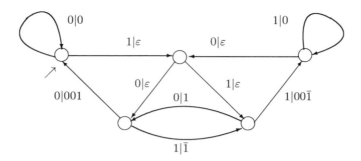

Fig. 3. The binary → "1 = 5/4 − 1/4" transformer

3.3 Minimal Representations

Motivated by algorithms on elliptic curves, in [9] representations of an integer n in base $q \geq 2$ with arbitrary digits are again studied; set

$$R_q(n) := \left\{ \varepsilon = (\varepsilon_0, \ldots, \varepsilon_l) \;\middle|\; l \in \mathbb{N},\; \varepsilon_i \in \mathbb{Z},\; n = \sum_{i=0}^{l} \varepsilon_i q^i \right\}.$$

We define the *cost of a representation* $\varepsilon \in R_q(n)$ as

$$c(\varepsilon) = c(\varepsilon_0, \ldots, \varepsilon_l) := l + 1 + \sum_{i=0}^{l} |\varepsilon_i|$$

and look for representations of n with minimum cost. The modified cost function, which we will call *relaxed costs*, namely

$$c'(\varepsilon) = c'(\varepsilon_0, \ldots, \varepsilon_l) := \sum_{i=0}^{l} |\varepsilon_i|$$

is more natural and easier to study.

We note that in general there is no unique minimal representation: Both $(-1, 2) \in R_3(5)$ and $(2, 1) \in R_3(5)$ are representations of 5 in base 3 with cost $c(-1, 2) = c(2, 1) = 5$. We will describe a special minimal representation ("reduced representation"), which is unique. In [9] there is a full set of conversion rules between the two minimal representations relative to the two types of costs. Although they have a language theoretic flavour, they cannot be included here.

In [9] we are able to derive an explicit formula for the digits of the reduced representation in the case of odd and even q, respectively. These formulæ are very useful for the computation of the frequencies of digits and the expected costs of the representations. Let $q \geq 3$ be odd and n be a positive integer. Let

$$\varepsilon_r := b_q^{(r)}(2n) - b_q^{(r)}(n), \qquad r \geq 0,$$

where $b_q^{(r)}(n)$ is the r-th digit of the "usual" q-adic expansion of n, i.e.,

$$b_q^{(r)}(n) = \left\lfloor \frac{n}{q^r} \right\rfloor - q \left\lfloor \frac{n}{q^{r+1}} \right\rfloor,$$

and $l := \max\{r \geq 0 \mid \varepsilon_r \neq 0\}$. Then $\varepsilon = (\varepsilon_0, \ldots, \varepsilon_l)$ is the relaxed reduced minimal representation. The formula for $q \geq 2$ even is more complicated: For $r \geq 0$ define

$$\varepsilon_r := \sum_{i=0}^{q/2-1} \left(\sum_{j=0}^{q/2-1} \left\lfloor \frac{n}{q^{r+2}} + \frac{1 + \frac{q}{2} + (q+1)(iq+j)}{q^2(q+1)} \right\rfloor \right.$$
$$- (q-1) \left\lfloor \frac{n}{q^{r+2}} + \frac{\frac{q}{2} + (q+1)(iq+q/2)}{q^2(q+1)} \right\rfloor$$
$$+ \sum_{j=q/2+1}^{q-1} \left\lfloor \frac{n}{q^{r+2}} + \frac{\frac{q}{2} + (q+1)(iq+j)}{q^2(q+1)} \right\rfloor \right)$$
$$+ \sum_{i=q/2}^{q-1} \left(\sum_{j=0}^{q/2-2} \left\lfloor \frac{n}{q^{r+2}} + \frac{1 + \frac{q}{2} + (q+1)(iq+j)}{q^2(q+1)} \right\rfloor \right.$$
$$- (q-1) \left\lfloor \frac{n}{q^{r+2}} + \frac{1 + \frac{q}{2} + (q+1)(iq+q/2-1)}{q^2(q+1)} \right\rfloor$$
$$+ \sum_{j=q/2}^{q-1} \left\lfloor \frac{n}{q^{r+2}} + \frac{\frac{q}{2} + (q+1)(iq+j)}{q^2(q+1)} \right\rfloor \right).$$

Let $l := \max\{r \geq 0 \mid \varepsilon_r \neq 0\}$. Then $\varepsilon = (\varepsilon_0, \ldots, \varepsilon_l) \in R_q(n)$ is the relaxed reduced minimal representation. We make that explicit for $q = 2$:

$$\varepsilon_r = \left\lfloor \frac{n}{2^{r+2}} + \frac{5}{6} \right\rfloor - \left\lfloor \frac{n}{2^{r+2}} + \frac{4}{6} \right\rfloor - \left\lfloor \frac{n}{2^{r+2}} + \frac{2}{6} \right\rfloor + \left\lfloor \frac{n}{2^{r+2}} + \frac{1}{6} \right\rfloor,$$

so we see that it is the representation studied by Reitwiesner and others.

It is also possible to formulate a simple algorithm to create the representation of an integer m: Consider $a \equiv m \bmod q$, and take a representative in the interval $-q/2 \leq a \leq q/2$. In case, a decision has to be made whether to take $a = -q/2$ or $a = q/2$. It is based on the fractional part of m/q^2: if $\{m/q^2\} < 1/2$ then take $a = q/2$. Continue with $m \leftarrow (m-a)/q$. This rule makes the system balanced.

We are also interested to count how often a given digit i occurs in the relaxed reduced expansion among the numbers $1, \ldots, n$; call that $\#_i(n)$. In the case q odd, the reduced representation is actually the (q, d) representation, with $d = -\frac{q-1}{2}$, and digits $d, d+1, \ldots, d+q-1$. (For example, $32\bar{1}01$ with base 4 represents $3 \cdot 4^4 + 2 \cdot 4^3 - 4^2 + 1 = 881$.) Digit counting in this representation is well known, see [10]. Every digit occurs with the same frequency $\frac{1}{q}$, and

$$\#_i(n) = \frac{1}{q} n \log_q n + \mathcal{O}(n).$$

Actually, more is known about the error term. It can be made fully explicit and is basically given by n times a periodic function in $\log_q n$. From that we get the average value of $c'(\varepsilon)$. It is

$$\left(\frac{q}{4} - \frac{1}{4q}\right) \log_q n + \mathcal{O}(1).$$

In the instance q even, things are a bit more complicated; digit 0 occurs with frequency $\frac{q+2}{q(q+1)}$, digits $\pm 1, \ldots, \pm \frac{q}{2} - 1$ with frequency $\frac{1}{q}$, digits $\pm \frac{q}{2}$ each with frequency $\frac{1}{2(q+1)}$. So we see that digit 0 appears a bit more often than the other digits, and the digits $\pm \frac{q}{2}$ (combined) appear a bit less often than the other digits. This leads to

$$\#_0(n) = \frac{q+2}{q(q+1)} n \log_q n + \mathcal{O}(n),$$

$$\#_i(n) = \frac{1}{q} n \log_q n + \mathcal{O}(n), \qquad \text{for } i = \pm 1, \ldots, \pm \frac{q}{2} - 1,$$

$$\#_i(n) = \frac{1}{2(q+1)} n \log_q n + \mathcal{O}(n), \qquad \text{for } i = \pm \frac{q}{2}.$$

Summing this, (multiplied by $|i|$), over all possible digits i, we get the average value of $\sum |\varepsilon_i|$:

$$\left(\frac{q}{4} - \frac{1}{2(q+1)}\right) \log_q n + \mathcal{O}(1).$$

3.4 Representations in the Sense of Morain and Olivos

The representation of n in base 2, with digits $-1, 0, 1$, where no two adjacent nonzero digits are adjacent, has the lowest number of nonzero digits (roughly speaking, the probability that a random digit is different from zero, is $\frac{1}{3}$). As we have seen, it can be obtained by repeated applications of replacement rules. Morain and Olivos in [16], see also [26], consider the simpler version, where all occurrences 1^a ($a \geq 2$, and the block is rendered by zeros) are simultaneously replaced by $10^{a-1}\bar{1}$. For example, 11011 is transformed into $10\bar{1}10\bar{1}$. This Morain/Olivos representation of natural numbers does not take care of carries (but they also consider them, in a second algorithm). However, it is also possible to give explicit formulæ for the digits:

$$a_0(n) = \left\lfloor \frac{n}{4} + \frac{3}{4} \right\rfloor - \left\lfloor \frac{n}{4} + \frac{2}{4} \right\rfloor - \left\lfloor \frac{n}{4} + \frac{1}{4} \right\rfloor + \left\lfloor \frac{n}{4} + \frac{0}{4} \right\rfloor,$$

$$a_1(n) = \left\lfloor \frac{n}{8} + \frac{6}{8} \right\rfloor - \left\lfloor \frac{n}{8} + \frac{5}{8} \right\rfloor - \left\lfloor \frac{n}{8} + \frac{2}{8} \right\rfloor + \left\lfloor \frac{n}{8} + \frac{1}{8} \right\rfloor,$$

and for $k \geq 2$

$$a_k(n) = \left\lfloor \frac{n}{2^{k+2}} + \frac{13}{16} \right\rfloor - \left\lfloor \frac{n}{2^{k+2}} + \frac{10}{16} \right\rfloor$$
$$+ \left\lfloor \frac{n}{2^{k+2}} + \frac{5}{16} \right\rfloor - 2\left\lfloor \frac{n}{2^{k+2}} + \frac{4}{16} \right\rfloor + \left\lfloor \frac{n}{2^{k+2}} + \frac{2}{16} \right\rfloor.$$

A rigorous proof of that is unpleasant but doable as in [9].

From this it follows that the number of nonzero digits $\sigma(n)$ in the Morain/Olivos representation is given by

$$\sigma(n) = \left\lfloor \frac{n}{4} + \frac{3}{4} \right\rfloor - \left\lfloor \frac{n}{4} + \frac{2}{4} \right\rfloor + \left\lfloor \frac{n}{4} + \frac{1}{4} \right\rfloor - \left\lfloor \frac{n}{4} + \frac{0}{4} \right\rfloor$$
$$+ \left\lfloor \frac{n}{8} + \frac{6}{8} \right\rfloor - \left\lfloor \frac{n}{8} + \frac{5}{8} \right\rfloor + \left\lfloor \frac{n}{8} + \frac{2}{8} \right\rfloor - \left\lfloor \frac{n}{8} + \frac{1}{8} \right\rfloor$$
$$+ \sum_{k\geq 2} \left\{ \left\lfloor \frac{n}{2^{k+2}} + \frac{13}{16} \right\rfloor - \left\lfloor \frac{n}{2^{k+2}} + \frac{10}{16} \right\rfloor + \left\lfloor \frac{n}{2^{k+2}} + \frac{5}{16} \right\rfloor - \left\lfloor \frac{n}{2^{k+2}} + \frac{2}{16} \right\rfloor \right\},$$

and therefore

$$\sum_{n<N} \sigma(n) = \left\lfloor \frac{N}{2} \right\rfloor + \left\lfloor \frac{N+1}{4} \right\rfloor$$
$$+ \sum_{n<N} \sum_{k=2}^{l+2} \left\{ \left\lfloor \frac{n}{2^{k+2}} + \frac{13}{16} \right\rfloor - \left\lfloor \frac{n}{2^{k+2}} + \frac{10}{16} \right\rfloor + \left\lfloor \frac{n}{2^{k+2}} + \frac{5}{16} \right\rfloor - \left\lfloor \frac{n}{2^{k+2}} + \frac{2}{16} \right\rfloor \right\},$$

with $l = \log_2 N$. Now we use Delange's approach [4] to develop that further (this leads to a more elementary computation than Thuswaldner's in [26]). Define

$$g(x) = \int_0^x \left\{ \left\lfloor t + \frac{13}{16} \right\rfloor - \left\lfloor t + \frac{10}{16} \right\rfloor + \left\lfloor t + \frac{5}{16} \right\rfloor - \left\lfloor t + \frac{2}{16} \right\rfloor - \frac{3}{8} \right\} dt,$$

then

$$\sum_{n<N} \sigma(n) = \frac{3N}{4} + \mathcal{O}(1) + \frac{3}{8}(\lfloor l \rfloor + 1) + \sum_{k=2}^{\lfloor l \rfloor + 2} 2^{k+2} g\left(\frac{N}{2^{k+2}} \right)$$
$$= \frac{3N}{4} + \mathcal{O}(1) + \frac{3}{8} \log_2 N + \frac{3}{8}(1 - \{\log_2 N\})$$
$$+ N 2^{2-\{\log_2 N\}} \sum_{r\geq 0} 2^{-r} g(2^{\{\log_2 N\}-2+r}).$$

Defining

$$H(x) = \frac{3}{4} + \frac{3}{8}(1 - \{x\}) + 2^{2-\{x\}} \sum_{r\geq 0} 2^{-r} g(2^{\{x\}-2+r}),$$

which is periodic with period 1, we find that the average of $\sigma(n)$ is given by

$$\frac{1}{N} \sum_{n<N} \sigma(n) = \frac{3}{8} \log_2 N + H(\log_2 N) + \mathcal{O}\left(\frac{1}{N}\right).$$

One can compute the Fourier coefficients of $H(x)$, which we omit because of lack of space. The formula should be compared with the $\sim \frac{1}{3} \log_2 N + F(\log_2 N)$ formula in the Reitwiesner instance. So the more sophisticated algorithm brings the constant down from $\frac{3}{8}$ to $\frac{1}{3}$.

References

1. A. Avizienis. Signed–digit number representations for fast parallel arithmetic. *IRE Trans.*, EC-10: 389–400, 1961.
2. L. Carlitz. Permutations with prescribed pattern. *Math. Nachr.*, 58:31–53, 1973.
3. L. Carlitz. A combinatorial property of q–Eulerian numbers. *American Mathematical Monthly*, 82:51–54, 1975.
4. H. Delange. Sur la fonction sommatoire de la fonction somme des chiffres. *Enseignement Mathématique*, 21:31–47, 1975.
5. C. Frougny. On–line finite automata for addition in some numeration systems. *Theor. Inform. Appl.*, 33:79–101, 1999.
6. P. Grabner, P. Kirschenhofer, and H. Prodinger. The sum–of–digits function for complex bases. *J. London Math. Soc.*, 57:20–40, 1998.
7. R. L. Graham, D. E. Knuth, and O. Patashnik. *Concrete Mathematics (Second Edition)*. Addison Wesley, 1994.
8. U. Güntzer and M. Paul. Jump interpolation search trees and symmetric binary numbers. *Information Processing Letters*, 26:193–204, 1987/88.
9. C. Heuberger and H. Prodinger. On minimal expansions in redundant number systems: Algorithms and quantitative analysis. *Computing*, 66: 377–393, 2001.
10. P. Kirschenhofer and H. Prodinger. Subblock occurrences in positional number systems and Gray code representation. *Journal of Information and Optimization Sciences*, 5:29–42, 1984.
11. P. Kirschenhofer and H. Prodinger. The path length of random skip lists. *Acta Informatica*, 31:775–792, 1994.
12. A. Knopfmacher and H. Prodinger. Combinatorics of geometrically distributed random variables: Value and position of the rth left-to-right maximum. *Discrete Mathematics*, 226:255–267, 2001.
13. D. E. Knuth. *The Art of Computer Programming*, volume 1: Fundamental Algorithms. Addison-Wesley, 1968. Third edition, 1997.
14. D. E. Knuth. *The Art of Computer Programming*, volume 3: Sorting and Searching. Addison-Wesley, 1973. Second edition, 1998.
15. G. Louchard and H. Prodinger. Ascending runs of sequences of geometrically distributed random variables: a probabilistic analysis. *submitted*, 2001.
16. F. Morain and J. Olivos. Speeding up the computations on an elliptic curve using addition-subtraction chains. *RAIRO Inform. Théor. Appl.*, 24(6):531–543, 1990.
17. H. Prodinger. Combinatorial problems of geometrically distributed random variables and applications in computer science. In V. Strehl and R. König, editors, *Publications de l'IRMA (Straßbourg) (=Séminaire Lotharingien de Combinatoire)*, volume 30, pages 87–95, 1993.
18. H. Prodinger. Combinatorics of geometrically distributed random variables: Left-to-right maxima. *Discrete Mathematics*, 153:253–270, 1996.

19. H. Prodinger. Combinatorics of geometrically distributed random variables: Inversions and a parameter of Knuth. *Annals of Combinatorics*, 5: 241–250, 2001.
20. H. Prodinger. Combinatorics of geometrically distributed random variables: Lengths of ascending runs. *LATIN2000, Lecture Notes in Computer Science 1776*, pages 473–482, 2000.
21. H. Prodinger. Combinatorics of geometrically distributed random variables: New q–tangent and q–secant numbers. *International Journal of Mathematical Sciences*, 24:825–838, 2000.
22. H. Prodinger. On binary representations of integers with digits $-1, 0, 1$. *Integers*, pages A8, 14 pp. (electronic), 2000.
23. H. Prodinger. Combinatorics of geometrically distributed random variables: Value and position of large left-to-right maxima, *Discrete Mathematics, to appear*, 2002.
24. H. Prodinger and T. A. Tsifhumulo. On q–Olivier functions, *submitted*, 2001.
25. G. Reitwiesner. Binary arithmetic. *Vol. 1 of Advances in Computers, Academic Press*, pages 231–308, 1960.
26. J. Thuswaldner. Summatory functions of digital sums occurring in Cryptography. *Periodica Math. Hungarica*, 38:111–130, 1999.
27. H. Wilf. On the outstanding elements of permutations, 1995, http://www.cis.upenn.edu/~wilf.

Proof Complexity of Pigeonhole Principles

Alexander A. Razborov*

Steklov Mathematical Institute, Moscow, Russia
Institute for Advanced Study, Princeton, USA

Abstract. The pigeonhole principle asserts that there is no injective mapping from m pigeons to n holes as long as $m > n$. It is amazingly simple, expresses one of the most basic primitives in mathematics and Theoretical Computer Science (counting) and, for these reasons, is probably the most extensively studied combinatorial principle. In this survey we try to summarize what is known about its proof complexity, and what we would still like to prove. We also mention some applications of the pigeonhole principle to the study of efficient provability of major open problems in computational complexity, as well as some of its generalizations in the form of general matching principles.

1 Introduction

Propositional proof complexity is an area of study that has seen a rapid development over a couple of last decades. It plays as important a role in the theory of feasible proofs as the role played by the complexity of Boolean circuits in the theory of efficient computations. Propositional proof complexity is in a sense complementary to the (non-uniform) computational complexity; moreover, there exist extremely rich and productive relations between the two areas. Besides the theory of (first-order) feasible proofs and computational complexity, propositional proof complexity is tightly connected in many ways with other areas like automated theorem proving and cryptography, and these connections are numerous both at the level of motivations and when it comes to proof techniques.

In my talk at the conference I gave a general overview of some important concepts, ideas and proof techniques behind this fascinating theory. A substantial portion of that material was already covered in various surveys and monographs (see e.g. [1,2,3,4,5]). For one particular subject from my lecture, however, the situation is very different. I am talking about the research on the proof complexity of specific tautologies that express various forms of the so-called *pigeonhole principle*. This principle (asserting that there is no injective mapping from m pigeons to n holes whenever $m > n$) is probably the most extensively studied combinatorial principle in proof complexity. It is amazingly simple and at the same time captures one of the most basic primitives in mathematics and Theoretical Computer Science (counting). It might be for these reasons that the pigeonhole principle somehow manages to find itself in the center of events, and many other important principles studied in the proof complexity are related to it in one or another way.

---

* Supported by The von Neumann Fund

Surprisingly, the proof complexity of the pigeonhole principle essentially depends on the number of pigeons m (as the function of the number of holes n) and on subtle details of its representation as a propositional tautology. This leads to a rich structural picture, and results making the skeleton of this picture are amongst the most beautiful and important in the whole theory. Moreover, for a long time they have been determining its methods, machinery and ideology.

In the last couple of years several more important touches have been added to this picture, and these results do not seem to have been surveyed in the literature yet. For this reason, this written contribution is entirely devoted to the pigeonhole principle. It is organized as follows. In Section 2 we present the proof systems appearing in our survey. Our pace in this section will be slow, as it is primarily designed for the beginners; more experienced readers may take notice of the notation introduced there and skip all the rest. Section 3 contains formal definitions of the basic pigeonhole principle and its various modifications. The next section 4 is central: we give a survey of known lower and upper bounds on the proof complexity of the pigeonhole principle. With the help of some of these bounds, we show in Section 5 that such proof systems as Resolution and Polynomial Calculus do not possess efficient proofs of circuit lower bounds. In Section 6 we survey a few known results about the complexity of more general matching principles. Finally, we conclude with several open problems in Section 7.

2 Propositional Proof Systems

Let $x_1, \ldots, x_n, \ldots$ be propositional variables, and $\mathcal{C}$ be a certain class of propositional formulas in these variables. Denote by $\mathrm{TAUT}_\mathcal{C}$ the set of all tautologies in the class $\mathcal{C}$.

Informally speaking, a propositional proof system is any complete and sound calculus for generating members of $\mathrm{TAUT}_\mathcal{C}$. Given such a calculus, we can associate with it the "theorem-extracting" function P that for every binary string w encoding a legitimate proof in this calculus extracts the theorem $P(w)$ this proof proves.[1] Note that $\mathrm{im}(P) = \mathrm{TAUT}_\mathcal{C}$ (this is tantamount to saying that our calculus is complete and sound). Also, it is conceivable that for any reasonable calculus the function P will be polynomial time computable.

Cook and Reckhow [6] proposed to take these properties of the theorem-extracting function as a general *axiomatic definition* of a propositional proof system.

Definition 1 ([6]). *A* **propositional proof system** *(often abbreviated as p.p.s.) for a class $\mathcal{C}$ of propositional formulas is any polynomial time computable function*

$$P : \{0,1\}^* \xrightarrow{\text{onto}} \mathrm{TAUT}_\mathcal{C}.$$

*For a tautology $\phi \in \mathrm{TAUT}_\mathcal{C}$, any string w such that $P(w) = \phi$ is called a P-***proof of*** ϕ.

---

[1] If w is a string that does not make sense, we let $P(w)$ be any fixed tautology from $\mathrm{TAUT}_\mathcal{C}$.

Denote by $S_P(\phi)$ the minimal possible bit size $|w|$ of a P-proof w of ϕ.[2] Then the basic question of proof complexity can be formulated as simply as that: what can we say about $S_P(\phi)$ for "interesting" proof systems P and "interesting" tautologies ϕ?

Definition 1 paves way for relating propositional proof complexity to the rich structure developed in *computational* complexity. For example, let us call a p.p.s. P *p-bounded* if $S_P(\phi)$ is bounded from above by a polynomial in $|\phi|$. Then we have

Theorem 1 ([6]). $\mathbf{NP} = co - \mathbf{NP}$ *if and only if there exists a p-bounded propositional proof system.*

This easy result actually indicates that in its full generality Definition 1 is just a reformulation of the standard characterization of $co - \mathbf{NP}$ and has only little to do with the "real" proof theory. Propositional proof complexity really begins as a separate subject only when we are interested in the performance of concrete p.p.s. that are natural and that are of independent interest. Still, even in that case some standard structural concepts from computational complexity are extremely useful. As an example, let us introduce the following notion.

Definition 2. *A p.p.s. P is **p-simulated** by another p.p.s. Q for the same class C if there exists a polynomial time algorithm A that transforms every P-proof into a Q-proof of the same tautology. That is, we demand that $\forall w \in \{0,1\}^* (Q(A(w)) = P(w))$. This relation is reflexive and transitive. Hence we say that two p.p.s are **p-equivalent** if they p-simulate each other.*

The notion of p-simulation is used for comparing strength of different proof systems (clearly, if P is p-simulated by Q then $S_Q(\phi) \leq S_P(\phi)^{O(1)}$) and arranging them into a hierarchy.

After this necessary digression into general structural issues, let us take a closer look at the specific proof systems used in this survey. The most basic example of a p.p.s. is the *Frege proof system* which is essentially the ordinary Hilbert-style propositional calculus from your favourite textbook in mathematical logic. Fix any complete language of propositional connectives (say, $L_0 \stackrel{\text{def}}{=} \{\neg, \wedge, \vee\}$), and let C consist of all propositional formulas in this language. A Frege proof system is specified by finitely many *Frege rules* of the form

$$\frac{A_1, \ldots, A_k}{B} \quad (A_1, \ldots, A_k, B \in C),$$

called *axiom schemes* if $k = 0$. All these rules are required to be (implicationally) sound. A *Frege proof* consists of a sequence of applications of *substitutional instances* of Frege rules, defined as the result of substituting arbitrary formulas for variables. Finally, we demand that there are sufficiently many Frege rules meaning that the resulting Frege system must be (implicationally) complete.

At the first glance, this definition is rather arbitrary since we have a considerable freedom in choosing the set of Frege rules. It turns out, however, that the resulting Frege

---

[2] It should be noted that although this complexity measure is by far more important, sometimes researchers are interested in more sophisticated ones. We will see below one example, degree of algebraic proofs.

systems are all p-equivalent. More generally, Reckhow [7] proved the following (highly non-trivial) result.

Theorem 2 ([7]). *Let P, Q be two Frege systems considered as proof systems for tautologies in the language that consists of their common connectives. Then P and Q are p-equivalent.*

Thus, the Frege proof system is uniquely defined up to p-equivalence, and we jubilantly denote it by F. In fact, its definition is even more robust. Instead of a Hilbert-style calculus we may consider a Gentzen-style sequent calculus, and we will still get a proof system p-equivalent to F.

One more important remark to be made along these lines (robustness) is this. It is of absolutely no importance in the classical proof theory whether we represent proofs in a *tree-like* form or in the *sequential* (sometimes also called *dag-like*) form, when a proof is a sequence of formulas (sometimes called *lines*) in which every line is obtained from preceding lines via an inference rule. The situation is potentially very different in proof complexity: when we expand a sequential proof into a tree, the bit size may in general grow exponentially as every formula in the proof will repeat itself many times. It turns out, however, that this distinction does not matter for the Frege proof system, and its tree-like and sequential versions are p-equivalent.

All other propositional proof systems considered in this survey will be p-simulated by F (in fact, they will be even defined as Frege proofs of a special form). They will be represented in the form of a Gentzen-style sequent calculus, and they will be given in the sequential (as opposed to tree-like) form. Also, the propositional language will be from now on fixed to $L_0 \stackrel{\text{def}}{=} \{\neg, \wedge, \vee\}$. The connectives $\wedge$ and $\vee$ will be allowed to have an arbitrary number of arguments.[3] The (logical) *depth* of a propositional formula ϕ and the hierarchy Σ_d, Π_d of propositional formulas are defined in the standard way.

Fix any constant $d \geq 1$, and let $\mathcal{C} \stackrel{\text{def}}{=} \Sigma_d$. Every formula in $\mathcal{C}$ can be re-written as a sequent $A_1, \ldots, A_k \longrightarrow B_1, \ldots, B_\ell$ in which all formulas $A_1, \ldots, A_k, B_1, \ldots, B_\ell$ are of depth at most $d - 1$. The system F_d, by definition, is the fragment of F that operates with sequents of this form (that is, in which every formula is in $\Sigma_{d-1} \cup \Pi_{d-1}$!) and has all ordinary rules of a Gentzen-style calculus (structural rules, logical rules and the cut rule). With a slight abuse of notation, when one is not interested much in the exact value of the depth d, the term *bounded-depth Frege proofs* is used.

Remark 1. One of the main ingredients in the proof of Theorem 2 is to show that every Frege proof of size S can be transformed into another Frege proof of size $S^{O(1)}$ and logical depth $O(\log S)$. In particular (with the same abuse of notation), the systems F and $F_{O(\log n)}$ have the same polynomial size proofs.

Remark 2. The p-simulation of sequential Frege proofs by tree-like Frege to a certain extent can be carried over to the bounded-depth case. Namely, F_d is p-simulated by

---
[3] The attentive reader might have observed at this point that we did not specify this parameter while discussing general Frege systems above. The reason should be already clear by now: the two versions are p-equivalent.

tree-like F_{d+2} and, in fact, even by the tree-like version of some "symmetrized" variant of F_{d+1} in which we also allow Π_{d+1}-formulas (we did not make a special provision for Π_d-formulas in our definition of the *sequential* version of F_d as in that case they can be always w.l.o.g. broken up into a sequence of Σ_{d-1}-formulas).

Remark 3. Given the convenience of this representation in the Gentzen form, some authors even define the depth of a proof directly in terms of the sequent calculus (see e.g. [8]). It is important to remember that our notation (which reflects the semantical meaning of lines in a proof rather than the way they are syntactically represented) is off by one, and our F_d corresponds to depth $(d-1)$ sequent calculus proofs in their notation.

The system F_1 is called *Resolution* and denoted by R; this is one of the most important and frequently studied propositional proof systems. At the first glance there is something wrong about F_1 since it operates with Σ_1-formulas, and this class does not contain any non-trivial tautologies at all. This is easily circumvented by the following trick. We in fact prove tautologies in Σ_2 (i.e., in the *DNF form*), but instead of directly *proving* a tautology ϕ, we are *refuting* its negation $\bar{\phi} \in \Pi_2$. $\bar{\phi}$ is of the form $C_1 \wedge C_2 \wedge \ldots \wedge C_s$, where C_i are Σ_1 formulas (often called *clauses*), and the resolution proof system is trying to infer a contradiction (= the empty clause) from the set of axioms $\{C_1, C_2, \ldots, C_s\}$ operating with clauses only. By inspecting the rules of the Gentzen calculus, we see that the only non-trivial rule left in this situation is the atomic cut rule

$$\frac{C \vee x \qquad D \vee \bar{x}}{C \vee D}$$

that receives the special name *resolution rule*.

Let Σ_d^t, Π_d^t consist of those formulas in Σ_{d+1}, Π_{d+1} respectively for which the number of inputs of any connective at the bottom (closest to the variables) level is at most t. Clearly, $\Sigma_d \subseteq \Sigma_d^t \subseteq \Sigma_{d+1}$, and this intermediate class turned out to be very convenient in circuit complexity e.g. in the proof of *Håstad Switching Lemma* [9]. No wonder that it very naturally appears in the proof complexity, too. With a slight abuse of notation, we define $F_{d+0.5}$ as the fragment of F_{d+1} in which all occurring formulas are in fact in $\Sigma_d^{polylog(n)}$ (that is, in the corresponding sequents all formulas must be in $\Sigma_{d-1}^{polylog(n)} \cup \Pi_{d-1}^{polylog(n)}$), n the number of variables.

$F_{1.5}$ is in particular very close to resolution: it operates with sequents in which every formula is either a clause of polylogarithmic *width* (defined as the number of literals in the clause) or its negation (*elementary conjunction*). More generally, let $R(t)$ be the similarly defined extension of Resolution in which all formulas are in Σ_1^t (thus, $F_{1.5} = R(polylog)$). In this survey we will be particularly interested in $R(2)$ and $R(O(1))$.

Our last system is of completely different nature (at least, at the first glance). Fix any field $\mathbb{F}$, and let us interpret the logical constants TRUE and FALSE as the elements 0 and 1 in this field, respectively. Then the clause[4] $x_{i_1}^{\epsilon_1} \vee \ldots \vee x_{i_w}^{\epsilon_w}$ is satisfied by a

---
[4] we use here the convenient notation $x^1 \stackrel{\text{def}}{=} x$ and $x^0 \stackrel{\text{def}}{=} \bar{x}$

truth assignment if and only if the corresponding 0-1 vector satisfies the polynomial equation $(x_{i_1} - \epsilon_1) \cdot \ldots \cdot (x_{i_w} - \epsilon_w) = 0$. We already saw before that proving Σ_2-formulas (DNFs) is equivalent to showing that a set of clauses is unsatisfiable, and now we take one step further and replace this task by the task of proving that the system of polynomial equations constructed from this set of clauses as described above does not have 0-1 solutions.

Polynomial Calculus (introduced in [10] and sometimes abbreviated as PC) is specifically designed for this latter task; it operates with polynomial equations over the field $\mathbb{F}$ and has *default axioms* $x_1^2 - x_1 = \cdots = x_n^2 - x_n = 0$ (ensuring that x_is take on 0-1 values) and two inference rules

$$\frac{f=0 \qquad g=0}{\alpha f + \beta g = 0} \, (\alpha, \beta \in \mathbb{F}) \qquad \frac{f=0}{f \cdot g = 0}.$$

The purpose is to infer the contradiction $1 = 0$ from a given system of polynomial equations. The complexity of a polynomial calculus proof is traditionally measured by its *degree* (defined as the maximal degree of all polynomials occurring in it) rather than by the bit size. Note that the default axioms allow us to assume w.l.o.g. that all polynomials in the proof are multi-linear; in particular, there always exists a polynomial calculus proof of degree $\leq n$.

3 Pigeonhole Principle(s)

Denote $\{1, \ldots, t\}$ by $[t]$. Let $m > n$ and $\{x_{ij} \mid i \in [m], j \in [n]\}$ be propositional variables that will be called *pigeonhole variables*. The *basic pigeonhole principle* PHP_n^m is the following DNF:

$$PHP_n^m \equiv \bigvee_{i \in [m]} \bigwedge_{j \in [n]} \bar{x}_{ij} \vee \bigvee_{j \in [n]} \bigvee_{\substack{i_1, i_2 \in [m] \\ i_1 \neq i_2}} (x_{i_1 j} \wedge x_{i_2 j}).$$

It will be more convenient (and, as we remarked in the previous section, simply necessary when working with p.p.s. below F_2) to work with its negation that consists of the following groups of clauses ("axioms"):

$$Q_i \stackrel{\text{def}}{=} \bigvee_{j=1}^{n} x_{ij} \; (i \in [m]); \tag{1}$$

$$Q_{i_1, i_2; j} \stackrel{\text{def}}{=} (\bar{x}_{i_1 j} \vee \bar{x}_{i_2 j}) \, (i_1 \neq i_2 \in [m], j \in [n]). \tag{2}$$

These clauses express that a multi-valued mapping from $[m]$ ("pigeons") to $[n]$ ("holes") is both total (everywhere defined) and injective. Since no such mapping exists whenever $m > n$, PHP_n^m is indeed a tautology.

For the purpose of orientation let us see what will happen when the number of pigeons m increases. In that case the task of the propositional proof system becomes easier (unnecessary pigeons i can be just ignored), and the complexity in general may decrease. In other words, the principle becomes *weaker* (= easier to prove). If we are proving upper bounds on its complexity (work along with the proof system), our task is also easier, and lower bounds, on the contrary, are harder since we are fighting on the other side.

Many (if not all) results surveyed in the next section can be presented, if desired, as a smooth function in the two parameters m and n. It is more instructive, however, to let n tend to infinity and view m as a certain function in n. Then it turns out that there are several "critical points" at which most of the *qualitative* changes in the behaviour of the pigeonhole principle occur, and these are

$$m = n+1, 2n, n^2, \infty.$$

For this reason, our tour in the next section will make a stop only at these four distinguished points, and we will see how drastically will the landscape be changing while we are driving from one to another.

The term "weak pigeonhole principle" traditionally refers to the case $m \geq 2n$. In order to distinguish between the three degrees of weakness, we use terms *moderately weak* ($m = 2n$), *weak* ($m = n^2$) and *very weak* ($m = \infty$).

As noted above, the "basic" pigeonhole principle expresses the fact that there is no multi-valued total injective mapping from $[m]$ to $[n]$. Besides increasing the number of pigeons, another way of weakening this principle consists in adding optional axioms dual to either (1) or to (2) or both. Namely, let

$$Q_{i;j_1,j_2} \stackrel{\text{def}}{=} (\bar{x}_{ij_1} \vee \bar{x}_{ij_2}) \ (i \in [m],\ j_1 \neq j_2 \in [n]) \tag{3}$$

(this group of axioms additionally requires that the assumed mapping is actually an ordinary single-valued function), and

$$Q_j \stackrel{\text{def}}{=} \bigvee_{i=1}^{m} x_{ij} \ (j \in [n]) \tag{4}$$

(which requires that the mapping is onto). The pigeonhole principle with the axioms (3) present is called *functional*, and the axioms (4) supply the prefix *"onto"* to the name. Altogether, we have four possible versions: the "basic" version PHP_n^m, the *functional* version $FPHP_n^m$, the *onto version onto $- PHP_n^m$* and the *functional onto version onto $- FPHP_n^m$*. All these versions have been frequently studied in the literature, quite often interchangeably and sometimes confusingly, so the word "basic" should not be understood as an attempt to distinguish this particular version out in its family as the most important. Anyway, the thumb rule is the same as with the dependence of m on n: the longer the name, the weaker is the principle, the easier are upper bounds, and the harder are lower bounds.

In many cases the fact that the axioms (1) have large width is rather annoying (see e.g. Footnote 5 below). One more version of the pigeonhole principle called *extended pigeonhole principle* $EPHP_n^m$ was introduced in [11] as a convenient way of circumventing this. In order to avoid ambiguity in the original definition of $EPHP_n^m$, we present it here in more invariant framework of [12].

Namely, for every Boolean function $f(x_1, \ldots, x_n)$ in n variables and every pigeon $i \in [m]$ introduce a new *extension variable* y_{if}, and identify the pigeonhole variable

x_{ij} with y_{i,x_j}. Then $EPHP_n^m$ is obtained by replacing the axioms (1) with new *local extension axioms*

$$y_{if_1}^{\epsilon_1} \vee y_{if_2}^{\epsilon_2} \vee y_{if_3}^{\epsilon_3} \left(f_1^{\epsilon_1} \vee f_2^{\epsilon_2} \vee f_3^{\epsilon_3} \geq \bigvee_{j=1}^{n} x_j \right). \tag{5}$$

It is easy to see that (1) can be easily inferred from (5) (more generally, local extension axioms of any width can be easily inferred from the local extension axioms (5) of width 3). Thus, $EPHP_n^m$ is weaker than PHP_n^m. [13, Section 5] observed that $FPHP_n^m$ is *weaker* than $EPHP_n^m$: namely, the substitution

$$y_{if}^{\epsilon} \mapsto \bigvee \{ x_{ij} \mid f(\chi_j) = \epsilon \}$$

(where χ_j is the n-bit Boolean input with the only one in position j) transforms proofs of $EPHP_n^m$ into proofs of $FPHP_n^m$. The same argument shows that the naturally defined version $EFPHP_n^m$ (obtained by relaxing the condition in (5) to $f_1^{\epsilon_1} \vee f_2^{\epsilon_2} \vee f_3^{\epsilon_3} \geq \bigvee_{j=1}^{n} x_j \wedge \bigwedge_{j_1 \neq j_2} (\bar{x}_{j_1} \vee \bar{x}_{j_2}))$ is in fact equivalent to $FPHP_n^m$.

The last subtle point is that our previous observation "m is larger $\implies$ principle is weaker" is *no longer a priori valid* in the presence of the onto axioms (4). The reason is very simple: when we restrict our mapping to fewer pigeons, as one of the by-side results of this restriction, the axioms (4) can get falsified.

4 Survey of Results

In this section we give a survey of at least the most important results about the complexity of the pigeonhole principle. As we promised in the previous section, our tour will make four stops, $m = n+1, 2n, n^2, \infty$. When presenting lower bounds, we always try our best to identify the weakest version (the strongest for upper bounds) to which the result is applicable, even if the original paper did not elaborate on the issue.

4.1 Classical Case: $m = n + 1$ (Hard but Expected)

It was the pigeonhole principle for which the first resolution lower bounds were proven in [14]. Haken's result is viewed by many as *the* result that really started the area of propositional proof complexity off.

Theorem 3 ([14]). $S_R(onto - FPHP_n^{n+1}) \geq \exp(\Omega(n))$.

Buss [15] further confirmed that Frege is a very powerful proof system by showing that the pigeonhole principle (any version) is easy for it.

Theorem 4 ([15]). $S_F(PHP_n^m) \leq n^{O(1)}$.

One natural proper subsystem of Frege in which PHP_n^m still has polynomial size proofs was identified in [16]; it is called the *cutting planes* proof system.

The next major step was to analyze the complexity of the pigeonhole principle with respect to bounded-depth Frege proof systems. Ajtai [17] proved a superpolynomial lower bound using non-standard models of arithmetic, and Bellantoni, Pitassi and Urquhart [18] presented a combinatorial version of his argument, at the same time extracting from it the following explicit bound:

Theorem 5 ([17,18]). *For every fixed* $d > 0$, $S_{F_d}(onto - FPHP_n^{n+1}) \geq n^{\epsilon_d \log n}$, *where ϵ_d is a constant depending only on d.*

Finally, Pitassi, Beame, Impagliazzio [19] and, independently, Krajíček, Pudlak and Woods [20] improved this lower bound to truly exponential.

Theorem 6 ([19,20]). *For every fixed* $d > 0$, $S_{F_d}(onto - FPHP_n^{n+1}) \geq \exp(n^{\epsilon_d})$, *where ϵ_d is a constant depending only on d.*

Another proof system for which PHP was the first tautology to be shown to be hard is Polynomial Calculus.[5] The following lower bound proved by Razborov [21] is applicable to an arbitrarily weak pigeonhole principle.

Theorem 7 ([21]). *Every polynomial calculus proof of $FPHP_n^\infty$ (over an arbitrary field $\mathbb{F}$) must have degree $\Omega(n)$.*

The proof of Theorem 7 uses a rather specific combinatorial argument called *pigeon dance*. The original proof was somewhat simplified in [22], although even that simpler form essentially depended on the pigeon dance. [23] gave another proof of the same $\Omega(n)$ lower bound which almost immediately follows from some general theory, but it can be applied only to the stronger principle $EPHP_n^m$ and only when $m = O(n)$ (cf. Theorem 29 below).

Looking at the statement of Theorem 7 more closely, we see that unlike all other negative results in this section, the prefix "onto" is missing there. The following observation made by Riis [24] shows that there is a very good reason for this omission.

Theorem 8 ([24]). *Assume that the ground field $\mathbb{F}$ has characteristic p and that $\binom{m}{d} \not\equiv \binom{n}{d}$ (mod p) for some $d \geq 1$. Then $onto - PHP_n^m$ has degree d polynomial calculus proof over $\mathbb{F}$.*

If $m = n + p^\ell$ then $\binom{m}{d} \equiv \binom{n}{d}$ (mod p) for all $d < \min\{n+1, p^\ell\}$ and Theorem 8 is no longer applicable, say, when n and p^ℓ are of the same order. Moreover, there are good reasons to believe (see Question 7 in Section 7) that $onto - PHP_n^{n+p^\ell}$ is hard for the polynomial calculus in characteristic p. Therefore, Theorem 8 perfectly illustrates the point made in Section 3: for the onto version, we no longer can assume that the complexity will be anti-monotone in m; instead, it may oscillate, at least for algebraic proof systems.

---

[5] One has to be a little bit creative when translating the axioms (1) to the algebraic language since the straightforward translation from Section 2 produces polynomials of intolerably high degree n. Instead, we transform them into degree 1 equations $\sum_{j \in [n]} x_{ij} - 1 = 0$. An alternative (and equivalent) way is to consider at once the extended version $EFPHP_n^m$ mentioned in Section 3.

4.2 Moderately Weak PHP_n^m: $m = 2n$ (Mystery Begins)

Many results in Section 4.1 are hard and deep but all of them were a sort of expected. It will be just the opposite in this section (I mean of course only the last part, their hardness and depth are quite competitive).

There is no a priori reason to believe that Theorem 6 can not be generalized to larger values of m. Nonetheless, it is indeed the case, and in fact this had been first shown even *prior* to that theorem. Namely, Paris, Wilkie and Woods proved in [25] that $FPHP_n^{2n}$ *does* possess short bounded-depth proofs, and Krajíček [26] calculated the exact value of depth resulting from their proof.

Theorem 9 ([25,26]).

a) $S_{F_{2.5}}(FPHP_n^{2n}) \leq n^{O(\log n)}$;
b) $S_{F_{1.5}}(onto - FPHP_n^{2n}) \leq n^{O(\log n)}$.

Buss and Turán [27] showed that the situation with Theorem 3 is exactly the opposite (at this stop!), and it readily generalizes to the moderately weak principle.

Theorem 10 ([27]). $S_R(onto - FPHP_n^{2n}) \geq \exp(\Omega(n))$.

In another important development, Maciel, Pitassi and Woods [8] were able to generalize Theorem 9 b) (and improve Theorem 9 a)) to the case of the basic principle:

Theorem 11 ([8]). $S_{F_{1.5}}(PHP_n^{2n}) \leq n^{O(\log n)}$.

Finally, the following recent result by Atserias [28] shows that in this situation depth can be traded for size.

Theorem 12 ([28]). $S_{F_d}(PHP_n^{2n}) \leq n^{((\log n)^{O(1/d)})}$.

Theorems 10 and 11 still leave open the gap between $R = R(1)$ and $F_{1.5} = R(polylog)$. All our intuition from complexity theory and mathematical logic strongly suggests that the distance between 2 and polylog should be much shorter than between 2 and 1. The last mysterious result (in Section 4.2!) due to Atserias, Bonet and Esteban [29] indicates that for the pigeonhole principle the situation is exactly the opposite.

Theorem 13 ([29]). $S_{R(2)}(onto - FPHP_n^{2n}) \geq \exp(n/(\log n)^{O(1)})$.

The case of $R(3)$ is still open.

4.3 Weak PHP_n^m: $m = n^2$ (Mystery Becomes Hard Labour)

Theorem 12 gets significantly improved in this case. Like Theorem 9, this was explicitly extracted from the paper [25] by Krajíček [2].

Theorem 14 ([25,2]). $S_{F_d}(PHP_n^{n^2}) \leq n^{(\log^{(\Omega(d))} n)}$, where $\log^{(t)}$ *is the t-wise composition of* $\log$ *with itself.*

The most remarkable thing that happens at this stop, however, is that the proof method of Theorems 3, 10 also completely breaks down. The question on determining the resolution proof complexity for the weak PHP has been very intriguing since it became clear, and it has been solved only very recently. Section 4.3 is entirely devoted to the history of this new result, which can be essentially viewed as the history of accumulating the necessary techniques.

The first major contribution was made by Buss and Pitassi [30]. Firstly, they proposed an extremely convenient "normal form" for resolution proofs of the pigeonhole principle that was used in many subsequent papers on the subject (proofs in this normal form operate exclusively with *monotone*, i.e., negation-free clauses). As a rather surprising corollary of this normal form, they showed that PHP_n^m and $onto - PHP_n^m$ behave in the same way with respect to Resolution.

Theorem 15 ([30]).
$S_R(onto - PHP_n^m) \leq S_R(PHP_n^m) \leq S_R(onto - PHP_n^m)^{O(1)}$.

This phenomenon seems to be very unique: Theorem 7 and Theorem 8 in particular imply that this is certainly not the case for the polynomial calculus.

Secondly, [30] solved the (relatively easy) case of tree-like Resolution.

Theorem 16 ([30]). *Every tree-like resolution proof of $onto - FPHP_n^m$ must have size at least 2^n, for any $m > n$.*

The next contribution was made in the paper by Razborov, Wigderson and Yao [31]. They identified two subsystems of Resolution and proved the desired lower bound for each of them (using different methods). We mention here only one of these systems, *rectangular calculus*, although we skip its formal definition. Intuitively, the idea is to concentrate only on those monotone clauses that are of "rectangular shape" $\bigvee_{i \in I} \bigvee_{j \in J} x_{ij}$ with $I \subseteq [m]$, $J \subseteq [n]$, and write down appropriate sound rules for operating with such clauses.

Theorem 17 ([31]). *Every rectangular calculus proof of $FPHP_n^{n^2}$ must have size $\exp(\Omega(n/(\log n)))$.*

A resolution proof is called *regular* if along every path in this proof every literal is resolved at most once. Proofs in every one of the two subsystems of Resolution considered in [31] are in fact regular. The following result by Pitassi and Raz made a major improvement on [31]:

Theorem 18 ([32]). *Every regular resolution proof of $FPHP_n^{n^2}$ must have size $\exp(n/(\log n)^{O(1)})$.*

Shortly after Raz [33] came up with a complete solution for the basic version PHP_n^m. By Theorem 15, this immediately extends to the onto version.

Theorem 19 ([33]). $S_R(onto - PHP_n^{n^2}) \geq \exp(n/(\log n)^{O(1)})$.

Razborov [13] gave a simpler proof of the same result. In the next paper [34] the lower bound was extended to the functional case, and, finally, in [35] the weakest functional onto version was also analyzed.

Theorem 20 ([13,34,35]). $S_R(onto - FPHP_n^{n^2}) \geq \exp(\Omega(n/(\log n)^2))$.

4.4 Very Weak PHP: $m = \infty$ (Last Twinkle of Mystery)

The reader who feels uncomfortable with infinitely many pigeons, may think of m in this section as of a sufficiently large number.

One more surprise still awaits us at this last stop. Namely, it had been conjectured for a while that $S_R(PHP_n^m) = \exp(\Omega(n))$ for every m whatsoever, even if we are not smart enough to prove this. The paper [30] already mentioned above disproved the conjecture, and [31] analyzed their proof to show that it is in fact carried over in the rectangular calculus.

Theorem 21 ([30]). *There exists a rectangular calculus proof of PHP_n^∞ that has size $\exp(O((n \log n)^{1/2}))$.*

All lower bounds from Section 4.3 readily extend to the very weak case (in fact, all of them were originally stated in this form).

Theorem 22 ([31]). *Every rectangular calculus proof of $FPHP_n^\infty$ must have size $\exp(\Omega(n^{1/2}))$.*

Theorems 21 and 22 determine the rectangular calculus complexity of the very weak pigeonhole principle up to a logarithmic factor in the exponent.

Theorem 23 ([32]). *Every regular resolution proof of $FPHP_n^\infty$ must have size $\exp(n^{\Omega(1)})$.*

Theorem 24 ([33]). $S_R(onto - PHP_n^\infty) \geq \exp(n^{\Omega(1)})$.

[33] also estimates the constant assumed in the expression $n^{\Omega(1)}$ above as between $1/10$ and $1/8$.

Theorem 25 ([13,34,35]). $S_R(onto - FPHP_n^\infty) \geq \exp(\Omega(n^{1/3}))$.

5 Application: Circuit Lower Bounds Are Hard for Weak Proof Systems

As we mentioned in Introduction, one of the reasons for the popularity of the pigeonhole principle consists in its tight connections with many other things. This and the next sections illustrate the point.

In [36, Appendix] Razborov proposed to study the provability of (first-order or second-order) principles expressing in a particular way that a given Boolean function can not be computed by short Boolean circuits. Shortly after, J. Krajíček observed that this question possesses an adequate re-formulation in terms of propositional proofs (which is by far more convenient), and it is this framework that is followed here.

More specifically, let f_n be a Boolean function in n variables, and let $t \leq 2^n$. Denote by $Circuit_t(f_n)$ any natural CNF *of size* $2^{O(n)}$ encoding the description of a size-t fan-in 2 Boolean circuit presumably computing f_n. Then its negation is a tautology if and only if the circuit size of f_n is greater than t. We demand that every clause in $Circuit_t(f_n)$ is of constant width (5 is enough). Given our liberal $2^{O(n)}$ bound on the size of $Circuit_t(f_n)$, it can be easily constructed simply by introducing a separate propositional variable x_{av} for the Boolean value computed at the node v on the input string $a \in \{0,1\}^n$; for a precise definition see [21,35]. Raz [33] also proposed to consider the variant of this principle in which *unbounded fan-in* circuits are allowed; we will denote this version by $Circuit_t^+(f_n)$. It is stronger than $Circuit_t(f_n)$, and we can no longer demand that every axiom is of constant width.

One of the main motivations for proposing this framework was that *all known lower bound proofs in circuit complexity can be carried over in it*. That is, if we restrict the class of circuits used in the definition of $Circuit_t(f_n)$ to monotone circuits, bounded-depth circuits, depth-2 threshold circuits etc. then this principle *will* become provable within polynomial (that is, $2^{O(n)}$) size in the Frege system or, in the worst case, in its natural extension known as *Extended Frege*. On the other hand, it is known that $Circuit_t(f_n)$ is hard for every proof system possessing *Efficient Interpolation Theorem* (see e.g. [4] for definitions and discussion of this theorem), but only *provided strong one-way functions exist*. In particular, Resolution and Polynomial Calculus do have Efficient Interpolation, hence the latter conclusion applies to them.

The only known *unconditional* (i.e., without any unproven assumptions) lower bounds on the complexity of $Circuit_t(f_n)$ are based on some of the lower bounds for the pigeonhole principle surveyed in the previous section, and on a reduction from PHP to $Circuit_t(f_n)$ discovered in [21]. The underlying idea of this reduction is simple. Let $A \stackrel{def}{=} f_n^{-1}(1)$. Then every counterexample to $onto - FPHP_t^{|A|}$ (encoded by the pigeonhole variables $\{x_{aj} \mid a \in A, j \in [t]\}$) can be used to construct a short circuits for f_n; namely, $f_n \equiv \bigvee_{j \in [t]} K_j$, where K_j is the characteristic function of that input $a \in A$ for which $x_{aj} = 1$. Elaborating a little bit on this simple idea, we can get rid of the onto axioms (in the case of PC!) and show

Lemma 1 ([21]). *If $Circuit_t(f_n)$ has polynomial calculus proof of degree d for some function f_n, then $FPHP_{t/2n}^{2^n}$ also has a PC proof of the same degree d.*

Lemma 1 and Theorem 7 immediately imply

Corollary 1 ([21]). *Every polynomial calculus proof of $Circuit_t(f_n)$ (for any function f_n) must have degree $\Omega(t/n)$.*

To extend the reduction from Lemma 1 to the case of Resolution, we apparently must employ the "onto" axioms (4).[6] Denote $|A|$ by m.

Lemma 2. a) [33] $S_R(Circuit_t^+(f_n)) \geq S_R(onto - PHP_{t-1}^m)$;
b) [35] $S_R(Circuit_t(f_n)) \geq S_R(onto - FPHP_{t/2n}^m)$.

Combining Lemma 2 a) with Theorem 24, we immediately get

Theorem 26 ([33]). $S_R(Circuit_t^+(f_n)) \geq \exp(t^{\Omega(1)})$.

---

[6] The remark in [13, Section 5] that $FPHP_n^m$ would suffice for the purpose seems to be erroneous.

Combining Lemma 2 b) with Theorem 25, we get a similar bound for $Circuit_t(f_n)$. If we, however, take into account that in Lemma 2 we always have $m \leq 2^n$, we can do slightly better (cf. Theorem 31 in the next section) and actually prove

Theorem 27 ([35]). $S_R(Circuit_t(f_n)) \geq \exp(\Omega(t/n^3))$.

6 Generalization: Matching Principles

One natural way to interpret $FPHP_n^m$ [$onto-FPHP_n^m$] is by saying that the complete bipartite graph on two sets of vertices U and V with $|U| = m$, $|V| = n$ does not contain a matching from U to V [a perfect matching, respectively]. One very natural question is what can be said about the complexity of this (perfect) *matching principle* for other graphs, not necessarily bipartite, or, perhaps, even hypergraphs. In this section we will survey what is known along these lines.

Ben-Sasson and Wigderson [11] introduced the principle $G - PHP$ which says that a given bipartite graph G from U to V does not contain a multi-valued matching (thus, $PHP_n^m \equiv K_{m,n} - PHP$), and this definition readily generalizes to all other versions of the pigeonhole principle. They were able to relate the complexity of $G - FPHP$ to expansion properties of the graph G.

Theorem 28 ([11]). *For every bipartite graph G on (U, V) which is a constant-rate expander of bounded minimal degree*[7], $S_R(G - FPHP) \geq \exp(\Omega(|U|))$.

Alekhnovich and Razborov [21] used a general theory developed in that paper to show that the expansion properties of G also imply hardness with respect to the polynomial calculus.

Theorem 29 ([21]). *For every bipartite graph G on (U, V) which is a constant-rate expander of bounded minimal degree, every polynomial calculus proof of $G - EPHP$ must have degree* $\Omega(|U|)$.

For an arbitrary (not necessarily bipartite) graph G, let $PM(G)$ be the principle asserting that G does not contain a *perfect* matching (thus, $onto - FPHP_n^m \equiv PM(K_{m,n})$). As an example, let G be a $(2n \times 2n)$ grid with two opposite corners removed, then $PM(G)$ is called the *mutilated chessboard problem*. Dantchev and Riis [37] proved the following tight lower bound for it (independently, Alekhnovich [38] showed a somewhat weaker bound $\exp(\Omega(n^{1/2}))$):

Theorem 30 ([37]). *Every resolution proof of the mutilated chessboard problem must have size* $\exp(\Omega(n))$.

Razborov [35] considered the principle $PM(G)$ in full generality. Let $\delta(G)$ be the minimal degree of a vertex $v \in V(G)$.

Theorem 31 ([35]). $S_R(PM(G)) \geq \exp\left(\Omega\left(\frac{\delta(G)}{(\log |V(G)|)^2}\right)\right)$.

---

[7] the minimum is taken over the nodes in U

This is a far-going generalization of Theorems 20 and 25. Vice versa, it is worth noting that Theorem 31 is in fact proved by a sort of indirect reduction from $FPHP_n^m$, followed by applying methods from [34].

Finally, [35] also contains a further generalization of Theorem 31 to hypergraphs. We formulate it here only for the special case of *complete r-hypergraph* intensively studied in the literature. Namely, let $r \nmid n$ and the principle $Count_r^n$ assert that an n-element set can not be partitioned into r-sets.

Theorem 32 ([35]). $S_R(Count_r^n) \geq \exp(\Omega(n/(r^2(\log n)(r + \log n))))$.

7 Open Problems

1. Theorem 4 and Remark 1 imply that PHP_n^{n+1} has polynomial size proofs of logical depth $O(\log n)$, whereas Theorem 6 implies that no such proof exists in bounded depth. That would be nice to further narrow this gap. In particular, is it true that $S_{F_{O(\log \log n)}}(PHP_n^{n+1}) \leq n^{O(1)}$?
2. Is it true that for some absolute constant $d \geq 1$ and for some $m = m(n)$, $S_{F_d}(onto-FPHP_n^m) \leq n^{O(1)}$? Quasi-polynomial upper bounds are provided by Theorems 9, 11, 12, 14, with the latter result particularly close to a polynomial.
3. Is it true that for some absolute constant $t \geq 1$ and for some $m = m(n)$, $S_{R(t)}(onto - FPHP_n^m) \leq \exp((\log n)^{O(1)})$? This is not true if $t \leq 2$ (Theorem 13), but becomes true if t is allowed to grow polylogarithmically in n (Theorems 9 b), 11).
4. When $m \geq n^2$, we do not know how to answer the previous question even for $t = 2$. Is it true that $S_{R(2)}(onto - FPHP_n^{n^2}) \leq \exp((\log n)^{O(1)})$?
5. What is the value of $\limsup_{n \to \infty} \frac{\log_2 \log_2 S_R(PHP_n^\infty)}{\log_2 n}$? From Theorems 21 and 25 we know that it lies in the interval $[1/3, 1/2]$.
6. In order to make this survey more structured, we stopped only at $m = n + 1, 2n, n^2, \infty$ to record the changes that had had happened along the road. The question of identifying the "turning points" more exactly also seems to be of considerable interest. As an example, consider the case $m = n + n^{1/2}$. It is consistent with our current knowledge that $S_{F_d}(PHP_n^{n+n^{1/2}}) \geq \exp(n^{\epsilon_d})$ for any fixed d, and it is also consistent that $S_{F_{1.5}}(PHP_n^{n+n^{1/2}}) \leq n^{O(\log n)}$. Rule out one of these two possibilities.
7. Is it true that for a fixed prime p and $\ell, n \to \infty$, $onto - FPHP_n^{n+p^\ell}$ does not possess bounded-degree PC proofs over any field of characteristic p? This is known for the weaker *Nullstellensatz proof system* [39].
8. Let G be a constant-rate bounded-degree expander on the sets of vertices U, V. Is it true that every polynomial calculus proof of $G - FPHP$ must have degree $\Omega(|U|)$? Theorem 29 answers this question for the stronger version $G - EPHP$.

Acknowledgement. I am grateful to Paul Beame and Sam Buss for useful suggestions promptly made on the original draft of this survey. Paul Beame has also contributed several open problems.

References

1. Urquhart, A.: The complexity of propositional proofs. Bulletin of Symbolic Logic **1** (1995) 425–467
2. Krajíček, J.: Bounded arithmetic, propositional logic and complexity theory. Cambridge University Press (1995)
3. Razborov, A.: Lower bounds for propositional proofs and independence results in Bounded Arithmetic. In auf der Heide, F.M., Monien, B., eds.: Proceedings of the 23rd ICALP, Lecture Notes in Computer Science, 1099, New York/Berlin, Springer-Verlag (1996) 48–62
4. Beame, P., Pitassi, T.: Propositional proof complexity: Past, present and future. Technical Report TR98-067, Electronic Colloquium on Computational Complexity (1998)
5. Pudlák, P.: The lengths of proofs. In Buss, S., ed.: Handbook of Proof Theory. Elsevier (1998) 547–637
6. Cook, S.A., Reckhow, A.R.: The relative efficiency of propositional proof systems. Journal of Symbolic Logic **44** (1979) 36–50
7. Reckhow, R.A.: On the lengths of proofs in the propositional calculus. Technical Report 87, University of Toronto (1976)
8. Maciel, A., Pitassi, T., Woods, A.: A new proof of the weak pigeonhole principle. Manuscript (1999)
9. Håstad, J.: Computational limitations on Small Depth Circuits. PhD thesis, Massachusetts Institute of Technology (1986)
10. Clegg, M., Edmonds, J., Impagliazzo, R.: Using the Groebner basis algorithm to find proofs of unsatisfiability. In: Proceedings of the 28th ACM STOC. (1996) 174–183
11. Ben-Sasson, E., Wigderson, A.: Short proofs are narrow - resolution made simple. In: Proceedings of the 31st ACM STOC. (1999) 517–526
12. Alekhnovich, M., Ben-Sasson, E., Razborov, A., Wigderson, A.: Pseudorandom generators in propositional complexity. In: Proceedings of the 41st IEEE FOCS. (2000) 43–53
13. Razborov, A.: Improved resolution lower bounds for the weak pigeonhole principle. Technical Report TR01-055, Electronic Colloquium on Computational Complexity (2001) Available at ftp://ftp.eccc.uni-trier.de/pub/eccc/reports/2001/TR01-055/index.html .
14. Haken, A.: The intractability or resolution. Theoretical Computer Science **39** (1985) 297–308
15. Buss, S.R.: Polynomial size proofs of the propositional pigeonhole principle. Journal of Symbolic Logic **52** (1987) 916–927
16. Cook, W., Coullard, C.R., Turán, G.: On the complexity of cutting plane proofs. Discrete Applied Mathematics **18** (1987) 25–38
17. Ajtai, M.: The complexity of the pigeonhole principle. In: Proceedings of the 29th IEEE Symposium on Foundations of Computer Science. (1988) 346–355
18. Bellantoni, S., Pitassi, T., Urquhart, A.: Approximation of small depth Frege proofs. SIAM Journal on Computing **21** (1992) 1161–1179
19. Pitassi, T., Beame, P., Impagliazzo, R.: Exponential lower bounds for the pigeonhole principle. Computational Complexity **3** (1993) 97–140
20. Krajíček, J., Pudlák, P., Woods, A.R.: Exponential lower bounds to the size of bounded depth Frege proofs of the pigeonhole principle. Random Structures and Algorithms **7** (1995) 15–39
21. Razborov, A.: Lower bounds for the polynomial calculus. Computational Complexity **7** (1998) 291–324
22. Impagliazzo, R., Pudlák, P., Sgall, J.: Lower bounds for the polynomial calculus and the Groebner basis algorithm. Computational Complexity **8** (1999) 127–144
23. Alekhnovich, M., Razborov, A.: Lower bounds for the polynomial calculus: non-binomial case. In: Proceedings of the 42nd IEEE Symposium on Foundations of Computer Science. (2001) 190–199

24. Riis, S.: Independence in Bounded Arithmetic. PhD thesis, Oxford University (1993)
25. Paris, J.B., Wilkie, A.J., Woods, A.R.: Provability of the pigeonhole principle and the existence of infinitely many primes. Journal of Symbolic Logic **53** (1988) 1235–1244
26. Krajíček, J.: On the weak pigeonhole principle. Fundamenta Mathematicae **170** (2001) 123–140
27. Buss, S., Turán, G.: Resolution proofs of generalized pigeonhole principle. Theoretical Computer Science **62** (1988) 311–317
28. Atserias, A.: Improved bounds on the weak pigeonhole principle and infinitely many primes from weaker axioms. In J. Sgall, A. Pultr, P.K., ed.: Proceedings of the 26th International Symposium on the Mathematical Foundations of Computer Science (Marianske Lazne, August '01), Lecture Notes in Computer Science 2136, Springer-Verlag (2001) 148–158
29. Atserias, A., Bonet, M.L., Esteban, J.L.: Lower bounds for the weak pigeonhole principle beyond resolution. To appear in *Information and Computation* (2000)
30. Buss, S., Pitassi, T.: Resolution and the weak pigeonhole principle. In: Proceedings of the CSL97, Lecture Notes in Computer Science, 1414, New York/Berlin, Springer-Verlag (1997) 149–156
31. Razborov, A., Wigderson, A., Yao, A.: Read-once branching programs, rectangular proofs of the pigeonhole principle and the transversal calculus. In: Proceedings of the 29th ACM Symposium on Theory of Computing. (1997) 739–748
32. Pitassi, T., Raz, R.: Regular resolution lower bounds for the weak pigeonhole principle. In: Proceedings of the 33rd ACM Symposium on the Theory of Computing. (2001) 347–355
33. Raz, R.: Resolution lower bounds for the weak pigeonhole principle. Technical Report TR01-021, Electronic Colloquium on Computational Complexity (2001)
34. Razborov, A.: Resolution lower bounds for the weak functional pigeonhole principle. Manuscript, available at http://www.mi.ras.ru/~razborov/matching.ps (2001)
35. Razborov, A.: Resolution lower bounds for perfect matching principles. Manuscript, available at http://www.mi.ras.ru/~razborov/matching.ps (2001)
36. Razborov, A.: Bounded Arithmetic and lower bounds in Boolean complexity. In Clote, P., Remmel, J., eds.: Feasible Mathematics II. Progress in Computer Science and Applied Logic, vol. *13*. Birkhäuser (1995) 344–386
37. Dantchev, S., Riis, S.: "Planar" tautologies hard for Resolution. In: Proceedings of the 42nd IEEE Symposium on Foundations of Computer Science. (2001) 220–229
38. Alekhnovich, M.: Mutilated chessboard is exponentially hard for resolution. Manuscript (2000)
39. Beame, P., Riis, S.: More on the relative strength of counting principles. In Beame, P., Buss, S., eds.: Proof Complexity and Feasible Arithmetics: DIMACS workshop, April 21-24, 1996, DIMACS Series in Dicrete Mathematics and Theoretical Computer Science, vol. 39. American Math. Soc. (1997) 13–35

Words and Patterns*

Antonio Restivo and Sergio Salemi

University of Palermo, Dipartimento di Matematica ed Applicazioni,
Via Archirafi 34, 90123 Palermo, Italy
{restivo,salemi}@dipmat.math.unipa.it

Abstract. In this paper some new ideas, problems and results on patterns are proposed. In particular, motivated by questions concerning avoidability, we first study the set of binary patterns that can occur in one infinite binary word, comparing it with the set of factors of the word. This suggests a classification of infinite words in terms of the "difference" between the set of its patterns and the set of its factors. The fact that each factor in an infinite word can give rise to several distinct patterns leads to study the set of patterns of a single finite word. This set, endowed with a natural order relation, defines a poset: we investigate the relationships between the structure of such a poset and the combinatorial properties of the word. Finally we show that the set of patterns of the words in a regular language is a regular language too.

1 Introduction

The study of patterns goes back to the beginning of last century with the papers of Axel Thue (cf. [17], [18]) on repetitions on words, repetitions corresponding to *unary* patterns. Such study was extended in [4], and independently in [19], to arbitrary patterns, providing a very general framework for researches on avoidability, i.e. researches concerning words that do not follow, or avoid, certain patterns. A typical question ask whether or not there exists an infinite word, on a given alphabet, that avoids a given pattern. An important reference on this topic, including very recent contributions, is [7].

From a different perspective D. Angluin (cf [2]) used patterns for defining languages: a pattern is a word that contains special symbols, called variables, and the associated *pattern language* is obtained by replacing the variables with arbitrary non-empty words, with the condition that two occurrences of the same variable have to be replaced with the same word. Starting from this basic definition, a theory of language generating devices based on patterns has been developed (cf. [10] and the more recent [14] for an overview of this theory).

The notion of pattern also appears, more or less explicitly, in several areas of combinatorics on words and formal languages, such as, for instance, word equations, DOL systems, etc.

In this paper we propose some new ideas on the study of patterns. The partial results here reported give rise to new questions and suggest some new research directions.

---
* Partially supported by MURST projects: *Bioinformatica e Ricerca Genomica*

Starting from some remarks about the special role of the Thue-Morse and Fibonacci words in binary avoidability, we first investigate the set of patterns occurring in an infinite word, comparing it with the set of factors of the word. If the alphabet of patterns and that of ordinary words are the same binary alphabet, we can consider the set $Fact(w)$ of factors of an infinite word w as a subset of the set $P(Fact(w))$ of patterns occurring in w. We report non trivial examples of infinite words w for which the difference set $P(Fact(w)) \setminus Fact(w)$ is finite and, on the contrary, words w for which the difference set is infinite and, moreover, the complexity function of the language $P(Fact(w))$ grows more quickly than the complexity function of the language $Fact(w)$. This suggests to characterize (to classify) infinite words in terms of the relationships between their (factor) complexity and pattern complexity.

The above examples, in particular, show that each factor in an infinite word can give rise to several distinct patterns. This leads to study the patterns of a single finite word. Given two finite word u, v, we write $u \leq v$ if u is a pattern of v. The relation $\leq$ is a partial order relation and we associate to a finite word v the set $P(v)$ of all words u such that $u \leq v$. $P(v)$ with the relation $\leq$ is a *poset*. We give some partial results and propose problems concerning the relationships between the structure of the poset $P(v)$ and the combinatorial properties of the word v.

In the last section, for a given language L, we consider the language $P(L)$ of all patterns of words in L. The main result of this section shows that, if L is a regular language, then $P(L)$ is a regular language. From this we derive that, given a pattern p and a regular language L, it is decidable whether L avoids p. A natural questions is whether similar results can be stated for other levels of the Chomsky hierarchy.

2 Avoidability

In order to define patterns, in general one makes use of two distinct alphabets. The first one, A, is the usual alphabet on which ordinary words are constructed. The second alphabet, E, is used in patterns. Its elements are usually called *variables*, and words in E^* are called *patterns*. This distinction is meant to help the understanding of the roles of the different words used. However, in most of the present paper, we treat a pattern as an ordinary word, which amounts to take $A = E$.

The *pattern language* associated to a pattern $p \in E^*$ is the language on A containing all the words $h(p)$, where h is a non-erasing morphism from E^* to A^* that substitutes an arbitrary non-empty word to every variable. It is denoted by $L(p)$.

Given a word $w \in A^*$, a word $p \in E^*$ is a *pattern of w* if $w \in L(p)$. The set of patterns of w is denoted by $P(w)$. If $A = E$, then $w \in P(w)$. Given a language L, the *set of patterns of L* is the language $P(L) \subseteq E^*$ defined as follows: a pattern $p \in E^*$ belongs to $P(L)$ if $p \in P(w)$, for some $w \in L$. In other terms:

$$P(L) = \{p \in E^* | L \cap L(p) \neq \emptyset\}.$$

A word $w \in A^*$ is said to *encounter* the pattern p if it contains as factor an element of the pattern language associated to p, i.e. if $Fact(w) \cap L(p) \neq \emptyset$. Equivalently, we

say that *p occurs* in w, otherwise w is said to *avoid p*. Denote by $Av(p)$ the set of words that avoid p:
$$Av(p) = \{w \in A^* | Fact(w) \cap L(p) = \emptyset\}.$$
A language L is said to avoid p if all its elements avoid p, i.e. if $L \subseteq Av(p)$.

These definitions also apply to infinite words $w \in A^\omega$. In particular, an infinite word w avoids p if $Fact(w) \cap L(p) = \emptyset$, or, equivalently, if $Fact(w) \subseteq Av(p)$.

A pattern p is *avoidable on* A if there are infinitely many words in A^* that avoid p, i.e. if $Av(p)$ is an infinite language. This is equivalent, by the König's lemma, to the existence of one infinite word in A^ω avoiding p. If, on the contrary, $Av(p)$ is finite, then p is *unavoidable on* A.

In the sequel of this section we take $A = E = \{a, b\}$, and by "avoidable" we often mean "avoidable on a binary alphabet". In this case the classification of avoidable patterns is completely solved (cf. [6], [16]). We need the following definition. A binary avoidable pattern p is *minimal* if any proper factor of p is unavoidable.

Theorem 1. *Every binary pattern of length at least 6 is avoidable in the binary alphabet. Moreover the minimal avoidable patterns, up to the automorphism exchanging a and b, are*
$$aaa, ababa, aabba, aababb, abaab, abbaa, abbab.$$

In the following we show the special role of the Fibonacci and Thue-Morse infinite words in binary avoidability. Recall that the Fibonacci infinite word
$$w_F = abaababaabaababaababaabaababa...$$
is the fixed point of the morphism $\varphi : a \mapsto ab, b \mapsto a$, and that the Thue-Morse infinite word
$$w_T = abbabaabbaababbabaababbaabbaba...$$
is the fixed point of the morphism $\mu : a \mapsto ab, b \mapsto ba$. These two words are among the most simple obtained as fixed points of morphisms, and still they have endless number of interesting combinatorial properties and are ubiquitous in combinatorics on words. Our first result shows their role in binary avoidability.

Proposition 1. *Every binary pattern of length at least 11 either is avoided by the Fibonacci word or it is avoided by the Thue-Morse word.*

This means that, whereas $P(Fact(w_F))$ and $P(Fact(w_T))$ are infinite sets, their intersection $P(Fact(w_F)) \cap P(Fact(w_T))$ is a finite set (actually the maximal length of its words is 10).

Previous result suggests the following definition. Two infinite words w_1 and w_2 are *pattern independent* if $P(Fact(w_1)) \cap P(Fact(w_2))$ is a finite set. (w_F, w_T) is an example of a pair of pattern independent words. It would be interesting to search for other pairs of pattern independent words and to characterize them (for instance among words that are fixed points of morphisms).

In the last part of this section we consider avoidability of sets of patterns. Whereas in the case of single binary patterns the problem of avoidability is completely solved, the problem remains still open, to our knowledge, for a finite set of patterns.

Given a set $\Pi = \{p_1, p_2, ..., p_k\}$ of patterns, denote by $Av(\Pi)$ the set

$$Av(\Pi) = Av(p_1) \cap Av(p_2) \cap ... \cap Av(p_k).$$

A set Π is *avoidable on* A if $Av(\Pi)$ contains infinitely many words. Otherwise Π is said *unavoidable on* A.

Given two avoidable patterns p, q, i.e. such that both $Av(p)$ and $Av(q)$ are infinite sets, it may occur that $Av(p) \cap Av(q)$ is finite, i.e. that the set $\{p, q\}$ is unavoidable. In the following table, under the hypothesis $E = A = \{a, b\}$, we report some experimental results concerning sets of patterns obtained by combining the minimal avoidable patterns in Theorem 1.

Denote by $p_1 = aaa$, $p_2 = ababa$, $p_3 = aabba$, $p_4 = aababb$, $p_5 = abaab$, $p_6 = abbaa$, $p_7 = abbab$ the minimal avoidable patterns and denote by $m(\Pi)$ the maximal length of words in $Av(\Pi)$, if such set is unavoidable.

Π	$m(\Pi)$
$\{p_1, p_5\}$	10
$\{p_1, p_7\}$	10
$\{p_1, p_2, p_4\}$	12
$\{p_1, p_3\}$	14
$\{p_1, p_6\}$	14
$\{p_3, p_4, p_5\}$	24
$\{p_4, p_6, p_7\}$	24
$\{p_2, p_4, p_5\}$	28
$\{p_2, p_4, p_7\}$	28
$\{p_3, p_4, p_6\}$	55
$\{p_3, p_7\}$	114
$\{p_5, p_6\}$	114

Denote by $\mathcal{M}$ the set of minimal avoidable patterns in Theorem 1. The previous table reports subsets of $\mathcal{M}$ that are unavoidable. We leave open the problem whether they are the minimal ones, i.e. whether their proper subsets are avoidable.

3 Patterns Occurring in Infinite Words

Some results in previous section are obtained by analysing the patterns occurring in an infinite word w, i.e. the set $P(Fact(w))$. The hypothesis $E = A = \{a, b\}$, implies that $Fact(w) \subseteq P(Fact(w))$, i.e. the language $P(Fact(w))$ is an extension of the language $Fact(w)$. The problem is how large is this extension.

We start by reporting a surprising result about the binary patterns occurring in the Thue-Morse word (cf. [9]).

Theorem 2. *(D. Guaiana). If p is a binary pattern occurring in the Thue-Morse word w_T such that $p \neq aabaa$ (and, by symmetry, $p \neq bbabb$), then p is a factor of w_T. In other terms:*

$$P(Fact(w_T)) = Fact(w_T) \cup \{aabaa, bbabb\}.$$

One can ask the question whether a similar property holds true for the Fibonacci word w_F.

Theorem 3. *There exist infinitely many binary patterns occurring in w_F that are not factors of w_F, i.e.*

$$Card(P(Fact(w_F)) \setminus Fact(w_F)) = \infty.$$

Proof. Given a finite word $x = a_1...a_n$, denote by x^R the *reversal* of x, i.e. $x^R = a_n...a_1$. A word x is a *palindrome* if $x = x^R$. Let $\varphi : a \mapsto ab, b \mapsto a$ be the morphism generating w_F. By well known properties of the Fibonacci word (cf.[12]) one has that, for all $n \geq 0$:

$$\varphi^{2n+1}(a) = uab \tag{1}$$
$$\varphi^{2n+2}(a) = vba \tag{2}$$
$$\varphi^{2n+3}(a) = vbauab = uabvab, \tag{3}$$

where u and v are words such that $u = u^R$ and $v = v^R$. Moreover, from Corollary 28 in [13], one has that

$$bub \in A^* \setminus Fact(w_F). \tag{4}$$

By taking into account (1) and (2) one derives:

$$\varphi(uab) = \varphi(u)\varphi(ab) = \varphi(u)aba = \varphi(ub)ba.$$

It follows that $v = \varphi(ub)$ and then we can write:

$$av = \varphi(b)\varphi(ub) = \varphi(bub).$$

On the other hand, since v is a palindrome, one has:

$$(va)^R = a(v)^R = av = \varphi(bub).$$

It follows that:

$$va = (\varphi(bub))^R = \tilde{\varphi}((bub)^R) = \tilde{\varphi}(bub),$$

where $\tilde{\varphi} : a \mapsto ba, b \mapsto a$. Since, by equations (3), va is factor of $\varphi^{2n+3}(a)$, and then it is a factor of w_F, one derives that bub is a pattern occurring in w_F. By taking into account (4), one can conclude that $bub \in P(Fact(w_F)) \setminus Fact(w_F)$. □

Given an infinite word w, define

$$\Delta(w) = Card(P(Fact(w)) \setminus Fact(w)).$$

By previous results, $\Delta(w_T) = 2$ and $\Delta(w_F) = \infty$. Notice that there exist words w such that $\Delta(w) = 0$. It suffices to consider *complete* words, i.e. words such that $Fact(w) = A^*$.

In the study of the parameter $\Delta(w)$ of an infinite word w, it is of particular interest to consider special families of infinite words, as, for instance, words that are fixed points of morphisms, or words of "low" complexity (and, in particular, words with linear complexity).

Denote by $\mathcal{F}$ the family of binary infinite words that are fixed point of morphisms, and denote by $\mathcal{L}$ the family of binary infinite words that are of linear complexity (see below for a definition of complexity). Previous results lead in a natural way to ask the following questions.

- Does there exist a word w in $\mathcal{F}$ (or in $\mathcal{L}$) such that $\Delta(w) = 0$?
- Does the property $\Delta(w) = 2$ uniquely characterizes (among the words in $\mathcal{F}$, or in $\mathcal{L}$) the Thue-Morse word?
- Characterize (among the words in $\mathcal{F}$, or in $\mathcal{L}$) the words w such that $\Delta(w)$ is finite.

Remark 1. Given a language L over the alphabet $\{a,b\}$, denote by $\overline{L}$ the closure of L by the automorphism interchanging a and b. By definition, $P(L)$ is invariant under such automorphism, i.e. $\overline{P(L)} = P(L)$. There is then a lack of symmetry in comparing, for a given infinite word w over $\{a,b\}$, the languages $Fact(w)$ and $P(Fact(w))$. Indeed $\overline{P(Fact(w))} = P(Fact(w))$, whereas the same property does not hold, in general, for $Fact(w)$. One can erroneously suppose that the difference in the behaviours of the Thue-Morse word and the Fibonacci word, stated in Theorem 2 and Theorem 3 respectively, is a consequence of this asymmetry. In fact the set of factors of the Thue-Morse word w_T is invariant under the interchanging of a and b ($\overline{Fact(w_T)} = Fact(w_T)$), whereas the same property does not hold for the Fibonacci word ($\overline{Fact(w_F)} \neq Fact(w_F)$). However, a minor variation on the arguments used in the proof of Theorem 3 shows that $Card(P(Fact(w_F)) \setminus \overline{Fact(w_F)}) = \infty$.

Taking into account previous remark, it appears more appropriate to compare $P(Fact(w))$ and $\overline{Fact(w)}$, and to introduce, together with the parameter $\Delta(w)$, the parameter $\overline{\Delta}(w)$, defined as follows:

$$\overline{\Delta}(w) = Card(P(Fact(w)) \setminus \overline{Fact(w)}).$$

One can then reformulate previous problems for the parameter $\overline{\Delta}(w)$.

Since, in most cases, the parameter $\Delta(w)$ (the parameter $\overline{\Delta}(w)$, resp.) is infinite, for a more accurate analysis we take into account the notion of "complexity". Given an infinite word w, its (factor) *complexity* (cf. [1]) is the function

$$f_w(n) = Card(Fact(w) \cap A^n).$$

In a similar way, the *pattern complexity* of an infinite word w is defined as follows:

$$p_w(n) = Card(P(Fact(w)) \cap A^n).$$

By definition, $f_w(n) \leq p_w(n)$, for all $n \geq 0$.

In the comparison between factors and patterns, a complete word w, i.e. such that $Fact(w) = A^*$, corresponds to an extremal case: $\Delta(w) = 0$ and $f_w(n) = p_w(n)$. Another trivial case corresponds to *repetitive* words. Recall that an infinite word w is repetitive if it contains as factors arbitrarily large powers, i.e. if, for any positive integer n, there exists a finite word u such that $u^n \in Fact(w)$. If we allow, in the definition of pattern, non-injective morphisms, every words of A^* trivially occurs as pattern in an infinite repetitive word. This is a consequence of the fact that, in the binary case, if the morphism h used in the definition of the pattern is not injective, then the words $h(a)$ and $h(b)$ are both powers of the same word. In order to avoid such trivial case, in the remainder of this section we restrict ourselves to patterns defined by injective morphisms.

A non trivial example of infinite word having pattern complexity that grows more quickly than its complexity is found among *Sturmian* words. Recall (cf. [5]) that a Sturmian word w is, by definition, a word of complexity $f_w(n) = n + 1$. Let us denote by St the set of all factors of all Sturmian words, i.e. $v \in St$ iff there exists a Sturmian word w such that $v \in Fact(w)$. The following theorem states, informally, that there exists a single Sturmian word that contains as patterns all the factors of all Sturmian words.

Theorem 4. *There exists a Sturmian word s such that $St \subseteq P(Fact(s))$.*

In [11] has been explicitly computed the complexity of the language St, i.e. the function $f_{St}(n) = Card(St \cap A^n)$. Moreover it is shown that $f_{St}(n) = \Theta(n^3)$. One derives the following corollary.

Corollary 1. *There exists an infinite word w such that $f_w(n) = n + 1$ and $p_w(n) = \Theta(n^3)$.*

In order to prove Theorem 4, let us recall some known facts about Sturmian words (cf. [5]). A special subfamily of the family of Sturmian words is that of *characteristic* Sturmian words, having the property that, for any Sturmian word x, there exists a characteristic Sturmian word y such that $Fact(x) = Fact(y)$. So, for the proof of Theorem 4, we can limit ourselves to consider only characteristic Sturmian words.

The characteristic Sturmian words can be constructed by the *standard method*: one defines a family of finite words, called *standard words*, and every characteristic Sturmian word is the limit of a sequence of standard words.

Consider two functions ρ_1 and ρ_2 from $A^* \times A^*$ into $A^* \times A^*$:

$$\rho_1(u, v) = (u, uv)$$

$$\rho_2(u,v) = (vu, v).$$

ρ_1 and ρ_2 are also known as the *Rauzy rules*. Any word $d \in \{1,2\}^*$ defines recursively a mapping (denoted again by d) as follows:

$$\epsilon(u,v) = (u,v)$$
$$di(u,v) = \rho_i d(u,v) \qquad \text{where } i \in \{1,2\}.$$

Denote by (u_d, v_d) the pair of words $d(a,b)$. The words u_d, v_d are called *standard* words. The following lemma is an immediate consequence of previous definitions.

Lemma 1. *For $d \in \{1,2\}^*$ consider a factorization $d = ps$. Let (u_d, v_d), (u_p, v_p), (u_s, v_s) be the pairs of standard words corresponding to d, p and s respectively. Then $u_d = h_p(u_s)$, $v_d = h_p(v_s)$, where h_p denotes the morphism from A^* into A^* defined as follows: $h_p : a \mapsto u_p, b \mapsto v_p$.*

It is well known (cf. [5]) that any characteristic Sturmian word can be obtained as follows. Let $t \in \{1,2\}^\omega$ be an infinite word over the alphabet $\{1,2\}$ and denote by $t(j)$ its prefix of length j. The sequences of standard words $(u_{t(j)})_{j \geq 0}$ and $(v_{t(j)})_{j \geq 0}$ both converge to the same characteristic Sturmian word x. t is called the *rule-sequence* of x and it is denoted by $\delta(x)$.

We are now ready to prove Theorem 4.

Proof of Theorem 4. Let z be a characteristic Sturmian word such that its rule-sequence $\delta(z)$ is a complete word, i.e. $Fact(\delta(z)) = \{1,2\}^*$. Let w be an arbitrary word in St. This means that there exists a characteristic Sturmian word x such that $w \in Fact(x)$. We shall prove that $w \in P(Fact(z))$. Let $\delta(x) \in \{1,2\}^\omega$ be the rule-sequence of x and denote by d_w the smallest prefix of $\delta(x)$ such that $w \in Fact(u_{d_w}) \cup Fact(v_{d_w})$. Suppose that $w \in Fact(u_{d_w})$ (in the alternative case, the proof is analogous). One can write: $u_{d_w} = u_1 w u_2$. Since $Fact(\delta(z)) = \{1,2\}^*$, then $d_w \in Fact(\delta(z))$. It follows that there exists a prefix t_w of $\delta(z)$ such that

$$t_w = p d_w$$

for some word $p \in \{1,2\}^*$. By the Lemma 1, one has

$$u_{t_w} = h_p(u_{d_w}),$$

where $h_p : a \mapsto u_p, b \mapsto v_p$. It follows that

$$u_{t_w} = h_p(u_1 w u_2) = h_p(u_1) h_p(w) h_p(u_2).$$

Since $u_{t_w} \in Fact(z)$, one has that $w \in P(Fact(z))$, i.e. w is a pattern of z. This concludes the proof.

An interesting problem is to search for other non trivial infinite words having pattern complexity that grows more quickly than their complexity.

For more details on the results of this section see [15].

4 Patterns of a Finite Word

The results of previous section show that each factor in an infinite word can give rise to several distinct patterns. This leads to study the patterns of a single finite word.

As in Section 3, $E = A = \{a, b\}$. For $w \in A^*$, denote by w' the word obtained from w by interchanging a and b. If u is a pattern of v, then u is also pattern of v' and, moreover, u' is a pattern of both v and v'. By this invariance, we can restrict ourselves to consider words beginning with the letter a, i.e. words belonging to aA^*.

Given two words $u, v \in aA^*$, we write $u \leq v$ if u is a pattern of v, i.e. if there exists a non-erasing morphism h from $\{a, b\}^*$ to $\{a, b\}^*$ such that $u = h(v)$. One can easily verify that $\leq$ is a partial order relation on aA^*.

Proposition 2. *The ordering $\leq$ is not a well-ordering.*

Proof. The set $\{a^p b^p \mid p \text{ prime}\}$ is an infinite set of pairwise incomparable elements. Indeed, if $a^p b^p \leq a^q b^q$, with p, q primes, one has necessarily that $a^p \leq a^q$ and $b^p \leq b^q$, which is impossible by the primality of p and q. □

Denote by $P(v)$ the set of patterns of the word $v \in aA^*$:

$$P(v) = \{u \in aA^* \mid u \leq v\}.$$

The set $P(v)$, with the order relation $\leq$, is a *poset*. The structure of such a poset provides useful information about the combinatorial properties of the word v.

Example 1. Let $v = abaababa$. $P(v)$ is given in the following figure:

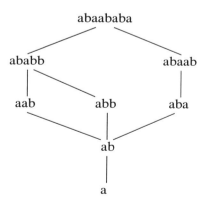

Proposition 3. *$P(v)$ is not in general a lattice.*

The proof is given by the following example.

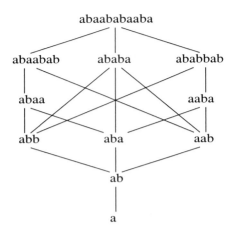

Example 2. Let $v = abaababaaba$. The reader can verify that the poset $P(v)$, given in the figure below, is not a lattice.

Problem: Characterize those words v such that $P(v)$ is a lattice.

Given a word v, we define the following parameters:

- $d(v)$: maximal cardinality of a chain in $P(v)$.
- $r(v)$: maximal cardinality of an antichain in $P(v)$.
- $m(v)$: number of maximal elements in $P(v)$.

Recall that a word u in $P(v)$ is *maximal* if, for $w \in P(v)$, $u \leq w$ implies that either $w = u$ or $w = v$.

EXAMPLE 10 (*continued*) The reader can verify that $d(abaababa) = 5$, $r(abaababa) = 3$, $m(abaababa) = 2$.

It is interesting to investigate, for a given word v, the relationships between previous parameter and the combinatorial properties of v. Let us consider the extremal cases, i.e. cases in which the parameters achieve their minimal or maximal values. Consider first the conditions of minimality.

Proposition 4. *There exist infinitely many words v such that $r(v) = m(v) = 1$.*

Proof. Let $\mu : a \mapsto ab, b \mapsto ba$ be the Thue-Morse morphism (cf. Section 2). One can verify that, for any $n \geq 0$, the poset $P(\mu^n(a))$ has the following structure:

$$\mu^n(a)$$
$$|$$
$$\mu^{n-1}(a)$$
$$\vdots$$
$$\mu(a)$$
$$|$$
$$a$$

It follows that $r(\mu^n(a)) = m(\mu^n(a)) = 1$. □

A natural question is whether there is a relation between Theorem 2 and the proof of Proposition 4. Another question is to search for other words v, different from those in the proof of Proposition 4, satisfying the condition $r(v) = 1$.

Proposition 5. *If $|v| \geq 5$, then $d(v) \geq 4$.*

Proof. Let $v \in aA^*$, with $|v| \geq 5$. For $|v| > 2$, one has trivially that $a < ab < v$, i.e. $d(v) \geq 3$. We show that, for $|v| \geq 5$, one has $a < ab < u < v$, where $u \in \{aab, aba, abb\}$. By definition, $v \in aA^*$, i.e. v begins with the letter a. If v ends with the letter a, then $u = aba < v$. If v ends with the letter b, consider the second letter of v. If such a letter is a, then $u = aab < v$, otherwise v begins with ab. In this last case, if v ends with bb, then $u = abb < v$, otherwise v begins and ends with ab, and then $u = aba < v$. This concludes the proof. □

We say that a word v is *2-primitive* if $d(v) = 4$. This definition is suggested by the classical notion of primitive word and it is motivated by the following remark. Apart some minor details, previous definitions can be given also in the case $E \neq A$. In the special case of unary patterns, i.e. $Card(E) = 1$, the parameter $d(v)$ achieves its minimal value $d(v) = 2$, for primitive words.

Proposition 6. *There exist infinitely many 2-primitive words.*

Proof. For any $n \geq 1$, the word $v_n = a(ab)^n b$ is 2-primitive, as one can easily verify. □

An interesting problem is to find an efficient algorithm to test whether a binary word is 2-primitive.

Let us now consider the maximal values of the parameters.

Proposition 7. *For any positive integer n, there is a word v such that $d(v) \geq n$.*

Proof. It suffices to consider the word $h^n(a)$, where h is a non-erasing morphism. □

Proposition 8. *For any positive integer n, there is a word v such that $r(v) \geq n$.*

Proof. Let k be an integer such that the number of primes dividing k is greater than $\sqrt{n}$. Let $v = a^k b^k$. The set
$$\{a^r b^s | r, s \text{ primes dividing } k\}$$
is an antichain of $P(v)$, by the argument in the proof of Proposition 2, and its cardinality is greater than n. □

All the examples we know, such as that in the proof of previous proposition, show that antichains of large cardinality occur for words containing powers of large order. This leads to propose the following conjecture. Recall that a word $v \in A^*$ is *k-power free* if $v = xy^k z$, with $x, y, z \in A^*$, implies that y is the empty word.

Conjecture: For any integer $k \geq 2$ there exists an integer $N(k)$ such that, if v is k-power free, then $r(v) \leq N(k)$.

5 Patterns of a Language

In this section the alphabet A of ordinary words and the alphabet E of patterns are arbitrary.

Recall from Section 2 that, given a language $L \subseteq A^*$, the set of patterns of L is the language
$$P(L) = \{p \in E^* | L(p) \cap L \neq \emptyset\}.$$
The main result of this section is the following theorem.

Theorem 5. *If L is a regular language, then $P(L)$ is a regular language. Moreover, one can effectively construct the automaton recognizing $P(L)$ from the automaton recognizing L.*

The proof is based on the following two lemma, reported here without proof.

Lemma 2. *Let L be a regular language. For each $p \in P(L)$, there exists a non-erasing morphism h_p from E^* to A^* such that $h_p(p) \in L$, with $|h_p(p)| \leq |p|C^{|p|}$, for some constant C depending only on L.*

Lemma 3. *Let L be a regular language. Then there exists a positive integer K, depending only on L, such that, for any pattern $p \in P(L)$, there exist a pattern $p' \in E^*$, with $|p'| \leq K$, and a non-erasing morphism h' from E^* to A^* such that $h'(p), h'(p') \in L$.*

Proof of Theorem 5. By definition
$$P(L) = \{p \in E^* | h(p) \in L, \ h \text{ non-erasing morphism from } E^* \text{ to } A^*\}.$$

Each element $p \in P(L)$ is identified by at least a morphism h. The idea of the proof is that it suffices to make use only of morphisms h taken from a finite set H. Indeed, by Lemma 3, in order to identify a given $p \in P(L)$, it suffices to consider a morphism h' which is used to identify another element $p' \in P(L)$ of length $|p'| \leq K$. Denote by P_K the set of elements of $P(L)$ of length smaller than or equal to K. Then, in definition of $P(L)$, one can make use only of the morphisms that identify the elements of P_K. By Lemma 2, any element $p \in P_K$ can be identified by using a morphism h_p such that
$$|h_p(p)| \leq |p|C^{|p|} \leq KC^K.$$

There is only a finite number of non-erasing morphisms h satisfying previous condition. Denote by H the set of such morphisms:
$$H = \{h_1, h_2, ..., h_t\}.$$

Define $R_i = h_i^{-1}(L), i = 1, 2, ..., t$. The languages R_i's are regular languages, since they are images by inverse morphisms of a regular language. Then
$$P(L) = R_1 \cup R_2 \cup ... \cup R_t$$
is a regular language.

Moreover, since the constants C and K are effectively related to the size of the automaton recognizing L, one can effectively construct the automaton recognizing $P(L)$. □

Recall from Section 2 that a language L avoids a pattern p if all the elements of L avoid p, i.e. if $Fact(L) \cap L(p) = \emptyset$, or, equivalently, if p does not belong to $P(Fact(L))$. If L is a regular language, by Theorem 5, $P(Fact(L))$ is a regular language too. A consequence is the following corollary.

Corollary 2. *Given a pattern p and a regular language L, it is decidable whether L avoids p.*

References

1. Allouche, J.P.: Sur la complexité des suites infinies. Bull. Belg. Math. Soc. **1** (1994) 133-143
2. Angluin, D.: Finding patterns common to a set of strings. Journal of Computer and System Sciences **21** (1980) 46-62
3. Béal, M.-P., Mignosi, F. Restivo, A., Sciortino, M.: Forbidden Words in Symbolic Dynamics. Advances in Appl. Math. **25** (2000) 163-193
4. Bean, D.R., Ehrenfeucht, A., McNulty, G.F.: Avoidable patterns in strings of symbols. Pacific J. Math. **85** (1979) 261-294
5. Berstel, J., Séébold, P.: Sturmian Words. In: Algebraic Combinatorics on Words, chapter 2, M.Lothaire (Ed.) Cambridge University Press, 2001
6. Cassaigne, J.: Unavoidable binary patterns. Acta Informatica **30** (1993) 385-395
7. Cassaigne, J.: Unavoidable patterns. In: Algebraic Combinatorics on Words, chapter 3, M. Lothaire (Ed.) Cambridge University Press, 2001
8. Choffrut, C., Karhumäki, J.: Combinatorics of Words. In: Handbook of Formal Languages, volume 1, chapter 6, 329-438, G.Rozenberg, A.Salomaa (Eds.) Springer, Berlin, 1997
9. Guaiana, D.: On the binary patterns of the Thue-Morse infinite word. Internal Report, University of Palermo, 1996
10. Mateescu, A., Salomaa, A.: Aspects of Classical Language Theory. In: Handbook of Formal Languages, volume 1, chapter 4, 175-251, G.Rozenberg, A.Salomaa (Eds.), Springer, Berlin, 1997
11. Mignosi, F.: On the number of factors of Sturmian words. Theoret. Comput. Sci. **82** (1991) 71-84
12. Mignosi, F., Pirillo, G.: Repetitions in the Fibonacci infinite word. RAIRO Theoretical Informatics and Applications **26**(3) (1992) 199–204
13. Mignosi, F., Restivo, A., Sciortino, M.: Words and Forbidden Factors. Theoret. Comput. Sci. **273**(1-2) (2001) 99–117
14. Mitrana, V.: Patterns and Languages: An overview. Grammars **2** (1999) 149-173
15. Restivo, A., Salemi, S.: Binary Patterns in Infinite Binary Words. Lecture Notes in Computer Science (to appear)
16. Roth, P.: Every binary pattern of length six is avoidable on the two-letter alphabet Acta Informatica **29** (1992) 95-107
17. Thue, A.: Über unendliche Zeichenreihen. Kra. Vidensk. Selsk. Skrifter. I. Mat.-Nat. Kl., Christiana **7** 1906
18. Thue, A.: Über die gegenseitige Lage gleicher Teile gewisser Zeichenreihen. Kra. Vidensk. Selsk. Skrifter. I. Mat.-Nat. Kl., Christiana **12** 1912
19. Zimin, A.I.: Blocking sets of terms. Math. USSR Sb **47** (1984) 353-364

A Short Introduction to Infinite Automata

Wolfgang Thomas

Lehrstuhl für Informatik VII, RWTH Aachen, 52056 Aachen, Germany
thomas@informatik.rwth-aachen.de

Abstract. Infinite automata are of interest not only in the verification of systems with infinite state spaces, but also as a natural (and so far underdeveloped) framework for the study of formal languages. In this survey, we discuss some basic types of infinite automata, which are based on the so-called prefix-recognizable, synchronized rational, and rational transition graphs, respectively. We present characterizations of these transition graphs (due to Muller/Schupp and to Caucal and students), mention results on their power to recognize languages, and discuss the status of central algorithmic problems (like reachability of given states, or decidability of the first-order theory).

1 Historical Background

In the early days of computer science, the first purpose of "finite automata" was to serve as an abstract model of circuits. In this context, a "state" is given by an association of boolean values to the latches of a circuit, i.e., by a bit vector $(b_1, \ldots, b_n)$ if there are n latches. A set of states is then given by a boolean function in n variables and can be defined by a boolean formula $\varphi(x_1, \ldots, x_n)$, and a transition relation (say via a fixed input) by a boolean formula $\psi(x_1, \ldots, x_n, x'_1, \ldots, x'_n)$ of $2n$ boolean variables. Only later in the emergence of computer science, when programming languages had entered the stage and the purpose of string processing was associated with finite automata, a more abstract view of automata became dominant: Now states and inputs were collected in abstract finite sets, often denoted Q and Σ, the transition relation being simply a subset of $Q \times \Sigma \times Q$.

In the framework of model checking (for the automatic verification of state-based programs, cf. [CGP99]), a reverse development took place. At first, the abstract view was pursued: A system, such as a reactive program or a protocol, is modelled by an annotated transition graph, whose vertices are the system's states. The task of model checking a temporal property (given by a temporal logic formula) is to compute the set of those vertices where the given formula is satisfied. However, as noted by McMillan [McM93], the efficiency of model checking is improved considerably when the so-called "symbolic approach" is adopted. This approach involves a return "back to the roots" of automata theory: States of transition graphs are bit vectors, and sets of states, as well as transition relations, are given by boolean functions. An essential point in symbolic model-checking to achieve efficiency is a particular representation of boolean functions, namely by the "ordered binary decision diagrams" (OBDD's). These OBDD's are a form of minimized acyclic finite automata accepting sets of 0-1-words of bounded length.

At this point, and aiming at the treatment of infinite systems, it is natural to cancel the constraint of bounded length. When admitting words of arbitrary finite length as representations of individual states, the OBDD's should be replaced by usual finite automata, thereby dropping the requirement of acyclicity. State properties will thus be captured by regular languages. For the specification of transition relations, one has to invoke automata theoretic definitions of word relations. There is a large reservoir of options here, by considering finite-state tranducers or word rewriting systems of different kinds. Presently, a lot of research is devoted to the study of infinite transition systems using these ideas, in an attempt to generalize the methodology of model checking to systems with infinite state spaces. (The reader can trace this research fairly well by consulting the LNCS-Proceedings of the annual conferences *CAV (Computer-Aided Verification)* and *TACAS (Tools and Algorithms in the Construction and Analysis of Systems)*). But also from the language theoretical point of view, infinite transition systems are interesting since they provide an alternative to standard models of automata for the definition of non-regular languages.

In all these investigations, the classical finite automata are still present: They enter as a tool to define the infinite structures (infinite state spaces, infinite transition graphs). Thus their dynamics is conceptually no more associated with the "progress in time" (as in the original circuit model), but rather with the "exploration of static data" (when the scanned words are considered as representations of individual states of an infinite system). In this role, finite automata will also be useful in the development of an infinitary but algorithmically oriented model theory.

In this paper we survey some basic classes of infinite transition systems, giving characterization results, remarks on their power as acceptors of languages, and discussing the status of algorithmic problems. Our aim is to help the reader to enter this fastly developing field and to give some references where more results and also detailed proofs can be accessed. We do not treat in any depth the many connections with process theory, where different means of specifying infinite transition systems are employed and where the semantics is based on bisimulation equivalence rather than on accepted languages. (The reader may start that subject with the survey [BE97], where the essential concepts and algorithmic results are presented.)

2 Some Basic Classes of Infinite Transition Graphs

We consider edge labelled transition graphs of the form $G = (V, (E_a)_{a \in A})$, where V is the set of vertices (sometimes considered as "states"), A the alphabet of edge labels, and E_a the set of a-labelled edges. Such a structure is extended to an automaton by the specification of initial and final vertices. In all cases considered below, where V is given as a regular language over some auxiliary alphabet Γ, the sets I, F of initial and final states will be assumed to be regular languages over Γ. For the classification of infinite automata it will thus suffice to look at their transition structure, i.e. at (the presentation of) the graph $(V, (E_a)_{a \in A})$.

2.1 Pushdown Graphs and Prefix-Recognizable Graphs

A graph $G = (V, (E_a)_{a \in A})$ is called *pushdown graph* if it is the transition graph of the reachable global states of some ϵ-free pushdown automaton. We consider pushdown automata in the format $\mathcal{P} = (Q, A, \Gamma, q_0, Z_0, \Delta)$, where Q is the finite set of control states, A the input alphabet, Γ the stack alphabet, q_0 the initial control state, $Z_0 \in \Gamma$ the initial stack symbol, and $\Delta \subseteq Q \times A \times \Gamma \times \Gamma^* \times Q$ the transition relation. As usual, a global state of the automaton is given by a control state and a stack content, i.e., by a word from $Q\Gamma^*$. The graph $G = (V, (E_a)_{a \in A})$ is now specified as follows:

- V is the set of configurations from $Q\Gamma^*$ which are reachable (via finitely many applications of transitions of Δ) from the initial global state $q_0 Z_0$.
- E_a is the set of all pairs $(p\gamma w, qvw)$ from V for which there is a transition (p, a, γ, v, q) in Δ.

Example 1. Consider the pushdown automaton over $A = \{a, b, c\}$ with the single state q_0, stack alphabet $\Gamma = \{X, Y\}$, and initial stack letter X, which recognizes the language $\{a^i b c^i \mid i \geq 0\}$ in the standard way (accepting by empty stack): By the transition (q_0, a, X, XY, q_0), it generates, starting from X, the stack content XY^i after reading i letters a, then upon reading b it cancels the top X from the stack by the transition $(q_0, b, X, \epsilon, q_0)$, and finally for an input letter c it deletes the top letter Y from the stack, using the transition $(q_0, c, Y, \epsilon, q_0)$. The infinite transition graph looks as follows:

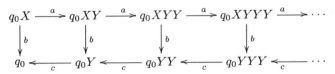

This view of pushdown automata is well-known; see e.g. [KS86, p.242].

By their definition, pushdown graphs are of bounded in- and out-degree. A more general class of graphs, which includes the case of vertices of infinite degree, has been introduced by Caucal [Cau96]. These graphs are introduced in terms of prefix rewriting systems in which "control states" (as they occur in pushdown automata) do no more enter and where a word on the top of the stack (rather than a single letter) may be rewritten. Thus, a rewriting step can be specified by a triple (u_1, a, u_2), describing a transition from a word $u_1 w$ via letter a to the word $u_2 w$. The feature of infinite degree is introduced by allowing generalized rewriting rules of the form $U_1 \to_a U_2$ with regular sets U_1, U_2 of words. Such a rule leads to the (in general infinite) set of rewrite triples (u_1, a, u_2) with $u_1 \in U_1$ and $u_2 \in U_2$. A graph $G = (V, (E_a)_{a \in A})$ is called *prefix-recognizable* if for some finite system $\mathcal{S}$ of such generalized prefix rewriting rules $U_1 \to_a U_2$ over an alphabet Γ, we have

- $V \subseteq \Gamma^*$ is a regular set,
- E_a consists of the pairs $(u_1 w, u_2 w)$ where $u_1 \in U_1$, $u_2 \in U_2$ for some rule $U_1 \to_a U_2$ from $\mathcal{S}$, and $w \in \Gamma^*$.

Example 2. Consider the prefix rewriting system over $\Gamma = \{X\}$ with the rules $\epsilon \to_a X$ and $X^+ \to_b \epsilon$ (here we identify a singleton set with its element). The corresponding graph $G = (V, E_a, E_b)$ over $V = \{X\}^*$ has the edge relations $E_a = \{(X^i, X^{i+1}) \mid i \geq 0\}$ and $E_b = \{(X^i, X^j) \mid i > j\}$; so the graph G is isomorphic to the structure of the natural numbers with the successor relation and the $>$-relation.

It is known that the pushdown graphs coincide with the prefix-recognizable graphs of finite in- and out-degree. A third class of graphs, located strictly between the pushdown graphs and the prefix-recognizable graphs, are the HR-equational graphs (sometimes just called equational or regular graphs), introduced by Courcelle [Cou89]. These graphs are generated by deterministic hyperedge replacement graph-grammars. In [CK01] a restriction of the prefix-recognizable graphs is presented which characterizes the HR-equational graphs.

2.2 Rational and Synchronized Rational Graphs

In many practical applications (of model checking), the restriction to prefix rewriting is too severe to allow the specification of interesting infinite-state systems. Exceptions are structures derived from recursive programs; in this case, prefix rewriting matches the structure of the state space to be described (see e.g. [ES01]).

From the classical literature on transducers ([EM65], [Eil74], [Ber79]), more powerful types of relations are known, in particular the rational and the synchronized rational relations. For simplicity we consider binary relations only, where moreover the alphabets of the two components coincide.

A relation $R \subseteq A^* \times A^*$ is *rational* if it can be defined by a regular expression starting from the atomic expressions $\emptyset$ (denoting the empty relation) and (u, v) for words u, v (denoting the relation $\{(u, v)\}$), by means of the operations union, concatenation (applied componentwise), and iteration of concatenation (Kleene star). An alternative characterization of these relations involves a model of nondeterministic automata which work one-way, but asynchronously, on the two components of an input $(w_1, w_2) \in A^* \times A^*$. To illustrate this model with a simple example, consider the suffix relation $\{(w_1, w_2) \mid w_1 \text{ is a suffix of } w_2\}$: A corresponding automaton would progress with its reading head on the second component w_2 until it guesses that the suffix w_1 starts; this, in turn, can be checked by moving the two reading heads on the two components simultaneously, comparing w_1 letter by letter with the remaining suffix of w_2.

A *rational transition graph* has the form $G = (V, (E_a)_{a \in A})$ where V is a regular set of words over an auxiliary alphabet Γ and where each E_a is a rational relation.

Example 3. ([Mor00]) Given an instance $(\overline{u}, \overline{v}) = ((u_1, \ldots, u_m), (v_1, \ldots, v_m))$ of the Post Correspondence Problem (PCP) over an alphabet Γ we specify a rational graph $G_{(\overline{u},\overline{v})} = (V, E)$. The vertex set V is Γ^*. The edge set E consists of the pairs of words of the form $(u_{i_1} \ldots u_{i_k}, v_{i_1} \ldots v_{i_k})$ where $i_1, \ldots, i_k \in \{1, \ldots, m\}$ and $k \geq 1$. Clearly, an asynchronously progressing nondeterministic automaton can check whether a word pair (w_1, w_2) belongs to E; basically the automaton has to guess successively the indices $i_1, \ldots, i_k$ and at the same time to check whether w_1 starts with u_{i_1} and w_2 starts with v_{i_1}, whether w_1 continues by u_{i_2} and w_2 by v_{i_2}, etc. So the graph $G_{(\overline{u},\overline{v})}$ is rational.

Clearly, in this graph there is an edge from some vertex w back to the same vertex w iff the PCP instance $(\overline{u}, \overline{v})$ has a solution (namely by the word w).

In *synchronized rational relations* a more restricted processing of an input (w_1, w_2) is required: the two components are scanned strictly letter by letter. Thus one can assume that the automaton reads letters from $A \times A$ if $w_1, w_2 \in A^*$. In order to cover the case that w_1, w_2 are of different length, one assumes that the shorter word is prolonged by dummy symbols \$ to achieve equal length. In this way a relation $R \subseteq A^* \times A^*$ induces a language L_R over the alphabet $(A \times A) \cup (A \times (A \cup \{\$\})) \cup ((A \cup \{\$\}) \times A)$. The relation R is called *synchronized rational* if the associated language L_R is regular. From this definition it is immediately clear that the synchronized rational relations share many good properties which are familiar from the theory of regular word languages. For example, one can reduce nondeterministic automata which recognize a word relation synchronously to deterministic automata, a fact which does not apply in the context of rational relations. For a comprehensive study of synchronized rational relations we recommend [FS93]. A graph $(V, (E_a)_{a \in A})$ is called synchronized rational if V is a regular language over an alphabet Γ and each edge relation $E_a \subset \Gamma^* \times \Gamma^*$ is synchronized rational.

Example 4. Consider the transition graph over $\Gamma = \{X_0, X, Y\}$ where there is an a-edge from X_0 to X and from X^i to X^{i+1} (for $i \geq 1$), a b-edge from $X^i Y^j$ to $X^{i-1} Y^{j+1}$ (for $i \geq 1, j \geq 0$), and a c-edge from Y^{i+1} to Y^i (for $i \geq 0$). We obtain the following synchronized rational graph:

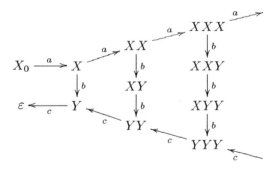

If X_0 is taken as initial and ε as final state, the resulting automaton recognizes the context-sensitive language consisting of the words $a^i b^i c^i$ for $i \geq 1$.

Example 5. The infinite two-dimensional grid $G_2 := (\mathbb{N} \times \mathbb{N}, E_a, E_b)$ (with E_a-edges $((i, j), (i, j + 1))$ and E_b-edges $((i, j), (i + 1, j))$) is a synchronized rational graph: It can be obtained using the words in $X^* Y^*$ as vertices, whence the edge relations become $E_a = \{(X^i Y^j, X^i Y^{j+1}) \mid i, j \geq 0\}$ and $E_b = \{(X^i Y^j, X^{i+1} Y^j) \mid i, j \geq 0\}$, which both are clearly synchronized rational relations.

In the literature, the synchronized rational relations appear under several names; in [EM65] they are called regular, in [Büc60] sequential, and in the context of combinatorial group theory [EC*92] they are named automatic. Here one speaks of automatic groups (see [EC*92]), more generally of "automatic structures" (see [BG00]).

We give another example which illustrates the power of synchronized rational relations:

Example 6. Given a Turing machine M with state set Q and tape alphabet Γ, we consider the graph G_M with vertex set $V_M = \Gamma^* Q \Gamma^*$, considered as the set of M-configurations. By an appropriate treatment of the blank symbol, we can assume that the length difference between two successive M-configurations is at most 1; from this it is easy to see that the relation E_M of word pairs which consist of successive M-configurations is synchronized rational. So the configuration graph $G_M = (V_M, E_M)$ is synchronized rational.

We now compare the four classes of graphs introduced so far.

Theorem 1. *The pushdown graphs, prefix-recognizable graphs, synchronized rational graphs, and rational graphs constitute, in this order, a strictly increasing inclusion chain of graph classes.*

For the proof, we first note that the prefix-recognizable graphs are clearly a generalization of the pushdown graphs and that the rational graphs generalize the synchronized rational ones. To verify that a prefix-recognizable graph is synchronized rational, we first proceed to an isomorphic graph which results by reversing the words under consideration, at the same time using suffix rewriting rules instead of prefix rewriting ones. Given this format of the edge relations, we can verify that it is synchronized rational: Consider a word pair (wu_1, wu_2) which results from the application of a suffix rewriting rule $U_1 \to_a U_2$, with regular U_1, U_2 and $u_1 \in U_1, u_2 \in U_2$. A nondeterministic automaton can easily check this property of the word pair by scanning the two components simultaneously letter by letter, guessing when the common prefix w of the two components is passed, and then verifying (again proceeding letter by letter) that the remainder u_1 of the first component is in U_2 and the remainder u_2 of the second component is in U_2.

The strictness of the inclusions may be seen as follows. The property of having bounded degree separates the pushdown graphs from the prefix-recognizable ones (see Example 2). To distinguish the other graph classes, one may use logical decidability results as presented in the next section. It will be shown there that the monadic second-order theory of a prefix-recognizable graph is decidable, which fails for the synchronized rational graphs, and that the first-order theory of a synchronized rational graph is decidable, which fails for the rational graphs.

2.3 External Characterizations

The introduction of classes of transition graphs as developed above rests on an explicit representation of vertices and edges in terms of words, respectively pairs of words. There is an alternative way to characterize transition systems by structural properties. In the terminology of Caucal, this amounts to proceed from an "internal" to an "external" presentation. We present here three of these characterization results, due to Muller and Schupp (in their pioneering paper [MS85]) and to Caucal and his students ([Cau96], [Mor00], [Ris01]).

Let $G = (V, (E_a)_{a \in A})$ be a graph of bounded degree and with designated "origin" vertex v_0. Let V_n be the set of vertices whose distance to v_0 is at most n (via paths formed from edges as well as reversed edges). Define G_n to be the subgraph of G induced by the vertex set $V \setminus V_n$, calling its vertices in $V_{n+1} \setminus V_n$ the "boundary vertices". The *ends* of G are the connected components (using edges in both directions) of the graphs G_n with $n \geq 0$. In [MS85], Muller and Schupp have established a beautiful characterization of pushdown graphs in terms of the isomorphism types of their ends (where an end isomorphism is assumed to respect the vertex property of being a boundary vertex):

Theorem 2. *A transition graph G of bounded degree is a pushdown graph iff the number of distinct isomorphism types of its ends is finite.*

A simple example is the full binary tree, which generates only a single isomorphism type by its ends (namely, the type of the full binary tree itself). In the subsequent characterizations, different structural properties of graphs are defined by different kinds of "reductions" to the structure of the binary tree. We consider the binary tree as the structure $T_2 = (\{0,1\}^*, E_0, E_1, \overline{E_0}, \overline{E_1})$, where E_0 (E_1) is the left (right) successor relation, while the edges of $\overline{E_0}, \overline{E_1}$ are the reverse ones of E_0, E_1, respectively. A finite path through the tree (which may move upwards as well as downwards, starting from some given vertex) can thus be described by a word over the alphabet $\{0, 1, \overline{0}, \overline{1}\}$. For example, starting from the vertex 10011, the word $\overline{1}\,\overline{1}\,0\,1$ describes the path leading up to 100 and then down to 10001.

We now can state Caucal's Theorem ([Cau96]) which says that the prefix-recognizable graphs can be obtained from the binary tree by a kind of regular modification, the edge sets E_a being described by regular sets of such path descriptions. (Thus prefix-recognizable graphs are rather "tree-like", which explains their limited use in modelling general transition systems.)

Theorem 3. *A transition graph is prefix-recognizable iff it can be obtained from the binary tree by an inverse regular substitution followed by a regular restriction.*

Let us explain the condition on the right-hand side, i.e. how it defines a graph $G = (V, (E_a)_{a \in A})$: By the regular restriction (to a regular set $V \subseteq \{0, 1\}^*$ of vertices of the binary tree), we are provided with the vertex set. The regular substitution maps each letter $a \in A$ to a regular language $L_a \subseteq \{0, 1, \overline{0}, \overline{1}\}^*$. The inverse application of this substitution yields the edges as follows: There is an a-edge from u to v iff there is a word $w \in L_a$ describing a path from u to v.

Another way of stating the theorem is to say that the prefix-recognizable graphs are those which can be obtained as "regular interpretations" in the binary tree. As shown in [Bar97] and [Blu00], this coincides with a logical reduction, namely via monadic second-order definable interpretations.

A slight extension of the notion of regular substitution in the above theorem yields a corresponding characterization of the rational graphs, established by Morvan [Mor00]. Here one considers "special linear languages" over the alphabet $\{0, 1, \overline{0}, \overline{1}\}$, which are defined to be generated by linear grammars with productions $A \to \overline{u}Bv$, $A \to \epsilon$, assuming that $\overline{u} \in \{\overline{0}, \overline{1}\}^*$ and $v \in \{0, 1\}^*$. By a "special linear substitution", each letter a is associated to a special linear language $L_a \subseteq \{0, 1, \overline{0}, \overline{1}\}^*$.

Theorem 4. *A transition graph is rational iff it can be obtained from the binary tree by an inverse special linear substitution followed by a regular restriction.*

Let us illustrate the theorem for the example of the $(\mathbb{N} \times \mathbb{N})$-grid. As regular representation of the vertex set we take the "comb-structure" 0^*1^* consisting of the leftmost branch *LB* of the binary tree (of the vertices in 0^*) together with all rightmost branches starting on *LB*. Then the a-edges simply lead from a vertex of this comb-structure to its right successor, and hence we can set $L_a = \{1\}$. The unique b-edge from a vertex $u \in 0^*1^*$, say $0^m 1^n$, of the binary tree is obtained by going back from u to *LB* (via n $\bar{1}$-edges), proceeding from the vertex 0^m reached on *LB* one step to the left, and from there proceeding down the rightmost path again n times. So we have $L_b = \{\bar{1}^i 01^i \mid i \geq 0\}$, which is a special linear language.

As shown by Rispal [Ris01], a suitable restriction of special linear substitutions yields also a characterization of the synchronized rational graphs.

2.4 Recognized Languages

Since the pushdown graphs are derived from pushdown automata it may not seem surprising that their use as automata (with regular sets of initial and final states) allows the recognition of precisely the context-free languages. For the proof, the reader should consult the fundamental paper [MS85]. Caucal [Cau96] has shown that the expressive power (regarding recognition of languages) of the automata does not increase when proceeding to prefix-recognizable graphs:

Theorem 5. *A language L is context-free iff L is recognized by a pushdown graph (with regular sets of initial and final states) iff L is recognized by a prefix-recognizable graph (with regular sets of initial and final states).*

Recently, this track of research was continued by surprising results regarding the rational and the synchronized rational graphs. Morvan and Stirling prove in [MoS01] that the rational graphs allow to recognize precisely the context-sensitive languages, and Rispal [Ris01] shows that the synchronized rational graphs reach the same expressive power:

Theorem 6. *A language L is context-sensitive iff L is recognized by a synchronized rational graph (with regular sets of initial and final states) iff L is recognized by a rational graph (with regular sets of initial and final states).*

We give some comments on the claim that rational transition graphs can only recognize context-sensitive languages. For this, a Turing machine has to check (with linear space in the length of the input) whether in the given rational transition graph G there is an path from an initial to a final vertex labelled by the input word. The Turing machine guesses this path nondeterministically, reusing the same tape segment for the different words that constitute the path through G. The main problem is the fact that the length of the individual words visited on the path through G may increase exponentially in the length of the path (which is the length of the input word). This may happen, for example, if the transducer defining the edge relation of G has transitions that involve rewriting a letter X by XX; this would allow to double the length of the word in each transition by an edge of G. The solution, which cannot be described in detail here, is to check the correctness of the word rewriting online while traversing the two considered words simultaneously; this can be done in constant space.

The converse direction is still harder; in [MoS01] strong normal forms of context-sensitive grammars due to Penttonen ([Pen74]) are applied.

It seems that Theorem 6 points to a deeper significance of the class of context-sensitive languages (which so far played its main role as a rather special space complexity class). In particular, it turns out that the context-sensitive languages can be recognized in a one-way read-only mode by (rational or synchronized rational) infinite automata, in contrast to the standard description by two-way Turing machines with read- and write-moves. It remains to be analyzed whether nontrivial facts on the context-sensitive languages (like the closure under complementation) have new (and maybe even simpler?) proofs in the framework of infinite automata. An important open question asks whether rational (or synchronized rational) graphs with a deterministic transition structure suffice for the recognition of the context-sensitive languages.

Caucal [Cau01] has extended the two theorems above, giving a class of infinite automata that accept precisely the recursively enumerable languages.

In process theory, one compares the expressive power of transition systems using sharper distinctions than by their ability to recognize languages: In this context, the semantics of transition systems (with designated initial state) is given by their bisimulation classes. With respect to bisimulation equivalence, the pushdown graphs are no more equivalent to the prefix-recognizable graphs (there are prefix-recognizable graphs which are not bisimilar to any pusdown graph); a corresponding proper inclusion with respect to bisimulation equivalence applies to the synchronized rational graphs in comparison with the rational graphs.

3 Algorithmic Problems

The infinite transition graphs considered above are all finitely presented, namely by finite automata defining their state properties, and by finite word rewriting systems or transducers defining their transition relations. Thus, decision problems concerning these graphs have a clear algorithmic meaning: as input one takes the finite automata which serve as presentations of transition graphs.

In this section we touch only a few (however central) decision problems on infinite transition systems. First we shall discuss the "reachability problem" which asks for the existence of paths with given start and end vertices. Secondly, we treat the "model-checking problem" regarding first-order and monadic second-order formulas, or, in a more classical terminology, the question of decidability of the first-order and the monadic second-order theory of a transition graph. (Here we assume that the reader is familiar with the fundamentals of first-order and monadic second-order logic, see e.g. [Tho97].)

3.1 Problems on Pushdown Graphs and Prefix-Recognizable Graphs

A main result of Büchi's paper [Büc65] on regular canonical systems (i.e., prefix rewriting systems) is the following theorem: The words which are generated from a fixed word by a prefix rewriting system form a regular language. As an application one obtains the well-known fact that the reachable global states of a pushdown automaton constitute a regular set. In the theory of program verification, the problem usually arises in a converse (but essentially equivalent) form: Here one starts with a set T of vertices as "target set", and the problem is to find those vertices (i.e., words) from which a vertex of T is reachable by a finite number of pushdown transition steps. More precisely, one deals with regular target sets. Given a pushdown graph G, the aim is to find from a finite automaton recognizing the vertex set T another finite automaton recognizing the set $pre^*(T)$ of those vertices from which there is a path through G to T. (The notation indicates that we are dealing with the transitive closure of the edge predecessor relation, starting with T.)

Theorem 7. *Given a pushdown graph $G = (V, E)$ (with $V \subseteq Q\Gamma^*$, where Q is the set of control states and Γ the stack alphabet of the underlying pushdown automaton), and given a finite automaton recognizing a set $T \subseteq Q\Gamma^*$, one can compute a finite automaton recognizing $pre^*(T)$. In particular, it can be decided for any vertex v whether it belongs to $pre^*(T)$ or not.*

The transformation of a given automaton $\mathcal{A}$ which recognizes T into the desired automaton $\mathcal{A}'$ recognizing $pre^*(T)$ works by a simple process of "saturation", which involves adding more and more transitions but leaves the set of states unmodified. It is convenient to work with Q as the set of initial states of $\mathcal{A}$; so a stack content qw of the pushdown automaton is scanned by $\mathcal{A}$ starting from state q and then processing the letters of w. The saturation procedure is based on the following idea: Suppose a pushdown transition allows to rewrite the stack content $p\gamma w$ into qvw, and that the latter one is accepted by $\mathcal{A}$. Then also the stack content $p\gamma w$ should be accepted. If $\mathcal{A}$ accepts qvw by a run starting in state q and reaching, say, state r after processing v, we enable the acceptance of $p\gamma w$ by adding a direct transition from p via γ to r. (For a detailed proof and complexity analysis of the construction see [EH*00]).

Let us embed the reachability problem into a more general question, that of model-checking monadic second-order properties. For this, we observe that reachability of a set T from a given vertex v can be expressed in the language of monadic second-order logic (MSO-logic). We indicate a corresponding formula $\varphi(x, Y)$ where x is a first-order variable (to be interpreted by single vertices v) and Y is a second-order variable (to be interpreted by target sets T). The formula $\varphi(x, Y)$ just has to express that x is in the smallest set containing Y and closed under taking "predecessor vertices". To express this more precisely, we assume a single edge relation E in the graph in question (which can be considered as the union of all edge relations E_a as considered before). The desired formula $\varphi(x, Y)$ says the following: "x belongs to each set Z containing Y, such that for any two vertices z, z' with $z' \in Z$ and $(z, z') \in E$ we have also $z \in Z$".

The MSO-definability of the reachability relation hints at a much more powerful decidability result than the previous theorem, which moreover extends to the class of prefix-recognizable graphs (cf. [Cau96]):

Theorem 8. *The monadic second-order theory of a prefix-recognizable graph is decidable.*

The proof works by a reduction to Rabin's Tree Theorem [Rab69] which states that the monadic second-order theory of the binary tree is decidable. The connection is provided by Theorem 3 above, which gives us a regular interpretation of any given prefix-recognizable graph G in the binary tree T_2. The interpretation of a graph G in T_2 by a regular substitution and a regular restriction is easily converted into an MSO-interpretation; this in turn allows to tranform any MSO-sentence φ into a new MSO-sentence φ' such that $G \models \varphi$ iff $T_2 \models \varphi'$. This provides a decision procedure for the MSO-theory of G, invoking the decidability of the MSO-theory of T_2.

3.2 Problems on Rational and Synchronized Rational Graphs

The decidability results above fail when we consider the much more extended class of synchronized rational graphs instead of the pushdown graphs or prefix-recognizable graphs.

Let us first consider the reachability problem. Here we use the fact that for a universal Turing machine M, say with a normalized halting configuration c_0, it is undecidable whether from a given start configuration c we can reach c_0. Hence, we know that for the synchronized rational graph G_M as introduced in Example 6 above, the reachability problem is undecidable.

Theorem 9. *There is a synchronized rational graph G such that the reachability problem over G (given two vertices, decide whether there is path from the first to the second) is undecidable.*

This result is one of the main obstacles in developing a framework for model checking over infinite systems: The synchronized rational graphs are a very natural framework for modelling interesting infinite systems, and most applications of model checking involve some kind of reachability analysis. Current research tries to find good restrictions of the class of synchronized rational graphs where the reachability problem is solvable.

Let us also look at a more ambitious problem than reachability: decidability of the monadic second-order theory of a given graph. Here we get undecidability for synchronized rational graphs with much simpler transition structure than that of the graphs G_M of the previous theorem, for example for the infinite two-dimensional grid (shown to be synchronized rational in Example 5). Note that the reachability problem over the grid is decidable.

Theorem 10. *The monadic second-order theory of the infinite two-dimensional grid G_2 is undecidable.*

The idea is to code the work of Turing machines in a more uniform way than in the previous result. Instead of coding a Turing machine configuration by a single vertex and capturing the Turing transitions directly by the edge relation, we now use a whole row of the grid for coding a configuration (by an appropriate coloring of its vertices with tape symbols and a Turing machine state). A Turing machine computation is thus represented by a sequence of colored rows, i.e., by a coloring of the grid. (We can assume that even a halting computation generates a coloring of the whole grid, by repeating the final configuration ad infinitum.) In this view, the horizontal edge relation is used to progress in space, while the vertical one allows to progress in time. A given Turing machine M (say with m states and n tape symbols) halts on the empty tape iff there is a coloring of the grid with $m + n$ colors which

- represents the initial configuration (on the empty tape) in the first row,
- respects the transition table of M between any two successive rows,
- contains a vertex which is colored by a halting state.

Such a coloring corresponds to a partition of the vertex set $\mathbb{N} \times \mathbb{N}$ of the grid into $m + n$ sets. One can express the existence of the coloring by saying "there exist sets $X_1, \ldots, X_{m+n}$ which define a partition and satisfy the requirements of the three items above". In this way one obtains effectively an MSO-sentence φ_M such that

$$M \text{ halts on the empty tape iff } G_2 \models \varphi_M.$$

When we restrict the logical framework to first-order logic (FO-logic), then we are back to decidability:

Theorem 11. *The first-order theory of a synchronized rational graph is decidable.*

The proof of this theorem is based on an idea which was first proposed by Büchi in [Büc60], as a method to show the decidability of Presburger arithmetic (the first-order theory of the structure $(\mathbb{N}, +)$). In both cases, Presburger arithmetic and the FO-theory of a synchronized rational graph, one works with representations of the structure's elements (numbers, respectively vertices) by words. For Presburger arithmetic one uses the representation by reversed binary numbers. An n-ary relation R is then described by a set of n-tuples of words. As in the definition of synchronized rational graphs, one codes R by a language L_R: For this, any n-tuple $(w_1, \ldots, w_n)$ of words, say over the alphabet Γ, is presented as a single word over the alphabet $(\Gamma \cup \{\$\})^n$, reading the words $w_1, \ldots, w_n$ simultaneously letter by letter and using the symbol $\$$ for filling up words at the end to achieve equal length. The words resulting from the n-tuples in R form the language L_R. The key step for the decidability proof is the following lemma: *If R is a first-order definable relation in a synchronized rational graph, then the language L_R is regular.* Given a synchronized graph $G = (V, (E_a)_{a \in A})$, this is easily proved by induction over the construction of first-order formulas: The case of relations defined by the atomic formulas $x = y$ and $(x, y) \in E_a$ is trivial by the assumption that G is synchronized rational. For the induction steps, regarding the connectives $\neg$, $\vee$, and the existential first-order quantifier, one uses the closure of the automata defined languages L_R under boolean operations and projection.

Let us finally show that the decidability of the first-order theory does not extend to rational graphs. We proceed in two steps. The first deals with a weaker result, concerning a uniform statement for the class of rational graphs, the second provides a single rational graph with an undecidable first-order theory.

Theorem 12. *([Mor00]) There is no algorithm which, given a presentation of a rational graph G and a first-order sentence φ, decides whether $G \models \varphi$.*

The proof is easy using the graphs $G_{(\overline{u}, \overline{v})}$ as introduced in Example 3 above, where $(\overline{u}, \overline{v})$ is an instance of the Post Correspondence Problem. As explained in Example 3, $(\overline{u}, \overline{v})$ has a solution iff $G_{(\overline{u}, \overline{v})} \models \exists x \, (x, x) \in E$. Thus a uniform algorithm to decide the first-order theory of any given rational graph would solve the Post Correspondence Problem.

In order to obtain a single rational graph with an undecidable first-order theory, we refine the construction. As for Theorem 9, we use a universal Turing machine M and the encoding of its undecidable halting problem (for different input words x) into a family of instances of the Post Correspondence Problem. For simplicity of exposition, we refer here to the standard construction of the undecidability of the PCP as one finds it in textbooks (see [HU79, Section 8.5]): A Turing machine M with input word x is converted into a PCP-instance $((u_1, \ldots, u_m), (v_1, \ldots, v_m))$ over an alphabet A whose letters are the states and tape letters of M and a symbol $\#$ (for the separation between M-configurations in M-computations). If the input word is $x = a_1 \ldots a_n$, then u_1 is set to be the initial configuration word $c(x) := \# q_0 a_1 \ldots a_n$ of M; furthermore we have always $v_1 = \#$, and $u_2, \ldots, u_m, v_2, \ldots, v_m$ only depend on M. Then the standard construction (of [HU79]) ensures the following:

M halts on input x

iff the PCP-instance $((c(x), u_2, \ldots u_m), (\#, v_2, \ldots, v_m))$ has a special solution.

Here a special solution is given by an index sequence $(i_2, \ldots, i_k)$ such that $c(x)u_{i_2} \ldots u_{i_k} = \# v_{i_2} \ldots v_{i_k}$.

Let G be the graph as defined from these PCP-instances in analogy to Example 3: The vertices are the words over A, and we have a single edge relation E with $(w_1, w_2) \in E$ iff there are indices $i_2, \ldots, i_k$ and a word x such that $w_1 = c(x)u_{i_2} \ldots u_{i_k}$ and $w_2 = \# v_{i_2} \ldots v_{i_k}$. Clearly G is rational, and we have an edge from a word w back to itself if it is induced by a special solution of some PCP-instance $((c(x), u_2, \ldots, u_m), (\#, v_2, \ldots, v_m))$.

In order to address the input words x explicitly in the graph, we add further vertices and edge relations E_a for $a \in A$. A $c(x)$-labelled path via the new vertices will lead to a vertex of G with prefix $c(x)$; if the latter vertex has an edge back to itself then a special solution for the the PCP-instance $((c(x), u_2, \ldots u_m), (\#, v_2, \ldots, v_m))$ can be inferred. The new vertices are words over a copy $\underline{A}$ of the alphabet A (consisting of the underlined versions of the A-letters). For any word $c(x)$ we shall add the vertices which arise from the underlined versions of the proper prefixes of $c(x)$, and we introduce an E_a-edge from any such underlined word w to $w\underline{a}$ (including the case $w = \epsilon$). There are also edges to non-underlined words: We have an E_a-edge from the underlined version of w to any non-underlined word which has wa as a prefix. Call the complete resulting graph G'. It is easy to see that G' is rational.

By construction of G', the PCP-instance $((c(x), u_2, \ldots u_m), (\#, v_2, \ldots, v_m))$ has a special solution iff

in G' there is a path, labelled with the word $c(x)$, from the vertex ϵ to a vertex which has an edge back to itself.

Note that the vertex ϵ is definable as the only one with outgoing E_a-edges but without an ingoing edge of one of the edge relations E_a. Thus the above condition is formalizable by a first-order sentence φ_x, using variables for the $|c(x)| + 1$ many vertices of the desired path. Altogether we obtain that the Turing machine M halts on input x iff $G' \models \varphi_x$. So we conclude the following:

Theorem 13. *There is a rational graph with an undecidable first-order theory.*

4 Concluding Remarks

The theory of infinite automata is still in its beginnings. We mention two interesting tracks of research originating in the results outlined above.

For the purposes of model checking, it would be important to obtain classes of transition graphs which are more general (or at least more flexible for modelling concrete systems) than prefix-recognizable graphs but still have good algorithmic properties (e.g. a decidable reachability problem). For example, one can analyze further options of word rewriting for the definition of transition relations (see for example [Bou01]). Or one can pursue the approach developed by C. Löding in [Löd01]: He represents vertices of transition graphs by finite trees (rather than words) and uses ground rewriting as the mechanism to define the transitions. In this way, interesting graphs are obtained which are not prefix-recognizable (the infinite two-dimensional grid being an example), but where the reachability problem remains decidable.

Another area is the study of language classes (beyond the regular languages) in terms of infinite automata. For example, the relation between nondeterministic and deterministic transition graphs as acceptors is not yet well understood, as well as their relation to the corresponding classical counterparts, pushdown automata and Turing machines. For the determinization problem, it may be useful to analyze the connection with uniformization results on rational relations. Also one may ask whether there are further language classes (besides the context-free and context-sensitive languages) which have appealing characterizations by finitely presented infinite automata. Finally, it would be interesting to see how much of the known general results on context-free, context-sensitive and other languages can be elegantly proved using the framework of infinite automata (instead of the standard approach by grammars, pushdown automata, and Turing machines).

Acknowledgment. Thanks a due to Didier Caucal, Thierry Cachat, Christof Löding, and Philipp Rohde for comments on a preliminary version of this paper.

References

[Bar97]   K. Barthelmann, On equational simple graphs, Tech. Rep. 9, Universität Mainz, Institut für Informatik 1997.

[Blu00]   A. Blumensath, Prefix-recognisable graphs and monadic second-order logic, Aachener Informatik-Berichte 2001-6, Dep. Of Computer Science, RWTH Aachen 2001.

[BG00]   A. Blumensath, E. Grädel, Automatic structures, Proc. 15th IEEE Symp. Logic in Computer Science (2000), 51-62.

[Ber79]   J. Berstel, *Transductions and Context-Free Languages*, Teubner, Stuttgart 1979.

[BE97]   O. Burkart, J. Esparza, More infinite results, *Bull. EATCS* **62** (1997), 138-159.

[Bou01]   A. Bouajjani, Languages, rewriting systems, and verification of infinite-state systems, in F. Orejas et al. (Eds.), Proc. ICALP 2001, LNCS 2076 (2001), Springer-Verlag 2001, pp. 24-39.

[Büc60]   J. R. Büchi, Weak second-order arithmetic and finite automata, *Z. Math. Logik Grundl. Math.* **6** (1960), 66-92.

[Büc65]   J. R. Büchi, Regular canonical systems, *Arch. Math. Logik u. Grundlagenforschung* **6** (1964), 91-111.

[Cau96]   D. Caucal, On transition graphs having a decidable monadic theory, in: F. Meyer auf der Heide, B. Monien (Eds.), Proc. 23rd ICALP, LNCS 1099 (1996), Springer-Verlag 1996, pp. 194-205.

[Cau01]   D. Caucal, On transition graphs of Turing machines, in: M. Margenstern, Y. Rogozhin (Eds.), Proc. Conf. on Machines, Computation and Universality, LNCS 2055 (2001), Springer-Verlag 2001, pp. 177-189.

[CK01]   D. Caucal, T. Knapik, An internal presentation of regular graphs by prefix-recognizable graphs, *Theory of Computing Systems* **34** (2001), 299-336.

[CGP99]   E.M. Clarke, O. Grumberg, D. Peled, *Model Checking*, MIT Press 1999.

[Cou89]   B. Courcelle, The monadic second-order logic of graphs II: Infinite graphs of bounded width, *Math. Systems Theory* **21** (1989), 187-221.

[EC*92]   D.B.A. Epstein, J.W. Cannon, D.F. Holt, S. Levy, M.S. Paterson, W. Thurston, *Word Processing in Groups*, Jones and Barlett 1992.

[Eil74]   S. Eilenberg, *Automata, Languages, and Machines*, Vol. A, Academic Press 1974.

[EM65]   C.C. Elgot, J. Mezei, On relations defined by generalized finite automata, *IBM J. Res. Dev.* **9** (1965), 47-68.

[EH*00]   J. Esparza, D. Hansel, P. Rossmanith, S. Schwoon, Efficient algorithms for model checking pushdown systems, Proc. CAV 2000, LNCS 1855 (2000), Springer-Verlag 2000, pp. 232-247.

[ES01]   J. Esparza, S. Schwoon, A BDD-based model checker for recursive programs, in: G. Berry et al. (Eds.), CAV 2001, LNCS 2102, Springer-Verlag 2001, pp. 324-336.

[FS93]   C. Frougny, J. Sakarovitch, Synchronized rational relations of finite and infinite words, *Theor. Comput. Sci.* **108** (1993), 45-82.

[HU79]   J.E. Hopcroft, J.D. Ullman, *Introduction to Automata Theory, Languages, and Computation*, Addison-Wesley 1979.

[KS86]   W. Kuich, A. Salomaa, *Semirings, Automata, Languages*, Springer 1986.

[Löd01]   C. Löding, Model-checking infinite systems generated by ground tree rewriting, to appear in Proc. FOSSACS 2002, LNCS, Springer-Verlag 2002.

[McM93]   K. L. McMillan, *Symbolic Model Checking*, Kluwer Academic Publishers 1993.

[Mor00]   C. Morvan, On rational graphs, in: J. Tiuryn (Ed.), Proc. FOSSACS 2000, LNCS 1784 (2000), Springer-Verlag, pp. 252-266.

[MoS01]   C. Morvan, C. Stirling, Rational graphs trace context-sensitive languages, in: A. Pultr, J. Sgall (Eds.), Proc. MFCS 2001, LNCS 2136, Springer-Verlag 2001, pp. 548-559

[MS85]   D. Muller, P. Schupp, The theory of ends, pushdown automata, and second-order logic, *Theor. Comput. Sci.* **37** (1985), 51-75.

[Pen74]   M. Penttonen, One-sided and two-sided context in formal grammars, *Inform. Contr.* **25** (1974), 371-392.

[Rab69]   M.O. Rabin, Decidability of second-order theories and automata on infinite trees, *Trans. Amer. Math. Soc.* **141** (1969), 1-35.

[Ris01]   C. Rispal, The synchronized graphs trace the context-sensitive languages, manuscript, IRISA, Rennes 2001.

[Tho97]   W. Thomas, Languages, automata, and logic, in *Handbook of Formal Language Theory* (G. Rozenberg, A. Salomaa, Eds.), Vol 3, Springer-Verlag, Berlin 1997, pp. 389-455.

The Power of One-Letter Rational Languages

Thierry Cachat

Lehrstuhl für Informatik VII, RWTH, D 52056 Aachen[**]
cachat@informatik.rwth-aachen.de

Abstract. For any language L, let $\mathrm{pow}(L) = \{u^j \mid j \geqslant 0,\ u \in L\}$ be the set of powers of elements of L. Given a rational language L (over a finite alphabet), we study the question, posed in [3], whether $\mathrm{pow}(L)$ *is rational or not*. While leaving open the problem in general, we provide an algorithmic solution for the case of one-letter alphabets. This case is still non trivial; our solution is based on Dirichlet's result that for two relatively prime numbers, their associated arithmetic progression contains infinitely many primes.

1 Introduction

A great amount of work was done in studying rational languages and finite automata since the origins of the theory in 1956 (see [8] and [11]). This theory is well developed and a lot of results are applicable via the effective properties of rational languages (see [7] or [10] for instance). The problem studied in this paper is easy to enunciate: We note

$$\mathrm{pow}(L) = \{u^j \mid j \geqslant 0,\ u \in L\} = \bigcup_{u \in L} u^* \subseteq L^*,$$

and consider the decision problem whether this language is rational if L is rational. More precisely, we are searching under which condition on L, $\mathrm{pow}(L)$ remains rational. This problem is far from trivial, it was first mentioned and left open in [3]. As a simple example, consider the language $L \in Rat(\{a, b\}^*)$ defined by the rational expression ab^+. Its power $pow(L) = \bigcup_{k > 0} (ab^k)^*$ is context-sensitive but it is not context-free and of course not rational.

We recall the background of [3]. The rational ω-languages are characterized by their ultimately periodic words, of the form uv^ω. Of course $uv^\omega = u(v^k)^\omega = uv^l(v^k)^\omega$, $l, k \geqslant 0$. Given $M \subseteq A^\omega$ one defines its "periods":

$$\mathrm{per}(M) = \{v \in A^+ \mid \exists u \in A^*,\ uv^\omega \in M\} \subseteq A^*\ .$$

An important result is that if M is rational, then $\mathrm{per}(M)$ is rational. Note that $\mathrm{pow}(\mathrm{per}(M)) = \mathrm{per}(M)$. We can also consider a partial representation $L \in \mathrm{Rat}(A^*)$ such that $\mathrm{pow}(L) = \mathrm{per}(M)$, and ask whether $\mathrm{pow}(L) \in \mathrm{Rat}(A^*)$.

For other classes of the Chomsky hierarchy than the regular languages, the question of the stability by the operation of power is easy and has a positive answer. We note the following facts:

[**] This work was done at IRISA, Rennes, France in 2000.

- The power of a recursively enumerable language $(u_i)_{i \geqslant 0}$ is recursively enumerable too: one can enumerate $\left((u_i)^j\right)_{i,j \geqslant 0}$ the same way as $\mathbb{N}^2$.
- The power of a recursive language L is also recursive: a Turing Machine can, in a finite time, look for all decompositions of a word $u \in A^*$, in the form of a power $u = v^k$, and test if $v \in L$ (on another part of the band).
- We have the same result for a context-sensitive language.

In this paper, we restrict ourselves essentially to the special case of a one-letter alphabet (say $A = \{a\}$), and give an effective solution in Theorem ??. With this restriction, the languages can be easily represented as sets of integers, and we can use some elementary facts of arithmetic. In Section 3, we expose our main theorem and prove it. After that we mention possible future work in the conclusion.

2 Rational Sets of Integers

In this section we recall some classical results to handle rational sets of integers and give examples.

We first recall the notion of rational languages over a *finite* alphabet A (see [10]): one denote $\mathrm{Rat}(A^*)$ the smallest set of languages which contains the finite languages and is closed under union ($\cup$), concatenation (.), and star (*) (the reflexive transitive closure of the concatenation). Since [8], it is known that $\mathrm{Rat}(A^*)$ is also the set of languages which are recognizable by finite automata, and is closed under complementation and intersection. The set of rational languages has very strong closure properties (under rational substitution, inverse substitution, residual, ...).

From now on we are considering a one-letter alphabet $A = \{a\}$ and remark that each word a^k of a^* is characterized by its length $k \in \mathbb{N}$. We identify a given $L \subseteq a^*$ with the set $\{k \geqslant 0 \mid a^k \in L\} \subseteq \mathbb{N}$. The structure $(a^*, \cup, \cdot, ^*)$ is isomorphic to $(\mathbb{N}, \cup, +, ^\circledast)$, where $\cup$ is the usual union, $M + N = \{m + n \mid m \in M, n \in N\}$, and $M^\circledast = \{0\} \cup M \cup (M + M) \cup \cdots$. The set of rational languages of integers is denoted by $\mathrm{Rat}(\mathbb{N})$:
$$M \in \mathrm{Rat}(\mathbb{N}) \iff \{a^k \mid k \in M\} \in \mathrm{Rat}(a^*) \,.$$

Note that over a one-letter alphabet, the context-free languages are exactly the rational languages. According to [Per 90, p.36], $\mathrm{Rat}(\mathbb{N})$ is the set of 1-recognizable languages of integers: their representation in basis 1 is rational.

The product of sets of integers is defined with the usual multiplication:
$$\forall M, N \subseteq \mathbb{N}, \quad M.N = \{m.n \mid m \in M, n \in N\} \,.$$

Note that the operations $+$ and . on sets of integers are still associative and commutative. We might omit the symbol (.) for the multiplication. For any $L \subseteq a^*$, $\mathrm{pow}(L)$ is isomorphic to $M\mathbb{N}$, with $M = \{k \geqslant 0 \mid a^k \in L\}$, just because $(a^k)^j = a^{kj}$ for $j \geqslant 0$. That is to say, multiplication over the integers corresponds to the power operation over words. Now we can formulate our main problem in terms of integers: given $M \in \mathrm{Rat}(\mathbb{N})$ we want to determine whether
$$M\mathbb{N} \in \mathrm{Rat}(\mathbb{N}) \,.$$

The previous remarks allow to characterize $\mathrm{Rat}(\mathbb{N})$ directly in an intuitive way.

Lemma 1 (rational [2] [4]) *For any $M \subseteq \mathbb{N}$, $M \in \mathrm{Rat}(\mathbb{N})$ iff M is ultimately periodic:*

$$\exists n > 0,\ \exists m \in \mathbb{N},\ I \subseteq [0, m),\ P \subseteq [m, m+n),\ M = I \cup (P + n\mathbb{N})\ .$$

By convention $n\mathbb{N} = \{n\}\mathbb{N} = \{nk \mid k \in \mathbb{N}\}$, and $[a, b)$ is the segment of integers between a and b (a included, b excluded). In the situation of this lemma, we will say that n is eligible as a period for M, and (n, m, I, P) is *a representation* of the rational language M. Let $[x]_n = (x + n\mathbb{Z}) \cap \mathbb{N}$ be the class of x modulo n in $\mathbb{N}$. The rationality of M is also equivalent to the following condition:

$$\exists n > 0,\ \forall x \in [0, n),\ [x]_n \cap M \text{ is finite or co-finite.}$$

Example 2 *The language $0 \cup (5 + 2\mathbb{N})$ is rational, represented by $(2, 5, \{0\}, \{5\})$.*

Note that for languages of integers, the well known pumping lemma becomes an equivalence:

$$L \in \mathrm{Rat}(\mathbb{N}) \Leftrightarrow \exists b \in \mathbb{N}, \forall l \in L, l \geqslant b, \exists\, 0 < r \leqslant b,\ l + r\mathbb{N} \subseteq L$$
$$\Leftrightarrow \exists b \in \mathbb{N}, \exists\, 0 < r \leqslant b, \forall l \in L, l \geqslant b,\ l + r\mathbb{N} \subseteq L\ .$$

This periodicity indicates that the "density" of a rational set is constant after a certain point, but we cannot always calculate it easily. The following example shows that it is not a sufficient condition of rationality.

Example 3 *Consider $M = \big((1 + 2\mathbb{N}) \cup \{2^k \mid k \geqslant 1\}\big) \setminus \{2^k + 1 \mid k \geqslant 1\}$.*
In each segment $[2m, 2m + 2)$ there is exactly one element of M, so we can say that the "density" is constant. But M is not ultimately periodic: its period would be greater than 2^k, for each $k \geqslant 1$.

We still have a proposition that allows to prove (by contraposition) that some sets are not rational (as used in Theorem **??**). We note $|M|$ the cardinal of a (finite) set $M \subseteq \mathbb{N}$.

Proposition 1 *For all $L \in \mathrm{Rat}(\mathbb{N})$,*

$$\lim_{t \to \infty} \frac{|L \cap [0, t)|}{t} \text{ is well defined, and if this limit is 0, then } L \text{ is finite.}$$

Consider some simple examples of the product of a rational set by $\mathbb{N}$:

$$\mathbb{N}\mathbb{N} = (1 + \mathbb{N})\mathbb{N} = (1 + 2\mathbb{N})\mathbb{N} = \mathbb{N} \in \mathrm{Rat}(\mathbb{N}),\ \text{because } \mathbb{N}\mathbb{N} \supseteq 1.\mathbb{N} = \mathbb{N}$$
$$(2 + 2\mathbb{N})\mathbb{N} = 2(1 + \mathbb{N})\mathbb{N} = 2\mathbb{N} \quad \in \mathrm{Rat}(\mathbb{N}),\ \text{but}$$
$$(3 + 2\mathbb{N})\mathbb{N} = \mathbb{N} \setminus 2^{\mathbb{N}} \qquad\qquad \notin \mathrm{Rat}(\mathbb{N}).$$

Here $2^{\mathbb{N}}$ is $\{2^k \mid k \in \mathbb{N}\}$. Indeed the set $(3 + 2\mathbb{N})\mathbb{N}$ consists of products of odd numbers greater than 1 and integers, so they can not be a power of 2; conversely, an integer that is not a power of 2 has at least one odd prime factor. And the language $2^{\mathbb{N}}$ is not (ultimately) periodic.

To compute $M\mathbb{N}$ for a given M, a natural approach is to calculate $M \cup 2M \cup 3M \cdots$. Possibly that after some steps the union remains the same. In this case $M\mathbb{N}$ is rational as a finite union of rationals. In the other case this may not be true. For example $M = 2 \cup (3 + 2\mathbb{N})$ has the property that $M\mathbb{N} = \mathbb{N} \setminus \{1\} \in \text{Rat}(\mathbb{N})$ (see Theorem ??), but the previous computation would be infinite (in brief $2^{n+1} \notin M.[0, 2^n)$).

In many cases we need the notions of division and prime number. The main idea of this paper is to use the basic properties of relative primality We write "$a \mid b$" for "a divides b". We assume that $a \mid 0$, even if $a = 0$. The set of prime numbers is denoted by $\mathcal{P}$.

We recall some results of arithmetic. They can be found e.g. in [1] or [5] (Chap.4). The main fact we have used is Theorem 2 below. It is well known that $\mathcal{P}$ is infinite and not rational. The idea of the classical proof will be reused in Lemma 4. In fact there are less and less prime numbers along the integers (see [5] or [12]), but the density is decreasing very slowly.

Proposition 2 For all $a, b \in \mathbb{N}$,
$a\mathbb{N} \cap b\mathbb{N} = \text{lcm}(a, b)\mathbb{N}$ and $ab = \text{lcm}(a, b) \gcd(a, b)$.

Theorem 3 (Dirichlet 1840) For all $a, b > 0$,

$$\gcd(a, b) = 1 \Leftrightarrow |\mathcal{P} \cap (a + b\mathbb{N})| = \infty \,.$$

The proof can be found in [6] or [12]. We just give the easy direction: let $a, b > 0$. If $\gcd(a, b) = d$ and $d > 1$ then all the integers of $(a + b\mathbb{N})$ are divisible by d, so they contain at most one prime number, namely d.

3 The Power of One-Letter Languages

In this section we will show our main result.

Theorem 4 For a given language $L \in \text{Rat}(\mathbb{N})$, one can decide algorithmically whether $L\mathbb{N} \in \text{Rat}(\mathbb{N})$. We denote

$$Inv(m, q) = \{x \in [m, m + q) \mid \gcd(x, q) = 1\} \,,$$

the set of integers relatively prime to q ("invertible" in $\mathbb{Z}/q\mathbb{Z}$), between m and $m+q-1$. To prove the theorem, we propose the following algorithm.

Algorithm: Rationality test for $L\mathbb{N}$, L rational

Input: $L \in \text{Rat}(\mathbb{N})$ represented by (q, m, I, P) (see Lemma 1) with
$L = I \cup (P + q\mathbb{N})$, where $q \geqslant 1$, $m \geqslant 0$, $I \subseteq [0, m)$, $P \subseteq [m, m + q)$
Output: "$L\mathbb{N} \in \text{Rat}(\mathbb{N})$" or "$L\mathbb{N} \notin \text{Rat}(\mathbb{N})$"

1 If $1 \in L$, then $L\mathbb{N} \in \text{Rat}(\mathbb{N})$, end.
2 Else, if $\emptyset \neq Inv(m, q) \cap P \neq Inv(m, q)$, then $L\mathbb{N} \notin \text{Rat}(\mathbb{N})$, end.
3 Else, if $Inv(m, q) \subseteq P$, obtain the answer with the equivalence:

$$L\mathbb{N} \in \text{Rat}(\mathbb{N}) \Leftrightarrow \forall p \leqslant m + q, \, p \in \mathcal{P}, \, \exists b \geqslant 1, \, p^b \in L \,,$$

end.

4 Else we have $\emptyset = Inv(m,q) \cap P$. Compute, for each prime divisor u of q such that $\exists x \in P, u \mid x$,

$$I_u = \left\{ \frac{x}{\gcd(u,x)} \;\middle|\; x \in I \right\},$$

$$P_u = \{x \in P \mid u \nmid x\} \cup \bigcup_{x \in P,\; u \mid x} \left\{ \frac{x}{u}, \frac{x+q}{u}, \cdots, \frac{x+q(u-1)}{u} \right\},$$

and $I'_u = (I_u \cup (P_u + q\mathbb{N})) \cap [0,m)$, $P'_u = (P_u + q\mathbb{N}) \cap [m, m+q)$.
Call recursively the algorithm with (q, m, I'_u, P'_u), to determine whether $L'_u\mathbb{N} \in \mathrm{Rat}(\mathbb{N})$, where $L'_u = I'_u \cup (P'_u + q\mathbb{N})$ and $L'_u\mathbb{N} = \frac{u\mathbb{N} \cap L\mathbb{N}}{u}$. Collect every answer. Then answer with:
$L\mathbb{N} \in \mathrm{Rat}(\mathbb{N}) \Leftrightarrow \forall u \in \mathcal{P}, u \mid q$, such that $\exists x \in P, u \mid x, L'_u\mathbb{N} \in \mathrm{Rat}(\mathbb{N})$.

As a preparation to the correctness proof, we show the two following lemmas related to points 2 and 3 of the algorithm.

Lemma 4 *If $1 \notin L$ and $\emptyset \neq Inv(m,q) \cap P \neq Inv(m,q)$, then $L\mathbb{N} \notin \mathrm{Rat}(\mathbb{N})$.*

Proof: By hypothesis, we can choose $k \in Inv(m,q) \setminus P$. We know by Dirichlet's Theorem (2) that $k + q\mathbb{N}$ contains infinitely many prime numbers ($\gcd(k,q) = 1$). They are not in $L = I \cup (P + q\mathbb{N})$, because $k \notin P$. Let $p \in k + q\mathbb{N}$ be a prime number, $p \notin L$, and by hypothesis $1 \notin L$, so $p \notin L\mathbb{N}$: the only way to write p as a product is $1.p$ or $p.1$. As a consequence, $M = [k]_q \setminus L\mathbb{N}$ is infinite. We will see that it is not rational. By hypothesis, we also have some $s \in Inv(m,q) \cap P$. Let $n > 0$ and consider

$$f = s(s+q) \cdots (s+qn).$$

The number f is relatively prime to q since s is, and so it is invertible in $\mathbb{Z}/q\mathbb{Z}$. Let $x > 0$ such that $x.f \equiv k - s \pmod{q}$. Remark that the integers $xf + s, xf + s + q, \cdots, xf + s + q.n$ are equivalent to k modulo q. But they are respectively divisible by $s, s+q, \cdots, s+qn$, since f is. So the numbers

$$\frac{xf+s}{s}, \frac{xf+s+q}{s+q}, \cdots, \frac{xf+s+qn}{s+qn}$$

are integers. Of course $s, s+q, \cdots, s+qn$ are in $(s+q\mathbb{N}) \subseteq L$, and so $xf+s, xf+s+q, \cdots, xf+s+qn$ are in $L\mathbb{N}$. As a consequence, the segment $[xf+s, xf+s+q(n+1))$ has an empty intersection with M.
This argument applies to any $n > 0$, and M infinite, so M is not ultimately periodic, hence $M \notin \mathrm{Rat}(\mathbb{N})$. We conclude that $L\mathbb{N}$ is not rational. ∎

Lemma 5 *If $1 \notin L$ and $Inv(m,q) \subseteq P$, then*

$$L\mathbb{N} \in \mathrm{Rat}(\mathbb{N}) \Leftrightarrow \forall p \leq m+q \; (p \in \mathcal{P} \Rightarrow \exists b \geq 1, p^b \in L).$$

Proof: If $Inv(m,q) \subseteq P$, then $P + q\mathbb{N}$ contains all numbers greater than m which are relatively prime to q. Thus it contains all prime numbers greater than $m+q$: if $p \in \mathcal{P}$, then p is relatively prime to q or p divides q; but if $p > q$, then $p \nmid q$. So only a finite number of primes is not in L. We denote $(p_k)_{k \geq 0}$ the *increasing* sequence of *all* prime numbers.

Let $n \geq 0$ be such that $p_{n+1} > m + q$. The language L contains the p_k for $k > n$. One fixes $M = \mathbb{N} \setminus L\mathbb{N}$. For each $k \leq n$, let $b_k = \min(\{b \geq 0 \mid p_k^b \in L\} \cup \{\infty\})$. One can prove that
$$M \subseteq \{p_0^{a_0} \cdots p_n^{a_n} \mid a_0 < b_0, \cdots, a_n < b_n\}.$$
Indeed, for each $x \in \mathbb{N}$, if x has a prime factor $p_j, j > n$, then $p_j \in L$, and $x \in p_j\mathbb{N} \subseteq L\mathbb{N} \Rightarrow x \notin M$. So, if $x \in M$, then x is $p_0^{a_0} \cdots p_n^{a_n}$, but if there is a k such that $a_k \geq b_k$, then $x \in p_k^{b_k}\mathbb{N} \subseteq L\mathbb{N}$.

In the case that $\forall p \leq m + q, p \in \mathcal{P}, \exists b \geq 1, p^b \in L$, each b_k is finite, and then M is finite (so rational), that is $L\mathbb{N}$ is co-finite, and rational. We conclude:
$$\forall p \leq m + q \, (p \in \mathcal{P} \Rightarrow \exists b \geq 1, p^b \in L) \Rightarrow L\mathbb{N} \in \mathrm{Rat}(\mathbb{N}).$$

Conversely, in the case that $\exists j \leq n, \forall b \geq 1, p_j^b \notin L$, we have $b_j = \infty$, and M is infinite: it contains $p_j^\mathbb{N}$ ($1 \notin L$, and the only factors of p_j^i are the p_j^k, $k \leq i$). We will prove that M is not rational with an argument of density inspired by [3]. We want to bound the number of elements of M lower than e^t, for $t \geq 1$:
$$p_0^{a_0} \cdots p_n^{a_n} < e^t \quad \Rightarrow \quad a_0 \ln p_0 + \cdots + a_n \ln p_n < t$$
$$\Rightarrow \quad \forall k \leq n, \, a_k < \frac{t}{\ln p_k} \leq \frac{t}{\ln 2} \quad \Rightarrow \quad a_0, \cdots, a_n \in \left[0, \frac{t}{\ln 2}\right).$$

That consists of at most $\left(\frac{t}{\ln 2} + 1\right)^{n+1}$ different (n+1)-tuples:
$$|M \cap [1, e^t]| < \left(\frac{t}{\ln 2} + 1\right)^{n+1} \quad \Rightarrow \quad \frac{|M \cap [1, e^t]|}{e^t} < \frac{\left(\frac{t}{\ln 2} + 1\right)^{n+1}}{e^t} \xrightarrow{t \to \infty} 0.$$

The density of M has the limit zero, and thanks to Proposition 2 (M is infinite), we conclude that $M \notin \mathrm{Rat}(\mathbb{N})$, then $L\mathbb{N} \notin \mathrm{Rat}(\mathbb{N})$. ∎

Now we prove the correctness of the algorithm for all cases.

Proof: (Theorem 4) We proceed in two steps: proof of partial correctness (under the assumption of termination), and proof of termination.

Proof of partial correctness
Let $L = I \cup (P + q\mathbb{N})$. We follow the algorithm step by step.

1- If $1 \in L$, then $L\mathbb{N} \supseteq 1\mathbb{N} = \mathbb{N}$, and $L\mathbb{N} = \mathbb{N} \in \mathrm{Rat}(\mathbb{N})$.
2- see Lemma 4
3- see Lemma 5
4- Else, we obtain $Inv(m, q) \cap P = \emptyset$. Each element of P has a (strict) common divisor with q, those of $P + q\mathbb{N}$ too, and those of $(P + q\mathbb{N})\mathbb{N}$ also.
$\forall x \in P, \exists u \in \mathcal{P}, u \mid q$ and $u \mid x$, consequently $x\mathbb{N} \subseteq u\mathbb{N}$
$$\Rightarrow P\mathbb{N} = \bigcup_{x \in P} x\mathbb{N} \subseteq \bigcup_{u \in \mathcal{P}, u \mid q, \exists x \in P, u \mid x} u\mathbb{N} = M.$$

We also have $(P + q\mathbb{N})\mathbb{N} \subseteq M$. The set M is rational, because it is a finite union of rational sets. Then

$$L\mathbb{N} \in \mathrm{Rat}(\mathbb{N}) \Leftrightarrow M \cap L\mathbb{N} \in \mathrm{Rat}(\mathbb{N}) \text{ and } (\mathbb{N} \setminus M) \cap L\mathbb{N} \in \mathrm{Rat}(\mathbb{N}).$$

But $(P + q\mathbb{N})\mathbb{N} \subseteq M$, and we already know that $(\mathbb{N} \setminus M) \cap L\mathbb{N} = (\mathbb{N} \setminus M) \cap I\mathbb{N}$ is rational (I is finite). So we have

$$L\mathbb{N} \in \mathrm{Rat}(\mathbb{N}) \Leftrightarrow M \cap L\mathbb{N} \in \mathrm{Rat}(\mathbb{N}).$$

We decompose into:

$$M \cap L\mathbb{N} = \left(\bigcup_{u \in \mathcal{P}, u \mid q, \exists x \in P, u \mid x} u\mathbb{N} \right) \cap L\mathbb{N} = \bigcup_{u \in \mathcal{P}, u \mid q, \exists x \in P, u \mid x} (u\mathbb{N} \cap L\mathbb{N}) \; ;$$

$$L\mathbb{N} \in \mathrm{Rat}(\mathbb{N}) \Leftrightarrow \forall u \in \mathcal{P}\Big(u \mid q \wedge \exists x \in P, u \mid x \Rightarrow (u\mathbb{N} \cap L\mathbb{N}) \in \mathrm{Rat}(\mathbb{N})\Big).$$

We still have to calculate the sets $u\mathbb{N} \cap L\mathbb{N}$.
Let $u \in \mathcal{P}, u \mid q$. For each $x \in P$, if $u \mid x$, then u divides each element of $x + q\mathbb{N}$, so $u\mathbb{N} \cap (x + q\mathbb{N})\mathbb{N} = (x + q\mathbb{N})\mathbb{N}$. Else $\gcd(u, x) = 1$ and u is also relatively prime to each integer of $x + q\mathbb{N}$, hence $u\mathbb{N} \cap (x + q\mathbb{N})\mathbb{N} = (x + q\mathbb{N})u\mathbb{N}$ (Prop. 2). In general

$$u\mathbb{N} \cap (P + q\mathbb{N})\mathbb{N} = \bigcup_{x \in P} (x + q\mathbb{N}) \frac{u}{\gcd(u, x)} \mathbb{N} \, .$$

Similarly for I,

$$u\mathbb{N} \cap I\mathbb{N} = \bigcup_{x \in I} \frac{xu}{\gcd(u, x)} \, .$$

The interest in calculating the intersection with $u\mathbb{N}$ is that all elements are now divisible by u:

$$\frac{u\mathbb{N} \cap L\mathbb{N}}{u} = \left(\bigcup_{x \in I} \frac{x}{\gcd(u, x)} \cup \bigcup_{x \in P} \left(\frac{x}{\gcd(u, x)} + \frac{q}{\gcd(u, x)} \mathbb{N} \right) \right) \mathbb{N} \, .$$

Consequently, $u\mathbb{N} \cap L\mathbb{N} \in \mathrm{Rat}(\mathbb{N})$ is equivalent to $L_u\mathbb{N} \in \mathrm{Rat}(\mathbb{N})$, with

$$L_u = I_u \cup (P_u + q\mathbb{N}) \, , \qquad I_u = \{x / \gcd(u, x) \mid x \in I\} \, ,$$

$$P_u = \{x \in P \mid u \nmid x\} \cup \bigcup_{x \in P,\ u \mid x} \left\{ \frac{x}{u}, \frac{x + q}{u}, \ldots, \frac{x + q(u-1)}{u} \right\} \, .$$

But this representation is not canonical with respect to Lemma 1, we must then consider:

$$I'_u = (I_u \cup (P_u + q\mathbb{N})) \cap [0, m) \, , \qquad P'_u = (P_u + q\mathbb{N}) \cap [m, m + q) \, ,$$

$$\text{and} \quad L'_u = I'_u \cup (P'_u + q\mathbb{N}) \, ,$$

as enunciated in the algorithm. One observes

$$L_u \mathbb{N} = L'_u \mathbb{N} = \frac{u\mathbb{N} \cap L\mathbb{N}}{u}.$$

Indeed, for each $x \in I$, x is replaced by x or x/u in L_u and L'_u. And for all $x \in P$, x is replaced by x or $x/u, (x+q)/u, \cdots, (x+q(u-1))/u$ in L_u. In the first case, x stay identically in L'_u, in the second each new element is smaller or equal than $(x + q(u-1))/u$, and

$$\frac{x+q(u-1)}{u} < \frac{m+q+q(u-1)}{u} \leqslant \frac{m}{2} + q \leqslant m+q$$

so the construction of L'_u does not omit any element, and $L_u = L'_u$. After that we can call the algorithm for (each) L'_u.

So we conclude the partial correctness of the algorithm.

Proof of termination
Most of the steps of the algorithm are clearly effective, using the representation (q, m, I, P) of L. We just have to justify two points: there is a finite number of recursive calls, and we can determine if

$$\forall p \leqslant m+q,\ p \in \mathcal{P},\ \exists b \geqslant 1,\ p^b \in L.$$

Let $p \in \mathcal{P}$, $p \leqslant m+q$. We compute the first powers of p: $p, p^2, \cdots, p^k$ with $p^k > m+q$. If one of them is in L, the condition is true for this p. Else, we can then calculate modulo q: $p^{k+1}, \cdots, p^{k+q} \pmod{q}$, since the following elements will not generate any new values modulo q. If one of them is in $P \pmod{q}$, the condition is true for p, else the condition is false (and $L\mathbb{N}$ is not rational).

For the recursive calls, we define the strict order $\prec$ on the (finite) sets of integers, by induction: $\forall A, B \subseteq \mathbb{N}$, $A, B \neq \emptyset$,

$$\emptyset \prec B$$
$$A \prec B \Leftrightarrow \begin{cases} \min(A) < \min(B), \text{ or} \\ \min(A) = \min(B), \text{ and } A \setminus \min(A) \prec B \setminus \min(B) \end{cases}$$

This is the lexicographical order over the "characteristic words" from $\{0, 1\}^\omega$ of A and B. The order $\prec$ is total.

We will prove that $I'_u \cup P'_u \prec I \cup P$, when $u \in \mathcal{P}, u \mid q, \exists x \in P, u \mid x$. Let y be the smallest of the $x \in I \cup P$ such that $u \mid x$. By construction, $\forall x \in I \cup P$, $x < y \Rightarrow u \nmid x$, we find x also (identically) in $I'_u \cup P'_u$, and it does not "generate" any other element in $I'_u \cup P'_u$. On the other side y generates $y/u \in I'_u \cup P'_u$, and of course $y/u < y$, so $I'_u \cup P'_u \prec I \cup P$.

The integers m and q remain the same in each recursive call, and for fixed m and q, there is only a finite number (2^{m+q}) of possible sets $I \cup P \subseteq [0, m+q)$. So the order $\prec$, restricted to these sets, is finite. It follows that the computation of the algorithm for a given L, i.e. (q, m, I, P), needs only (recursively) the computation of finitely many (q, m, I', P'), which proves the termination. ∎

To speed up the algorithm, it might be possible to factorize L by $\text{lcm}(I \cup P \cup \{q\})$: the greatest common divisor of (all) the elements of $I \cup P \cup \{q\}$. One can also try to minimize the period q and the basis m.

As an application of the algorithm we can conclude that very often the power of rational languages is *not* rational. For example

$$\forall a, b \in \mathbb{N}, \ (a + b\mathbb{N})\mathbb{N} \in \text{Rat}(\mathbb{N}) \ \Leftrightarrow \ a \mid b \text{ or } b \mid a \ .$$

After some eventually reductions of the languages through point 4 of the algorithm that can be done directly by a factorization with $\gcd(a, b)$, one obtains $a', b' \in \mathbb{N}, \gcd(a', b') = 1$. The cases a' or $b' \in \{0, 1\}$, corresponding to $a \mid b$ or $b \mid a$ are easy. Otherwise, if $b' = 2$, then $a' \geq 3$ and a' is odd, we conclude unrationality with point 3 and $\forall k \geq 0, 2^k \notin (a' + b'\mathbb{N})$. If $b' \geq 2$, point 2 shows that the language is unrational.

4 Conclusion

We have given an algorithmically solution to the question enunciated in Section 2:

given $M \in \text{Rat}(\mathbb{N})$, is $M\mathbb{N}$ rational?

Depending on the presence of relative prime numbers in M, our algorithm either states directly that M is not rational using Dirichlet's theorem, or it considers the representation of all prime numbers in M, or it decomposes M into "smaller" languages by intersection with simple periodic sets and use recursivity. It was necessary to use some facts of arithmetic. The runtime is mainly influenced by the bound 2^{m+q} for the number of recursive calls; we do not give here a precise computation of the time complexity.

As future work, one could try to extend the result of Theorem ?? to any product of two rational languages of integers (or even not necessarily rational). It is easy to remark that if $L\mathbb{N} \notin \text{Rat}(\mathbb{N})$ then $\forall t \geq 0, \ L(t + \mathbb{N}) \notin \text{Rat}(\mathbb{N})$, but the converse proposition is false: $(2 + \mathbb{N})(2 + \mathbb{N}) = (4 + \mathbb{N}) \setminus \mathcal{P} \notin \text{Rat}(\mathbb{N})$. In another direction, one can try to answer the question: "is $\text{pow}(L)$ rational?" for any rational language L (over any finite alphabet). After that, the most general question would be: given a finite alphabet A, two languages $L \in \text{Rat}(A^*)$ and $M \in \text{Rat}(\mathbb{N})$,

is $A^M = \{u^k \in A^* \mid u \in L, k \in M\}$ rational?

Acknowledgement. Great thanks to Didier Caucal that gives us the subject of this paper, and a lot of advice, to Wolfgang Thomas and Tanguy Urvoy for their helpful comments and suggestions, to the referees for their remarks.

References

1. Jean-Marie ARNAUDIÈS and Henri FRAYSSE, *Cours de mathématiques - 1, Algèbre*, classes préparatoires, Bordas, 1987.
2. J. Richard BÜCHI, *Weak second-order arithmetic and finite automata*, Zeit. für math. Logik und Grund. der Math. 6, pp. 66-92, 1960.

3. Hugues CALBRIX, *Mots ultimement périodiques des langages rationnels de mots infinis*, Thèse de l'Université Denis Diderot-Paris VII, Litp Th96/03, Avril 1996.
4. Samuel EILENBERG and Marcel-Paul SCHÜTZENBERGER, Rational sets in commutative monoids, Journal of Algebra 13, pp. 173-191, 1969.
5. Ronald L. GRAHAM, Donald E. KNUTH and Oren PATASHNIK, *Concrete mathematics*, Addison-Wesley, 1989, second edition 1994.
6. Godfrey H. HARDY and Edward M. WRIGHT, *An Introduction to the Theory of Numbers*, Clarendon Press, Oxford, 1938; fifth edition, 1979.
7. John E. HOPCROFT and Jeffrey D. ULLMAN, *Introduction to automata theory, languages and computation*, Addison-Wesley, 1979.
8. Stephen C. KLEENE, *Representation of events in nerv nets and finite automata*, In C.E.Shannon and J.McCarthy, editors, Automata Studies, Princeton University Press, 1956.
9. Armando B. MATOS, *Periodic sets of integers*, Theoretical Computer Science 127, Elsevier, 1994.
10. Dominique PERRIN, *Finite Automata*, Handbook of Theoretical Computer Science, J. van Leeuwen, Elsevier Science publishers B.V., 1990.
11. Dominique PERRIN, *Les débuts de la théorie des automates*, Technique et Science Informatique (vol. 14. 409-43), 1995.
12. Jean-Pierre SERRE, *Cours d'arithmétique*, Presses Universitaires de France, 1970. Or *A course in arithmetic*, Springer-Verlag, New York-Heidelberg, 1973.

The Entropy of Lukasiewicz-Languages

Ludwig Staiger

Martin-Luther-Universität Halle-Wittenberg
Institut für Informatik
von-Seckendorff-Platz 1, D–06099 Halle (Saale), Germany
staiger@informatik.uni-halle.de

Abstract. The paper presents an elementary approach for the calculation of the entropy of a class of context-free languages. This approach is based on the consideration of roots of a real polynomial and is also suitable for calculating the Bernoulli measure.

In the paper [Ku70] a remarkable information-theoretic property of Lukasiewicz' comma-free notation was developed. The languages of well-formed formulas of the implicational calculus with one variable and one n-ary operation ($n \geq 2$) in Polish parenthesis-free notation have generative capacity $h_2(\frac{1}{n})$ where h_2 is the usual Shannon entropy. In this paper we investigate the more general case of pure Lukasiewicz languages where the number of variables and n-ary operations is arbitrary, however, the arity of the operations is fixed. A further generalization results in substituting words for letters in pure Lukasiewicz languages. This is explained in more detail in Section 2.

The main purpose of our investigations is to study the same information theoretic aspect of languages as in [Ku70,Ei74,JL75], namely the generative capacity of languages. This capacity, in language theory called the *entropy of languages* resembles directly Shannon's channel capacity (cf. [Jo92]). It measures the amount of information which must be provided on the average in order to specify a particular symbol of a word in a language. For a connection of the entropy of languages to Algorithmic Information Theory see e.g. [LV93,St93].

After having investigated basic properties of generalized Lukasiewicz languages we first calculate their Bernoulli measures in Section 3. Here we derive and investigate in detail a basic real equation closely related to the measure of generalized Lukasiewicz languages.

These investigations turn out to be useful not only for the calculation of the measure but also for estimating the entropy of of generalized Lukasiewicz languages which will be carried out in Section 4. In contrast to [Ku70] we do not require the powerful apparatus of the theory of complex functions utilized there for the more general task of calculating the entropy of unambiguous context-free languages, we develop a simpler apparatus based on augmented real functions. Here we give an exact formula for the entropy of pure Lukasiewicz languages with arbitrary numbers of letters representing variables and n-ary operations.

The final section deals with the entropy of the star languages (submonoids) of generalized Lukasiewicz languages.

Next we introduce the notation used throughout the paper. By $\mathbb{N} = \{0, 1, 2, \ldots\}$ we denote the set of natural numbers. Let X be an alphabet of cardinality $\# X = r$. By X^* we denote the set (monoid) of words on X, including the *empty word* e. For $w, v \in X^*$ let $w \cdot v$ be their *concatenation*. This concatenation product extends in an obvious way to subsets $W, V \subseteq X^*$. For a language W let $W^* := \bigcup_{i \in \mathbb{N}} W^i$ be the *submonoid* of X^* generated by W, and by $W^\omega := \{w_1 \cdots w_i \cdots : w_i \in W \setminus \{e\}\}$ we denote the set of infinite strings formed by concatenating words in W. Furthermore $|w|$ is the *length* of the word $w \in X^*$ and $\mathbf{A}(B)$ is the set of all finite prefixes of strings in $B \subseteq X^* \cup X^\omega$.

As usual language $V \subseteq X^*$ is called a code provided $w_1 \cdots w_l = v_1 \cdots v_k$ for $w_1, \ldots, w_l, v_1, \ldots, v_k \in V$ implies $l = k$ and $w_i = v_i$. A language $V \subseteq X^*$ is referred to as an ω-code provided $w_1 \cdots w_i \cdots = v_1 \cdots v_i \cdots$ where $w_1, \ldots, w_i, \ldots, v_1, \ldots, v_i, \cdots \in V$ implies $w_i = v_i$. A code V is said to have a *finite delay of decipherability*, provided for every $w \in V$ there is an $m_w \in \mathbb{N}$ such that if $w \cdot v_1 \cdots v_{m_w}$ is a prefix of $w' \cdot u$ where $v_1, \ldots, v_{m_w}, w' \in V$ and $u \in V^*$ implies $w = w'$ (cf. [St85, DL94]).

1 Pure Lukasiewicz-Languages

In this section we consider languages over a finite or countably infinite alphabet A. Let $\{A_0, A_1\}$ be a partition of A into two nonempty parts and let $n \geq 2$. The *pure* $\{A_0, A_1\}$-n-**Lukasiewicz-language** is defined as the solution of the equation

$$\widetilde{\mathbf{L}} = A_0 \cup A_1 \cdot \widetilde{\mathbf{L}}^n . \tag{1}$$

It is a simple deterministic language (cf. [AB97, Section 6.7]) and can be obtained as $\bigcup_{i \in \mathbb{N}} \widetilde{\mathbf{L}}_i$ where $\widetilde{\mathbf{L}}_0 := \emptyset$ and $\widetilde{\mathbf{L}}_{i+1} := A_0 \cup A_1 \cdot \widetilde{\mathbf{L}}_i^n$.

$\{A_0, A_1\}$-n-Lukasiewicz-languages have the following properties.

Proposition 1. *1. $\widetilde{\mathbf{L}}$ is a prefix code.*
 2. If $w \in A^*$ and $a_0 \in A_0$ then $w \cdot a_0^{|w| \cdot n} \in \widetilde{\mathbf{L}}^*$.
 3. $\mathbf{A}(\widetilde{\mathbf{L}}^*) = A^*$

Along with $\widetilde{\mathbf{L}}$ it is useful to consider its *derived language* $\widetilde{\mathbf{K}}$ which is defined by the following equation.

$$\widetilde{\mathbf{K}} := A_1 \cdot \bigcup_{i=0}^{n-1} \widetilde{\mathbf{L}}^i \tag{2}$$

Proposition 2. *1. $\mathbf{A}(\widetilde{\mathbf{L}}) \setminus \widetilde{\mathbf{L}} = \widetilde{\mathbf{K}}^*$*
 2. Every $w \in A^*$ has a unique factorization $w = v \cdot u$ where $v \in \widetilde{\mathbf{L}}^*$ and $u \in \widetilde{\mathbf{K}}^*$.

Proposition 3. *$\widetilde{\mathbf{K}}$ is an ω-code having an infinite delay of decipherability.*

2 The Definition of Lukasiewicz Languages

Generalized Lukasiewicz-languages are constructed from pure Lukasiewicz-languages via composition of codes (cf. Section I.6 of [BP85]) as follows. We start with a code

$\tilde{C}$, $\#\tilde{C} \geq 2$, and an bijective morphism $\psi : A^* \to \tilde{C}^*$. Let $C := \psi(A_0) \subseteq \tilde{C}$ and $B := \psi(A_1) \subseteq \tilde{C}$. This partitions the code $\tilde{C}$ it into nonempty parts C and B.

Let $\tilde{\mathbf{L}}$ be the $\{A_0, A_1\}$-n-Lukasiewicz-language and $\mathbf{L} := \psi(\tilde{\mathbf{L}})$. Thus, $\mathbf{L}$ is the composition of the codes $\tilde{\mathbf{L}}$ and $\tilde{C}$ via ψ, $\mathbf{L} = \tilde{C} \circ_\psi \tilde{\mathbf{L}}$. Analogously to the previous section $\mathbf{L}$ is called $\{C, B\}$-n-*Lukasiewicz-language*. For the sake of brevity, we shall omit the prefix "$\{C, B\}$-n-" when there is no danger of confusion. Throughout the rest of the paper we suppose C and B to be disjoint sets for which $C \cup B$ is a code, and we suppose n to be the composition parameter described in Eq. (3) below.

Utilizing the properties of the composition of codes (cf. [BP85, Section 1.6]) from the results of the previous section one can easily derive that $\mathbf{L}$ has the following properties.

$$\mathbf{L} = C \cup B \cdot \mathbf{L}^n \qquad (3)$$

Property 1.   *1.* $\mathbf{L} \subseteq (C \cup B)^*$
   2. $\mathbf{L}$ is a code, and if $C \cup B$ is a prefix code then $\mathbf{L}$ is also a prefix code.
   3. If $w \in (B \cup C)^*$ and $v \in C$ then $w \cdot v^{|w| \cdot n} \in \mathbf{L}^*$.
   4. $\mathbf{A}(\mathbf{L}^*) = \mathbf{A}((C \cup B)^*)$

It should be mentioned that $\mathbf{L}$ might be a prefix code, in case if B is not a prefix code.

Example 1. Let $X = \{a, b\}$, $C := \{baa\}$ and $B := \{b, ba\}$. Then it is easily verified that $\mathbf{L}$ is a prefix code. ☐

In the same way as above we define the *derived language* K as $K := \tilde{C} \circ_\psi \tilde{K}$, and we obtain the following.

$$K := B \cdot \bigcup_{i=0}^{n-1} \mathbf{L}^i \qquad (4)$$

Propositions 2.2 and 3 prove that the language K is related to $\mathbf{L}$ via the following properties.

Theorem 1.   *1.* $\mathbf{A}(\mathbf{L}) = K^* \cdot \mathbf{A}(C \cup B)$.
   2. Every $w \in (C \cup B)^*$ has a unique factorization $w = v \cdot u$ where $v \in \mathbf{L}^*$ and $u \in K^*$.
   3. K is a code having an infinite delay of decipherability.

3 The Measure of Lukasiewicz Languages

In this section we consider the measure of Lukasiewicz languages. Measures of languages were considered in Chapters 1.4 and 2.7 of [BP85]. In particular, we will consider so-called Bernoulli measures.

3.1 Valuations of Languages

As in [Fe95] we call a morphism $\mu : X^* \to (0, \infty)$ of the monoid X^* into the multiplicative monoid of the positive real numbers a *valuation*. A valuation μ such that $\mu(X) = \sum_{x \in X} \mu(x) = 1$ is known as *Bernoulli measure* on X^* (cf. [BP85, Chap. 1.4]).

A valuation is usually extended to a mapping $\mu : 2^{X^*} \to [0, \infty]$ via $\mu(W) := \sum_{w \in W} \mu(w)$.

Now consider the measure $\mu(\mathbf{L})$ for a valuation μ on X^*. Since the decomposition $\mathbf{L} = C \cup B \cdot \mathbf{L}^n$ is unambiguous, we obtain

$$\mu(\mathbf{L}) = \mu(C) + \mu(B) \cdot \mu(\mathbf{L})^n \ .$$

The representation $\mathbf{L} = \bigcup_{i=1}^{\infty} \mathbf{L}_i$ where $\mathbf{L}_1 := C$ and $\mathbf{L}_{i+1} := C \cup B \cdot \mathbf{L}_i$ allows us to approach the measure $\mu(\mathbf{L})$ by the sequence

$$\mu_1 := \mu(C)$$
$$\mu_{i+1} := \mu(C) + \mu(B) \cdot \mu_i{}^n$$
$$\mu(\mathbf{L}) = \lim_{i \to \infty} \mu_i \ .$$

We have the following

Theorem 2. *If the equation $\lambda = \mu(C) + \mu(B) \cdot \lambda^n$ has a positive solution then $\mu(\mathbf{L})$ equals the smallest positive number such that $\mu(\mathbf{L}) = \mu(C) + \mu(B) \cdot \mu(\mathbf{L})^n$, otherwise $\mu(\mathbf{L}) = \infty$.*

In order to give a more precise evaluation of $\mu(\mathbf{L})$, in the subsequent section we take a closer look to our basic equation

$$\lambda = \gamma + \beta \cdot \lambda^n \ , \tag{5}$$

where $\gamma, \beta > 0$ are positive reals.

In order to estimate $\mu(K)$ we observe that the unambiguous representation of Eq. (4) yields the formula $\mu(K) = \mu(B) \cdot \sum_{i=0}^{n-1} \mu(\mathbf{L})^i$. Then the following connection between the valuations $\mu(\mathbf{L})$ and $\mu(K)$ is obvious.

Property 2. It holds $\mu(\mathbf{L}) = \infty$ iff $\mu(K) = \infty$.

3.2 The Basic Equation $\lambda = \gamma + \beta \cdot \lambda^n$

This section is devoted to a detailed investigation of the solutions of our basic equation (5). As a result we obtain estimates for the Bernoulli measures of $\mathbf{L}$ and K as well as a useful tool when we are going to calculate the entropy of Lukasiewicz languages in the subsequent sections.

Let $\bar{\lambda}$ be an arbitrary positive solution of Eq. (5). Then we have the following relationship to the value $\gamma + \beta$.

$$\begin{aligned} \bar{\lambda} &< 1 \Leftrightarrow \bar{\lambda} < \gamma + \beta \ , \\ \bar{\lambda} &= 1 \Leftrightarrow \bar{\lambda} = \gamma + \beta \ , \text{ and} \\ \bar{\lambda} &> 1 \Leftrightarrow \bar{\lambda} > \gamma + \beta \end{aligned} \tag{6}$$

In order to study positive solutions it is convenient to consider the positive zeroes of the function.

$$f(\lambda) = \gamma + \beta \cdot \lambda^n - \lambda \tag{7}$$

The graph of the function f reveals that f has exactly one minimum at $\lambda_{\min}$, $0 < \lambda_{\min} = \frac{1}{\sqrt[n-1]{\beta n}} < \infty$ on the positive real axis. Thus it has at most two positive roots $\lambda_0, \hat{\lambda}$ which satisfy $0 < \lambda_0 \leq \lambda_{\min} \leq \hat{\lambda}$. Since $f(0) = \gamma > 0$, $f(1) = \gamma + \beta - 1$ and $f(\gamma + \beta) = \beta((\gamma + \beta)^n - 1)$, f has at least one positive root provided $\gamma + \beta \leq 1$ and these positive roots satisfy the equivalences

$$\begin{aligned}\gamma + \beta < 1 &\Leftrightarrow \lambda_0 < \gamma + \beta < 1 < \hat{\lambda}, \text{ and} \\ \gamma + \beta = 1 &\Leftrightarrow \lambda_0 = 1 \vee \hat{\lambda} = 1.\end{aligned} \quad (8)$$

If $\gamma + \beta > 1$ the function f has a positive root if and only if $f(\lambda_{\min}) = \gamma - \frac{n-1}{n} \cdot \frac{1}{\sqrt[n-1]{\beta n}} \leq 0$, that is, $\gamma^{n-1} \cdot \beta \leq \frac{(n-1)^{n-1}}{n^n}$.

In connection with λ_0 we consider the value $\kappa_0 := \beta \cdot \sum_{i=0}^{n-1} \lambda_0^i$. This value is related to $\mu(K)$ in the same way as λ_0 to $\mu(\mathbf{L})$. In view of $\beta(\lambda_0^n - 1) = \beta \cdot \lambda_0^n + \gamma - \gamma - \beta = \lambda_0 - (\gamma + \beta)$, we have

$$\kappa_0 = \begin{cases} n \cdot \beta & \text{, if } \lambda_0 = 1 \text{ and} \\ \dfrac{\lambda_0 - (\gamma + \beta)}{\lambda_0 - 1} & \text{, otherwise.} \end{cases} \quad (9)$$

As a corollary to Eq. (9) we obtain the following.
Corollary 1. $(\kappa_0 - 1) \cdot (\lambda_0 - 1) = 1 - (\gamma + \beta)$
We obtain our main result on the dependencies between th coefficients of our basic equation (5) and the values of λ_0 and κ_0.
Theorem 3. *If $f(\lambda) = \gamma + \beta \cdot \lambda^n - \lambda$ has a positive root then its minimum positive root λ_0 and the value $\kappa_0 := \beta \cdot \sum_{i=0}^{n-1} \lambda_0^i$ depend in the following way from the coefficients γ and β.*

$$\lambda_0 < 1 \wedge \kappa_0 < 1 \Leftrightarrow \gamma + \beta < 1 \quad (10)$$
$$\lambda_0 < 1 \wedge \kappa_0 = 1 \Leftrightarrow \gamma + \beta = 1 \wedge \beta > \tfrac{1}{n} \quad (11)$$
$$\lambda_0 < 1 \wedge \kappa_0 > 1 \Leftrightarrow \gamma + \beta > 1 \wedge \beta > \tfrac{1}{n} \wedge f(\lambda_{\min}) \leq 0 \quad (12)$$
$$\lambda_0 = 1 \wedge \kappa_0 < 1 \Leftrightarrow \gamma + \beta = 1 \wedge \beta < \tfrac{1}{n} \quad (13)$$
$$\lambda_0 = 1 \wedge \kappa_0 = 1 \Leftrightarrow \gamma + \beta = 1 \wedge \beta = \tfrac{1}{n} \quad (14)$$
$$\lambda_0 > 1 \wedge \kappa_0 < 1 \Leftrightarrow \gamma + \beta > 1 \wedge \beta < \tfrac{1}{n} \wedge f(\lambda_{\min}) \leq 0 \quad (15)$$

The remaining cases $\lambda_0 = 1 \wedge \kappa_0 > 1$ and $\lambda_0 > 1 \wedge \kappa_0 \geq 1$ are impossible.

From Theorem 3 and Eq. (6) we obtain the following.
Corollary 2. *If $\lambda_0 > 1$ then $\lambda_0 > \gamma + \beta > 1$.*
Finally, we consider the case $\lambda_0 = \lambda_{\min}$, that is, when f has a positive root of multiplicity two. It turns out that in this case we have some additional restrictions.
Lemma 1. $f(\lambda) = \gamma + \beta \cdot \lambda^n - \lambda$ *has a positive root of multiplicity two iff $\gamma^{n-1} \cdot \beta = \frac{(n-1)^{n-1}}{n^n}$. Moreover in this case $\lambda_{\min} = \frac{n}{n-1} \cdot \gamma$, $\gamma + \beta \geq 1$, and $\gamma + \beta = 1$ if and only if $\beta = \frac{1}{n}$ or $\gamma = \frac{n-1}{n}$.*
Comparing with the equivalences of Theorem 3 we observe that multiple positive roots are possible only in the cases of Eqs. (12), (14) and (15), and that in the case of Eq. (14) we have necessarily multiple positive roots.

3.3 The Bernoulli Measure of Lukasiewicz Languages

The last part of Section 3 is an application of the results of the previous subsections to Bernoulli measures. As is well known, a code of Bernoulli measure 1 is maximal (cf. [BP85]). The results of the previous subsection show the following necessary and sufficient conditions.

Theorem 4. *Let* $\mathbf{L} = C \cup B \cdot \mathbf{L}$, $C, B \subseteq X^*$ *be a Lukasiewicz language,* K *its derived language and* $\mu : X^* \to (0,1)$ *a Bernoulli measure. Then* $\mu(\mathbf{L}) = 1$ *iff* $\mu(C \cup B) = 1$ *and* $\mu(B) \leq \frac{1}{n}$, *and* $\mu(\mathrm{K}) = 1$ *iff* $\mu(C \cup B) = 1$ *and* $\mu(B) \geq \frac{1}{n}$.

Thus Theorem 4 proves that pure Lukasiewicz languages $\widetilde{\mathbf{L}}$ and their derived languages $\widetilde{\mathrm{K}}$ are maximal codes.

Resuming the results of Section 3 one can say that in order to achieve maximum measure for both Lukasiewicz languages $\mathbf{L}$ and K it is necessary and sufficient to distribute the measures $\mu(C)$ and $\mu(B)$ as $\mu(C) = \frac{n-1}{n}$ and $\mu(B) = \frac{1}{n}$, thus respecting the composition parameter n in the defining equation (1). A bias in the measure distribution results in a measure loss for at least one of the codes $\mathbf{L}$ or K.

4 The Entropy of Lukasiewicz Languages

In [Ku70] Kuich introduced a powerful apparatus in terms of the theory of complex functions to calculate the entropy of unambiguous context-free languages.

For our purposes it is sufficient to consider real functions admitting the value ∞. The coincidence of Kuich's and our approach for Lukasiewicz languages is established by Pringsheim's theorem which states that a power series $\mathfrak{s}(t) = \sum_{i=0}^{\infty} s_i t^i$, $s_i \geq 0$, with finite radius of convergence rad $\mathfrak{s}$ has a singular point at rad $\mathfrak{s}$ and no singular point with modulus less than rad $\mathfrak{s}$. For a more detailed account see [Ku70, Section 2].

4.1 Definition and Simple Properties

The notion of entropy of languages is based on counting words of equal length. Therefore, from now on we assume our alphabet X to be finite of cardinality $\# X = r$, $r \geq 2$.

For a language $W \subseteq X^*$ let $\mathsf{s}_W : \mathbb{N} \to \mathbb{N}$ where $\mathsf{s}_W(n) := \# W \cap X^n$ be its *structure function*, and let

$$\mathsf{H}_W = \limsup_{n \to \infty} \frac{\log_r(1 + \mathsf{s}_W(n))}{n}$$

be its *entropy* (cf. [Ku70]).

Informally, this concept measures the amount of information which must be provided on the average in order to specify a particular symbol of a word in a language.

The *structure generating function* corresponding to s_W is

$$\mathfrak{s}_W(t) := \sum_{i \in \mathbb{N}} \mathsf{s}_W(i) \cdot t^i. \tag{16}$$

$\mathfrak{s}_W$ is a power series with convergence radius

$$\operatorname{rad} W := \limsup_{n \to \infty} \frac{1}{\sqrt[n]{\mathsf{s}_W(n)}}.$$

As it was explained above, it is convenient to consider $\mathfrak{s}_W$ also as a function mapping $[0, \infty)$ to $[0, \infty) \cup \{\infty\}$, where we set

$$\mathfrak{s}_W(\operatorname{rad} W) := \sup\{\mathfrak{s}_W(\alpha) : \alpha < \operatorname{rad} W\}, \text{ and} \tag{17}$$
$$\mathfrak{s}_W(\alpha) := \infty, \text{ if } \alpha > \operatorname{rad} W. \tag{18}$$

Having in mind this variant of $\mathfrak{s}_W$, we observe that $\mathfrak{s}_W$ is a nondecreasing function which is continuous in the interval $(0, \operatorname{rad} W)$ and continuous from the left in the point $\operatorname{rad} W$.

Then the entropy of languages satisfies the following property[1].

Property 3.

$$\mathsf{H}_W := \begin{cases} 0 & \text{, if } W \text{ is finite, and} \\ -\log_r \operatorname{rad} W & \text{, otherwise.} \end{cases}$$

Before we proceed to the calculation of the entropy of Lukasiewicz languages we mention still some properties of the entropy of languages which are easily derived from the fact that $\mathfrak{s}_W$ is a positive series (cf. [Ei74, Proposition VIII.5.5]).

Property 4. Let $V, W \subseteq X^*$. Then $0 \leq \mathsf{H}_W \leq 1$ and, if W and V are nonempty, we have $\mathsf{H}_{W \cup V} = \mathsf{H}_{W \cdot V} = \max\{\mathsf{H}_W, \mathsf{H}_V\}$.

For the entropy of the star of a language we have the following (cf. [Ei74,Ku70]).

Property 5. If $V \subseteq X^*$ is a code then $\mathfrak{s}_{V^*}(t) = \sum_{i \in \mathbb{N}} (\mathfrak{s}_V(t))^i = \frac{1}{1 - \mathfrak{s}_V(t)}$ and

$$\mathsf{H}_{V^*} = \begin{cases} \mathsf{H}_V & \text{, if } \mathfrak{s}_V(\operatorname{rad} V) \leq 1 \text{, and} \\ -\log_r \inf\{\gamma : \mathfrak{s}_V(\gamma) = 1\} & \text{, otherwise.} \end{cases}$$

In general $\sum_{i \in \mathbb{N}} (\mathfrak{s}_W(t))^i$ is only an upper bound to $\mathfrak{s}_{W^*}(t)$. Hence only in case $\mathfrak{s}_W(t) < 1$ one can conclude that $\mathfrak{s}_{W^*}(t) \leq \sum_{i \in \mathbb{N}} (\mathfrak{s}_W(t))^i < \infty$ and, consequently, $t \leq \operatorname{rad} W^*$. Thus we obtain a sufficient condition for the equality $\mathsf{H}_W = \mathsf{H}_{W^*}$ depending on the value of $\mathfrak{s}_W(\operatorname{rad} W)$.

Corollary 3. *Let $W \subseteq X^*$. We have $\mathsf{H}_W = \mathsf{H}_{W^*}$ if $\mathfrak{s}_W(\operatorname{rad} W) \leq 1$, and if W is a code $W \subseteq X^*$ it holds $\mathsf{H}_W < \mathsf{H}_{W^*}$ if and only if $\mathfrak{s}_W(\operatorname{rad} W) > 1$.*

4.2 The Calculation of the Convergence Radius

Property 3 showed the close relationship between H_W and $\operatorname{rad} W$, and Corollary 3 proved that the value of $\mathfrak{s}_W$ at the point $\operatorname{rad} W$ is of importance for the calculation of the entropy of the star language of W, H_{W^*}.

Therefore, in this section we are going to estimate the convergence radius of the power series $\mathfrak{s}_\mathbf{L}(t)$ and, simultaneously, the values $\mathfrak{s}_\mathbf{L}(\operatorname{rad} \mathbf{L})$ and $\mathfrak{s}_\mathbf{K}(\operatorname{rad} \mathbf{L})$ (Observe that $\operatorname{rad} \mathbf{L} = \operatorname{rad} \mathbf{K}$ in view of Eq. (4) and Property 4). We start with the equation

$$\mathfrak{s}_\mathbf{L}(t) = \mathfrak{s}_C(t) + \mathfrak{s}_B(t) \cdot \mathfrak{s}_\mathbf{L}(t)^n \tag{19}$$

---

[1] If, in the definition of H_W, the additional term $1+$ in the numerator is omitted one obtains $\mathsf{H}_W = -\infty$ for finite languages.

which follows from the unambiguous representation in Eq. (3) and the observation that $\mathrm{rad}\,\mathbf{L} = \sup\{t : \mathfrak{s}_\mathbf{L}(t) < \infty\} = \inf\{t : \mathfrak{s}_\mathbf{L}(t) = \infty\}$, because the function $\mathfrak{s}_\mathbf{L}(t)$ is nondecreasing (even increasing on $[0, \mathrm{rad}\,\mathbf{L}])$.

From Section 3.2 we know that, for fixed t, $t < \mathrm{rad}\,\mathbf{L}$, the value $\mathfrak{s}_\mathbf{L}(t)$ is one of the solutions of Eq. (5) with $\gamma = \mathfrak{s}_C(t)$ and $\beta = \mathfrak{s}_B(t)$. Similarly to Theorem 2 one can prove the following.

Theorem 5. *Let* $t > 0$. *If Eq. (5) has a positive solution for* $\gamma = \mathfrak{s}_C(t)$ *and* $\beta = \mathfrak{s}_B(t)$ *then* $\mathfrak{s}_\mathbf{L}(t) = \lambda_0$, *and if Eq. (5) has no positive solution then* $\mathfrak{s}_\mathbf{L}(t)$ *diverges, that is,* $\mathfrak{s}_\mathbf{L}(t) = \infty$.

This yields an estimate for the convergence radius of $\mathfrak{s}_\mathbf{L}(t)$.

$$\mathrm{rad}\,\mathbf{L} = \inf\left\{\mathrm{rad}\,(C \cup B)\right\} \cup \left\{t : \mathfrak{s}_C(t)^{n-1} \cdot \mathfrak{s}_B(t) > \tfrac{(n-1)^{n-1}}{n^n}\right\} \qquad (20)$$

We obtain some corollaries to Theorem 5 and Eq. (20) which allow us to estimate $\mathrm{rad}\,\mathbf{L}$.

Corollary 4. *If* $\tfrac{(n-1)^{n-1}}{n^n} \le \mathfrak{s}_C(\mathrm{rad}\,(C \cup B))^{n-1} \cdot \mathfrak{s}_B(\mathrm{rad}\,(C \cup B))$ *then* $\mathrm{rad}\,\mathbf{L}$ *is the solution of the equation* $\mathfrak{s}_C(t)^{n-1} \cdot \mathfrak{s}_B(t) = \tfrac{(n-1)^{n-1}}{n^n}$.

Corollary 5. *We have* $\mathrm{rad}\,\mathbf{L} = \mathrm{rad}\,(C \cup B)$ *if and only if* $\mathrm{rad}\,(C \cup B) < \infty$ *and* $\tfrac{(n-1)^{n-1}}{n^n} \ge \mathfrak{s}_C(\mathrm{rad}\,(C \cup B))^{n-1} \cdot \mathfrak{s}_B(\mathrm{rad}\,(C \cup B))$.

Observe that Corollary 4 covers the case $\mathrm{rad}\,(C \cup B) = \infty$. Thus, Corollaries 4 and 5 show that Eq. (19) has always a finite solution when $t = \mathrm{rad}\,\mathbf{L}$.

In the case of Corollary 4 when $\mathfrak{s}_C(\mathrm{rad}\,\mathbf{L})^{n-1} \cdot \mathfrak{s}_B(\mathrm{rad}\,\mathbf{L}) = \tfrac{(n-1)^{n-1}}{n^n}$ the value $\mathfrak{s}_\mathbf{L}(\mathrm{rad}\,\mathbf{L})$ is a double root of Eq. (20). Then Lemma 1 applies and we obtain the following.

Lemma 2. *If* $\mathfrak{s}_C(\mathrm{rad}\,\mathbf{L})^{n-1} \cdot \mathfrak{s}_B(\mathrm{rad}\,\mathbf{L}) = \tfrac{(n-1)^{n-1}}{n^n}$ *then* $\mathfrak{s}_{C \cup B}(\mathrm{rad}\,\mathbf{L}) \ge 1$, *and* $\mathfrak{s}_\mathbf{L}(\mathrm{rad}\,\mathbf{L}) = \tfrac{n}{n-1} \cdot \mathfrak{s}_C(\mathrm{rad}\,\mathbf{L})$. *Furthermore, in this case, we have*

1. $\mathfrak{s}_{C \cup B}(\mathrm{rad}\,\mathbf{L}) = \mathfrak{s}_\mathbf{L}(\mathrm{rad}\,\mathbf{L}) = \mathfrak{s}_K(\mathrm{rad}\,\mathbf{L}) = 1$, $\mathfrak{s}_B(\mathrm{rad}\,\mathbf{L}) = \tfrac{1}{n}$, *and* $\mathfrak{s}_C(\mathrm{rad}\,\mathbf{L}) = \tfrac{n-1}{n}$, *or*
2. $\mathfrak{s}_{C \cup B}(\mathrm{rad}\,\mathbf{L}) > 1$, $\mathfrak{s}_\mathbf{L}(\mathrm{rad}\,\mathbf{L}) < 1$ *and* $\mathfrak{s}_K(\mathrm{rad}\,\mathbf{L}) > 1$, *or*
3. $\mathfrak{s}_{C \cup B}(\mathrm{rad}\,\mathbf{L}) > 1$, $\mathfrak{s}_\mathbf{L}(\mathrm{rad}\,\mathbf{L}) > 1$ *and* $\mathfrak{s}_K(\mathrm{rad}\,\mathbf{L}) < 1$.

We give an example that all three cases are possible.

Example 2. Let $m \ge 1$ and $C \cup B \subseteq X^m$. Then $\mathfrak{s}_C(t)^{n-1} \cdot \mathfrak{s}_B(t) = (\#C \cdot t^m)^{n-1} \cdot \#B \cdot t^m = \tfrac{(n-1)^{n-1}}{n^n}$ has the minimum positive solution

$$\mathrm{rad}\,\mathbf{L} = \sqrt[m \cdot n]{\left(\tfrac{n-1}{n \cdot \#C}\right)^{n-1} \cdot \tfrac{1}{n \cdot \#B}},$$

and, according to Lemma 2,

$$\mathfrak{s}_\mathbf{L}(\mathrm{rad}\,\mathbf{L}) = \tfrac{n}{n-1} \cdot \mathfrak{s}_C(\mathrm{rad}\,\mathbf{L}) = \tfrac{n}{n-1} \cdot \#C \cdot (\mathrm{rad}\,\mathbf{L})^m = \sqrt[n]{\tfrac{\#C}{(n-1) \cdot \#B}}\,.$$

Choosing n, $\#C$ and $\#B$ appropriately yields all three cases in Lemma 2.

If we set $m := 1$, $\#B := 1$, and $C := X \setminus B$, whence $\#C = r - 1$, we obtain a slight generalization of Kuich's example [Ku70, Example 1] (see also [JL75, Example 4.1]) to alphabets of cardinality $\#X = r \geq 2$, yielding $\mathsf{H_L} = h_r(\frac{n-1}{n})$ where $h_r(p) = -(1-p) \cdot \log_r(1-p) - p \cdot \log_r \frac{p}{r-1}$ is the r-ary entropy function well-known from information theory (cf. [Jo92, Section 2.3]). This function satisfies $0 \leq h_r(p) \leq 1$ for $0 \leq p \leq 1$ and $h_r(p) = 1$ iff $p = \frac{r-1}{r}$. ❑

In the case of Corollary 5 when $\mathfrak{s}_C(\text{rad }\mathbf{L})^{n-1} \cdot \mathfrak{s}_B(\text{rad }\mathbf{L}) < \frac{(n-1)^{n-1}}{n^n}$ the value $\mathfrak{s_L}(\text{rad }\mathbf{L})$ is a single root of Eq. (20). Then the results of Section 3.2 show that $\mathfrak{s}_B(t) = \frac{1}{n}$ and simultaneously $\mathfrak{s}_C(t) = \frac{n-1}{n}$ is impossible for $t \leq \text{rad }\mathbf{L}$. The other cases listed in Theorem 3 are possible. This can be shown using Lukasiewicz languages constructed in Example 2 as basic codes $C \cup B$ and splitting them appropriately.

Example 3. Let $X := \{a, b, d\}$ and define the following Lukasiewicz languages:

$$\mathbf{L}_1 := \{a\} \cup b \cdot \mathbf{L}_1^2, \quad \mathbf{L}_2 := \{a, d\} \cup b \cdot \mathbf{L}_2^2, \text{ and } \mathbf{L}_3 := \{a\} \cup \{b, d\} \cdot \mathbf{L}_3^2$$

Then $\text{rad }\mathbf{L}_1 = \frac{1}{2}$, $\text{rad }\mathbf{L}_2 = \text{rad }\mathbf{L}_3 = \frac{1}{2\sqrt{2}}$, $\mathfrak{s}_{L_1}(\text{rad }\mathbf{L}_1) = 1$, $\mathfrak{s}_{L_2}(\text{rad }\mathbf{L}_2) = \sqrt{2}$ and $\mathfrak{s}_{L_3}(\text{rad }\mathbf{L}_3) = \frac{1}{\sqrt{2}}$.

Let $C_4 := \{a, baa\}$, $B_4 := \mathbf{L}_1 \setminus C_4$ and $\mathbf{L}_4 := C_4 \cup B_4 \cdot \mathbf{L}_4^2$. Since $\mathfrak{s}_{C_4}(\text{rad }\mathbf{L}_1) = \frac{5}{8}$ and $\mathfrak{s}_{B_4}(\text{rad }\mathbf{L}_1) = \frac{3}{8} < \frac{1}{2}$ we have $\mathfrak{s}_{C_4}(\text{rad }\mathbf{L}_1) \cdot \mathfrak{s}_{B_4}(\text{rad }\mathbf{L}_1) < \frac{1}{4}$. Thus Eq. (13) applies.

Interchanging the rôles of C_4 and B_4 yields Eq. (11) for $\mathbf{L}_4' := B_4 \cup C_4 \cdot \mathbf{L}_4'^2$.

Next, split $\mathbf{L}_2$ into the union of C_5 and B_5 in such a way that $\mathfrak{s}_{C_5}(\text{rad }\mathbf{L}_5) > \frac{1+\sqrt{2}}{2}$. Then $\mathfrak{s}_{B_5}(\text{rad }\mathbf{L}_5) < \frac{\sqrt{2}-1}{2} < \frac{1}{2}$ and $\mathfrak{s}_{C_5}(\text{rad }\mathbf{L}_2) \cdot \mathfrak{s}_{B_5}(\text{rad }\mathbf{L}_2) < \frac{1}{4}$, and Eq. (15) is satisfied for $\mathbf{L}_5 := C_5 \cup B_5 \cdot \mathbf{L}_5^2$.

Likewise, interchanging the rôles of C_5 and B_5 yields that for $\mathbf{L}_5' := B_5 \cup C_5 \cdot \mathbf{L}_5'^2$ the hypothesis of Eq. (12) is satisfied.

Finally, choosing $C_6 := \{a\}$, $B_6 := \mathbf{L}_3 \setminus C_6$, for $\mathbf{L}_6 := C_6 \cup B_6 \cdot \mathbf{L}_6^2$ the hypothesis of Eq. (10) obtains. ❑

We conclude this subsection with some useful observations about the function $\mathfrak{s_L}$.

$$\mathfrak{s}_C(t) \leq \mathfrak{s_L}(t) \leq \frac{1}{\sqrt[n]{n \cdot \mathfrak{s}_B(t)}} \text{ for all } t \leq \text{rad }\mathbf{L}$$

Lemma 3. *Let t_1 be a solution of $\mathfrak{s}_{C \cup B}(t) = 1$ and let $\mathfrak{s}_B(t_1) > \frac{1}{n}$. Then $\text{rad }\mathbf{L} \geq t_1$ and $\mathfrak{s_L}(\text{rad }\mathbf{L}) < 1$.*

4.3 The Entropies of $\mathbf{L}^*$ and $\mathbf{K}^*$

The previous part of Section 4 was mainly devoted to explain how to give estimates on the entropy of $\mathbf{L}$ and $\mathbf{K}$ on the basis of the structure generating functions of the basic codes C and B. As a byproduct we could sometimes achieve some knowledge about $\mathfrak{s_L}(\text{rad }\mathbf{L})$ and, consequently, also about $\mathfrak{s_K}(\text{rad }\mathbf{L})$ useful in the light of Property 5.

We are going to explore this situation in more detail in this section.

We distinguish the two cases when $\mathsf{H_L} = \mathsf{H_K} = \mathsf{H}_{C \cup B}$ (which is in fact the case rad $\mathbf{L}$ = rad $(C \cup B)$) and $\mathsf{H_L} = \mathsf{H_K} < \mathsf{H}_{C \cup B}$. Note that these two cases do not fully coincide with the respective cases of Corollaries 4 and 5.

Theorem 6. *Let* $\mathsf{H_L} = \mathsf{H_K} = \mathsf{H}_{C \cup B}$. *Then the following three implications hold.*

1. *If* $\mathfrak{s}_{C \cup B}(\mathrm{rad}\,(C \cup B)) \leq 1$ *then*
$$\mathsf{H_L} = \mathsf{H_K} = \mathsf{H_{L^*}} = \mathsf{H_{K^*}} = \mathsf{H}_{(C \cup B)^*}.$$

2. *If* $\mathfrak{s}_{C \cup B}(\mathrm{rad}\,(C \cup B)) > 1$ *and* $\mathfrak{s}_B(\mathrm{rad}\,(C \cup B)) > \frac{1}{n}$ *then*
$$\mathsf{H_L} = \mathsf{H_K} = \mathsf{H_{L^*}} < \mathsf{H_{K^*}} = \mathsf{H}_{(C \cup B)^*}.$$

3. *If* $\mathfrak{s}_{C \cup B}(\mathrm{rad}\,(C \cup B)) > 1$ *and* $\mathfrak{s}_B(\mathrm{rad}\,(C \cup B)) < \frac{1}{n}$ *then*
$$\mathsf{H_L} = \mathsf{H_K} = \mathsf{H_{K^*}} < \mathsf{H_{L^*}} = \mathsf{H}_{(C \cup B)^*}.$$

Theorem 7. *Let* $\mathsf{H_L} = \mathsf{H_K} > \mathsf{H}_{C \cup B}$. *Then the following three implications hold.*

1. *If* $\mathfrak{s}_{C \cup B}(\mathrm{rad}\,\mathbf{L}) = 1$ *then*
$$\mathsf{H_L} = \mathsf{H_K} = \mathsf{H_{L^*}} = \mathsf{H_{K^*}} = \mathsf{H}_{(C \cup B)^*}.$$

2. *If* $\mathfrak{s}_{C \cup B}(\mathrm{rad}\,\mathbf{L}) > 1$ *and* $\mathfrak{s}_B(\mathrm{rad}\,\mathbf{L}) > \frac{1}{n}$ *then*
$$\mathsf{H_L} = \mathsf{H_K} = \mathsf{H_{L^*}} < \mathsf{H_{K^*}} = \mathsf{H}_{(C \cup B)^*}.$$

3. *If* $\mathfrak{s}_{C \cup B}(\mathrm{rad}\,\mathbf{L}) > 1$ *and* $\mathfrak{s}_B(\mathrm{rad}\,\mathbf{L}) < \frac{1}{n}$ *then*
$$\mathsf{H_L} = \mathsf{H_K} = \mathsf{H_{K^*}} < \mathsf{H_{L^*}} = \mathsf{H}_{(C \cup B)^*}.$$

In conclusion, one should remark that in the case of entropy of Lukasiewicz languages a similar situation as in the case of their Bernoulli measures appears. In order to achieve maximum possible entropy for both Lukasiewicz languages **L** and K it is necessary and sufficient to choose basic codes C and B whose power series $\mathfrak{s}_C(t)$ and $\mathfrak{s}_B(t)$ behave in agreement with the composition parameter n of the Lukasiewicz language.

Acknowledgement. My thanks are due to Jeanne Devolder who helped me in simplifying my first rather lengthy proof of Proposition 2.

References

[AB97] J.-M. Autebert, J. Berstel and L. Boasson, Context-Free Languages and Pushdown Automata, in: *Handbook of Formal Languages*, G. Rozenberg and A. Salomaa (Eds.), Vol. 1, Springer-Verlag, Berlin 1997, 111–174.
[BP85] J. Berstel and D. Perrin. *Theory of Codes*. Academic Press, Orlando 1985.

[DL94] J. Devolder, M. Latteux, I. Litovski and L. Staiger, Codes and Infinite Words, *Acta Cybernetica*, **11** (1994), 241–256.
[Ei74] S. Eilenberg, *Automata, Languages and Machines*. Vol. A, Academic Press, New York 1974.
[Fe95] H. Fernau, Valuations of Languages, with Applications to Fractal Geometry, *Theoret. Comput. Sci.*, **137** (1995), 177–217.
[Jo92] R. Johannesson, *Informationstheorie*, Addison-Wesley, 1992.
[JL75] J. Justesen and K. Larsen, On Probabilistic Context-Free Grammars that Achieve Capacity, *Inform. Control* **29** (1975), 268–285.
[Ku70] W. Kuich, On the Entropy of Context-Free Languages, *Inform. Control* **16** (1970), 173–200.
[LV93] M. Li and P.M.B. Vitányi, *An Introduction to Kolmogorov Complexity and its Applications*. Springer-Verlag, New York, 1993.
[St85] L. Staiger, On infinitary finite length codes, RAIRO–*Inform. Théor.*, **20** (1986), 483–494.
[St93] L. Staiger, Kolmogorov Complexity and Hausdorff Dimension. *Inform. and Comput.* **103** (1993), 159–194.

Collapsing Words vs. Synchronizing Words

D.S. Ananichev and M.V. Volkov*

Department of Mathematics and Mechanics,
Ural State University, 620083 Ekaterinburg, RUSSIA
{Dmitry.Ananichev,Mikhail.Volkov}@usu.ru

Abstract. We investigate the relationships between two types of words that have recently arisen in studying "black-box" versions of the famous Černý problem on synchronizing automata. Considering the languages formed by words of each of these types, we verify that one of them is regular while the other is not, thus showing that the two notions in question are different. We also discuss certain open problems concerning words of minimum length in these languages.

1 Background and Motivation

Let $\mathcal{A} = \langle Q, A, \delta \rangle$ be a finite automaton, where, as usual, Q denotes the state set, A stands for the input alphabet, and $\delta : Q \times A \to Q$ is the transition function defining an action of the letters in A on Q. The action extends in a unique way to an action $Q \times A^+ \to Q$ of the free A-generated semigroup A^+; the latter action is still denoted by δ. The automaton $\mathcal{A}$ is called *synchronizing* (or *directable*) if there exists a word $w \in A^+$ whose action resets $\mathcal{A}$, that is, brings all its states to a particular one: $\delta(q, w) = \delta(q', w)$ for all $q, q' \in Q$. Any word w with this property is said to be a *reset* word for the automaton. It is rather natural to ask how long such a word may be. We refer to the question of determining the length of the shortest reset word as to the *Černý problem*. Černý conjectured in [1]—that is, almost 40 years ago—that for any synchronizing automaton $\mathcal{A}$ there exists a reset word (clearly, depending on the structure of $\mathcal{A}$) of length $(|Q| - 1)^2$. Although being confirmed in some special cases (cf. [2,8,3,4], to mention a few most representative papers only), this simply looking conjecture still constitutes an open problem.

Now suppose that $\mathcal{A}$ is a black-box automaton with the input alphabet A. This means that we have no information of internal structure of $\mathcal{A}$ except the number n of its states. We want to construct an input word $w \in A^+$ which resets $\mathcal{A}$ in the above sense, provided the automaton is synchronizing. Does such an n-*synchronizing* word exist? The answer becomes pretty obvious as soon as one realizes that the desired property of w amounts to claim that w should reset **every** synchronizing automaton with n states and with the input alphabet A. Since the number of such automata is finite, we can take a reset word for each of them and concatenate all these words.

Given A and n, how long may be an n-synchronizing word? It can be easily shown that its length cannot be less than $|A|^{n-1}$ [7, Theorem 7]. Apparently, the word obtained

---

* The authors acknowledge support from the Russian Education Ministry through the Grant Center at St Petersburg State University, grant E00-1.0-92, and from the INTAS through the Network project 99-1224 "Combinatorial and Geometric Theory of Groups and Semigroups and its Applications to Computer Science".

from the simple reasoning above is much longer, and it is also true that the reasoning provides no construction in the proper sense of the word. The next natural idea (modulo the Černý conjecture) is to concatenate all words in A^+ of length $(n-1)^2$: the resulting word of length $(n-1)^2|A|^{(n-1)^2}$ will contain as a factor a reset word for each synchronizing automaton with n states and with the input alphabet A, and thus, it will be n-synchronizing. With a bit more effort, one can extract from this idea a construction for an n-synchronizing word of length $O(|A|^{(n-1)^2})$ (cf. [5, Theorem 3.1] or [7, Corollary 2]). In order to avoid relying on the still unproved conjecture, one can instead employ a result by Pin [11] so far yielding the best approximation to the length of the shortest reset words for synchronizing automata: the same construction then produces a somewhat longer, but non-conditional n-synchronizing word of length $O(|A|^{\frac{1}{6}(n^3-n)})$ [7, Corollary 6].

A surprising observation in [7] is that the idea discussed in the previous paragraph does not lead to the best possible estimation of the length of an n-synchronizing word. It turns out that the shortest n-synchronizing word does not need consisting of all shortest "individual" reset words somehow put together. Using a combinatorial approach by Sauer and Stone [12], the authors of [7] have found a transparent recursive construction for an n-synchronizing word with length $O(|A|^{\frac{1}{2}(n^2-n)})$ [7, Corollary 14]. We see that the word obtained this way is indeed essentially shorter than any word arising from the "concatenation" approach outlined above.

The recursion used in [12] and then in [7] is based on a property of words which appears to be of independent interest. In order to introduce this property we recall a few basic definitions concerning transformations. By a *transformation* of a set X we mean an arbitrary function f whose domain is X and whose range (denoted by $\mathsf{Im}(f)$) is a non-empty subset of X. The *rank* $\mathsf{rk}(f)$ of the function f is the cardinality of its range. Transformations of X form a semigroup under the usual composition of functions; the semigroup is called *the full transformation semigroup over* X and is denoted by $T(X)$. If the set X is finite with n elements, the semigroup $T(X)$ is also denoted by T_n.

Given a transformation f of a finite set X, we denote by $\mathsf{df}(f)$ its *deficiency*, that is, the difference $|X| - \mathsf{rk}(f)$. For a homomorphism

$$\varphi : A^+ \to T(X),$$

we denote by $\mathsf{df}(\varphi)$ the maximum of the deficiencies $\mathsf{df}(v\varphi)$ where v runs over A^+. Now we say that a word $w \in A^+$ is *k-collapsing* (*has property* Δ_k or *witnesses for deficiency* k in terminology of [12] and respectively [7]), provided that **for all homomorphisms** $\varphi : A^+ \to T(X)$ where X is a finite set, $\mathsf{df}(w\varphi) \geq k$ whenever $\mathsf{df}(\varphi) \geq k$.

We note that since the cardinality of the set X is **not** fixed in the definition of a k-collapsing word, the existence of such words for every k is not completely obvious. However, it should be clear that if $A = \{a_1, \ldots, a_t\}$, then the product $w = a_1 \cdots a_t$ is 1-collapsing. (Indeed, if $\mathsf{df}(\varphi) \geq 1$ for a homomorphism $\varphi : A^+ \to T(X)$, then at least one of the letters $a_1, \ldots, a_t$ should be evaluated at a transformation which is not a permutation whence $w\varphi$ is not a permutation as well). Using this observation as the induction basis, one can then prove that k-collapsing words exist for all k, see [12, Theorem 3.3] or [7, Theorem 13] for details.

Even though the notion of a collapsing word has arisen in a purely combinatorial environment, it is (as observed in [7]) rather natural from the standpoint of automata theory because it is tightly connected with Pin's extension of the Černý problem (see [9, 10]). Recall that the transition function $\delta : Q \times A^+ \to Q$ of the automaton $\mathcal{A} = \langle Q, A, \delta \rangle$ defines a natural homomorphism $\varphi : A^+ \to T(Q)$ via the rule

$$\begin{aligned} \varphi : v &\mapsto v\varphi : Q \to Q \\ v\varphi : q &\mapsto \delta(q, v) \end{aligned} \quad (1)$$

Suppose that the deficiency of this homomorphism is at least k, where $1 \leq k < |Q|$. Then the *Pin problem* is to determine the length of the shortest word $w \in A^+$ verifying $\mathrm{df}(w\varphi) \geq k$. Clearly, the aforementioned Černý problem corresponds to the case $k = |Q| - 1$. (Pin also generalized the Černý conjecture in the following natural way: if $\mathrm{df}(\varphi) \geq k$, then there exists a word $w \in A^+$ of length k^2 for which $\mathrm{df}(w\varphi) \geq k$. In [9, 10] he proved this generalized conjecture for $k \leq 3$, but recently J. Kari [6] exhibited a counter example in the case $k = 4$.)

A comparison between the Pin problem and the problem of determining the shortest k-collapsing word immediately reveals an obvious similarity in them. In fact, the only difference between the two situations is that in the former case we look for the shortest rank-decreasing word for a particular homomorphism of deficiency $\geq k$, while in the latter case we are interested in an object with the same property but with respect to an arbitrary homomorphism of deficiency $\geq k$, in other words, with respect to a black-box automaton. Thus, collapsing words relate to the Pin problem exactly in the same way in which synchronizing words relate to the Černý problem. As for the connection between these two types of words, we immediately observe the following fact:

Proposition 1. *If a word is k-collapsing, then it is $(k + 1)$-synchronizing.*

So far the best constructions of synchronizing words have been based on Proposition 1. Intuitively, however, the property of being k-collapsing looks stronger than that of being $(k + 1)$-synchronizing—recall that, for a given k, the cardinality of the state set is fixed in the latter case while the former property applies to arbitrarily large automata. Thus, studying the relationship between the two classes of words is natural if one aims to clarify if there exists a way to a more economic construction for synchronizing words than those using collapsing words.

2 Languages of Collapsing and Synchronizing Words

We fix a finite alphabet A. Let $\mathcal{C}_k$ and $\mathcal{S}_n$ denote the languages of all k-collapsing words and respectively of all n-synchronizing words over A ($k = 0, 1, 2, \ldots, n = 1, 2, 3, \ldots$). With this notation, Proposition 1 can be written as the inclusion

$$\mathcal{C}_k \subseteq \mathcal{S}_{k+1} \quad (2)$$

It is easy to show that $\mathcal{C}_1 = \mathcal{S}_2$ (the language consists of all words involving every letter in A); another simple case is $|A| = 1$ when the languages $\mathcal{C}_k$ and $\mathcal{S}_{k+1}$ are readily seen to coincide for all $k = 0, 1, 2, \ldots$. It turns out that the inclusion (2) is strict in all other cases.

Theorem 1. *For alphabets with at least two letters and for any $k \geq 2$, the language $\mathcal{C}_k$ is strictly contained in the language $\mathcal{S}_{k+1}$.*

We obtain Theorem 1 comparing the two following propositions:

Proposition 2. *The language $\mathcal{S}_n$ is regular for all $n = 1, 2, 3, \ldots$.*

Proposition 3. *For alphabets with at least two letters and for any $k \geq 2$, the language $\mathcal{C}_k$ is not regular.*

Proposition 2 is basically known. Indeed, if $\mathcal{A} = \langle Q, A, \delta \rangle$ is a synchronizing automaton, then the language of all words that reset $\mathcal{A}$ is regular (see, for example, [13, Theorem 9.9])—it is recognized by the power automaton $\mathcal{P}(\mathcal{A})$ whose states are non-empty subsets of the set Q and whose transition function δ' is defined via

$$\delta'(P, a) = \{\delta(q, a) \mid q \in P \subseteq Q, \ a \in A\},$$

provided one takes Q as the initial state of $\mathcal{P}(\mathcal{A})$ and the singleton subsets as its final states. Thus, the language $\mathcal{S}_n$ is regular as the intersection of the "reset" languages of all synchronizing automata with n states and with the input alphabet A (recall that the number of such automata is finite).

In order to illustrate the basic ideas of the proof of Proposition 3 without going into bulky detail, we restrict here to its partial case when $k = 2$. Take two different letters $a, b \in A$ and a positive integer ℓ and consider the word

$$w_\ell(a, b) = b^\ell a b^{\ell+1} a.$$

Lemma 1. *Let $\mathcal{A} = \langle Q, A, \delta \rangle$ be an automaton and let $\varphi : A^+ \to T(Q)$ be the homomorphism defined via (1). If $\mathsf{df}(\varphi) \geq 2$, then at least one of the following three conditions holds true:*

(i) *there exist two different letters $a, b \in A$ such that $\mathsf{df}(aba\varphi) \geq 2$;*
(ii) *there exist two different letters $a, b \in A$ such that $\mathsf{df}(ab^2 a\varphi) \geq 2$ and $\mathsf{df}(aw_m(a, b)w_\ell(a, b)\varphi) \geq 2$ whenever m and ℓ are coprime;*
(iii) *there exist three different letters $a, b, c \in A$ such that $\mathsf{df}(abca\varphi) \geq 2$.*

Proof. Let $P = \{a \in A \mid a\varphi \text{ is a permutation of } Q\}$, $D = A \setminus P$. Clearly, $D \neq \varnothing$. Take an arbitrary letter $a \in D$ and denote the transformation $a\varphi$ by α. We may assume that $\mathsf{df}(\alpha) = \mathsf{df}(\alpha^2) = 1$—otherwise (ii) holds in an obvious way (with a in the role of b and an arbitrary other letter in the role of a). Since $\mathsf{df}(\alpha) = 1$, there exists a unique state $x_a \in Q \setminus \mathsf{Im}(\alpha)$. Since $\mathsf{df}(\alpha) = \mathsf{df}(\alpha^2)$, α restricted to $\mathsf{Im}(\alpha)$ is a permutation. This implies the existence of a unique state $y_a \in \mathsf{Im}(\alpha)$ such that $x_a \alpha = y_a \alpha$.

In the course of the proof we shall often refer to the following easy observation:

Lemma 2. *Let $\alpha = a\varphi$ be as above, π an arbitrary permutation of the set Q. Then $\mathsf{df}(\alpha \pi \alpha) \geq 2$ if and only if $x_a \pi \notin \{x_a, y_a\}$.*

Proof. It is clear that $\mathsf{df}(\alpha \pi \alpha) \geq 2$ if and only if $x_a, y_a \in \mathsf{Im}(\alpha \pi) = \mathsf{Im}(\alpha)\pi$. Since $Q = \{x_\alpha\} \cup \mathsf{Im}(\alpha)$ and π is a permutation, the latter condition is equivalent to $x_a \pi \notin \{x_a, y_a\}$, as required. □

Returning to the proof of Lemma 1, first suppose that $|D| = 1$. Denote the only letter in D by a and let $x = x_a$, $y = y_a$. Since $\mathsf{df}(\varphi) \geq 2$, there must exist a letter $b \in P$ such that the permutation $\beta = b\varphi$ does not fix the state x. Besides that, there must be a letter $c \in P$ such that the permutation $\gamma = c\varphi$ does not preserve the set $\{x,y\}$. If there is no letter that combines both the properties, then β transposes x and y while $x\gamma = x$ and $y\gamma \notin \{x,y\}$. Then $x\beta\gamma = y\gamma \notin \{x,y\}$, and we can apply Lemma 2 to the permutation $\beta\gamma$. Thus, $\mathsf{df}(abca\varphi) = \mathsf{df}(\alpha\beta\gamma\alpha) \geq 2$, and (iii) holds.

We may therefore assume that there exits a letter $b \in P$ such that both $x\beta \neq x$ and $\{x,y\}\beta \neq \{x,y\}$. For each positive integer s, consider the word $ab^s a$. By Lemma 2 $\mathsf{df}(ab^s a\varphi) \geq 2$ if and only if $x\beta^s \notin \{x,y\}$. In particular, if $x\beta \notin \{x,y\}$, then $\mathsf{df}(aba\varphi) \geq 2$ and (i) holds. Now suppose that $x\beta \in \{x,y\}$, that is, $x\beta = y$ while $y\beta \notin \{x,y\}$. Then $x\beta^2 \notin \{x,y\}$ and $\mathsf{df}(ab^2 a\varphi) \geq 2$. In order to show that (ii) holds, it remains to verify that for each pair (m,ℓ) of coprime integers, $\mathsf{df}(aw_m(a,b)w_\ell(a,b)\varphi) \geq 2$. If $x\beta^m \notin \{x,y\}$, then $\mathsf{df}(ab^m a\varphi) \geq 2$, and we are done because $ab^m a$ is a factor of the word $aw_m(a,b)w_\ell(a,b)$. If $x\beta^m = y$, then $x\beta^{m+1} \notin \{x,y\}$ whence $\mathsf{df}(ab^{m+1}a) \geq 2$, again yielding the desired conclusion. Thus, we may assume that $x\beta^m = x$, and similarly, $x\beta^\ell = x$. But m and ℓ are coprime whence $rm + q\ell = 1$ for some integers r, q. Using this, we easily deduce that

$$x\beta = x\beta^{rm+q\ell} = x(\beta^m)^r(\beta^\ell)^q = x$$

which contradicts our choice of the letter b. Thus, (ii) holds indeed.

Now consider the situation when $|D| > 1$. Take two arbitrary letters $a, b \in D$. If $\{x_b, y_b\} \subseteq \mathsf{Im}(\alpha)$, then $\mathsf{df}(ab\varphi) \geq 2$, and (i) holds. Therefore we may assume that for all $a, b \in D$, $x_a \in \{x_b, y_b\}$ and $x_b \in \{x_a, y_a\}$. Under this assumption, there are three possible cases.

Case 1: there exists a state $x \in Q$ such that $x_a = x$ for all $a \in D$ but $y_a \neq y_{a'}$ for some $a, a' \in D$.

Since $\mathsf{df}(\varphi) \geq 2$, there must exist a letter $b \in P$ such that the permutation $\beta = b\varphi$ does not fix the state x. We can also choose a letter $a \in D$ such that $y_a \neq x\beta$. Then $x\beta \notin \{x, y_a\}$, whence $\mathsf{df}(aba\varphi) \geq 2$ by Lemma 2, and (i) holds.

Case 2: there exists a subset $\{x,y\} \subseteq Q$ such that $\{x_a, y_a\} = \{x,y\}$ for all $a \in D$ and there are two letters $c, d \in D$ such that $x_c = y_d$, $x_d = y_c$.

Since $\mathsf{df}(\varphi) \geq 2$, there must exist a letter $b \in P$ such that the permutation $\beta = b\varphi$ does not preserve the set $\{x,y\}$. Therefore, for some letter $a \in \{c,d\}$, we have $x_a\beta \notin \{x,y\}$. Again, Lemma 2 implies that $\mathsf{df}(aba\varphi) \geq 2$, and (i) holds.

Case 3: there exist states $x, y \in Q$ such that $x_a = x$, $y_a = y$ for all $a \in D$.

This case is completely analogous to the case of an automaton with a unique non-permutation letter which we already have analyzed. □

Now we fix a pair (m, ℓ) of positive integers and a pair (a, b) of different letters in A. Consider the word

$$W[m, \ell] = Vaba^2 w_m(a,b)w_\ell(a,b)$$

in which $V = V_1 V_2 V_3 V_4$ with

- V_1 being a product of all words of the form cdc, cd^2c, $cdec$ for all collections of different letters $c, d, e \in A \setminus \{a, b\}$;
- V_2 being a product of all words of the form cdc, cd^2c, $cdec$ for all collections of different letters $c, d, e \in A \setminus \{a\}$;
- V_3 being a product of all words of the form $abcabacb$, $acbabcab$ for all letters $c \in A \setminus \{a, b\}$;
- V_4 being a product of all words of the form cdc, cd^2c, $cdec$ for all collections of different letters $c, d, e \in A \setminus \{b\}$.

Lemma 3. *Let p be an odd prime and $\ell \geq p$. The word $W[p, \ell]$ belongs to the language $\mathcal{C}_2$ if and only if p does not divide ℓ.*

Proof. **Necessity.** Consider the automaton $\mathcal{C}_n$ whose states are the residues modulo n and whose transition function is defined by $\delta(0, a) = 1$, $\delta(m, a) = m$ for $0 < m < n$; $\delta(m, b) = m + 1 \pmod{n}$, $\delta(m, c) = m$ for all $c \in A \setminus \{a, b\}$ and for all $m = 0, 1, \ldots, n-1$ (cf. Figure 1 in which we have omitted the letters whose action is trivial). This automaton is known to be synchronizing (in fact, it was $\mathcal{C}_n$ that Černý used in his pioneering paper [1] as the example of a synchronizing automaton with n states whose shortest reset word is of length $(n - 1)^2$.) Therefore if $\varphi : A^+ \to T_n$ is the homomorphism induced by the function δ as in (1), then $\mathrm{df}(\varphi) = n - 1$. Letting $n = p$, an odd prime, we obtain $\mathrm{df}(\varphi) \geq 2$, but it amounts to a straightforward calculation to check that if p divides ℓ, then $\mathrm{df}(W[p, \ell]\varphi) = \mathrm{df}(b^r(ab)^q a^s a w_p(a, b) w_\ell(a, b)\varphi) = 1$. This shows that $W[p, \ell] \notin \mathcal{C}_2$.

Sufficiency readily follows from the construction of the word $W[p, \ell]$ and Lemma 1. □

Proof of Proposition 3 (for $k = 2$). Suppose that the language $\mathcal{C}_2$ is recognized by the automaton $\mathcal{A} = \langle Q, A, \delta \rangle$ with the initial state q_0. Since the state set Q is finite, there exist two odd primes p and r such that $p < r$ and

$$\delta(q_0, Vaba^2 w_p(a, b)) = \delta(q_0, Vaba^2 w_r(a, b)). \tag{3}$$

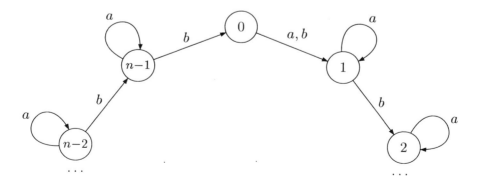

Fig. 1. The automaton $\mathcal{C}_n$

Let s be the least integer such that $ps > r$. Then applying the word $w_{ps}(a,b) = b^{ps}ab^{ps+1}a$ to the state in (3), we obtain

$$\delta(q_0, W[p, ps]) = \delta(q_0, W[r, ps]). \tag{4}$$

Since ps is divisible by p, but not by r, Lemma 3 implies that the automaton $\mathcal{A}$ accepts the word $W[r, ps]$ but rejects the word $W[p, ps]$. This means that the right-hand side of the equality (4) represents a final state, while the left-hand side does not, a contradiction. □

The proof of Proposition 3 for the general case will be published elsewhere.

3 Words of Minimum Length in $\mathcal{S}_3$ and $\mathcal{C}_2$

When studying collapsing or synchronizing words, we are especially interested in the shortest possible words within these classes. From this point of view, Theorem 1 is not yet sufficient because in spite of the inequality $\mathcal{C}_k \neq \mathcal{S}_{k+1}$, the set $\min \mathcal{C}_k$ of words of minimum length in $\mathcal{C}_k$ may well coincide with the set $\min \mathcal{S}_{k+1}$. This happens, for example, for the languages $\mathcal{C}_2$ and $\mathcal{S}_3$ over $\{a, b\}$. Indeed, in Section 2 we proved that $\mathcal{C}_2 \subsetneq \mathcal{S}_3$. On the other hand, it can be easily checked that the sets of words of minimum length in these two languages coincide. Namely, one can prove (via constructing appropriate 3-state automata) that every 2-collapsing or 3-synchronizing word of minimum length over $\{a,b\}$ should contain as a factor every word of length 3 except a^3 and b^3. This immediately shows that the minimum length of a word in $\mathcal{C}_2$ or in $\mathcal{S}_3$ is 8. Then searching through words of length 8 yields that both $\min \mathcal{C}_2$ and $\min \mathcal{S}_3$ consists of the words aba^2b^2ab, $abab^2a^2b$, ab^2a^2bab, ab^2aba^2b and their mirror images. In contrast, here we exhibit an example indicating that over three letters one should expect $\min \mathcal{C}_2 \neq \min \mathcal{S}_3$ (and giving an alternative proof that $\mathcal{C}_2 \neq \mathcal{S}_3$ over three letters).

Proposition 4. *The word*

$$w_{20} = abc^2 a^2 b \cdot cbabc \cdot b^2 (ca)^2 ab$$

of length 20 is 3-synchronizing but not 2-collapsing.

The word was first discovered by a computer program written by our students Ilya Petrov and Anton Savin. (In fact, their program has made an exhaustive search through all the words of length 20 over $\{a, b, c\}$ which start with ab and contain none of the factors a^3, b^3, or c^3 and has found 22 3-synchronizing words among them.) We have then managed to check that w_{20} is 3-synchronizing by hand. Clearly, this proof requires a lengthy case-by-case analysis, so we cannot reproduce it here. In a rough outline, the proof works as follows. We have constructed 96 regular languages $\mathcal{L}^{(i)}$ over $\{a, b, c\}$ such that a word $w \in \{a, b, c\}^+$ is 3-synchronizing if and only if for each $i = 1, \ldots, 96$, w has a factor from the language $\mathcal{L}^{(i)}$ (the factors corresponding to different languages may overlap and even coincide). Then we have verified that the word w_{20} indeed contains a factor from each of these test languages. For instance, the factor $cbabc$ (which we have distinguished in the above definition of w_{20}) "represents" 4 test languages:

$$c(b+a^2)^*ab(a+b^2+bab^*ab)^*c,$$
$$c(a+b^2)^*ba(b+a^2+aba^*ba)^*c,$$
$$c(b+ab^*a)((a+b)(b+ab^2a))^*c,$$
$$cb^*(a^2+ab)(b^2+ba+ab^*(a^2+ab))^*c.$$

In order to verify that the word w_{20} fails to be 2-collapsing, consider the following automaton:

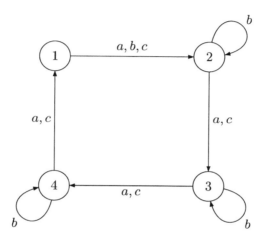

Fig. 2. The automaton showing that w_{20} is not 2-collapsing

The automaton is synchronizing as it is obtained from the Černý automaton C_4 by doubling a letter, but a straightforward calculation shows that the word w_{20} does not collapse its state set to a 2-element subset.

Another program by Petrov and Savin has checked all words of length 19 over $\{a,b,c\}$ and found no 3-synchronizing words among them. Even though this result so far has been confirmed neither by any proof done by hand nor by an alternative computer calculation, it is very tempting to conjecture that w_{20} is a 3-synchronizing word of minimum length. If this is the case, then indeed $\min C_2 \neq \min S_3$.

Sauer and Stone formulate in [12] the following open problem: for a given alphabet with t letters, determine for each positive integer k the minimum length $\mu_k(t)$ of k-collapsing words. Obviously $\mu_1(t) = t$ for any t; besides that, the only value of the function $\mu_k(t)$ known so far is $\mu_2(2) = 8$. From [7, Theorem 13] one can extract the estimation $\mu_2(t) \leq t^3 + 3t^2 + 2t$ which for $t = 3$ gives $\mu_2(3) \leq 60$. The following result shows that in fact $\mu_2(3) \leq 27$.

Proposition 5. *The word*
$$w_{27} = abc^2b^2c \cdot bca^2c^2a \cdot cab^2a^2b \cdot (abc)^2$$
of length 27 *is 2-collapsing.*

Again we omit the proof. Recently our student Pavel Martjugin has constructed a 2-collapsing word of length 58 over four letters.

Acknowledgments. We thank the anonymous referees for their valuable remarks.

References

1. Černý, J.: Poznámka k homogénnym eksperimentom s konecnými avtomatami. Mat.-Fyz. Cas. Slovensk. Akad. Vied. **14** (1964) 208–216 [in Slovak].
2. Černý, J., Pirická, A., Rosenauerova, B.: On directable automata. Kybernetika (Praha) **7** (1971) 289–298.
3. Dubuc, L.: Les automates circulaires biases verifient la conjecture de Černý. RAIRO, Informatique Théorique et Applications **30** (1996) 495–505 [in French].
4. Goehring, W.: Minimal initializing word: A contribution to Černý conjecture. J. Autom. Lang. Comb. **2** (1997) 209–226.
5. Ito, M., Duske, J.: On cofinal and definite automata. Acta Cybernetica **6** (1983) 181–189.
6. Kari, J.: A counter example to a conjecture concerning synchronizing words in finite automata. EATCS Bulletin **73** (2001) 146.
7. Margolis, S., Pin, J.-E., Volkov, M. V.: Words guaranteeing minimal image. Proc. III Internat. Colloq. on Words, Languages and Combinatorics, submitted.
8. Pin, J.-E.: Sur un cas particulier de la conjecture de Černý. Automata, Languages, Programming; 5th Colloq., Udine 1978, Lect. Notes Comput. Sci. **62** (1978) 345–352 [in French].
9. Pin, J.-E.: Le Problème de la Synchronisation. Contribution à l'Étude de la Conjecture de Černý. Thèse de 3éme cycle. Paris, 1978 [in French].
10. Pin, J.-E.: Sur les mots synchronisants dans un automate fini. Elektronische Informationverarbeitung und Kybernetik **14** (1978) 283–289 [in French].
11. Pin, J.-E.: On two combinatorial problems arising from automata theory. Ann. Discrete Math. **17** (1983) 535–548.
12. Sauer, N., Stone, M. G.: Composing functions to reduce image size. Ars Combinatoria **31** (1991) 171–176.
13. Starke, P. H.: Abstrakte Automaten. Berlin, 1969 [in German].

A Note on Synchronized Automata and Road Coloring Problem*

Karel Culik II[1], Juhani Karhumäki[2], and Jarkko Kari[3]

[1] Department of Computer Science
University of South Carolina
Columbia S.C. 29208, USA
culik@cs.sc.edu

[2] Department of Mathematics and Turku Centre for Computer Science
University of Turku
FIN-20014, Turku, Finland
karhumak@cs.utu.fi

[3] Department of Computer Science
14 MLH University of Iowa
Iowa City, IA 52242, USA
jjkari@cs.uiowa.edu

Abstract. We consider a problem of labeling a directed multigraph so that it becomes a synchronized finite automaton, as an ultimate goal to solve the famous Road Coloring Conjecture, cf. [1,2]. We introduce a relabeling method which can be used for a large class of automata to improve their "degree of synchronization". This allows, for example, to formulate the conjecture in several equivalent ways.

1 Introduction

Synchronization properties in automata theory are fundamental, and often at the same time very challenging. Two examples of such problems are as follows. Let us call a finite automaton $\mathcal{A}$ *synchronized* if there exists a word w which takes each state of $\mathcal{A}$ to a single special state s. Such a w is called *synchronizing* word for $\mathcal{A}$.

Now, so-called *Cerny's Conjecture* [3,11] claims that each synchronized automaton possesses a synchronizing word of length at most $(n-1)^2$, where n is the cardinality of the state set of $\mathcal{A}$. Despite many attempts the conjecture in the general case is still unsolved, the best upper bound being cubic in n, cf. [6]. However, recently in [7] a natural extension of Cerny's Conjecture stated in [12] was shown to be false.

Cerny's Conjecture asks something about synchronized automata. *Road Coloring Problem*, in turn, asks for a dual task: change, if possible, an automaton to a synchronized one. More precisely, given a deterministic complete and strongly connected automaton, can it be relabelled to a synchronized automaton.

It is well known, cf. Lemma 2, that the Road Coloring Problem has a negative answer if the greatest common divisor of the lengths of all loops in $\mathcal{A}$ is larger than 1. In the

---

* Supported by the Academy of Finland under grant 44087 and by NSF under grant CCR 97-33101.

opposite case - which due to the strong connectivity is equivalent to the existence of two loops of coprime lengths - the answer is conjectured to be affirmative. This is the *Road Coloring Conjecture, RC-conjecture* for short. In terms of graphs it is formulated as follows: Let us call a directed graph G *acceptable* if it is of uniform outdegree and strongly connected (i.e. for any pair (p, q) of vertices there is a path from p to q) and *primitive* if the greatest common divisor of lengths of its loops is one. The conjecture claims that each acceptable primitive graph can be labeled to a synchronized finite automaton.

Intuitively the above means that if a traveler in the network of colored roads modeled by an acceptable primitive graph gets lost, he can find a way back home by following a single instruction, the synchronized word.

The Road Coloring Conjecture has attracted a lot of attention over the past 20 years. However, it has been established only in a very limited cases, cf. [4,9,10], and it is stated as a "notorious open problem" in [8].

We attempt to solve the problem by analyzing properties of different labelings of finite automata. In particular, we describe a method to increase the synchronization degree of an automaton by relabeling it in a suitable way. Here the synchronization degree $n_{\mathcal{A}}$ of an automaton $\mathcal{A}$ is the minimal number of states of $\mathcal{A}$ such that there exists a word w taking each state of $\mathcal{A}$ to one of these $n_{\mathcal{A}}$ states. Unfortunately, our method does not work for all labelings, but it does work for a quite large subclass of labelings, and, moreover, allows to formulate the RC-conjecture in two equivalent ways.

This paper is organized as follows. In Section 2 we fix our terminology and formulate several conjectures connected to synchronization properties including the Road Coloring Conjecture. In Section 3 we introduce an automaton which, for a given automaton $\mathcal{A}$, computes all words having the maximal synchronizing effect, in particular all synchronizing words, if the automaton is synchronized. In Section 4 we relate the synchronization to certain equivalence relations. Section 5 introduces our method to improve the synchronization. Finally, in Section 6 we point out the equivalencies of some of our conjectures, as well as show that any primitive acceptable graph G can be commutatively (resp. lengthwise) synchronized, i.e. G can be labeled to a finite automaton $\mathcal{A}$ such that there exists a state s and a word w so that, for each state q, there exists a word w_q satisfying: w_q takes q to s and w_q is commutatively equivalent to w (resp. w_q is of the same length as w). The proofs of the results are illustrated by examples.

2 Preliminaries

In this section we fix the necessary terminology, cf. [5], and formulate several conjectures including the Road Coloring Conjecture.

Let $G = (V, E)$ be a directed graph with vertex set V and the edge set E, where multiple edges are allowed. We consider only graphs which are

(i) *strongly connected*, and
(ii) of *uniform outdegree*, i.e. all vertices have the same outdegree, say n.

Such a graph is called *acceptable*. Clearly, each acceptable graph can be labeled by an n-letter alphabet to yield a deterministic strongly connected and complete automaton without initial and final states. By a *labeling* of an acceptable graph we mean such a labeling, or also an automaton defined by such a labeling.

Let G be an acceptable graph and L_G the set of its all loops. We call G *primitive* if the greatest common divisor of the lengths of the loops in L_G is equal to 1; otherwise G is *imprimitive*. Further we call G *cyclic* if there exist an $n \geq 2$ and a partition of the vertex set V of G into the classes $V_0, \ldots, V_{n-1}$ such that whenever $p \longrightarrow q$ is an edge in G, then $p \in V_i$ and $q \in V_{i+1 \pmod{n}}$ for some $i = 0, \ldots, n-1$. Otherwise G is *acyclic*. We have an easy connection:

Lemma 1. *An acceptable graph G is imprimitive iff it is cyclic.*

In order to formulate our conjectures we recall some terminology of automata. Let $\mathcal{A} = (Q, \Sigma, \delta)$ be complete deterministic finite automaton without final and initial states. We say that the automaton $\mathcal{A}$ is *synchronized* at state $s \in Q$ if there exists a word $w \in \Sigma^*$ that takes each state q of Q into s, i.e. $\delta(q, w) = s$ for all $q \in Q$. The word w is called a *synchronizing* word for $\mathcal{A}$. Clearly, if a strongly connected automaton is synchronized at a given state it is so at any of its states.

Now, we extend the above to graphs. We say that an acceptable graph is *synchronized* if it has a labeling making it a synchronized automaton. Note that originally in [1] the word "collapsible" was used instead of "synchronized". Now, the conjecture can be stated as follows:

Road Coloring Conjecture. Each primitive acceptable graph is synchronized.

The conjecture, if true, is optimal:

Lemma 2. *Each imprimitive graph is not synchronized.*

Next we define a few special types of labelings of an acceptable primitive graph. Let δ be such a labeling. We say that two vertices p and q are *reducible*, in symbols $p \sim q$, if there exists a word w such that $\delta(p, w) = \delta(q, w)$, i.e. word w takes p and q to the same state. Accordingly such a δ is called (p, q)-*synchronized*. Clearly, the reducibility defines a relation on Q, which is symmetric and reflexive, but not necessarily transitive. If it is also transitive, i.e. an equivalence relation, and nondiscrete, i.e. some $p \neq q$ are reducible, then the labeling δ is called *strong*. The nondiscreteness is to avoid some trivial exceptions in our later considerations. Finally, the labeling δ is called *stable* (resp. (p, q)-*stable* for the pair $(p, q) \in Q^2$), if the reducibility relation is consistent with δ, i.e. for all $s, t \in Q$ and $u \in \Sigma^*$, whenever $s \sim t$, then also $\delta(s, u) \sim \delta(t, u)$ (resp. $\delta(p, u) \sim \delta(q, u)$ for all $u \in \Sigma^*$).

Now we formulate several conjectures. The first one is a weaker version of the Road Coloring Conjecture.

Conjecture A. *Let G be acceptable primitive graph. For each pair (p, q) of vertices of G there exists a (p, q)-synchronized labeling.*

Conjecture A seems to be much weaker than the RC-conjecture but it might be equivalent to it. The two other conjectures we formulate are, as we shall show, equivalent to the Road Coloring Conjecture.

Conjecture B. For each acceptable primitive graph there exists a strong labeling.

Conjecture C. Let G be an acceptable primitive graph. Then, there exist vertices p and q, with $p \neq q$, and a labeling δ such that δ is (p,q)-stable.

The following example illustrates our conjectures

Example 1. Consider the automata shown in Figure 1.

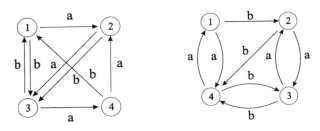

Fig. 1. Automata $\mathcal{A}$ and $\mathcal{B}$

The automaton $\mathcal{A}$ possesses a synchronizing word, for example $baaab$, while the automaton $\mathcal{B}$ does not possess any. The reducibility relations are the full relation and the relation $\{(2,3),(3,2),(1,4),(4,1)\}$, respectively. Consequently also in the latter case the reducibility is an equivalence relation, and hence the labeling is strong.

We conclude this section by generalizing the notion of a synchronized automaton in a number of ways. Let G be acceptable graph and $\mathcal{A}$ the automaton defined by the labeling δ of G. We say that the *synchronization degree* of $\mathcal{A}$ is

$$n_\mathcal{A} = \min_{w \in \Sigma^*} \{\text{card}(P) \mid \delta(Q, w) = P\},$$

and that the *synchronization degree* of G is

$$n_G = \min_\delta \{n_\mathcal{A} \mid \mathcal{A} \text{ is a labeling of } G \text{ via } \delta\}.$$

Consequently, $\mathcal{A}$ or G is synchronized iff $n_\mathcal{A}$ or n_G is equal to 1, respectively.

Finally, we say that $\mathcal{A}$ (or G) is *commutatively* (resp. *lengthwise*) synchronized if there exists a word w such that, for any vertex q, there exists a word w_q commutatively equivalent to w (resp. of the same length as w) such that $\delta(q, w_q) = \delta(p, w_p)$ for any p, $q \in Q$.

In all examples and technical considerations we consider only binary alphabets Σ. Indeed, this case seems to capture the difficulty of the problem.

3 An Automaton for Synchronizing Words

Let G be an acceptable graph and $\mathcal{A}$ the automaton obtained from it via the labeling δ. Hence $\mathcal{A} = (Q, \Sigma, \delta)$ is a complete deterministic automaton without initial and final states. We define another automaton $\mathcal{A}_s$ as follows:

$$\mathcal{A}_s = (2^Q, \Sigma, \delta_s, Q),$$

where Q is the initial state and the transition function δ_s is defined by

$$\delta_s(P, q) = \bigcup_{p \in P} \delta(p, a) \quad \text{for } P \subseteq Q, a \in \Sigma. \tag{1}$$

Clearly $\mathcal{A}_s$ is complete and deterministic, and, moreover, a word w is synchronizing for $\mathcal{A}$ if and only if $\delta_s(Q, w)$ is singleton. Hence, we have

Lemma 3. *The set of all synchronizing words for $\mathcal{A}$ is computed by the automaton $(2^Q, \Sigma, \delta_s, Q, F)$, where the set F of final states consists of all singletons of the power set 2^Q.*

Next we recall the notion of the synchronization degree $n_\mathcal{A}$ of the automaton $\mathcal{A}$:

$$n_\mathcal{A} = \min\{\text{card}(P) \mid P \text{ is reachable state of } \mathcal{A}_s\},$$

and define

$$Q_{\min} = \{P \mid \text{card}(P) = n_\mathcal{A} \text{ and } P \text{ is reachable in } \mathcal{A}_s\}.$$

Using these notions we define another automaton $\mathcal{A}_{\min}$ as follows:

$$\mathcal{A}_{\min} = (Q_{\min}, \Sigma, \delta_s).$$

The automaton $\mathcal{A}_{\min}$ plays an important role in our subsequent considerations.

Lemma 4. *The automaton $\mathcal{A}_{\min}$ is deterministic, complete and strongly connected.*

To illustrate the above notions we return to our example.

Example 1. *(continued).* Consider automata $\mathcal{A}$ and $\mathcal{B}$ of Example 1. The automata $\mathcal{A}_s$, $\mathcal{A}_{\min}$, $\mathcal{B}_s$ and $\mathcal{B}_{\min}$ are shown in Figure 2, where the min-automata are shown by the dash lines. It follows that $\mathcal{A}$ is synchronized and its shortest synchronizing word is $baaab$. On the other hand, $\mathcal{B}$ is not synchronized - its synchronization degree is 2.

Let us analyze a bit more the above automata $\mathcal{A}_s$ and $\mathcal{A}_{\min}$. Let w be a word leading the initial state Q of $\mathcal{A}_s$ into $Q_{\min}$, in other words

$$\delta_s(Q, w) = \{p_1, \ldots, p_{n_\mathcal{A}}\}. \tag{2}$$

Then w defines an equivalence relation $\sim_w$ on Q, where the equivalence classes consist of subsets $\{q \mid \delta(q, w) = p_i\}$ for $i = 1, \ldots, n_\mathcal{A}$. Of course, in general, such partitions need not be unique. However, the uniqueness is characterized by the strongness of the labeling, cf. Theorem 1 in the next section.

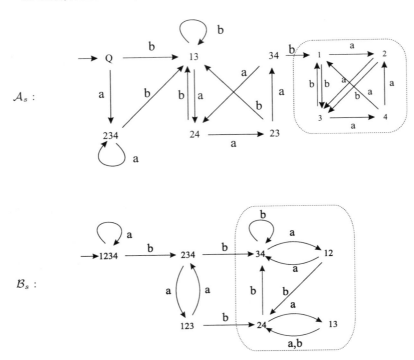

Fig. 2. Automata $\mathcal{A}_s$, $\mathcal{B}_s$, $\mathcal{A}_{\min}$, and $\mathcal{B}_{\min}$, the latter ones being those shown by dashlines.

4 Equivalence Relations vs. Synchronization

Let $\mathcal{A} = (Q, \Sigma, \delta)$ be an automaton obtained from an acceptable primitive graph via a labeling δ. From the point of the synchronization a fundamental notion is that of reducibility relation $\sim$: $p \sim q$ iff there exists a word w such that $\delta(p, w) = \delta(q, w)$. As we already noted this relation is reflexive and symmetric, but not necessarily transitive, as shown by the next example.

Example 2. Consider an automaton $\mathcal{C}$ and its variants $\mathcal{C}_s$ and $\mathcal{C}_{\min}$ shown in Figure 3. In this case states 1 and 2 are reducible by b, and states 1 and 3 by ab, but 2 and 3 are not reducible by any word. Hence, the relation $\sim$ is not an equivalence relation. This is connected to the fact that the partitions of Q defined by different words (as was explained in Section 3) need not coincide. Indeed, the word b defines the partition $\{\{1, 2\}, \{3, 4\}\}$, while the word ab defines the partition $\{\{1, 3\}, \{2, 4\}\}$.

Actually, we have the following important characterization.

Theorem 1. *The labeling δ is strong if and only if the partitions of (2) in Section 3 are independent of w. Moreover, δ is consistent with respect to the reducibility relation $\sim$.*

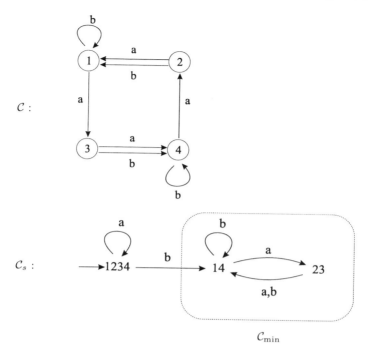

Fig. 3. Automata $\mathcal{C}$ and $\mathcal{C}_s$ as well as $\mathcal{C}_{\min}$ obtained from the latter.

Proof. Assume first that the labeling δ is strong, i.e., the reducibility relation $\sim$ is an equivalence relation. Since the states of the automaton $\mathcal{A}_{\min}$ are pairwise irreducible, the number of equivalence classes under $\sim$ is $n_\mathcal{A}$. On the other hand, by the construction of $\mathcal{A}_{\min}$, any partition in (2) induced by a fixed word w is of the same cardinality. So the independence of these partitions follows since any class in these partitions is a subset of an equivalence class of the reducibility relation.

Conversely, if the partitions in (2) are independent of w, then clearly all states which are reducible belong to a same equivalence class in (2). Hence, the reducibility matches with the unique relation, and consequently is an equivalence relation.

The second sentence follows directly from the uniqueness of the partitions (2). □

Automata like in Example 2 are problematic for our general approach. Indeed, as we shall see, whenever the labeling is strong (and hence the reducibility is an equivalence relation) we can improve the synchronization degree of a nonsynchronized automaton.

Note also that in the previous example the transitive closure of the reducibility relation is the full relation. Our next result shows that for any acceptable primitive graph such a labeling can be found.

Theorem 2. *For any acceptable primitive graph G there exists a labeling such that the transitive closure of the reducibility relation is the full relation.*

We are not able to use Theorem 2 to solve the RC-conjecture, but it can be used to prove a weaker result, as we shall see in Section 6. The proof of Theorem 2 resembles that of Theorem 3, so we postpone it after the proof of Theorem 3.

5 Improving the Synchronization

In this section we introduce our central tool to improve the synchronization degree of an automaton, or more precisely of an automaton having a strong labeling.

Let $\mathcal{A} = (Q, \Sigma, \delta)$ be a complete deterministic automaton with a *strong* labeling δ. We introduce still another automaton $\mathcal{A}_P$, referred to as a partition automaton of $\mathcal{A}$, as follows: Let $w \in \Sigma^*$ be a word such that

$$\delta(Q, w) \in Q_{\min} = \{p_1, \ldots, p_{n_{\mathcal{A}}}\}, \qquad (3)$$

and $P_1, \ldots, P_{n_{\mathcal{A}}}$ the partition of Q defined by w, i.e.

$$P_i = \{q \mid \delta(q, w) = p_i\}.$$

Since δ is strong, by Theorem 1, the partition is independent of w (if it only satisfies (3)). Now the *partition automaton* of $\mathcal{A}$ is

$$\mathcal{A}_P = (P_{\min}, \Sigma, \delta_p),$$

where $P_{\min} = \{P_1, \ldots, P_{n_{\mathcal{A}}}\}$ and δ_p is defined, for all $P, P' \in P_{\min}$ and $a \in \Sigma$, by the condition

$$P' \in \delta_p(P, a) \text{ iff there exists } p \in P, p' \in P' : \delta(p, a) = p'.$$

Let us call $P_{\min}$ the δ-partition of $\mathcal{A}$.

We obtain

Lemma 5. *The automaton $\mathcal{A}_P$ is complete, deterministic and co-deterministic.*

As a corollary of the determinism above we obtain the following crucial fact.

Lemma 6. *The partition $P_{\min}$ is consistent with the transition function δ, i.e. whenever p and p' are in the same $P_{\min}$-class so are $\delta(p, a)$ and $\delta(p', a)$ for any $a \in \Sigma$.*

Before continuing let us return to Example 1.

Example 1. *(continued)* For the automaton $\mathcal{B}$ the min-automaton $\mathcal{B}_{\min}$ and the partition automaton $\mathcal{B}_P$ are as shown in Figure 4.

Now we are ready for our crucial lemma

Theorem 3 (Relabeling lemma). *Let $\mathcal{A} = (Q, \Sigma, \delta)$ be a strong automaton of synchronization degree $n_{\mathcal{A}} \geq 2$, and $P_{\min}$ its δ-partition. Then, if the partition automaton $\mathcal{A}_P$ is not cyclic, there exists a relabeling of $\mathcal{A}$ into an automaton $\mathcal{A}' = (Q, \Sigma, \delta')$ having the synchronization degree strictly smaller than $n_{\mathcal{A}}$.*

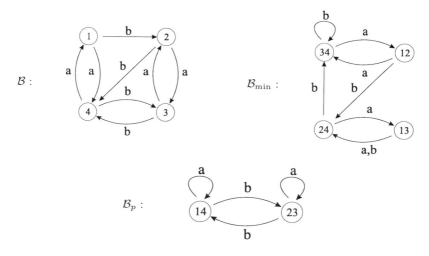

Fig. 4. Automaton $\mathcal{B}$, its min-automaton $\mathcal{B}_{\min}$, and its partition automaton $\mathcal{B}_p$.

For illustration we apply the above construction to automaton $\mathcal{B}$ of Example 1.

Example 1. *(Continued)* Consider the partition automaton $\mathcal{B}_P$ computed above. Now, we switch a and b in the labels starting from the state 23. Then the corresponding automata $\mathcal{B}'$, $\mathcal{B}'_s$ are as in Figure 5. Hence, $\mathcal{B}'$ indeed is synchronizing.

In theorem 3 we excluded the case when $\mathcal{A}_P$ is cyclic. The following simple result, which is a direct consequence of Lemma 6, takes care of that case.

Lemma 7. *If the automaton $\mathcal{A}_P$ is cyclic so is the underlying graph G.*

6 Applications

As applications of our previous lemmas and ideas we can prove the following results.

Theorem 4. *Road Coloring Conjecture is equivalent to Conjecture B.*

Proof. Clearly, the RC-conjecture implies Conjecture B. The other direction is proved by induction on the number of vertices of the graph G. Let G be an acceptable primitive graph with $n+1$ vertices. By our assumption G can be labeled to an automaton $\mathcal{A}$ with strong labeling δ. Hence, the partition automaton $\mathcal{A}_P$ exists, and it contains at most n states. Let the partition of the state set of $\mathcal{A}$ be $P_{\min}$. If $P_{\min}$ consists of only one class we are done: then G is synchronized.

Consider the other case. Since G is primitive, so is the underlying graph of $\mathcal{A}_P$. Hence, by induction hypothesis, it can be relabeled, say by δ', to an automaton having a synchronized word, say u. Let $\mathcal{A}'$ be the automaton obtained from $\mathcal{A}$ by the corresponding relabeling. Then the word u brings each class of $P_{\min}$ to a single class, say $\mathcal{P}_u$. But, by

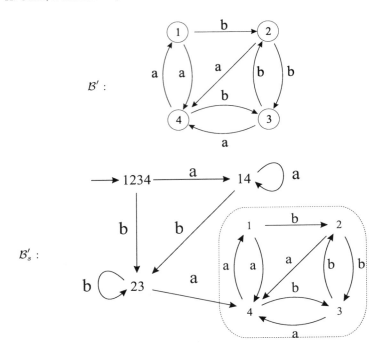

Fig. 5. Automaton $\mathcal{B}'$ obtained from $\mathcal{B}$ by relabeling and its variant $\mathcal{B}'_s$.

the construction of $\mathcal{P}_{\min}$, there exists a word v which takes the class $\mathcal{P}_u$ into a single state of $\mathcal{A}$. Now, let v' be the label of the same path under the labeling δ'. Then, due to the δ-consistency of $P_{\min}$ and the fact that the relabeling is done in the partition automaton $\mathcal{A}_P$, the similar argumentation as used in the proof of Theorem 3 shows that the word uv' is synchronizing for $\mathcal{A}$. □

Similarly, even using similar argumentation, we can show

Theorem 5. *Road coloring conjecture is equivalent to Conjecture C.*

Finally, using again very much the same ideas introduced above, in particular in Theorem 2, we can prove the following unconditional result.

Theorem 6. *Each acceptable primitive graph can be labeled to yield commutatively (or lengthwise) synchronized automaton.*

Theorem 6 gives a concrete example of the applicability of our approach.

References

1. R. L. Adler, L. W. Goodwyn and B. Weiss, *Equivalence of Topological Markov Shifts*, Israel J. Math. 27, 49-63, (1977).

2. M.-P. Beal and D. Perrin, *Symbolic Dynamics and Finite automata*, in: G. Rozenberg and A. Salomaa (eds.) Handbook of Formal Languages Vol II, 49-63, (1997).
3. J. Černý, *Poznámka k homogennym experimenton s konečnými automatmi*, Mat. fyz. čas. SAV 14, 208-215, (1964).
4. J. Friedman *On the road coloring problem*, Math. Soc. 110, 1133-1135, (1990).
5. J. E. Hopcroft and J. D. Ullman, *Introduction to Automata Theory, Languages and Computation*, (Addison-Wesley 1979).
6. A. Kljachko, I. Rystsow and K. Spivak, *Extended combinatorial problem concerning the length of the reset word in a finite automaton*, Cybernetics 23, 165-170, (1987).
7. J. Kari, *A counter example to a conjecture concerning synchronized words in finite automata*, EATCS Bull. 73, 146, (2001).
8. D. Lind and B. Marcus, *An Introduction of Symbolic Dynamics and Coding*, (Cambridge Univ. Press, 1995).
9. G. L. O'Brien, *The road coloring problem*, Israel J. Math. 39, 145-154, (1981).
10. D. Perrin and M.P. Schüzenberger, *Synchronizing prefix codes and automata and the road coloring problem*, in: P. Walters (ed.) Symbolic Dynamics and its Applications, Contemporary Mathematics 135, 295-318, (1991).
11. J.-E. Pin, *Le probleme de la conjecture de Černý*, These de 3^e cycle, Universite de Paris VI (1978).
12. J.-E. Pin, *On two combinatorial problems arising from automata theory*, Annals of Discrete Mathematics 17, 535–548, (1983).

Shuffle Quotient and Decompositions*

C. Câmpeanu[1], K. Salomaa[1], and S. Vágvölgyi[2]

[1] Department of Computing and Information Science
Queen's University
Kingston, Ontario K7L 3N6, Canada
[2] Department of Applied Informatics
University of Szeged
Szeged, Árpád tér 2, H-6720 Hungary

Abstract. We introduce a right congruence relation that is the analogy of the Nerode congruence when catenation is replaced by shuffle. Using this relation we show that for certain subclasses of regular languages the shuffle decomposition problem is decidable. We show that shuffle decomposition is undecidable for context-free languages.

1 Introduction

A language L is said to have a non-trivial shuffle decomposition if L can be expressed as a shuffle of two languages such that neither one of them is the singleton language consisting of the empty word. It is not known whether for a given regular language L we can effectively decide whether or not L has a non-trivial shuffle decomposition.

Here we introduce the notion of shuffle quotient and using it, given an arbitrary shuffle decomposition for a language L, we obtain the corresponding normalized decomposition for L. Thus any language that has a non-trivial shuffle decomposition has also a normalized decomposition that is given in terms of maximal languages. We show that the components of any normalized decomposition of L are closed under a right congruence ϱ_L that is defined in a somewhat similar way as the Nerode congruence of L when we think that catenation is replaced by shuffle. Using this result we show that the shuffle decomposition problem is decidable for commutative regular languages and for locally testable languages. On the other hand, we show that shuffle decomposition for context-free languages is undecidable in both of the cases if we allow the component languages to be arbitrary or require that the component languages are context-free.

Finally let us mention some related recent work. Assume that the minimal DFA accepting the language L_i has n_i states, $i = 1, 2$. By the state complexity of shuffle we mean the worst case size of the minimal DFA for the language $L_1 ⧢ L_2$. A tight lower bound for the state complexity of shuffle is given in [2]. The result uses incomplete DFA's and it is not known whether the bound can be reached with complete DFA's. The state complexities of many basic operations for regular languages are given in [16, 17] and in the references listed there. The complexity of accepting shuffle languages,

---

* Work supported by Natural Sciences and Engineering Research Council of Canada Grant OGP0147224 and the Bolyai János Research Grant of the Hungarian Academy of Sciences. All correspondence to K. Salomaa, ksalomaa@cs.queensu.ca

that is, languages defined by regular expressions that additionally use the shuffle and shuffle closure operations is investigated in [7,8]. Extensions of the shuffle operation are considered in [5,6,11,12].

2 Preliminaries

We assume the reader to be familiar with the basic notions of formal languages and with finite automata in particular, cf. e.g. [4,14,16]. Here we briefly recall some definitions and introduce notation.

The set of words over a finite alphabet Σ is denoted Σ^*, the empty word is λ. The length of a word $w \in \Sigma^*$ is denoted $|w|$. When there is no confusion, a singleton language $\{w\}$ is denoted simply as w. The power set of a set A is denoted 2^A. If L is a finite language, the length of the longest word in L is denoted $\text{length}(L)$. The mirror-image (or reversal) of $w \in \Sigma^*$ is $\text{mir}(w)$.

The Nerode (right-)congruence of $L \subseteq \Sigma^*$ is the relation $\gamma_L \subseteq \Sigma^* \times \Sigma^*$ defined by

$$u \gamma_L v \text{ iff } (\forall w \in \Sigma^*)(uw \in L \Leftrightarrow vw \in L)$$

A language L is regular iff γ_L has a finite index.

A language L is *commutative* if always when $w \in L$ any permutation of w is also in L.

We recall the definition of locally testable languages, cf. e.g. [13,16]. For $k \geq 1$ and $w \in \Sigma^*$ such that $|w| \geq k$ we denote by $\text{pre}_k(w)$ and $\text{suf}_k(w)$, respectively, the prefix and the suffix of length k of w, and by $\text{int}_k(w)$ the set of interior subwords of length k of w (subwords that occur in a position other than the prefix or the suffix). A language $L \subseteq \Sigma^*$ is said to be *k-testable* if for any words $w_1, w_2 \in \Sigma^*$ of length at least k the conditions

$$\text{pre}_k(w_1) = \text{pre}_k(w_2), \ \text{suf}_k(w_1) = \text{suf}_k(w_2), \ \text{int}_k(w_1) = \text{int}_k(w_2)$$

imply that $w_1 \in L$ iff $w_2 \in L$. A language is said to be *locally testable* if it is k-testable for some $k \geq 1$.

The *shuffle* of words $w_1, w_2 \in \Sigma^*$ is the set

$$w_1 \shuffle w_2 = \{ u_1 v_1 u_2 v_2 \cdots u_m v_m \mid$$
$$u_i, v_i \in \Sigma^*, \ i = 1, \ldots, m, \ w_1 = u_1 \cdots u_m, \ w_2 = v_1 \cdots v_m \}.$$

For example, $a \shuffle bc = \{abc, bac, bca\}$ and

$$ab \shuffle bc = \{abbc, abcb, babc, bacb, bcab\}.$$

The shuffle operation is associative and commutative. Alternatively, the shuffle operation can be defined recursively.

Definition 1. *We define the shuffle operation $\shuffle$ recursively by the following conditions:*

1. $u \shuffle \lambda = \lambda \shuffle u = \{u\}$ *for any* $u \in \Sigma^*$.
2. *For* $u, v \in \Sigma^*$ *and* $a, b \in \Sigma$ *we define*

$$au \shuffle bv = a(u \shuffle bv) \cup b(au \shuffle v).$$

The shuffle operation is extended in the natural way for languages. The shuffle of two languages L_1 and L_2 is

$$L_1 \shuffle L_2 = \bigcup_{w_1 \in L_1, w_2 \in L_2} w_1 \shuffle w_2.$$

A nondeterministic finite automaton (NFA) is denoted as a five-tuple

$$\mathbf{A} = (A, \Sigma, \delta, A_0, A_f) \tag{1}$$

where A is the finite set of states, Σ is the finite input alphabet, $\delta : A \times \Sigma \longrightarrow 2^A$ is the state transition relation, $A_0 \subseteq A$ is the set of initial states and $A_f \subseteq A$ is the set of final states.

In the standard way, the state transition relation δ is extended to a function $\hat{\delta} : A \times \Sigma^* \longrightarrow 2^A$. We denote also $\hat{\delta}$ simply by δ. The language accepted by $\mathbf{A}$ is

$$L(\mathbf{A}) = \{w \in \Sigma^* \mid (\exists q \in A_0)\ \delta(q, w) \cap A_f \neq \emptyset\}.$$

A five-tuple $\mathbf{A} = (A, \Sigma, \delta, A_0, A_f)$ as in (1) is said to be a deterministic finite automaton (DFA) if (i) A_0 consists of exactly one state of A, and (ii) for each $a \in A$ and $z \in \Sigma$, $\delta(a, z)$ is either a singleton set or the empty set (in which case we say that the transition is not defined). A DFA is said to be *complete* if $\delta(a, z)$ is always defined. Both NFA's and DFA's accept exactly the regular languages.

The following result is well known.

Theorem 1. *If L_1 and L_2 are regular languages (over an alphabet Σ), then also $L_1 \shuffle L_2$ is regular.*

3 Shuffle Decomposition

We say that a language L has a *(non-trivial) shuffle decomposition* if we can write

$$L = L_1 \shuffle L_2 \tag{2}$$

where neither one of the languages L_1 or L_2 is $\{\lambda\}$. Note that any language L has always the so called trivial decompositions $L = L \shuffle \{\lambda\} = \{\lambda\} \shuffle L$. When speaking of a shuffle decomposition as in (2), unless otherwise mentioned, we always assume that $L_1 \neq \{\lambda\} \neq L_2$.

Example 1. A language $L \subseteq \Sigma^*$ does not have a shuffle decomposition if we can write $L = \alpha L_0$ where $\alpha \in \Sigma^*$ contains occurrences of two distinct letters.

To see this, write $\alpha = a^k \beta$, where $k \geq 1$, $a \in \Sigma$ and the first letter of $\beta \in \Sigma^*$ is not a. Now assume that we could write $L = L_1 \shuffle L_2$ where $L_i \neq \{\lambda\}$, $i = 1, 2$. Choose non-empty words $u_i \in L_i$, $i = 1, 2$, and write $u_i = a^{k_i} v_i$, $k_i \geq 0$, where the first letter of v_i is not a. Since $u_1 \shuffle u_2 \subseteq \alpha L_0$ it follows that $k_1 + k_2 \leq k$ and $k_1 = k = k_2$. This is impossible. □

The answer to the following question is still open. Given a regular language L (for instance in terms of an NFA accepting L), can we algorithmically decide whether or not L has a shuffle decomposition. In spite of the general assumption that "everything should be decidable for regular languages", so far the shuffle decomposition problem has resisted all efforts to solve it. The difficulty is caused by the fact that if we have a decomposition as in (2) it is possible that the minimal automata for L_1 and L_2 are much larger than the minimal automaton for L. For instance, Σ^* has shuffle decompositions in terms of regular languages where the components can have arbitrarily high state-complexity.

Thus it is not clear whether we could determine the existence of a shuffle decomposition for L even by a brute force search of all DFA's up to a given size. Results on shuffle decompositions of finite languages are given in [1] but hardly anything is known about the question for general regular languages.

Next we introduce the operation of shuffle quotient. It is the analogy of left/right quotient when catenation is replaced by shuffle.

Definition 2. *Let L and L_1 be languages over an alphabet Σ. The* shuffle quotient *of L with respect to L_1 is defined as*

$$L \bowtie L_1 = \{w \in \Sigma^* \mid L_1 \shuffle w \subseteq L\}.$$

Note that the shuffle quotient of L with respect to itself, $L \bowtie L$, is the shuffle residual of L as defined in [6]. We have the following result.

Theorem 2. *Assume that $L = L_1 \shuffle L_2$. We define*

$$L'_1 = L \bowtie L_2, \text{ and}$$

$$L'_2 = L \bowtie L'_1.$$

Then L has the following maximal decomposition

$$L = L'_1 \shuffle L'_2. \tag{3}$$

Proof. By the definition of shuffle quotient, $L \supseteq (L \bowtie L_2) \shuffle L_2$. On the other hand, $L_1 \subseteq L \bowtie L_2$ and thus $L = (L \bowtie L_2) \shuffle L_2 = L'_1 \shuffle L_2$. Using commutativity and the same argument again we see that $L = L'_1 \shuffle L'_2$. □

We say that the decomposition (3) $L'_1 \shuffle L'_2$ is the *normalized decomposition* of L corresponding to the given decomposition $L = L_1 \shuffle L_2$. Note that corresponding to an arbitrary shuffle decomposition there always exists a normalized decomposition. A language L may naturally have different normalized decompositions.

Note that $L_1 \subseteq L'_1$ and $L_2 \subseteq L'_2$. In the above result actually we have also

$$L'_1 = L \bowtie L'_2 \tag{4}$$

and thus the definitions of L'_1 and L'_2 can be seen as symmetric. The equation (4) is established as follows. Since $L_2 \subseteq L'_2$ it follows that $L'_1 = L \bowtie L_2 \supseteq L \bowtie L'_2$. On the other hand, by (3) we have

$$L'_1 \subseteq (L'_1 \shuffle L'_2) \bowtie L'_2 = L \bowtie L'_2.$$

Thus a normalized decomposition $L = L'_1 \shuffle L'_2$ satisfies the conditions

$$\begin{cases} L'_1 = L \bowtie L'_2 \\ L'_2 = L \bowtie L'_1 \end{cases}$$

Example 2. Let $\Sigma = \{a, b, c\}$. Choose $L_1 = \{c\}$ and $L_2 = \Sigma^* ab \Sigma^*$. Then

$$L = L_1 \shuffle L_2 = \Sigma^* c \Sigma^* ab \Sigma^* \cup \Sigma^* acb \Sigma^* \cup \Sigma^* ab \Sigma^* c \Sigma^*.$$

It can be verified that

$$L_1' = L \bowtie L_2 = \{a, b\}^* c (\{a, b\}^+ c)^* \{a, b\}^*$$

and

$$L_2' = L \bowtie L_1' = L_2. \quad \square$$

The proof of the following lemma is given in the appendix.

Lemma 1. *If L and L_1 are regular languages then also $L \bowtie L_1$ is regular.*

4 Decidability Results

Using the shuffle quotient operation we can now decide the shuffle decomposition problem for certain subclasses of regular languages.

Theorem 3. *The following question is decidable.*

Input: *A regular language L and $k \in \mathbb{N}$.*
Output: *Does there exist a finite language L_1 with $\text{length}(L_1) \leq k$ and L_2 such that $L = L_1 \shuffle L_2$.*

Proof. Assume that $L = L_1 \shuffle L_2$. Denote $M = L \bowtie L_1$. Then $L_1 \shuffle M = L$ because $L_2 \subseteq M$ and, on the other hand, by the definition of shuffle quotient $L_1 \shuffle M \subseteq L$.

Thus by Lemma 1 we can decide the existence of the required decomposition by going through the finitely many possibilities for the pair of languages L_1 and M. $\quad \square$

The following definition can be viewed as an analogy of the Nerode congruence when we replace catenation of words by shuffle.

Definition 3. *Let $L \subseteq \Sigma^*$. We define the binary relation ϱ_L on Σ^* by setting $(u, v \in \Sigma^*)$:*

$$u \varrho_L v \text{ iff } (\forall x \in \Sigma^*) \ L \bowtie ux = L \bowtie vx.$$

Note that always $\varrho_L \subseteq \gamma_L$ where γ_L is the Nerode congruence.

Example 3. Let $\Sigma = \{a, b\}$, $L = L_1 \shuffle L_2$ where $L_1 = (ab)^*$ and $L_2 = a$. It is easy to verify that $L \bowtie a = (ab)^*$, $L \bowtie (ab)^m = a$ when $m \geq 1$, $L \bowtie \lambda = L$, $L \bowtie w = \lambda$ for $w \in L - \{a\}$, and $L \bowtie w = \emptyset$ in all other cases.

Using the above observations we see that the equivalence classes of ϱ_L are the following:

$C_0 = \{\lambda\}, C_1 = \{a\}, C_2 = (ab)^+, C_3 = (ab)^+ a,$
$C_4 = (L - (ab)^* a) a \cup \{aa\}, C_5 = L - (ab)^* a, C_6 = \Sigma^* - \bigcup_{i=0}^{5} C_i.$

Note that C_6 consists of exactly the words $u \in \Sigma^*$ such that

$$(\forall x \in \Sigma^*) \ \{w \in \Sigma^* \mid w \shuffle ux \subseteq L\} = \emptyset.$$

The "transitions" between the classes are as follows:

$C_0 a \subseteq C_1, C_0 b \subseteq C_6, C_1 a \subseteq C_4, C_1 b \subseteq C_2,$
$C_2 a \subseteq C_3, C_2 b \subseteq C_6, C_3 a \subseteq C_4, C_3 b \subseteq C_2,$
$C_4 a \subseteq C_6, C_4 b \subseteq C_5, C_5 a \subseteq C_4, C_5 b \subseteq C_6, C_6 a \subseteq C_6, C_6 b \subseteq C_6.$ $\quad \square$

For a regular language L the relation ϱ_L does not necessarily have a finite index.
Example 4. Let $\Sigma = \{a, b, c\}$ and define $L = \Sigma^* ab \Sigma^*$. Now for any $i < j$, c^i is not ϱ_L equivalent to c^j because

$$(ab)^{i+1} \in L \bowtie c^i - L \bowtie c^j.$$

Furthermore, ϱ_L may have infinitely many classes also in cases where the regular language L has a non-trivial shuffle decomposition. Choose $\Gamma = \{a, b, c, d\}$ and let

$$L = (\Gamma^* ab \Gamma^*) \shuffle d^* = \Gamma^* a d^* b \Gamma^*.$$

Exactly as above we see that any words c^i and c^j where $i \neq j$ are not ϱ_L equivalent. □

Theorem 4. (a) ϱ_L *is a right congruence.*
(b) *If $L = L_1' \shuffle L_2'$ is a normalized shuffle decomposition of L (that is, $L_1' = L \bowtie L_2'$, $L_2' = L \bowtie L_1'$) then each of the languages L, L_1', L_2' is closed with respect to ϱ_L.*

Proof. Claim (a) follows directly from the definition of ϱ_L. For (b), assume that $u \in L_1'$ and $u \varrho_L v$. Since $L_1' = \{w \mid w \shuffle L_2' \subseteq L\}$, we have

$$L_2' \subseteq \{w \mid w \shuffle u \subseteq L\} = \{w \mid w \shuffle v \subseteq L\}$$

and thus $v \in L_1'$. By a symmetric argument we see that L_2' is closed with respect to ϱ_L. Finally to see that L is closed with respect to ϱ_L, consider $u \in L$ and $u \varrho_L v$. Then in particular, $L \bowtie u = L \bowtie v$ and since $L \bowtie u$ contains the empty word, so does $L \bowtie v$. This means that $v \in L$. □

Theorem 5. *Given a commutative regular language L we can effectively decide whether or not L has a shuffle decomposition.*

Proof. If L has a shuffle decomposition, by Theorem 2 it has a normalized decomposition $L = L_1' \shuffle L_2'$. By Theorem 4, L_i' is closed with respect to ϱ_L, $i = 1, 2$. Since L is commutative, for any $u, v \in \Sigma^*$,

$$u \gamma_L v \text{ iff } L \bowtie u = L \bowtie v, \qquad (5)$$

(here γ_L is the Nerode congruence). We show that for all $u, v \in \Sigma^*$

$$L \bowtie u = L \bowtie v \text{ iff } u \varrho_L v. \qquad (6)$$

The implication from right to left follows from the definition of ϱ_L. Conversely if $L \bowtie u = L \bowtie v$, we can write

$$(\forall y \in \Sigma^*) \; \{xy \mid u \shuffle xy \subseteq L\} = \{xy \mid v \shuffle xy \subseteq L\}.$$

Since L is commutative it follows that

$$(\forall y \in \Sigma^*) \; \{xy \mid uy \shuffle x \subseteq L\} = \{xy \mid vy \shuffle x \subseteq L\}.$$

This implies that

$$(\forall y \in \Sigma^*) \; \{x \mid uy \shuffle x \subseteq L\} = \{x \mid vy \shuffle x \subseteq L\}$$

and thus $u \varrho_L v$.

Now we have $\varrho_L = \gamma_L$ by (5) and (6). By Theorem 4 the components of a normalized decomposition are closed under ϱ_L. For a regular language L, the Nerode congruence γ_L is of finite index and its classes can be effectively computed. Thus there exist only finitely many possibilities for a normalized decomposition and we can effectively determine whether or not a decomposition exists. □

Theorem 6. *For a locally testable language L we can effectively decide whether or not L has a shuffle decomposition.*

Proof. Assume that L is k-testable, $k \geq 1$. Given an arbitrary regular language L we can effectively determine whether or not L is locally testable and in the case of a positive answer we can find $k \geq 1$ such that L is k-testable, [9,10,15]. Let $u_1, u_2 \in \Sigma^*$ be arbitrary. By the definition of ϱ_L we have $u_1 \varrho_L u_2$ iff

$$(\forall x \in \Sigma^*) \quad w \sqcup\!\sqcup u_1 x \subseteq L \text{ iff } w \sqcup\!\sqcup u_2 x \subseteq L.$$

Since L is k-testable, the relation $w \sqcup\!\sqcup u_i x \subseteq L$ depends only on the k-prefix and k-suffix of the words w and $u_i x$ and the sets $\text{int}_k(u_i x)$, $\text{int}_k(w)$. Thus given k we have an upper bound for the number of ϱ_L classes and the classes are regular languages that can be effectively computed.

By Theorem 4 the components of a normalized decomposition of L are closed under ϱ_L. Thus we can effectively determine whether or not a normalized decomposition exists.
□

Note that for a given DFA **A** we can decide in polynomial time whether or not $L(\mathbf{A})$ is locally testable but the problem of finding the order of local testability is NP-hard [3, 9,10,15].

5 Undecidability Results

Contrasting the results of the previous section we show that it is undecidable whether or not a context-free language has a shuffle decomposition.

It should be noted that context-free languages are not even closed under shuffle. For instance, if $L = \{a^n b^n \mid n \geq 1\} \sqcup\!\sqcup \{c^n d^n \mid n \geq 1\}$ then $L \cap a^* c^* b^* d^* = \{a^n c^m b^n d^m \mid m, n \geq 1\}$ which is not context-free. We can consider different variants of the context-free shuffle decomposition problem depending on whether or not we require that the component languages are also context-free.

The proof of the following result uses a reduction to the Post Correspondence Problem (PCP) that is a modification of the well known construction used to show that context-free language equivalence is undecidable.

Theorem 7. *Given a context-free language L it is undecidable in general whether or not we can write $L = L_1 \sqcup\!\sqcup L_2$ where*

1. *$L_1, L_2 \neq \{\lambda\}$ are arbitrary languages,*
2. *$L_1, L_2 \neq \{\lambda\}$ are context-free languages,*
3. *$L_1, L_2 \neq \{\lambda\}$ are regular languages.*

Proof. Let $\mathbf{M} = (u_1, \ldots, u_k; v_1, \ldots, v_k)$, $k \geq 1$, $u_i, v_i \in \{0,1\}^*$, $i = 1, \ldots, k$, be an arbitrary PCP instance. We construct a context-free language $L_\mathbf{M}$ such that $L_\mathbf{M}$ has a shuffle decomposition in terms of arbitrary languages iff $L_\mathbf{M}$ has a shuffle decomposition in terms of regular languages iff the PCP instance $\mathbf{M}$ does not have a solution.

Denote $\Sigma = \{0, 1, \#\}$ and let $\bar{\Sigma} = \{\bar{x} \mid x \in \Sigma\}$. For $w \in \Sigma^*$ we also denote by $\bar{w} \in \bar{\Sigma}^*$ the word that is obtained from w by replacing each symbol $x \in \Sigma$ by $\bar{x}$.

We define $L_0 \subseteq (\Sigma \cup \bar{\Sigma} \cup \{\$, \%\})^*$ as follows

$$L_0 = (\%\Sigma^*) \shuffle (\bar{\Sigma}^*\$).$$

Let $L_1 = L_0 - L_1'$ where

$$L_1' = \{\%01^{i_1} \cdots 01^{i_n} \# u_{i_n} \cdots u_{i_1} \# \# \operatorname{mir}(\bar{v}_{j_1}) \cdots \operatorname{mir}(\bar{v}_{j_m}) \# \bar{1}^{j_m} \bar{0} \cdots \bar{1}^{j_1} \bar{0} \$ \mid$$
$$i_1, \ldots, i_n, j_1, \ldots, j_m \in \{1, \ldots, k\}, m, n \geq 1\}.$$

Now L_0 is regular and a nondeterministic pushdown automaton can verify that the input is not in L_1'. Thus L_1 is context-free.

Further let $L_2 = L_0 - L_2'$ where

$$L_2' = \{\%01^{i_1} \cdots 01^{i_n} \# w \# \# \operatorname{mir}(\bar{w}) \# \bar{1}^{i_n} \bar{0} \cdots \bar{1}^{i_1} \bar{0} \$ \mid$$
$$w \in \Sigma^*, i_1, \ldots, i_n \in \{1, \ldots, k\}, n \geq 1\}.$$

Similarly as above it is seen that L_2 is context-free. We define

$$L_\mathbf{M} = L_1 \cup L_2$$

and thus also $L_\mathbf{M}$ is context-free. From the definition of the languages L_1 and L_2 it is seen that $L_\mathbf{M} = L_0$ iff the PCP instance $\mathbf{M}$ does not have a solution.

Thus if $\mathbf{M}$ does not have a solution then $L_\mathbf{M}$ has a shuffle decomposition where the components are regular languages. To complete the proof it is sufficient to show that if $\mathbf{M}$ has a solution then $L_\mathbf{M}$ does not have a non-trivial shuffle decomposition (where the components can be arbitrary languages $\neq \{\lambda\}$).

Assume that $\mathbf{M}$ has a solution

$$u_{i_1} \cdots u_{i_n} = v_{i_1} \cdots v_{i_n}, \quad i_1, \ldots, i_n \in \{1, \ldots, k\}. \tag{7}$$

For the sake of contradiction assume that we can write

$$L_\mathbf{M} = D_1 \shuffle D_2, \quad D_1, D_2 \neq \{\lambda\}. \tag{8}$$

Let $\{x, y\} = \{1, 2\}$ and assume that some word in D_x contains the symbol $\%$. Since

$$L_\mathbf{M} \subseteq (\%\Sigma^*) \shuffle (\bar{\Sigma}^*\$)$$

this implies that no word of D_y contains an occurrence of a symbol of Σ. Similarly, if some word in D_x contains the symbol $\$$ it follows that no word of D_y contains an occurrence of a symbol of $\bar{\Sigma}$.

Since neither one of the languages D_1, D_2 is $\{\lambda\}$, it follows that exactly one of the languages contains occurrences of the symbol $\%$ and exactly the other one of the languages contains occurrences of the symbol $. Thus without loss of generality (the other possibility being symmetric) we can assume that

$$D_1 \subseteq (\{\%\} \cup \Sigma)^* \text{ and } D_2 \subseteq (\{\$\} \cup \bar{\Sigma})^*. \tag{9}$$

Now using the notations of (7)

$$\%01^{i_1} \cdots 01^{i_n} \# u_{i_n} \cdots u_{i_1} \#\$ \in L_\mathbf{M}$$

and thus (9) implies that

$$\%01^{i_1} \cdots 01^{i_n} \# u_{i_n} \cdots u_{i_1} \# \in D_1.$$

Similarly we have

$$\%\bar{\#}\mathrm{mir}(\bar{v}_{i_1}) \cdots \mathrm{mir}(\bar{v}_{i_n}) \bar{\#} \bar{1}^{i_n} \bar{0} \cdots \bar{1}^{i_1} \bar{0}\$ \in L_\mathbf{M}$$

and again (9) implies that

$$\bar{\#}\mathrm{mir}(\bar{v}_{i_1}) \cdots \mathrm{mir}(\bar{v}_{i_n}) \bar{\#} \bar{1}^{i_n} \bar{0} \cdots \bar{1}^{i_1} \bar{0}\$ \in D_2.$$

This means that the word

$$w = \%01^{i_1} \cdots 01^{i_n} \# u_{i_n} \cdots u_{i_1} \#\bar{\#}\mathrm{mir}(\bar{v}_{i_1}) \cdots \mathrm{mir}(\bar{v}_{i_n}) \bar{\#} \bar{1}^{i_n} \bar{0} \cdots \bar{1}^{i_1} \bar{0}\$$$

where $i_1, \ldots, i_n$ are as in (7) is in $D_1 \sqcup D_2 = L_\mathbf{M}$. On the other hand, by (7), $w \in L'_1 \cap L'_2$ which implies that $w \notin L_1$ and $w \notin L_2$. This is a contradiction since $L_1 \cup L_2 = L_\mathbf{M}$. □

Acknowledgement. We thank Heechul Lim for implementing the algorithm from the proof of Lemma 1. His software enabled us to construct large numbers of shuffle quotient examples.

References

1. J. Berstel, L. Boasson. Shuffle factorization is unique. Research report November 1999. http://www-igm.univ-mlv.fr/ berstel/Recherche.html
complexity
2. C. Câmpeanu, K. Salomaa, S. Yu. Tight lower bound for the state complexity of shuffle of regular languages. Accepted for publication in *Journal of Automata, Languages and Combinatorics.*
3. P. Caron. Families of locally testable languages. *Theoretical Computer Science* **242** (2000) 361–376.
4. J.E. Hopcroft and J.D. Ullman. *Introduction to Automata Theory, Languages and Computation.* Addison-Wesley, 1979.
5. B. Imreh, M. Ito, M. Katsura. On shuffle closures of commutative regular languages. In: *Combinatorics, Complexity & Logic,* Proc. of DMTCS'96, Springer-Verlag, 1996, pp. 276–288.
6. M. Ito, L. Kari, G. Thierrin. Shuffle and scattered deletion closure of languages. *Theoretical Computer Science* **245** (2000) 115–133.

7. M. Jantzen. Extending regular expressions with iterated shuffle. *Theoretical Computer Science* **38** (1985) 223–247.
8. J. Jedrzejowicz, A. Szepietowski. Shuffle languages are in P. *Theoretical Computer Science* **250** (2001) 31–53.
9. S. Kim, R. McNaughton, R. McCloskey. A polynomial time algorithm for the local testability problem of deterministic finite automata. IEEE *Trans. Comput.* **40** (1991) 1087–1093.
10. S. Kim, R. McNaughton. Computing the order of a locally testable automaton. *SIAM Journal of Computing* **23** (1994) 1193–1215.
11. A. Mateescu, G.D. Mateescu, G. Rozenberg, A. Salomaa. Shuffle-like operations on ω-words. In: *New Trends in Formal Languages,* Lecture Notes in Computer Science 1218, Springer-Verlag, 1997, 395-411.
12. A. Mateescu, G. Rozenberg, A. Salomaa. Shuffle on trajectories: Syntactic constraints. *Theoretical Computer Science* **197** (1998) 1–56.
13. R. McNaughton, S. Papert. *Counter-Free Automata.* MIT Press, Cambridge, Mass. 1971.
14. A. Salomaa. *Formal Languages.* Academic Press, 1973.
15. A.N. Trahtman. Optimal estimation on the order of local testability of finite automata. *Theoretical Computer Science* **231** (2000) 59–74.
16. S. Yu. Regular languages. In: *Handbook of Formal Languages, Vol. I,* G. Rozenberg and A. Salomaa, eds., Springer-Verlag, pp. 41–110, 1997.
17. S. Yu, Q. Zhuang, K. Salomaa. The state complexities of some basic operations on regular languages. *Theoretical Computer Science* **125** (1994) 315–328.

Appendix

Proof of Lemma 1. Let L and L_1 be as in the statement of Lemma 1. Assume that L_1 is accepted by a DFA $\mathbf{A} = (A, \Sigma, \delta_A, a_0, A_f)$ and that L is accepted by a complete DFA $\mathbf{B} = (B, \Sigma, \delta_B, b_0, B_f)$. We construct a complete DFA $\mathbf{C} = (C, \Sigma, \delta, c_0, C_f)$ such that $L(\mathbf{C}) = L \bowtie L_1$.

Each state of $\mathbf{C}$ will be a subset of $A \times B$. For the sake of easier readability we denote elements of $A \times B$ using square brackets in the form $[a, b]$, where $a \in A, b \in B$.

Then $\mathbf{C} = (C, \Sigma, \delta, c_0, C_f)$ is defined as follows:

(i) $C = 2^{A \times B}$,
(ii) $c_0 = \{\ [\delta_A(a_0, u), \delta_B(b_0, u)] \mid u \in \Sigma^*\ \}$,
(iii) $C_f = \{\ X \subseteq A \times B \mid \text{if } ([a,b] \in X,\ a \in A_f) \text{ then } b \in B_f\}$, in other words, C_f consists of subsets of $(A_f \times B_f) \cup ((A - A_f) \times B)$,

and the transition relation δ is defined as follows. Let $X \subseteq A \times B$ be arbitrary, and $z \in \Sigma$ be an arbitrary input symbol. Then

(iv) $\delta(X, z) = \{\ [\delta_A(a, u), \delta_B(\delta_B(b, z), u)] \mid u \in \Sigma^*, [a, b] \in X\ \}$.

Intuitively, the construction works as follows. After reading an input word w the automaton is in a state $X \subseteq A \times B$.

The second components of the pairs in X consist of all elements of B that the automaton $\mathbf{B}$ can reach after reading an arbitrary word belonging to $w \shuffle w_1$, where $w_1 \in \Sigma^*$ is arbitrary.

On the other hand, the first components of the pairs belonging to X are the result of the computation of $\mathbf{A}$ that is performed only on the word w_1 (that is shuffled with w). Thus the first components of the pairs check that the inserted word is accepted by $\mathbf{A}$.

Assume now that after reading an input word w the automaton **C** reaches an accepting state $X \in C_f$. The second components of the pairs in X are all the states where the automaton **B** can be after reading an arbitrary word belonging to $w \sqcup\!\sqcup w_1$ (where $w_1 \in \Sigma^*$ is arbitrary). The first component in each pair is the state that **A** reaches after reading the same shuffled word w_1. Thus the definition of C_f means that always when **A** accepts the shuffled word w_1 (that is, $w_1 \in L_1$) then the result of the shuffle (any word in $w \sqcup\!\sqcup w_1$) has to be accepted by **B** (that is, $w \sqcup\!\sqcup w_1 \subseteq L(\mathbf{B}) = L$).

Thus **C** accepts an input w iff $w \sqcup\!\sqcup L_1 \subseteq L$, that is, $L(\mathbf{C}) = L \bowtie L_1$. □

The Growing Context-Sensitive Languages Are the Acyclic Context-Sensitive Languages

Gundula Niemann[1] and Jens R. Woinowski[2]*

[1] Fachbereich Mathematik/Informatik
Universität Kassel, D–34109 Kassel, Germany
niemann@theory.informatik.uni-kassel.de
[2] Fachbereich Informatik
TU Darmstadt, D–64289 Darmstadt, Germany
woinowski@iti.informatik.tu-darmstadt.de

Abstract. The growing context-sensitive languages have been defined by Dahlhaus and Warmuth using strictly monotone grammars, and they have been characterized by Buntrock and Loryś by weight-increasing grammars. The acyclic context-sensitive languages are defined by context-sensitive grammars the context-free kernels of which contain no cycles of chain rules, which is equivalent to being context-sensitive and weight-increasing at the same time.
In this paper we show that these two language classes coincide, that is, for each weight-increasing grammar there exists an equivalent one that is weight-increasing and context-sensitive at the same time.

1 Introduction

Dahlhaus and Warmuth [10] considered the class **GCSL** of *growing context-sensitive languages*. These languages are generated by monotone grammars each production rule of which is strictly length-increasing, that is, they are produced by strictly monotone grammars. Dahlhaus and Warmuth proved that these languages have membership problems that are decidable in polynomial time.

Although it might appear from the definition that **GCSL** is not an interesting class of languages, Buntrock and Loryś showed that **GCSL** is an *abstract family of languages* [6], that is, this class of languages is closed under union, concatenation, iteration, intersection with regular languages, ε-free homomorphisms, and inverse homomorphisms. Exploiting these closure properties Buntrock and Loryś characterized the class **GCSL** through various other classes of grammars that are less restricted [6,7]. Of these the most important ones in our context are the weight-increasing grammars. A grammar is called weight-increasing if there exists a homomorphic mapping relating each symbol of the grammar to a natural number such that for each rule the weight of the left-hand side is smaller than that of the right-hand side, that is, the weight of a sentential form increases with every application of a rule. Using these grammars Buntrock and Otto [8] obtained a characterization of the class **GCSL** by a nondeterministic machine model, the so-called *shrinking pushdown automaton with two pushdown stores* (**sTPDA**).

* current address: sd&m AG, software design & management, D–81737 München, Germany, jens.woinowski@sdm.de

The class GCSL can be seen as an insertion into the Chomsky hierarchy [13], as the definition is natural and simple and it lies strictly between CFL and CSL. Also it reappears in a central position in the hierarchy of McNaughton classes of languages as defined by Beaudry, Holzer, Niemann, and Otto [2]. These classes are obtained as generalizations of the Church-Rosser languages, which were introduced in [14] by defining languages via string rewriting systems using additional nonterminal symbols and border markers.

Due to the well-known characterization of the class of context-sensitive languages CSL by monotone grammars [9], the class GCSL defined by strictly monotone grammars has been given the name "growing *context-sensitive*". Thus it arises as a natural question, whether we obtain the same language class if we use context-sensitive grammars as defined by [9] that are additionally length-increasing or weight-increasing.

A context-sensitive rule replaces one symbol by a nonempty string depending on some context to the left and/or to the right of it. Therefore we can define the context-free kernel of a context-sensitive grammar as the grammar we obtain by omitting the context in every rule. A context-sensitive grammar is acyclic, if its context-free kernel contains no cycle of chain rules. These grammars have been defined under the name 1_A-grammars by Parikh [15]. We denote the corresponding language class of acyclic context-sensitive languages by ACSL. The acyclic context-sensitive grammars are exactly those that are context-sensitive and weight-increasing at the same time [5].

By GACSL we denote the class of growing acyclic context-sensitive languages which are defined by length-increasing context-sensitive grammars. Obviously, these grammars are acyclic. This class was defined by Parikh under the name 1_B-grammars [15], and it also has been investigated by Brandenburg [4], who defined the class of context grammars with normal kernel. It turns out that these grammars characterize the same language class GACSL [5]. We give the exact definitions in Section 2.

It is easily seen that GACSL ⊆ ACSL ⊆ GCSL. Buntrock investigates these three classes in his Habilitationsschrift [5], and he conjectures that both inclusions are strict. In this paper we show that the language classes GCSL and ACSL actually coincide, that is, for a weight-increasing grammar we can find an equivalent one that is also weight-increasing and at the same time context-sensitive.

It remains as an open question whether GCSL also coincides with GACSL. We conjecture that this is not the case.

2 Preliminaries

Let Σ be a finite alphabet. Then Σ^* denotes the set of strings over Σ including the empty string ε, and $\Sigma^+ := \Sigma^* \smallsetminus \{\varepsilon\}$. A *grammar* is a quadruple $G = (N, T, S, P)$, where N and T are finite disjoint alphabets of nonterminal and terminal symbols, respectively, $S \in N$ is the start symbol and $P \subseteq (N \cup T)^* N (N \cup T)^* \times (N \cup T)^*$ is a finite set of productions [9,12].

A grammar $G = (N, T, S, P)$ is *context-sensitive* if each production $(\alpha \to \beta) \in P$ is of the form $(xAz \to xyz)$, where $x, y, z \in (N \cup T)^*$, $A \in N$, and $y \neq \varepsilon$. We denote the corresponding language class by CSL.

A grammar $G = (N, T, S, P)$ is *monotone* if the start symbol S does not appear on the right-hand side of any production of G, and $|\alpha| \leq |\beta|$ holds for all productions $(\alpha \to \beta) \in P$ satisfying $\alpha \neq S$.

The class CSL is characterized by monotone grammars [9].

A function $\varphi : \Sigma \to \mathbb{N}_+$ is called a *weight-function*. Its extension to Σ^*, which we will also denote by φ, is defined inductively through $\varphi(\varepsilon) := 0$ and $\varphi(wa) := \varphi(w) + \varphi(a)$ for all $w \in \Sigma^*$ and $a \in \Sigma$. A particular weight-function is the *length-function* $|.| : \Sigma \to \mathbb{N}_+$, which assigns each letter the weight (length) 1.

Definition 1. *A language is called* growing context-sensitive *if it is generated by a strictly monotone grammar $G = (N, T, S, P)$, that is, the start symbol S does not appear on the right-hand side of any production of G, and $|\alpha| < |\beta|$ holds for all productions $(\alpha \to \beta) \in P$ satisfying $\alpha \neq S$. By* GCSL *we denote the class of growing context-sensitive languages.*

Note that we refer to this class as growing *context-sensitive* for historical reasons, although it is defined by strictly *monotone* grammars. As shown by Dahlhaus and Warmuth [10] the membership problem for each growing context-sensitive language can be solved in polynomial time. The class of context-free languages CFL is strictly contained in GCSL, and GCSL is strictly contained in the class CSL, e.g., the Gladkij language $L_{G1} := \{w\mathcal{c}w^\sim\mathcal{c}w \mid w \in \{a,b\}^*\}$, where $w^\sim$ denotes the reversal of w, is a context-sensitive language that is not growing context-sensitive [11,3,8]. Further, the class GCSL has many nice closure properties [6,7]. A detailed presentation of this class can be found in Buntrock's Habilitationsschrift [5].

Definition 2. *A grammar $G = (N, T, S, P)$ is* weight-increasing *if there exists a weight-function $\varphi : (N \cup T)^* \to \mathbb{N}^+$ such that $\varphi(\alpha) < \varphi(\beta)$ holds for all productions $(\alpha \to \beta) \in P$.*

Lemma 3. *[6,5] The class* GCSL *is characterized by weight-increasing grammars.*

Acyclic context-sensitive grammars have been defined by Parikh who called them type 1_A-grammars [15]. Like [5] we follow Aarts in the presentation of the definition [1]. A context-free grammar is called *acyclic* if there exists no nonterminal A such that $A \Rightarrow_G^+ A$, that is, if it has no cycle of chain rules. For a context-sensitive grammar $G = (N, T, S, P)$ the *context-free kernel* $G' = (N, T, S, P')$ is defined by $P' = \{(A \to \beta) \mid \exists x, y \in (N \cup T)^* : (xAy \to x\beta y) \in P\}$.

Definition 4. *A context-sensitive grammar $G = (N, T, S, P)$ is* acyclic, *if its context-free kernel is acyclic. We denote the set of acyclic context-sensitive grammars by* ACSG *and the corresponding set of languages by* ACSL.

Lemma 5. *[5]* ACSG *is the set of weight-increasing context-sensitive grammars.*

Parikh also introduces the class of context-sensitive grammars that are length-increasing at the same time. Note that such a grammar is also acyclic. He refered to them as 1_B-grammars.

Definition 6. *A context-sensitive grammar is* growing acyclic *if it is length-increasing, that is,* $|\alpha| < |\beta|$ *holds for all productions* $(\alpha \to \beta) \in P$ *satisfying* $\alpha \neq S$. *By* GACSG *we denote this class of grammars, and by* GACSL *we refer to the corresponding language class.*

Brandenburg investigates this language class [4] by using so-called context grammars with normal kernel. These are defined as follows.

Definition 7. *[4] A context-sensitive grammar is a* context grammar with normal kernel *if its context-free kernel is in* ε-*free Chomsky normal form.*

It is easily seen that such a grammar is growing acyclic as well. On the other hand for each growing acyclic context-sensitive grammar the kernel does not contain chain rules. So we see that an equivalent context grammar with normal kernel can be constructed (see [5]). Brandenburg [4] shows that GACSL contains CFL and is strictly contained in CSL. Buntrock [5] gives an example for a language in GACSL $\setminus$ CFL.

So the following chain of inclusions holds:

$$\text{CFL} \subsetneq \text{GACSL} \subseteq \text{ACSL} \subseteq \text{GCSL} \subsetneq \text{CSL} \quad .$$

These inclusions obviously raise the question of whether GACSL is strictly contained in ACSL, and whether ACSL is strictly contained in GCSL. Here we will answer the latter question by showing that for each weight-increasing grammar there exists an equivalent grammar that is weight-increasing and context-sensitive at the same time. It follows that ACSL and GCSL coincide.

3 The Main Result

In this section we construct a weight-increasing context-sensitive grammar from an arbitrary length-increasing one. Here, we combine two techniques: The well-known construction of a context-sensitive grammar from a monotone one given by Chomsky [9], and a new technique given by the second author in [17] called weight-spreading. Here, the sentential form is compressed into compression symbols. The weight of these compression symbols is defined via the length of their contents. In the length-increasing grammar, each derivation step increases the length of the sentential form. So the weight of the compression symbols touched in the simulation increases by a certain amount ρ. As the compression symbols touched are not changed all at once but one after another, we divide ρ into portions introducing dummy symbols that are inserted into the sentential form. Each compression symbol receives one of the weight portions when changed (encoded in the length of its content, which now also contains dummy symbols), and thus we spread the weight ρ over the different compression symbols involved in the simulation of this derivation step.

As by Lemma 3 for any weight-increasing grammar a strictly monotone grammar can be constructed, our construction implies that the weight-increasing grammars and the weight-increasing context-sensitive grammars define the same language class, that is, GCSL = ACSL.

Now we look at the construction of a weight-increasing context-sensitive grammar from a strictly monotone one in detail. Let $G = (N, T, S, P)$ be a strictly monotone grammar. We define a set

$$W_\# = \#^{\leq 1} \cdot ((N \cup T) \cdot \#)^* \cdot (N \cup T) \cdot \#^{\leq 1} \ .$$

The set of compression symbols is now defined as follows:

$$N_1 = \{\zeta_w \mid w \in W_\# \wedge 2\mu + 1 \leq |w| \leq 4\mu + 1\},$$

where $\mu = \max\{|r| : (\ell \to r) \in P\}$. That is, for each compression symbol ζ_w, w stores at least μ letters and we can always store up to 2μ letters in w indepent of the position of dummy symbols. We define the natural morphism $\widehat{} : W_\# \to (N \cup T)^+$ that deletes the dummy symbols $\#$ from a word $w \in W_\#$. We also define a set of blocking symbols that will carry the complete information of the derivation step simulated.

$$N_t = \left\{ \zeta_t \;\middle|\; \begin{array}{l} t = (w_1', w_1'', w_2', w_2'', (\ell \to r)), \\ \text{where } w_1', w_1'', w_2', w_2'' \in W_\#, \\ |w_1'| + |w_1''| \leq 4\mu + 1, |w_2'| + |w_2''| \leq 4\mu + 1, \\ (\ell \to r) \in P, \text{ and } \widehat{w_1'' w_2'} = \ell \end{array} \right\}$$

Define the set of nonterminals N' by

$$N' = N_1 \cup \{S'\} \cup N_t \ ,$$

where S' is another new symbol. In the simulation, that is, in the application of a rule from P, at most 2 compression symbols can be touched. P' contains the following rules: customize

Start rules:

$S' \to v$ for $v \in L(G), |v| \leq \mu$
$S' \to \zeta_w$ for $\mu + 1 \leq |\widehat{w}| \leq 2\mu, w \in ((N \cup T) \cdot \#)^+$ and $S \to_P^* \widehat{w}$

Simulation rules: If only one compression symbol is touched, that is, $w = w_1 w_2 w_3$, $\widehat{w_2} = \ell$ for some $(\ell \to r) \in P$, $w_1 \in \#^{\leq 1} \cdot ((N \cup T) \cdot \#)^*$, $w_2 \in ((N \cup T) \cdot \#)^+$, $w_3 \in ((N \cup T) \cdot \#)^* \cdot (N \cup T) \cdot \#^{\leq 1} \cup \{\varepsilon\}$, and if $w_4 \in ((N \cup T) \cdot \#)^+$ such that $\widehat{w_4} = r$, then we have two cases:

1.1 If the resulting content string fits into one compression symbol, that is, $|w_1 w_4 w_3| \leq 4\mu + 1$:

$$(\zeta_{w_1 w_2 w_3} \to \zeta_{w_1 w_4 w_3}) \in P' \ .$$

1.2 If the resulting content string does not fit into one compression symbol, that is, $|w_1 w_4 w_3| > 4\mu + 1$:

$$(\zeta_{w_1 w_2 w_3} \to \zeta_{z_1} \zeta_{z_2}) \in P' \ ,$$

where $|z_1| = 2\mu + k$, $|z_2| = 2\mu + 1$, $k := |w_1 w_4 w_3| - 4\mu - 1$, and $z_1 z_2 = w_1 w_4 w_3$.

If two compression symbols are touched, that is, if $w_1 = w_1'w_1''$, $w_2 = w_2'w_2''$, $\widetilde{w_1''w_2'} = \ell$ for some $(\ell \to r) \in P$, $w_1' \in \#^{\leq 1}((N \cup T) \cdot \#)^+$, $w_1''w_2' \in ((N \cup T) \cdot \#)^+$, $w_2'' \in ((N \cup T) \cdot \#)^* \cdot (N \cup T) \cdot \#^{\leq 1}$, then let $t = (w_1', w_1'', w_2', w_2'', (\ell \to r))$. Here we distinguish the following cases:

2.1 If the encoded lefthand side is split directly after an original symbol, that is, if $w_1'' = \ell_1 \# \ldots \ell_k$ and $w_2' = \#\ell_{k+1}\# \ldots \ell_{|\ell|}\#$, then the following rule belongs to P':

$$\zeta_{w_1}\zeta_{w_2} \to \zeta_{w_1}\zeta_t \quad .$$

The encoded righthand side of the rule is split as follows: $r_1\# \ldots r_k\#$ is put into the first compression symbol, replacing w_1'', and the second part $r_{k+1}\# \ldots r_{|r|}\#$ is put into the second compression symbol, replacing w_2'.

2.1.1 If the first part of the encoded righthand side fits into one compression symbol, that is, if $|w_1'r_1\# \ldots r_k\#| \leq 4\mu + 1$, then the following rule also belongs to P':

$$\zeta_{w_1}\zeta_t \to \zeta_{w_1'r_1\#\ldots r_k\#}\zeta_t \quad .$$

2.1.2 On the other hand, if the first part of the encoded righthand side does not fit into one compression symbol, that is, if we have $|w_1'r_1\# \ldots r_k\#| > 4\mu + 1$, then the following rule belongs to P', where $w_1' = w_{1,1}'w_{1,2}'$ such that $|w_{1,1}'| = 2\mu + 1$:

$$\zeta_{w_1}\zeta_t \to \zeta_{w_{1,1}'}\zeta_{w_{1,2}'r_1\#\ldots r_k\#}\zeta_t \quad .$$

In each of these two cases there are two subcases to consider. We name the first part of the righthand side of the rule above by z, that is, $z = \zeta_{w_1'r_1\#\ldots r_k\#}$ considering Case 2.1.1, and $z = \zeta_{w_{1,1}'}\zeta_{w_{1,2}'r_1\#\ldots r_k\#}$ considering Case 2.1.2. Now, the respective subcases can be denoted as follows (for $i = 1, 2$):

2.1.i.1 If the second part of the encoded righthand side fits into one compression symbol, that is, if $|r_{k+1}\# \ldots r_{|r|}\#w_2''| \leq 4\mu + 1$, then the following rule belongs to P':

$$z\zeta_t \to z\zeta_{r_{k+1}\#\ldots r_{|r|}\#w_2''} \quad .$$

2.1.i.2 On the other hand, if the second part of the encoded righthand side does not fit into one compression symbol, that is, if $|r_{k+1}\# \ldots r_{|r|}\#w_2''| > 4\mu + 1$, then the following rule belongs to P', where $w_2'' = w_{2,1}''w_{2,2}''$ such that $|r_{k+1}\# \ldots r_{|r|}\#w_{2,1}''| = 2\mu + 1$:

$$z\zeta_t \to z\zeta_{r_{k+1}\#\ldots r_{|r|}\#w_{2,1}''}\zeta_{w_{2,2}''} \quad .$$

2.2 If the encoded lefthand side is split directly in front of an original symbol, that is, if $w_1'' = \ell_1\# \ldots \ell_k\#$ and, accordingly, $w_2' = \ell_{k+1}\# \ldots \ell_{|\ell|}\#$, then the following rule belongs to P':

$$\zeta_{w_1}\zeta_{w_2} \to \zeta_{w_1}\zeta_t \quad .$$

Here, the encoded righthand side of the rule is split as follows: $r_1\# \ldots r_k\#r_{k+1}$ is put into the first compression symbol and the second part $\#r_{k+2}\# \ldots r_{|r|}\#$ is put into the second compression symbol.

2.2.1 If the first part of the encoded righthand side fits into one compression symbol, that is, if $|w'_1 r_1 \# \ldots r_{k+1}| \leq 4\mu + 1$, then the following rule also belongs to P':

$$\zeta_{w_1} \zeta_t \to \zeta_{w'_1 r_1 \# \ldots r_{k+1}} \zeta_t \quad .$$

2.2.2 On the other hand, if the first part of the encoded righthand side does not fit into one compression symbol, that is, if we have $|w'_1 r_1 \# \ldots r_{k+1}| > 4\mu + 1$, then the following rule belongs to P', where $w'_1 = w'_{1,1} w'_{1,2}$ such that $|w'_{1,1}| = 2\mu + 1$:

$$\zeta_{w_1} \zeta_t \to \zeta_{w'_{1,1}} \zeta_{w'_{1,2} r_1 \# \ldots r_{k+1}} \zeta_t \quad .$$

Again, in each of these two cases there are two subcases to consider. We name the first part of the righthand side of the rule above by z, that is, $z = \zeta_{w'_1 r_1 \# \ldots r_{k+1}}$ considering Case 2.2.1, and $z = \zeta_{w'_{1,1}} \zeta_{w'_{1,2} r_1 \# \ldots r_{k+1}}$ considering Case 2.2.2. Now, the respective subcases can be denoted as follows (for $i = 1, 2$):

2.2.i.1 If the second part of the encoded righthand side fits into one compression symbol, that is, if $|\# r_{k+2} \# \ldots r_{|r|} \# w''_2| \leq 4\mu + 1$, then the following rule belongs to P':

$$z \zeta_t \to z \zeta_{\# r_{k+2} \# \ldots r_{|r|} \# w''_2} \quad .$$

2.2.i.2 On the other hand, if the second part of the encoded righthand side does not fit into one compression symbol, that is, if $|\# r_{k+2} \# \ldots r_{|r|} \# w''_2| > 4\mu + 1$, then the following rule belongs to P', where $w''_2 = w''_{2,1} w''_{2,2}$ such that $|\# r_{k+2} \# \ldots r_{|r|} \# w''_{2,1}| = 2\mu + 1$:

$$z \zeta_t \to z \zeta_{\# r_{k+2} \# \ldots r_{|r|} \# w''_{2,1}} \zeta_{w''_{2,2}} \quad .$$

Ending rules:

$$\zeta_w \to a_1 \ldots a_m \text{ for } \zeta_w \in N_1 \text{ with } \widehat{w} = a_1 \ldots a_m \in T^* \quad .$$

Define $G' = (N', T, S', P')$.

It is easily seen that G is context-sensitive.

By an examination of the rules we see that $L(G') = L(G)$, as G is simulated step by step in the compression symbols and t uniquely determines the rule applied and the position where it is applied.

We define a weight function as follows:

$$\varphi(S') = 1,$$
$$\varphi(\zeta_w) = 2 \cdot |w| \text{ for } \zeta_w \in N_1,$$
$$\varphi(\zeta_t) = 2 \cdot |w'_2 w''_2| + 1 \text{ for } t = (w'_1, w''_1, w'_2, w''_2, (\ell \to r)),$$
$$\varphi(a) = 10 \text{ for } a \in T.$$

It follows that $|w| < |v|$ implies $\varphi(\zeta_w) + 1 < \varphi(\zeta_v)$, and by definition it holds that $\varphi(\zeta_{w'_2 w''_2}) < \varphi(\zeta_{(w'_1, w''_1, w'_2, w''_2, (\ell \to r))})$. From this it is easily seen that G' is weight-increasing. Thus by Lemma 5 G' is an acyclic context-sensitive grammar.

So, for each growing context-sensitive grammar there exists an equivalent acyclic context-sensitive grammar, which by the trivial inclusion ACSL ⊆ GCSL can be stated as follows.

Theorem 8. GCSL = ACSL.

4 Conclusion

Although intuitively it seems not to be the case, we have seen that acyclic context-sensitive grammars and strictly monotone grammars describe the same language class, namely the class of growing context-sensitive languages GCSL.

As GCSL is recognized by a certain machine model, the so-called shrinking two-pushdown automaton, from this characterization various restrictions for this machine model can be derived. The similar construction for the class of Church-Rosser languages CRL in [16,17], where a normal form for length-reducing string rewriting systems describing Church-Rosser languages is built, implies similar restrictions also for deterministic two-pushdown automata. In fact, by rebuilding the construction a normal form for deterministic as well as for nondeterministic shrinking two-pushdown automata can be obtained. The exact description and proof for this restricted normal form is the matter of research in the near future.

It remains as an open question whether length-increasing context-sensitive grammars also characterize GCSL, that is, whether GACSL and GCSL coincide. On the one hand, intuitively we think that this is not the case. On the other hand, currently we have no candidate for a language that might separate these classes, and also the fact that GCSL and ACSL coincide is counterintuitive.

References

1. Erik Aarts. Recognition for acyclic context-sensitive grammars is probably polynomial for fixed grammars. In *Proceedings of the 14th International Conference on Computational Linguistics*, 1992.
2. Martin Beaudry, Markus Holzer, Gundula Niemann, and Friedrich Otto. McNaughton Languages. Mathematische Schriften Kassel 26/00, Universität Kassel, November 2000.
3. Ronald V. Book. *Grammars with Time Functions*. Phd thesis, Harvard University, Cambridge, Massachusetts, February 1969.
4. Franz-Josef Brandenburg. Zur Verallgemeinerung von Grammatiken durch Kontext. Seminarberichte des Instituts für Theorie der Automaten und Schaltnetzwerke 73, Gesellschaft für Mathematik und Datenverarbeitung mbH, Bonn, 1974.
5. Gerhard Buntrock. *Wachsende kontextsensitive Sprachen*. Habilitationsschrift, Fakultät für Mathematik und Informatik, Universität Würzburg, July 1996. English title: Growing context-sensitive languages.
6. Gerhard Buntrock and Krzysztof Loryś. On growing context-sensitive languages. In W. Kuich, editor, *Proceedings of the 19th International Colloquium on Automata, Languages and Programming*, number 623 in Lecture Notes in Computer Science, pages 77–88, Berlin/New York, 1992. Springer.

7. Gerhard Buntrock and Krzysztof Loryś. The variable membership problem: Succinctness versus complexity. In P. Enjalbert, E. W. Mayr, and K.W. Wagner, editors, *Proceedings of the 11th Symposium on Theoretical Aspects of Computer Science*, number 775 in Lecture Notes in Computer Science, pages 595–606, Berlin/New York, 1994. Springer.
8. Gerhard Buntrock and Friedrich Otto. Growing context-sensitive languages and Church-Rosser languages. *Information and Computation*, 141:1–36, 1998.
9. Noam Chomsky. On certain formal properties of grammars. *Information and Control*, 2(2):137–167, June 1959.
10. Elias Dahlhaus and Manfred K. Warmuth. Membership for growing context–sensitive grammars is polynomial. *Journal of Computer and System Sciences*, 33:456–472, 1986.
11. Aleksey Vsevolodovich Gladkij. On the complexity of derivations in context-sensitive grammars. *Algebra i Logika, Seminar*, 3(5-6):29–44, 1964. In Russian.
12. John E. Hopcroft and Jeffrey D. Ullman. *Introduction to Automata Theory, Languages and Computation*. Series in Computer Science. Addison-Wesley, Reading, MA, 1979.
13. Robert McNaughton. An insertion into the chomsky hierarchy? In Juhani Karhumäki, Hermann A. Maurer, Gheorghe Păun, and Grzegorz Rozenberg, editors, *Jewels are Forever, Contributions on Theoretical Computer Science in Honor of Arto Salomaa*, pages 204–212. Springer, 1999.
14. Robert McNaughton, Paliath Narendran, and Friedrich Otto. Church-Rosser Thue systems and formal languages. *Journal of the Association Computing Machinery*, 35:324–344, 1988.
15. Rohit J. Parikh. On context-free languages. *Journal of the Association Computing Machinery*, 13:570–581, 1966.
16. Jens R. Woinowski. A normal form for Church-Rosser language systems (with appendix). Technical Report TI-7/00, Technische Universität Darmstadt, June 2000.
17. Jens R. Woinowski. A normal form for Church-Rosser language systems. In A. Middeldorp, editor, *Proceedings of the 12th International Conference on Rewriting Techniques and Applications*, number 2051 in Lecture Notes in Computer Science, pages 322–337, Berlin, 2001. Springer.

Recognizable Sets of N-Free Pomsets Are Monadically Axiomatizable

Dietrich Kuske

Department of Mathematics and Computer Science
University of Leicester
LEICESTER
LE1 7RH, UK
D.Kuske@mcs.le.ac.uk

Abstract. It is shown that any recognizable set of finite N-free pomsets is axiomatizable in counting monadic second order logic. Differently from similar results by Courcelle, Kabanets, and Lapoire, we do not use MSO-transductions (i.e., one-dimensional interpretations), but two-dimensional interpretations of a generating tree in an N-free pomset. Then we have to deal with the new problem that set-quantifications over the generating tree are translated into quantifications over binary relations in the N-free pomset. This is solved by an adaptation of a result by Potthoff & Thomas on monadic antichain logic.

1 Introduction

In 1960, Büchi showed that a set of finite words is recognizable iff it is axiomatizable in monadic second order logic. This result has been extended in several directions: To infinite words where it is used to prove the decidability of the monadic second order theory of ω by Büchi, to finite ordered trees of bounded degree by Doner, and to infinite ordered trees of bounded degree by Rabin. Similar relations were shown for aperiodic sets of finite and infinite words and first-order logic by McNaughton & Papert as well as natural fragments of first-order logic by Thomas.

Courcelle initiated the consideration of recognizable sets of graphs introducing algebras of graphs. A set of graphs is recognizable in such an algebra iff it is the union of equivalence classes of a congruence of finite index. He could show that any set of graphs that is axiomatized in monadic second order logic MSO is recognizable [1]. This result holds even for the extension of MSO to counting monadic second order logic (where one can make statements on the size of a set modulo some number n). The inverse is false in general, but true for sets of graphs of bounded tree-width (Lapoire [11]).

Lodaya & Weil aimed at an extension of the equivalence between rational and recognizable sets of words to N-free pomsets [12,13]. In [10], I showed that any set of N-free pomsets that is axiomatizable in monadic second order logic (first-order logic, resp.) is recognizable (aperiodic, resp.). The converse implications were shown for sets of bounded width [9,10] and left open in general.

In this paper, it is shown that any recognizable set of N-free pomsets is axiomatizable in counting monadic second order logic. The general strategy of the proof is similar to proofs by Courcelle for series-parallel graphs [2], for partial k-paths by Kabanets [8], and for graphs of bounded tree width by Lapoire [11]: Given an N-free pomset p, we interpret a generating tree in p. The set of generating trees of a recognizable set of N-free pomsets is recognizable and therefore CMSO-axiomatizable. Now the interpretation of a generating tree in p is used to show that the recognizable set of N-free pomsets is CMSO-axiomatizable. Differently from [2,8,11], our interpretation is two-dimensional, i.e., nodes of the generating tree are represented by pairs of nodes of the N-free pomset, in particular, our interpretation is no MSO-transduction. This causes the problem that the quantifications over sets of nodes of the generating tree lead to quantifications over binary relations in the N-free pomset. We solve this problem by two observations: In the counting monadic second order logic for trees, it suffices to allow set quantifications over sets of leaves (is a straightforward extension of a result by Potthoff & Thomas [15]). And leaves of the generating tree are interpreted by pairs of the form (x,x), i.e., sets of leaves can be seen as subsets of the N-free pomset. Hence we obtain that a recognizable set of N-free pomsets is axiomatizable in counting monadic second order logic. Together with results from [4,10], recognizability and axiomatizability in CMSO are therefore equivalent for N-free pomsets. This equivalence holds for *finite* N-free pomsets, only. [9, Example 3] shows that recognizable sets of infinite N-free pomsets are not necessarily axiomatizable in CMSO. Conversely, the methods from [10] can be used to show that CMSO-axiomatizable sets of infinite N-free pomsets are recognizable.

This result does not follow from Lapoire's result on sets of (hyper)graphs of bounded tree-width since N-free pomsets have unbounded tree width. Neither does our result follow from the equivalence between recognizability and CMSO-definability of sets of series-parallel graphs [2] since these graphs have source and sink nodes. The lack of these source and sink nodes requires the use of a two-dimensional interpretation in our setting instead of MSO-transductions.[1]

The equivalence between first-order axiomatizable and aperiodic sets does not hold: From an example by Potthoff [14], it follows that aperiodic sets need not be elementarily axiomatizable.

2 Preliminaries

2.1 N-Free Pomsets

In this paper, any partially ordered set is assumed to be nonempty and finite. Let $(P, \leq)$ be a partially ordered set. We write x co y for elements $x, y \in P$ if they are incomparable. An *N-free poset* $(P, \leq)$ is a finite and nonempty partially ordered set such that the partially ordered set $(N, \leq_N)$ cannot be embedded into $(P, \leq)$. We fix an alphabet Σ, i.e., a nonempty finite set. Then NF denotes the set of all Σ-labeled N-free posets $(P, \leq, \lambda)$. These labeled posets are called *N-free pomsets*.

The poset $(N, \leq_N)$

Next, we define the sequential and the parallel product of N-free pomsets: Let $p_1 = (P_1, \leq_1, \lambda_1)$ and $p_2 = (P_2, \leq_2, \lambda_2)$

---

[1] Only recently I have learnt that similar methods were used by Hoogeboom and ten Pas in their investigation of text languages [7].

be N-free pomsets with $P_1 \cap P_2 = \emptyset$. The *sequential product* $p_1 \cdot p_2$ of p_1 and p_2 is the N-free pomsets $(P_1 \cup P_2, \leq_1 \cup P_1 \times P_2 \cup \leq_2, \lambda_1 \cup \lambda_2)$. Thus, in $p_1 \cdot p_2$, the pomset p_2 is put on top of the pomset p_1. On the contrary, the *parallel product* $p_1 \parallel p_2$ is defined to be $(P_1 \cup P_2, \leq_1 \cup \leq_2, \lambda_1 \cup \lambda_2)$, i.e. here the two partial orders are set side by side. It is a result in the folklore of order theory that a poset is N-free iff it can be constructed from the singletons by sequential and parallel product (cf. [5]). This equivalence is the reason for the alternative name of N-free pomsets: series-parallel or sp-pomsets. By NF $\parallel$ NF $\cup \Sigma$ we denote the set of all non-connected or singleton N-free pomsets.

In this paper, we will use the following two properties of N-free pomsets.

Lemma 1. *Let $p = (P, \leq, \lambda)$ be an N-free pomset.*

- *p is connected if and only if any two elements $x, y \in P$ are bounded from above or from below.*
- *If p is connected and $|P| \geq 2$, there are uniquely determined N-free pomsets $p_0 \in$ NF and $p_1 \in$ (NF $\parallel$ NF) $\cup \Sigma$ such that $p = p_0 \cdot p_1$.*

Next we want to define recognizable sets of N-free pomsets. Let S be a set that is equipped with two associative binary operations $\cdot$ and $\parallel$. We assume, in addition, $\parallel$ to be commutative. Then $(S, \cdot, \parallel)$ is an *sp-algebra* [12]. Note that the set of N-free pomsets is an sp-algebra. A set L of N-free pomsets is *recognizable* if there exists a finite sp-algebra $(S, \cdot, \parallel)$ and a homomorphism $\eta :$ NF $\to S$ such that $L = \eta^{-1}\eta(L)$.

2.2 Logic

In this section, we will define counting monadic second order formulas and their interpretations over N-free pomsets. *CMSO-formulas* involve first order variables $x, y, z \ldots$ for vertices and monadic second order variables $X, Y, Z, \ldots$ for sets of vertices. They are built up from the atomic formulas $\lambda(x) = a$ for $a \in \Sigma$, $x \leq y$, $x \in X$, and $\mathrm{mod}_{p,q}(X)$ with $0 \leq p < q$ by means of the boolean connectives $\neg, \vee, \wedge, \to, \leftrightarrow$ and quantifiers $\exists, \forall$ (both for first order and for second order variables). Formulas without free variables are called sentences. An elementary formula (sentence) is a formula (sentence) without set variables. The atomic formula $\mathrm{mod}_{p,q}(X)$ states that the set X contains p mod q elements. Then the satisfaction relation $\models$ between N-free pomsets $p = (P, \leq, \lambda)$ and CMSO-sentences φ is defined canonically with the understanding that first order variables range over vertices of P and second order variables over subsets of P.

Let φ be a CMSO-sentence and let $L = \{p \in$ NF $\mid p \models \varphi\}$ denote the set of N-free pomsets that satisfy φ. Then we say that the sentence φ *axiomatizes the set L* or that L is *CMSO-axiomatizable*[2]. If φ is an elementary sentence, we say that L is *elementarily axiomatizable*.

A set of N-free pomsets *of bounded width* is MSO-axiomatizable if and only if it is recognizable [9]. Without assuming bounded width, this is clearly false since the set of antichains of even size is recognizable but not MSO-axiomatizable. The following lemma follows from [4]; alternatively, it can be shown using the techniques from [10]:

---

[2] "1-definable" in Courcelle's terminology, see, e.g., [3]. In [2,8,11], "2-definable" sets are considered where one can in addition quantify over sets of edges.

Lemma 2. *Any CMSO-axiomatizable set of N-free pomsets is recognizable.*

In this paper, we will show the converse implication. This result strengthens one obtained in [9] by dropping the condition "bounded width".

2.3 Reduced Terms

Definition 1. *A* tree *is a structure* (V, son, ρ) *where V is a finite set,* $\text{son} \subseteq V^2$ *is an acyclic relation, and* $\rho : V \to \{\|, \cdot\} \cup \Sigma$ *is a labeling function. We require in addition that there is precisely one node (the* root*) without a father node, and that any other node has precisely one father.*

A reduced term *is a structure* $(V, \text{son}, \rho, \text{firstson})$ *such that* (V, son, ρ) *is a tree,* $\text{firstson} \subseteq \text{son}$ *is a binary relation on V, and for any $v \in V$ the following hold:*

- $\rho(v) \in \Sigma$ *iff v is a leaf.*
- *If* $\rho(v) = \cdot$, *then v has precisely two sons v_1 and v_2. For these sons, we have* $(v, v_1) \in \text{firstson}$, $(v, v_2) \notin \text{firstson}$, *and* $\rho(v_2) \neq \cdot$.
- *If* $\rho(v) = \|$, *then v has at least two sons and any son w of v satisfies $\rho(w) \neq \|$ and* $(v, w) \notin \text{firstson}$.

The set of reduced terms is denoted by RTerm.

Let $t = (V, \text{son}, \rho, \text{firstson})$ be a reduced term and let $v \in V$ be some node of t. Then the restriction of t to the ancestors of v is a reduced term. Hence we can extend any mapping $\eta : \Sigma \to (S, \cdot, \|)$ into some sp-algebra by induction:

- $\overline{\eta}(t) = \eta(a)$ if t is a singleton term whose only node is labeled by a,
- $\overline{\eta}(\cdot(t_1, t_2)) = \overline{\eta}(t_1) \cdot \overline{\eta}(t_2)$, and
- $\overline{\eta}(\|(t_1, t_2, \ldots, t_n)) = \overline{\eta}(t_1) \| \overline{\eta}(t_2) \cdots \| \overline{\eta}(t_n)$.

In particular, the extension of the mapping $\Sigma \to (\text{NF}, \cdot, \|) : a \mapsto a$ associates an N-free pomset to any reduced term t. This pomset will be denoted by $\text{val}(t)$, i.e., val is a mapping from RTerm to NF. One can show that val is even bijective. Hence the set of reduced terms is another incarnation of the free sp-algebra over the set Σ (with suitably defined operations $\cdot$ and $\|$).

Example 1. The following picture shows an N-free pomset p (on the left) and a reduced term $t(p)$ with $\text{val}(t(p)) = p$. In the reduced term $t(p)$, the sons of a $\cdot$-labeled node are ordered such that the left one is the first son.

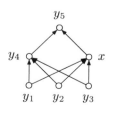
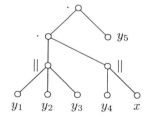

Since reduced terms are relational structures, counting monadic second order logic can be defined along the lines of CMSO for N-free pomsets above. The only difference is that the atomic formulas now are of the form $\rho(x) = \alpha$ for $\alpha \in \{\cdot, \|\} \cup \Sigma$, $(x, y) \in$ son, $(x, y) \in$ firstson, $x \in X$, and $\mathrm{mod}_{p,q}(X)$. A set X of reduced terms is *CMSO-axiomatizable* iff there exists a formula φ such that $t \models \varphi$ iff $t \in X$ for any reduced term t.

Lemma 3. *Let $L \subseteq \mathrm{NF}$ be a recognizable set of N-free pomsets. Then the set $\mathrm{val}^{-1}(L) \subseteq \mathrm{RTerm}$ is CMSO-axiomatizable.*

Proof. Let $\eta : \mathrm{NF} \to S$ be a homomorphism that recognizes L. The mapping $\eta \upharpoonright \Sigma$ can be extended as described above to a mapping $\overline{\eta} : \mathrm{RTerm} \to S$. We mimic the evaluation of $\overline{\eta}$ on a reduced term $t = (V, \mathrm{son}, \rho, \mathrm{firstson})$ by a CMSO-formula. This formula states the existence of sets X_s for $s \in S$ that form a partition of the sets of nodes. Let r be the corresponding mapping from the set of nodes into S. Then, some local conditions have to be satisfied, in particular, $r(x)$ has to be the parallel product of $\{r(y) \mid (x, y) \in \mathrm{son}\}$ for any $\|$-labeled node x. Since the number of sons of a $\|$-labeled node is not bounded, this cannot be translated into MSO. To overcome this problem, one uses the fact that there are $m, k \in \mathbb{N}$ such that $s^{n+m} = s^n$ for $n \geq k$ in the commutative semigroup $(S, \|)$. Thus, to determine $r(x)$, for any $s \in S$, we only need to know

- if there are more than k sons y of x satisfying $r(y) = s$ and
- the number of sons y of x satisfying $r(y) = s$, decreased by k, modulo m.

This can be expressed in CMSO (but not in pure monadic second order logic) which finishes the sketch of proof. □

3 Construction of the Generating Term $t(p)$

It is easy to show inductively that any N-free pomset p is the value of some reduced term $t(p)$ and that this reduced term is even uniquely determined. Therefore, we will call $t(p)$ the generating term of p. In this section, we define the generating term $t(p)$ explicitly from the N-free pomset p without using induction. This will be the basis for our interpretation of the generating term $t(p)$ in the N-free pomset p in the following section. The foundation for the construction of the set of vertices of $t(p)$ is layed down by the following definition:

Definition 2. *Let $p = (P, \leq, \lambda)$ be an N-free pomset and let $x, y \in P$. Then $\mathrm{lf}_p(x, y)$, the least factor of p containing x and y, is defined by*

$$\mathrm{lf}_p(x, y) = \begin{cases} \{z \in P \mid \forall h \in P : (x, y < h \to z < h) \text{ and} \\ \qquad \forall h \in P : (x, y > h \to z > h)\} & \text{if } x \text{ co } y \\ \{z \in P \mid \neg(z < x), \neg(y < z), \text{ and } \neg(x \text{ co } z \text{ co } y)\} & \text{if } x \leq y \\ \mathrm{lf}_p(y, x) & \text{if } y < x. \end{cases}$$

Example 2. Let p be the N-free pomset from Example 1. Then $\mathrm{lf}_p(x,x) = \{x\}$, $\mathrm{lf}_p(x,y_4) = \{x,y_4\}$, and $\mathrm{lf}_p(x,y_1) = \{y_1,y_2,y_3,y_4,x\} = \mathrm{lf}_p(x,y_2)$. You might check that any node in the reduced term $t(p)$ corresponds to some set $\mathrm{lf}_p(z_1,z_2)$, but the inverse is not true: $\mathrm{lf}_p(x,y_5) = \{x,y_4,y_5\}$ is not the value of any subtree of $t(p)$.

Let $p = (P, \leq, \lambda)$ be an N-free pomset and $X \subseteq P$. Then the restriction of p to X is an N-free pomset. In this sense, we can speak of the "connected components of X" and of "sequential prefixes of X" as we do in the following definition.

Definition 3. *Let $p = (P, \leq, \lambda)$ be an N-free pomset. We define three sets of subsets of P as follows:*
$V_0 = \{\mathrm{lf}_p(x,y) \mid x \text{ co } y\} \cup \{\{x\} \mid x \in P\}$,
$V' = $ *the set of connected components of elements of $V_0 \cup \{P\}$, and*
$V_1 = $ *the set of sequential prefixes of elements of V'.*

Note that $V' \subseteq V_1$ since we consider any connected N-free pomset as a sequential prefix of itself. The set $V = V_0 \cup V_1$ will be the set of nodes of the generating term $t(p)$.

The following lemma relates the definitions of the sets V_0 and V_1 to the recursive construction of an N-free pomset. This relation is the basis for the recursive proofs of Lemmas 6 and 7.

Lemma 4. *1. Let p_i be connected N-free pomsets, $n > 1$, and $p = p_1 \parallel p_2 \parallel \cdots \parallel p_n$. Then we have*
$$V_0(p) = \{p\} \cup \bigcup_{1 \leq i \leq n} V_0(p_i), \text{ and } V_1(p) = \bigcup_{1 \leq i \leq n} V_1(p_i).$$
2. Let p_0, p_1 be N-free pomsets with $p_1 \in (\mathrm{NF} \parallel \mathrm{NF}) \cup \Sigma$. Then $V_0(p_0 \cdot p_1) = V_0(p_0) \cup V_0(p_1)$ and $V_1(p_0 \cdot p_1) = V_1(p_0) \cup V_1(p_1) \cup \{p_0, p_0 \cdot p_1\}$.

Next we define the edges of the generating term $t(p)$.

Definition 4. *Let p be an N-free pomset. We define three binary relations on $V_0 \cup V_1$ as follows:*

$E_1 = \{(v_0, v_1) \in V_0 \times V_1 \mid v_1 \text{ is a connected component of } v_0 \text{ and } v_0 \neq v_1\}$,
$E_2 = \{(v, v') \in V_1 \times V_1 \mid v' \text{ is a maximal proper sequential prefix of } v\}$, and
$E_3 = \{(v_1, v_0) \in V_1 \times V_0 \mid v_0 \text{ is a minimal proper sequential suffix of } v_1\}$.

Similarly to Lemma 4, the following lemma prepares the recursive proof of Lemmas 6 and 7 below.

Lemma 5. *1. Let p_i be connected N-free pomsets, $n > 1$, and $p = p_1 \parallel p_2 \parallel \cdots \parallel p_n$. Then we have*
$$E_1(p) = \{(p, p_i) \mid 1 \leq i \leq n\} \cup \bigcup_{1 \leq i \leq n} E_1(p_i),$$
$$E_2(p) = \bigcup_{1 \leq i \leq n} E_2(p_i), \text{ and } E_3(p) = \bigcup_{1 \leq i \leq n} E_3(p_i).$$

2. Let p_0, p_1 be N-free pomsets with $p_1 \in (\text{NF} \parallel \text{NF}) \cup \Sigma$. Then

$$E_1(p_0 \cdot p_1) = E_1(p_0) \cup E_1(p_1),$$
$$E_2(p_0 \cdot p_1) = \{(p_0 \cdot p_1, p)\} \cup E_2(p_0) \cup E_2(p_1), \text{ and}$$
$$E_3(p_0 \cdot p_1) = \{(p_0 \cdot p_1, p_1)\} \cup E_3(p_0) \cup E_3(p_1).$$

Now we construct the generating term $t(p)$ from the N-free pomset p using the sets and relations defined above:

Definition 5. *Let p be an N-free pomset. Let $V = V_0 \cup V_1$, son $= E_1 \cup E_2 \cup E_3$, and firstson $= E_2$. The labeling $\rho : V \to \Sigma \cup \{\parallel, \cdot\}$ is defined by $\rho(X) = \cdot$ if $X \in V_1 \setminus V_0$, $\rho(\{x\}) = \lambda(x)$, and $\rho(X) = \parallel$ otherwise. Then the generating term $t(p)$ is given by $t(p) = (V, \text{son}, \rho, \text{firstson})$.*

By induction, one can infer the following from Lemmas 4 and 5.

Lemma 6. *Let p be an N-free pomset. Then the generating term $t(p)$ is a reduced term with $\text{val}(t(p)) = p$.*

4 Interpretation of the Generating Term $t(p)$ in p

In this section, we will show how we can interpret the generating term $t(p)$ in the N-free pomset $p = (P, \leq, \lambda)$. Recall that the nodes of the generating term are subsets of P. We start by showing that these sets can be represented by pairs of elements of P (which is clear for the elements of V_0):

Lemma 7. *Let $p = (P, \leq, \lambda)$ be an N-free pomset and $X \in V$ (i.e., X is a node of the generating term $t(p)$). Then there exist $x, y \in P$ with $\text{lf}_p(x, y) = X$.*

Next we show the existence of some particular formulas that single out those pairs (x, y) which stand for a node in the generating term (formula $\nu_0 \vee \nu_1$), that are labeled by α (formula label$_\alpha$) as well as those pairs of pairs that are connected by an edge (formula $\eta_1 \vee \eta_2 \vee \eta_3$). Note that different pairs of elements can stand for the same node in the generating term. Therefore, we also need the formula eq that expresses precisely this:

Lemma 8. *There are elementary formulas $\text{eq}(x_1, x_2, y_1, y_2)$, $\nu_i(x_1, x_2)$ ($i = 0, 1$), $\eta_i(x_1, x_2, y_1, y_2)$ ($i = 1, 2, 3$), and $\text{label}_\alpha(x_1, x_2)$ ($\alpha \in \Sigma \cup \{\cdot, \parallel\}$) such that for any N-free pomset $p = (P, \leq, \lambda)$ and any $x_i, y_i \in P$, we have*

1. *$p \models \text{eq}(x_1, x_2, y_1, y_2)$ iff $\text{lf}_p(x_1, x_2) = \text{lf}_p(y_1, y_2)$,*
2. *$p \models \nu_i(x_1, x_2)$ iff $\text{lf}_p(x_1, x_2) \in V_i$ ($i = 0, 1$),*
3. *$p \models \eta_i(x_1, x_2, y_1, y_2)$ iff $(\text{lf}_p(x_1, x_2), \text{lf}_p(y_1, y_2)) \in E_i$ ($i = 1, 2, 3$), and*
4. *$p \models \text{label}_\alpha(x_1, x_2)$ iff $\rho(\text{lf}_p(x_1, x_2)) = \alpha$ for $\alpha \in \Sigma \cup \{\cdot, \parallel\}$.*

Proof. Since we can express that an element $z \in P$ belongs to $\text{lf}_p(x_1, x_2)$, the formula eq exists as required. The formula ν_0 is obviously $x_1 \text{ co } x_2 \vee x_1 = x_2$. By Lemma 1, we can express that $\text{lf}_p(x_1, x_2)$ is connected by an elementary formula. Hence there exists a formula $\nu'(x_1, x_2)$ which is satisfied by x_1, x_2 iff $\text{lf}_p(x_1, x_2) \in V'$. A subset of an

N-free pomset is a sequential prefix iff any of its elements is below any element of its complement. Hence, from ν', we can build the formula ν_1.

Using Lemma 1 again, we can state that $\mathrm{lf}_p(y_1, y_2)$ is a connected component of $\mathrm{lf}_p(x_1, x_2)$ which gives us the formula η_1. Since any connected component of a least factor $\mathrm{lf}_p(x_1, x_2)$ is of the form $\mathrm{lf}_p(z_1, z_2)$ for some z_1, z_2, we can formulate that $\mathrm{lf}_p(y_1, y_2)$ is a maximal proper prefix of $\mathrm{lf}_p(x_1, x_2)$ which gives us the formula η_2. To obtain η_3 note that a subset of an N-free pomset is a minimal proper suffix iff its complement is a maximal proper prefix. □

With $\nu = \nu_0 \vee \nu_1$, son $= \eta_1 \vee \eta_2 \vee \eta_3$, and firstson $= \eta_2$, the lemma defines an interpretation of $t(p)$ in p. But this interpretation is 2-dimensional (i.e., nodes of the reduced term are represented by [sets of] pairs of elements of the N-free pomset). Let $L \subseteq \mathrm{NF}$ be recognizable. Then we know that $\mathrm{val}^{-1}(L) \subseteq \mathrm{RTerm}$ is CMSO-axiomatizable by Lemma 3. Suppose this set were elementarily axiomatizable, i.e., there were an elementary sentence φ such that $\mathrm{val}^{-1}(L) = \{t \in \mathrm{RTerm} \mid t \models \varphi\}$. Then we obtain that L is elementarily axiomatizable by a general result on interpretations (cf. [6, Theorem 5.3.2]). The idea of the proof is to replace any elementary variable in φ (that ranges over nodes in t) by two elementary variables (that stand for elements of p and together represent the least factor containing them and therefore a node of the generating term). Hence, we obtain an elementary sentence $\overline{\varphi}$ from φ by the following inductive procedure:

$$\overline{\exists x \psi} = \exists x_1, x_2 (\nu(x_1, x_2) \wedge \overline{\psi}), \quad \overline{\rho(x) = \alpha} = \mathrm{label}_\alpha(x_1, x_2),$$
$$\overline{\mathrm{son}(x, y)} = \eta(x_1, x_2, y_1, y_2), \quad \overline{\mathrm{firstson}(x, y)} = \eta_2(x_1, x_2, y_1, y_2), \text{ and}$$
$$\overline{(x = y)} = \mathrm{eq}(x_1, x_2, y_1, y_2)$$

where $\alpha \in \{\cdot, \|\} \cup \Sigma$. Then $t(p) \models \varphi$ iff $p \models \overline{\varphi}$ for any N-free pomset p by Lemma 8.

In general the set $\mathrm{val}^{-1}(L)$ is not elementarily, but only CMSO-axiomatizable (see Lemma 3). Using the idea above, set variables would be replaced by binary relations which is impossible in CMSO. In order to solve this problem, we have to make a short detour and consider monadic logics on trees:

In [15], Potthoff & Thomas consider monadic second order logic on proper binary trees. They show that the restriction of set variables to sets of leaves does not reduce the expressive power. As explained in [16, p. 403], the idea is the following: "... the inner nodes can be mapped injectively into the set of leaves: From a given inner node, we follow the path which first branches right and then always branches left until a leaf is reached. Thus a set of inner nodes can be coded by a set of leaves ... Using this idea, quantifiers over subsets of proper binary trees can be simulated by quantifiers over" sets of leaves.

In order to make a similar idea go through, we extend CMSO to LCMSO by allowing in addition the atomic formula $(x, y) \in \mathrm{son}^*$ which states "x is an ancestor of y". The satisfaction relation $\models_L$ for formulas of LCMSO and reduced terms t is defined canonically with the understanding that set variables range over sets of leaves, only (which explains the prefix L in LCMSO). The proof of the following lemma uses the idea from Potthoff and Thomas, but defines three instead of just one injection of the inner nodes of a reduced term into its leaves. These three injections are necessary since the sons of a $\|$-labeled node are not ordered.

Lemma 9. *Let φ be a CMSO-formula. Then there exists an LCMSO-formula ψ such that for any reduced term $t \in \text{RTerm}$ we have $t \models \varphi$ iff $t \models_L \psi$.*

In the reduced term $t(p)$, leaves are of the form $\{(x,x)\}$ for some $x \in P$. Hence sets of leaves in $t(p)$ can be seen as subsets of P. This allows us to axiomatize in CMSO any recognizable set of N-free pomsets:

Theorem 1. *Let L be a set of N-free pomsets. Then L is recognizable if and only if it is CMSO-axiomatizable.*

Proof. Let L be recognizable and let $T = \{t \in \text{RTerm} \mid \text{val}(t) \in L\}$. Then, by Lemmas 3 and 9, there is an LCMSO-sentence φ such that $T = \{t \in \text{RTerm} \mid t \models_L \varphi\}$. We extend the construction of $\overline{\varphi}$ from above by setting $\overline{(x,y) \in \text{son}^\star} = (\text{lf}_p(x_1, x_2) \supseteq \text{lf}_p(y_1, y_2))$, $\overline{\exists X \psi} = \exists X \overline{\psi}$, and $\overline{x \in X} = (x_1 = x_2 \wedge x_1 \in X)$. Then $p \models \overline{\varphi}$ iff $t(p) \models_L \varphi$ which is equivalent to $t(p) \in T$ and therefore to $p \in L$. The inverse implication is Lemma 2. □

The proof of the above theorem uses what is known as "relativized, 2-dimensional, first-order interpretation" (cf. Hodges [6, pp. 212 ff]): The nodes of $t(p)$ are represented by equivalence classes of pairs of elements of p; therefore, the interpretation is "two-dimensional". Since not all pairs (x_1, x_2) give rise to a node of $t(p)$ (i.e., since there are pairs of nodes that violate ν), the interpretation is "relativized". It is "first-order" since our formulas ν etc. are elementary formulas.

In theoretical computer science, a similar concept is known as MSO-transduction (Courcelle [2]). An MSO-transduction can be seen as a relativized and parametrized one-dimensional monadic second order interpretation that differs in three aspects from the interpretation we consider here: (1) They are one-dimensional. (2) The formula eq that defines an equivalence relation is simply $x = x$, i.e., the equivalence relation eq^p is trivial there.[3] (3) The formulas are not elementary but monadic formulas. Courcelle shows that the preimage of a CMSO-axiomatizable set under an MSO-transduction is CMSO-axiomatizable. This property of MSO-transductions is the basis for the proofs in [2,8,11] that recognizable sets of graphs of bounded tree width are CMSO-axiomatizable.

In [3, p. 65], Courcelle discusses the possibility of more general MSO-transductions by allowing multidimensional interpretations. But, as he observes, in general this results in transductions that do not reflect CMSO-axiomatizability in general.

Thus, the main contribution of this paper is the proof that in certain situations multidimensional interpretations can be used instead of MSO-transductions.

5 Elementarily Axiomatizable Sets of N-Free Pomsets

An sp-algebra $(S, \cdot, \|)$ is *aperiodic* iff the semigroups $(S, \cdot)$ and $(S, \|)$ are aperiodic. A set of N-free pomsets L is *aperiodic* iff it can be accepted by a homomorphism into some finite aperiodic sp-algebra. In [10], I showed for any set L of N-free pomsets:

---

[3] Nontrivial equivalence relations can be introduced without affecting the valitity of Courcelle's results at the expense of an additional parameter, cf. [2, Lemma 2.4].

Theorem 2 ([10]). *Let L be a set of N-free pomsets.*
- *If L is elementarily axiomatizable, then L is aperiodic.*
- *If L is aperiodic and width-bounded, then L is elementarily axiomatizable.*

In this section, it will be shown that the condition on L to be width-bounded in the second statement cannot be dropped.

Let $\Sigma = \{\wedge, \vee, 0, 1\}$. We consider $\wedge$ and $\vee$ as binary operations and 0 and 1 as constants. Then, any term over this signature can naturally be considered as a Σ-labeled tree and therefore as an N-free pomset. Let $L \subseteq \mathrm{NF}$ be the set of all "terms" over Σ that evaluate to 1. By [14, Theorem 4.2], the set L is not elementarily axiomatizable. We will show that L is aperiodic:

Let $S = \{p, p_0, p_1, s_\wedge, s_\vee, s_0, s_1, \bot\}$. The two operations $\cdot$ and $\|$ are defined as follows: $s_0 \| s_1 = s_1 \| s_0 = p$, $\quad s_0 \| s_0 = p_0$, $s_1 \| s_1 = p_1$,
$p_0 \cdot s_\wedge = p \cdot s_\wedge = s_0, \quad p_1 \cdot s_\wedge = s_1,$
$p_0 \cdot s_\vee = s_0, p_1 \cdot s_\vee = p \cdot s_\vee = s_1,$
and $x \| y = \bot$ as well as $x \cdot y = \bot$ in any other case. Then $(S, \cdot, \|)$ is easily seen to be an aperiodic sp-algebra since $x \cdot x \cdot y = x \| x \| y = \bot$ for any $x, y \in S$. A homomorphism from NF onto S is defined by $\eta(\alpha) = s_\alpha$ for $\alpha \in \{\wedge, \vee, 0, 1\}$. Then, indeed, $L = \eta^{-1}(s_1)$ is aperiodic, i.e., we showed

Theorem 3. *There is an aperiodic set of N-free pomsets which is not elementarily axiomatizable.*

References

1. B. Courcelle. The monadic second-order logic of graphs. I: Recognizable sets of finite graphs. *Information and Computation*, 85:12–75, 1990.
2. B. Courcelle. The monadic second-order logic of graphs. V: on closing the gap between definability and recognizability. *Theoretical Comp. Science*, 80:153–202, 1991.
3. B. Courcelle. Monadic second-order definable graph transductions: a survey. *Theoretical Comp. Science*, 126:53–75, 1994.
4. B. Courcelle and J.A. Makowsky. VR and HR graph grammars: A common algebraic framework compatible with monadic second order logic. In *Graph transformations*, 2000. cf. tfs.cs.tu-berlin.de/gratra2000/proceedings.html.
5. J.L. Gischer. The equational theory of pomsets. *Theoretical Comp. Science*, 61:199–224, 1988.
6. W. Hodges. *Model Theory*. Cambridge University Press, 1993.
7. H.J. Hoogeboom and P. ten Pas. Monadic second-order definable text languages. *Theoretical Comp. Science*, 30:335–354, 1997.
8. V. Kabanets. Recognizability equals definability for partial k-paths. In *ICALP'97*, Lecture Notes in Comp. Science vol. 1256, pages 805–815. Springer, 1997.
9. D. Kuske. Infinite series-parallel posets: logic and languages. In J. D. P. Rolim U.Montanari and E. Welzl, editors, *ICALP 2000*, Lecture Notes in Comp. Science vol. 1853, pages 648–662. Springer, 2000.
10. D. Kuske. A model theoretic proof of Büchi-type theorems and first-order logic for N-free pomsets. In A. Ferreira and H. Reichel, editors, *STACS 2001*, Lecture Notes in Comp. Science vol. 2010, pages 443–454. Springer, 2001.

11. D. Lapoire. Recognizability equals monadic second order definability, for sets of graphs of bounded tree width. In *STACS'98*, Lecture Notes in Comp. Science vol. 1373, pages 618–628. Springer, 1998.
12. K. Lodaya and P. Weil. Series-parallel languages and the bounded-width property. *Theoretical Comp. Science*, 237:347–380, 2000.
13. K. Lodaya and P. Weil. Rationality in algebras with a series operation. *Information and Computation*, 2001. To appear.
14. A. Potthoff. Modulo-counting quantifiers over finite trees. *Theoretical Comp. Science*, 126:97–112, 1994.
15. A. Potthoff and W. Thomas. Regular tree languages without unary symbols are star-free. In Z. Ésik, editor, *Fundamentals of Computation Theory*, Lecture Notes in Comp. Science vol. 710, pages 396–405. Springer, 1993.
16. W. Thomas. Languages, automata, and logic. In G. Rozenberg and A. Salomaa, editors, *Handbook of Formal Languages*, pages 389–455. Springer Verlag, 1997.

Automata on Series-Parallel Biposets*

Z. Ésik and Z.L. Németh

Dept. of Computer Science
University of Szeged
P.O.B. 652
6701 Szeged, Hungary
{esik, zlnemeth}@inf.u-szeged.hu

Abstract. We provide the basics of a 2-dimensional theory of automata on series-parallel biposets. We define recognizable, regular and rational sets of series-parallel biposets and study their relationship. Moreover, we relate these classes to languages of series-parallel biposets definable in monadic second-order logic.

1 Introduction

Finite automata process words, i.e., elements of a finitely generated free semigroup. In this paper, we define automata whose input structure is a finitely generated free bisemigroup equipped with two associative operations. The elements of the free bisemigroup may be represented by labelled series-parallel biposets. We introduce recognizable, regular and rational sets of series-parallel biposets and study their relationship. Moreover, by relying on the main result of Hoogeboom and ten Pas [16], we relate these classes to languages of series-parallel biposets definable in monadic second-order logic. All of our results can be generalized to higher dimensions, i.e., to any finite number of associative operations.

Our study owes much to the work of Hoogeboom and ten Pas [15,16] on text languages, and to the recent work of Lodaya and Weil [19,20] and Kuske [17,18] on languages of series-parallel posets that may be seen as a two-dimensional extension of the classical theory to a situation where one of the two associative operations is commutative. We believe that the case that none of the two operations is commutative is more fundamental. An independent study of automata and languages over free bisemigroups was also initiated by Hashiguchi et. al. [14]. However, the approach taken in *op. cit.* is very syntactic. See the last section for a comparison.

2 Biposets

Let n denote a positive integer and let Σ denote a finite alphabet. A Σ-labelled n-poset, or n-poset, for short, is a finite nonempty set P of vertices equipped with n (irreflexive) partial orders $<_i$, $i \in [n] = \{1, \ldots, n\}$, and a labelling function $\lambda : P \to \Sigma$. A Σ-labelled biposet, or biposet, is an n-poset for $n = 2$. The two partial orders of a biposet

---

* Research supported by grant no. T30511 from the National Foundation of Hungary for Scientific Research.

are called the *horizontal* and the *vertical order*. Accordingly, we write $<_h$ and $<_v$. A *morphism* between n-posets P and Q is a function on the vertices that preserves the partial orders and the labelling. An *isomorphism* is a bijective morphism whose inverse is also a morphism. Below we will identify isomorphic n-posets.

Suppose that $P = (P, <_1^P, \ldots, <_n^P, \lambda_P)$ and $Q = (Q, <_1^Q, \ldots, <_n^Q, \lambda_Q)$ are Σ-labelled n-posets. Without loss of generality assume that P and Q are disjoint. For each $i \in [n]$, we define the $\circ_i$-*product* $P \circ_i Q$ to be the n-poset with underlying set $P \cup Q$, partial orders

$$<_j^{P \circ_i Q} = \begin{cases} <_j^P \cup <_j^Q & \text{if } j \neq i \\ <_i^P \cup <_i^Q \cup (P \times Q) & \text{if } j = i \end{cases}$$

and labelling $\lambda_{P \circ_i Q} = \lambda_P \cup \lambda_Q$. When $n = 2$, the product operations $\circ_1$ and $\circ_2$ are called the *series product* or *horizontal product* and the *parallel product* or *vertical product*, respectively. It is clear that the product operations $\circ_i$ are associative.

Each letter $a \in \Sigma$ may be identified with the singleton n-poset labelled a. Let $\mathrm{SP}_n(\Sigma)$ denote the collection of n-posets that can be generated from the singletons by the n product operations.

Theorem 1. [9] *An n-poset $P = (P, <_1, \ldots, <_n, \lambda_P)$ is in $\mathrm{SP}_n(\Sigma)$ iff the following conditions hold.*

1. *For every $u, v \in P$ with $u \neq v$ there is exactly one $i \in [n]$ such that $u <_i v$ or $v <_i u$ holds.*
2. *Each poset $(P, <_i)$, $i \in [n]$ is N-free, i.e., it does not have an induced subposet isomorphic to the poset $([4], <)$ with $1 < 3, 2 < 3, 2 < 4$.*
3. *P satisfies the following* triangle condition: *If u, v, w are different vertices of P, then u, v, w are related by at most 2 of the partial orders $<_i$ (i.e., there is no triangle whose sides have different "colours").*

Note that when $n = 1, 2$, the last condition holds automatically as does the second for $n = 1$. Thus, when $n = 1$, an n-poset is in $\mathrm{SP}_n(\Sigma)$ iff it is a labelled linear order, i.e., a word. An immediate consequence of Theorem 1 is the fact that any "induced sub-n-poset" of an n-poset in $\mathrm{SP}_n(\Sigma)$ is also in $\mathrm{SP}_n(\Sigma)$.

Proposition 1. [9] *$\mathrm{SP}_n(\Sigma)$ is freely generated by Σ in the variety of algebras equipped with n associative operations.*

Call an n-poset P in $\mathrm{SP}_n(\Sigma)$ $\circ_i$-*irreducible*, where $i \in [n]$, if P has no decomposition into the $\circ_i$-product of two or more n-posets (in $\mathrm{SP}_n(\Sigma)$). If this condition does not hold, call P $\circ_i$-*reducible*. Proposition 1 relies on the fact that each n-poset P in $\mathrm{SP}_n(\Sigma)$ is either a singleton or there is a unique i such that P is $\circ_i$-reducible. Moreover, in that case, P has, up to associativity, a unique *maximal decomposition* into a $\circ_i$-product of $\circ_i$-irreducible n-posets. We call the biposets in $\mathrm{SP}_2(\Sigma)$ *series-parallel*.

Remark 1. Theorem 1 is a particular instance of a more general result proved in [9] which concerns Σ-labelled sets equipped with n partial orders $<_i$ and m symmetric irreflexive relations $\sim_j$. The relations $<_i$ define n associative operations and the relations $\sim_j$ define

m associative and commutative operations. The general result is a common extension of the geometric characterization of series-parallel partial orders by Grabowski [13] and Valdes et al. [25], and the characterization of cographs by Corneil et al. [2]. For the case that $n = 1$ and $m = 2$, see also Boudol and Castellani [1].

Labelled n-posets satisfying the first condition of Theorem 1 correspond to those (labelled) reversible antisymmetric *2-structures* of Ehrenfeucht and Rozenberg [6] which are transitive. The third condition is the angularity property [7] for these 2-structures. The n-posets satisfying both the first and the third condition correspond to the T-structures of [8], while the n-posets satisfying all three conditions correspond to the uniformly nonprimitive T-structures. Uniformly nonprimitive 2-structures are studied in detail in Engelfriet et al. [4].

In the subsequent sections we will only consider biposets, and in particular series-parallel biposets, or *sp-biposets*, for short. All of our results can be generalized, in a straightforward way, to n-posets in $\mathrm{SP}_n(\Sigma)$. We will denote $\mathrm{SP}_2(\Sigma)$ by $\mathrm{SPB}(\Sigma)$ and write $\cdot$ for the horizontal and $\circ$ for the vertical product.

Remark 2. Labelled biposets with the property that any two elements are related by exactly one of the two partial orders are called *texts* by Ehrenfeucht and Rozenberg in [8]. The sp-biposets are the uniformly nonprimitive, or *alternating texts*. Suppose that $P = (P, <_h, <_v, \lambda_P)$ is a labelled biposet which is a text. Then the relations $\sqsubset_1 = <_h \cup <_v$ and $\sqsubset_2 = <_h \cup <_v^{-1}$ are strict linear orders on P, where $<_v^{-1}$ is the reverse of the relation $<_v$. Moreover, the relations $<_h$ and $<_v$ can be recovered from these linear orders. In fact, this correspondence defines a bijection between texts and finite nonempty labelled sets equipped with two not necessarily different strict linear orders, see [8]. The operations of horizontal and parallel product on texts correspond to natural operations on (isomorphism classes of) labelled biposets equipped with two strict linear orders. It follows that $\mathrm{SPB}(\Sigma)$ can be represented as an algebra of isomorphism classes of such biposets satisfying a condition ("primitive quartet-freeness") corresponding to N-freeness. See [7] and [5] for details.

3 Recognizable and Regular Languages

The concept of recognizable sp-biposet languages, i.e., recognizable subsets of $\mathrm{SPB}(\Sigma)$, where Σ is a finite alphabet, can be derived from standard general notions, cf. [10]. Recall that a *bisemigroup* is an algebra $B = (B, \cdot, \circ)$ equipped with two associative binary operations $\cdot$ and $\circ$. Homomorphisms and congruences of bisemigroups are defined as usual. A congruence, or equivalence relation of a bisemigroup is of *finite index* if the partition induced by the relation has a finite number of blocks.

Definition 1. *A language $L \subseteq \mathrm{SPB}(\Sigma)$ is recognizable if there is a finite bisemigroup B and a homomorphism $h : \mathrm{SPB}(\Sigma) \to B$ such that $L = h^{-1}(h(L))$.*

It is clear that $L \subseteq \mathrm{SPB}(\Sigma)$ is recognizable iff there is a finite index congruence ϑ of $\mathrm{SPB}(\Sigma)$ which *saturates* L, i.e., L is the union of some blocks of the partition induced by ϑ. It follows by standard arguments that the class Rec of recognizable sp-biposet languages is (effectively) closed under the boolean operations and inverse homomorphisms, so that if h is a homomorphism $\mathrm{SPB}(\Sigma) \to \mathrm{SPB}(\Sigma')$ and $L \subseteq \mathrm{SPB}(\Sigma')$ is recognizable, then so is $h^{-1}(L)$. Other closure properties will be given later.

Regular sets of sp-biposets will be defined using parenthesising automata that process sp-biposets in a sequential manner. The definition below involves a finite set Ω of parentheses. *We assume that Ω is partitioned into opening and closing parentheses that are in a bijective correspondence.*

Definition 2. *A (nondeterministic)* parenthesizing automaton *is a 9-tuple* $\mathbf{S} = (S, H, V, \Sigma, \Omega, \delta, \gamma, I, F)$, *where S is the nonempty, finite set of states, H and V are the sets of* horizontal *and* vertical states, *which give a disjoint decomposition of S, Σ is the* input alphabet, *Ω is a finite set of parentheses, moreover,*

- $\delta \subseteq (H \times \Sigma \times H) \cup (V \times \Sigma \times V)$ *is the* labelled transition relation,
- $\gamma \subseteq (H \times \Omega \times V) \cup (V \times \Omega \times H)$ *is the* parenthesizing relation,
- $I, F \subseteq S$ *are the sets of* initial *and* final *states, respectively.*

Definition 3. *Suppose that $P \in \mathrm{SPB}(\Sigma)$ and $p, q \in S$. We say that $\mathbf{S}$ has a run on P from p to q, denoted $(p, P, q)_{\mathbf{S}}$ if one of the following conditions holds.*

(Base) $P = a \in \Sigma$ *and* $(p, a, q) \in \delta$.
(HH) $p, q \in H$ *and P has maximal horizontal decomposition $P = P_1 \cdot \ldots \cdot P_n$, where $n \geq 2$, and $\exists r_1, \ldots, r_{n-1} \in S$, $r_0 = p$, $r_n = q$ such that $(r_{i-1}, P_i, r_i)_{\mathbf{S}}$, for all $i \in [n]$.*
(VV) $p, q \in V$ *and P has maximal vertical decomposition $P = P_1 \circ \ldots \circ P_n$, where $n \geq 2$, and $\exists r_1, \ldots, r_{n-1} \in S$, $r_0 = p$, $r_n = q$ such that $(r_{i-1}, P_i, r_i)_{\mathbf{S}}$ for all $i \in [n]$.*
(HV) $p, q \in H$ *and P has maximal vertical decomposition $P = P_1 \circ \ldots \circ P_n$, where $n \geq 2$, and $\exists (_k,)_k \in \Omega$, $p', q' \in V$ and $(p, (_k, p'), (q',)_k, q) \in \gamma$ such that $(p', P, q')_{\mathbf{S}}$ holds.*
(VH) $p, q \in V$ *and P has maximal horizontal decomposition $P = P_1 \cdot \ldots \cdot P_n$, where $n \geq 2$, and $\exists (_k,)_k \in \Omega$, $p', q' \in H$ and $(p, (_k, p'), (q',)_k, q) \in \gamma$ such that $(p', P, q')_{\mathbf{S}}$ holds.*

Remark 3. Note that $(p, P, q)_{\mathbf{S}}$ implies $p, q \in H$ or $p, q \in V$. So in (HH) we have $r_1, r_2, \ldots r_{n-1} \in H$, and similarly, in (VV) we have $r_1, r_2, \ldots r_{n-1} \in V$.

The sp-biposet *language $L(\mathbf{S})$ accepted by the automaton $\mathbf{S}$* is defined as the set of all labels of a run from an initial state to a final state. Formally,

$$L(\mathbf{S}) = \{P \in \mathrm{SPB}(\Sigma) \mid \exists i \in I, \, f \in F \, : \, (i, P, f)_{\mathbf{S}}\}.$$

We say that two automata are *equivalent* if they accept the same sp-biposet language.

Definition 4. *An sp-biposet language $L \subseteq \mathrm{SPB}(\Sigma)$ is said to be* regular *if it is accepted by a parenthesizing automaton. We denote the class of all regular sp-biposet languages by* Reg.

Theorem 2. Reg = Rec, *i.e., an sp-biposet language* $L \subseteq \mathrm{SPB}(\Sigma)$ *is recognizable iff L is regular.*

In our proof, we show how to construct a parenthesizing automaton from a finite bisemigroup, and conversely, how to construct a finite bisemigroup from a parenthesizing automaton.

Remark 4. Each language $L \subseteq \mathrm{SPB}(\Sigma)$ can be recognized by a smallest bisemigroup, called the *syntactic bisemigroup* of L. This bisemigroup B_L, unique up to isomorphism, corresponds to the syntactic semigroup [22] of a word language, and to the syntactic algebra of a tree language, cf. [23]. For our present purpose it is sufficient to define B_L as the quotient of $\mathrm{SPB}(\Sigma)$ with respect to the largest congruence $\sim_L$ that saturates L. We clearly have that L is recognizable iff B_L is finite.

4 Rationality

There are several meaningful definitions of rational sets of sp-biposets. Here we will only consider the simplest of them: series rational, parallel rational, birational and generalized rational sets.

Definition 5. *Let $L_1, L_2 \subseteq \mathrm{SPB}(\Sigma)$, where Σ is an alphabet. We define the following operations, called* horizontal *(or* series product*),* vertical *(or* parallel product*),* horizontal iteration *(or* series iteration*) and* vertical iteration *(or* parallel iteration*).*

$$L_1 \cdot L_2 := \{P \cdot Q \mid P \in L_1,\ Q \in L_2\}$$
$$L_1 \circ L_2 := \{P \circ Q \mid P \in L_1,\ Q \in L_2\}$$
$$L_1^{\cdot+} := \{P_1 \cdot \ldots \cdot P_n \mid P_i \in L_1,\ n \geq 1\}$$
$$L_1^{\circ+} := \{P_1 \circ \ldots \circ P_n \mid P_i \in L_1,\ n \geq 1\}.$$

Moreover, if $L_1 \in \mathrm{SPB}(\Sigma)$, $\xi \notin \Sigma$, $L_2 \in \mathrm{SPB}(\Sigma \cup \{\xi\})$, then the sp-biposet language in $\mathrm{SPB}(\Sigma)$ obtained by substituting (non uniformly) biposets in L_1 for ξ in the members of L_2 is denoted by $L_2[L_1/\xi]$. We refer to this operation as ξ-substitution.

Definition 6. *The class of* birational languages *is the least class* BRat *of sp-biposet languages containing the finite sp-biposet languages in $\mathrm{SPB}(\Sigma)$, for all Σ, and closed under union, sequential and parallel product and sequential and parallel iteration. The class of* generalized birational languages *is the least class* GRat *of sp-biposet languages containing the finite languages and closed under the above operations and complementation.*

Clearly, BRat $\subseteq$ GRat. Our first result is that GRat $\subseteq$ Rec.

Lemma 1. *Every parenthesizing automaton $\mathbf{S}$ is equivalent to a parenthesizing automaton $\mathbf{S}^h$ ($\mathbf{S}^v$, resp.) with a single initial and final state, both horizontal (vertical, resp.), such that the initial state is not the target and the final state is not the origin of any (labelled or parenthesising) transition.*

We use the above Lemma to prove:

Proposition 2. *The class of regular (i.e., recognizable) languages is closed under ξ-substitution.*

We have already noted that Rec is closed under the Boolean operations and inverse homomorphisms. Using Proposition 2, we can immediately derive some further closure conditions of Rec.

Corollary 1. *The class* Rec *of recognizable (i.e., regular) sp-biposet languages is (effectively) closed under the Boolean operations, horizontal and vertical product, horizontal and vertical iteration, homomorphism and inverse homomorphism. Thus, since every finite language is recognizable, we have that* GRat $\subseteq$ Rec.

Definition 7. *Define the* alternation depth $\mathrm{ad}(P)$ *of an sp-biposet* $P \in \mathrm{SPB}(\Sigma)$ *inductively as follows:*

- *if P is a letter in Σ, then $\mathrm{ad}(P) = 0$,*
- *if $P = P_1 \cdot \ldots \cdot P_n$, then $\mathrm{ad}(P) = \max\{\mathrm{ad}(P_1), \ldots, \mathrm{ad}(P_n)\} + 1$,*
- *if $P = P_1 \circ \ldots \circ P_n$, then $\mathrm{ad}(P) = \max\{\mathrm{ad}(P_1), \ldots, \mathrm{ad}(P_n)\} + 1$,*

where the decompositions are maximal and $n \geq 2$. The alternation depth *of an sp-biposet language L is defined as the supremum of the alternation depths of its elements:* $\mathrm{ad}(L) := \sup\{\mathrm{ad}(P) \mid P \in L\}$.

Note that $\mathrm{ad}(L)$ may be ∞. We denote by $\mathrm{BD}^{\leq n}$ the class of sp-biposet languages L with $\mathrm{ad}(L) \leq n$, and by BD the class of *bounded alternation depth sp-biposet languages* $\bigcup_{n < \infty} \mathrm{BD}^{\leq n}$.

Theorem 3. BRat = Rec $\cap$ BD.

Two subclasses of BRat are also of interest. The class SRat of *series rational languages* is the least class containing the finite languages closed under union, series and parallel product and series iteration. Call a language $L \subseteq \mathrm{SPB}(\Sigma)$ *series bounded* (SB) if there is a constant K such that for all $P \in L$, the length of each $<_h$-chain in P is bounded by K. *Parallel bounded languages* (PB) and *parallel rational languages* (PRat) are defined symmetrically. We clearly have that SB $\cup$ PB $\subseteq$ BD.

Corollary 2. SRat = Rec $\cap$ PB *and* PRat = Rec $\cap$ SB.

In the full version of the paper we will prove that it is decidable for a recognizable sp-biposet language whether it is birational, series rational or parallel rational. We do not know the answer for generalized rational languages.

Suppose that B is a bisemigroup. An *elementary $\cdot$-translation* on B is a function $f : B \to B$ of the form $f(x) = b \cdot x$ or $f(x) = x \cdot b$, where b is a fixed element of B. *Elementary $\circ$-translations* are defined in the same way. An *elementary translation* is an elementary $\cdot$-translation or an elementary $\circ$-translation. (Note that the same function can be both an elementary $\cdot$-translation and an elementary $\circ$-translation.) An *alternating* *translation* is any composition of elementary translations $f_1, \ldots, f_k$ such that for some i, f_i is an elementary $\cdot$-translation, and for some j, f_j is an elementary $\circ$-translation.

The following proposition provides a necessary condition for a language to be generalized rational. This result is related to Schützenberger's well-known characterization of star-free word languages, see [22].

Proposition 3. *If a language* $L \subseteq \mathrm{SPB}(\Sigma)$ *is in* GRat *then its syntactic bisemigroup* B_L *satisfies the following condition: For all alternating translations $p(x)$ and $b \in B_L$, if $p^n(b) = b$ for some $n \geq 1$, then $p(b) = b$.*

Proposition 4. SRat $\cup$ PRat $\subset$ BRat $\subset$ GRat $\subset$ Rec, *where each inclusion is proper. Moreover,* SRat *and* PRat *are incomparable with respect to set inclusion.*

Indeed, it is clear that SRat $\cup$ PRat $\subseteq$ BRat $\subseteq$ GRat $\subseteq$ Rec and that the first two inclusions are proper. As for the last inclusion, let $\Sigma = \{a\}$, and define $P_0 = a$, $P_{n+1} = a \cdot (P_n \circ a)$, $n \geq 0$. Then let $L = \{P_{2i} : i \geq 0\}$, the set consisting of every second of the P_n. The syntactic bisemigroup of L is finite but does not satisfy the condition given in Proposition 3, so that $L \in$ Rec $-$ GRat.

Open problem. Does the converse of Proposition 3 hold?

If so, then it is decidable for a recognizable language whether or not it is generalized rational.

5 Logical Definability

In this section we relate monadic second-order definable (MSO-definable) sp-biposet languages to recognizable languages.

Suppose that Σ is an alphabet. An *atomic formula* is of the form $P_a(x)$, $X(x)$, $x <_h y$ or $x <_v y$, where a is any letter in Σ, x, y are first-order variables ranging over vertices in an sp-biposet, and X is a (monadic) second-order variable ranging over subsets of the vertex set of an sp-biposet. Here, $P_a(x)$ means that vertex x is labelled by a and $X(x)$ means that x belongs to X. The atomic formulas $x <_h y$ and $x <_v y$ have their expected meanings. (We assume a fixed countable set of first-order, and a fixed countable set of second-order variables.) *Formulas* are composed from atomic formulas by the boolean connectives $\vee$ and $\neg$ and first- and second-order existential quantifiers $\exists x$ and $\exists X$. We define in the usual way when a *closed formula (sentence)* φ holds in, or is satisfied by an sp-biposet P, denoted $P \models \varphi$. The *language* L_φ *defined by* φ is $\{P \in \mathrm{SPB}(\Sigma) \mid P \models \varphi\}$.

Definition 8. *We say that a language* $L \subseteq \mathrm{SPB}(\Sigma)$ *is* MSO-definable *if there is sentence φ with $L = L_\varphi$.*

We let MSO denote the class of MSO-definable languages in $\mathrm{SPB}(\Sigma)$, for all alphabets Σ.

It is not hard to show that MSO $\subseteq$ Rec. We can argue by formula induction. In order to do that, we first associate a language $L_\varphi \subseteq \mathrm{SPB}(\Sigma \times \mathcal{V} \times \mathcal{P}(\mathcal{W}))$ to any formula φ whose free variables are contained in the finite sets $\mathcal{V}$ of first-order and $\mathcal{W}$ of second-order variables, where $\mathcal{P}(\mathcal{W})$ denotes the powerset of $\mathcal{W}$. Our definition parallels that in [24], and makes use of the closure properties of recognizable languages given in Corollary 1. (An alternative way of proving MSO $\subseteq$ Rec would be through a compositionality property of the monadic theories of sp-biposets. See Kuske [18] for a general outline of this method.)

Recognizable and MSO-definable text languages, with texts defined as isomorphism classes of nonempty finite labelled sets equipped with two strict linear orders, were studied by Hoogeboom and ten Pas in [16]. (See Remark 2.) The notion of recognizability clearly does not depend on the concrete representation of the free bisemigroups. On the other hand, the equivalence of the representations of free bisemigroups by texts and as labelled sp-biposets can be established within the language of first-order logic. Thus, from the (more general) equivalence results proved in [16], we immediately have:

Theorem 4. Rec = MSO.

The inclusion Rec $\subseteq$ MSO is shown for texts in [16] by interpreting the "structure" of a text within the text. This method originates in [3].

Corollary 3. *The following conditions are equivalent for a language $L \subseteq$ SPB(Σ) of bounded alternation depth: 1. L is recognizable. 2. L is regular. 3. L is birational. 4. L is generalized birational. 5. L is MSO-definable. When L is parallel-bounded, the above conditions are further equivalent to the condition that L is series rational.*

6 Comparison with Other Work

Our investigations have been influenced to a great extent by the work of Hoogeboom and ten Pas [15,16] on text languages, in particular on logical definability, and the recent work of K. Lodaya and P. Weil, and subsequently by D. Kuske, on automata (and logic) on series-parallel posets (sp-posets), i.e., finite nonempty labelled sets equipped with a single partial order subject to the N-free condition. These posets, equipped with the series product and the parallel product, where the parallel product is now just disjoint union (hence commutative), form the free "semi-commutative" bisemigroups, cf. [13].[1] In [19,20], Lodaya and Weil defined recognizable languages of sp-posets as well as regular languages accepted by "branching automata", and rational languages. They showed that a language of sp-posets is regular iff it is rational, and that the recognizable languages form a proper subclass of the regular languages. Aside from semi-commutativity, their notion of recognizability corresponds to ours, and the one in [15] (actually this notion is well established in a very general setting, just as the the notion of equational set, see below). On the other hand, their notion of rationality is much more general than our birationality, and although our parenthesizing automata owe much to their branching automata, they are not a non-commutative version of branching automata. The above differences, together with the well-known fact that rationality and recognizability do not

---

[1] Grabowski called a bisemigroup with a neutral element a double monoid.

coincide for free commutative semigroups explain why the above mentioned results of Lodaya and Weil are so different from ours.

Nevertheless Lodaya and Weil also obtained several results similar to ours. They studied bounded width poset languages that correspond to our parallel bounded biposet languages and showed in [20] that for such languages, the concepts of recognizability, regularity and series rationality are all equivalent. Moreover, Kuske proved in [18] that for bounded width poset languages, these conditions are equivalent to MSO-definability. These equivalences correspond to our Corollary 3, the parallel bounded case.

What we called a birational sp-biposet language corresponds to the series-parallel rational sp-poset languages of Lodaya and Weil. In [17,18], Kuske showed that any series rational poset language is MSO-definable and that every MSO-definable poset language is recognizable. On the other hand, there easily exist recognizable but not MSO-definable sp-poset languages. In an earlier version of this paper we proposed as an open problem whether Rec is included in MSO. We have since learned that the equality Rec = MSO has been established by Hoogeboom and ten Pas in [16] for text languages, from which Theorem 4 follows immediately.

By a generalized rational sp-poset language Lodaya and Weil understood a language that would elsewhere be called equational. In the realm of associative operations, they correspond to the much broader class of context-free languages. We have not studied context-free biposet languages.

The main object of study in [21] is the extension of the classical framework to automata over free algebras with a single associative operation and a collection of operations not satisfying any nontrivial equations. It is shown that a suitably adapted version of branching automata captures recognizable languages, and that there exists a corresponding notion of rationality. Lodaya and Weil also discuss, in a rather indirect way, the situation when at least one of the additional operations is associative. In this case they find that the recognizable languages form a proper subclass of the regular languages which coincide with the rational languages. Their "asymmetric" notion of regularity is different from ours (which is "symmetric"), and their notion of rationality they show to correspond to regularity is much more general than ours. Corollary 2 also appears in [21].

Automata and languages over free bisemigroups (more precisely, free bisemigroups with identity) have also been studied in Hashiguchi et al. [14]. The elements of the free bisemigroup are represented by ordinary words (involving parentheses) in "standard form". Accordingly, ordinary finite automata are used to accept sp-biposet languages. More precisely, they define two kinds of acceptance modes: the free binoid mode and the free monoid mode. The free monoid mode is rather restricted, since the language accepted in the ordinary sense by the finite automaton should consist of only such words that are standard forms of sp-biposets. The free binoid mode is closer to our approach. We suspect that it corresponds to those parenthesising automata having a single pair of parentheses. No notion related to our recognizability, rationality, or logical definability is considered. On the other hand, they define phrase structure grammars (B-grammars) generating sp-biposet languages in standard form. (The definition takes a full page and consists of 31 items!) In particular, they define left and right linear B-grammars and show that they determine different language classes that lie somewhere between finite automata in the free monoid, and the free binoid mode.

A different two-dimensional generalization of the classical framework is provided by the *picture languages*. Pictures themselves are labeled biposets with a very regular structure. They come with two operations, corresponding to horizontal and vertical product, but these are only partially defined, cf. [11]. The notion of recognizability is based on tilings and behaves differently, since recognizable picture languages are not closed under complement and their emptiness problem is undecidable. For the description of picture languages using formal logic we refer to [12,26].

Acknowledgment. Parts of the results were obtained during the first author's visit at BRICS and the Department of Computer Science of the University of Aalborg. This author is indebted to Anna Ingólfsdóttir, Luca Aceto and Kim G. Larsen for their kind hospitality.

References

1. Boudol, G., Castellani, I.: Concurrency and atomicity. *Theoret. Comput. Sci.* **59** (1988) 25–34.
2. Corneil, D.G., Lerchs, H., Burlinham, L.S.: Complement reducible graphs. *Discr. Appl. Math.* **3** (1981) 163–174.
3. Courcelle, B.: The monadic second-order logic on graphs V: on closing the gap between definability and recognizability. *Theoret. Comput. Sci.* **80** (1991) 153–202.
4. Engelfriet, J., Harju, T., Proskurowski, A., Rozenberg, G.: Characterization and complexity of uniformly nonprimitive labeled 2-structures. *Theoret. Comput. Sci.* **154** (1996) 247–282.
5. Ehrenfeucht, A., ten Pas, P., Rozenberg, G.: Combinatorial properties of texts. *Theor. Inf. Appl.* **27** (1993) 433–464.
6. Ehrenfeucht, A., Rozenberg, G.: Theory of 2-structures, Part 1: clans, basic subclasses, and morphisms. Part 2: representation through labeled tree families. *Theoret. Comput. Sci.* **70** (1990) 277–303, 305–342.
7. Ehrenfeucht, A., Rozenberg, G.: Angular 2-structures. *Theoret. Comput. Sci.* **92** (1992) 227–248.
8. Ehrenfeucht, A., Rozenberg, G.: T-structures, T-functions and texts. *Theoret. Comput. Sci.* **116** (1993) 227–290.
9. Ésik, Z.: Free algebras for generalized automata and language theory. *RIMS Kokyuroku 1166*, Kyoto University, Kyoto (2000) 52–58.
10. Gécseg, F., Steinby, M.: *Tree Automata*. Akadémiai Kiadó, Budapest (1984).
11. Giammarresi, D., Restivo, A.: Two-dimensional finite state recognizability. *Fund. Inform.* **25** (1996) 399-422.
12. Giammarresi, D., Restivo, A., Seibert, S., Thomas, W.: Monadic second order logic over pictures and recognizability by tiling systems. In: proc. *STACS 94, Caen*, LNCS, Vol. 775. Springer (1994) 365–375.
13. Grabowski, J.: On partial languages. *Fund. Inform.* **4** (1981) 427–498.
14. Hashiguchi, K., Ichihara, S., Jimbo, S.: Formal languages over free binoids. *J. Autom. Lang. Comb.* **5** (2000) 219–234.
15. Hoogeboom, H.J., ten Pas, P.: Text languages in an algebraic framework. *Fund. Inform.* **25** (1996) 353–380.
16. Hoogeboom, H.J., ten Pas, P.: Monadic second-order definable text languages. *Theory Comput. Syst.* **30** (1997) 335–354.

17. Kuske, D.: Infinite series-parallel posets: logic and languages. In: proc. *ICALP 2000*, LNCS, Vol. 1853. Springer (2001) 648–662.
18. Kuske, D.: Towards a language theory for infinite N-free pomsets. (to appear).
19. Lodaya, K., Weil, P.: Kleene iteration for parallelism. In: proc. *FST & TCS 98*, LNCS, Vol. 1530. Springer-Verlag (1998) 355–366.
20. Lodaya, K., Weil, P.: Series-parallel languages and the bounded-width property. *Theoret. Comput. Sci.* **237** (2000) 347–380.
21. Lodaya, K., Weil, P.: Rationality in algebras with series operation. Inform. and Comput. (to appear).
22. Pin, J.-E.: *Varieties of Formal Languages*. Plenum Publishing Corp., New York (1986).
23. Steinby, M.: General varieties of tree languages. *Theoret. Comput. Sci.* **205** (1998) 1–43.
24. Straubing, H.: *Automata, Formal Logic and Circuit Complexity*. Birkhauser, Boston (1994).
25. Valdes, J., Tarjan, R.E., Lawler, E.L.: The recognition of series-parallel digraphs. *SIAM J. Comput.* **11** (1982) 298–313.
26. Wilke, Th.: Star-free picture expressions are strictly weaker than first-order logic. In: proc. *ICALP 97*, LNCS, Vol. 1256. Springer-Verlag (1997) 347–357.

Hierarchies of String Languages Generated by Deterministic Tree Transducers

Joost Engelfriet and Sebastian Maneth

Leiden University, LIACS, PO Box 9512, 2300 RA Leiden, The Netherlands
{engelfri, maneth}@liacs.nl

Abstract. The composition of total deterministic macro tree transducers gives rise to a proper hierarchy with respect to the generated string languages (these are the languages obtained by taking the yields of the output trees). The same holds for attributed tree transducers, for controlled EDT0L systems, and for YIELD mappings (which shows properness of the IO-hierarchy).

1 Introduction

Macro tree transducers [EV85,FV98] serve as a model of syntax-directed semantics. They combine top-down tree transducers and macro grammars, that is, they are finite state transducers, the states of which are equipped with parameters that allow to handle context information.

A macro tree transducer can be used as a string language generator by applying its translation to a tree language, and then taking the yields of the resulting output trees. It can also be viewed as a controlled (tree) grammar, where the generation of the output trees is controlled by the input trees. Then, the iteration of control corresponds to the composition of translations. The string languages generated by the composition closure of (nondeterministic) macro tree transducers form a large full AFL, while still having decidable membership, emptiness, and finiteness problems [DE98]. Because of their special relevance to syntax-directed semantics we here investigate total deterministic macro tree transducers only, a combination of total deterministic top-down tree transducers and IO (inside-out) macro grammars.

The question arises, whether composition of total deterministic macro tree transducers (MTTs) gives rise to a proper hierarchy with respect to the generated string languages. A hierarchy $X(n)$ is proper if, for every n, $X(n) \subsetneq X(n+1)$, and it is infinite if, for every n, $X(n) \subsetneq X(m)$ for some $m > n$. For the two ingredients of MTTs the situation is as follows. Since (total deterministic) top-down tree transducers are closed under composition [Rou70], they do not form a hierarchy. The iteration of IO macro grammars by the concept of level-n grammars gives rise to a proper and to an infinite hierarchy [Dam82], for the generated tree and string languages, respectively: the so-called IO-hierarchies (see, e.g., [ES78]). With respect to the translations it is shown in [EV85] that composition of MTTs yields a proper hierarchy, based on the fact that the height of the output tree of an MTT is exponentially bounded by the height of the input tree. In [Dam82] it is proved that also the generated tree languages form a proper hierarchy. For the generated string languages, the composition of MTTs yields an infinite hierarchy; the proof in [Dam82] combines the above exponential bound with the concept of rational index [BCN81]. To prove properness of this hierarchy we use instead a "bridge theorem" (cf. [Eng82], and the section on translational techniques in [Gre81]).

The bridge theorem we use was proved in [Man99]. It states, for two languages L' and L such that L' is obtained from L by some kind of string insertion: if L' is generated by an MTT, then L can be generated by a top-down tree transducer. Since MTTs are closed under composition with top-down tree transducers, this theorem allows us to step down from the composition of $n+1$ MTTs to that of n MTTs. We apply the bridge theorem to two different types of insertion and obtain these results:

(1) Composition of MTTs yields a proper hierarchy with respect to the generated string languages, i.e., there is a language L' generated by the composition of $n+1$ MTTs, which cannot be generated by the composition of n MTTs. Here L' is obtained from a language L by inserting a sequence of b's before each symbol of a string in L (for a new symbol b), viz., b^i before the i-th symbol from the right. In fact, L' can already be generated by the $(n+2)$-fold iteration of controlled EDT0L systems. This implies properness of the EDT0L-hierarchy. The EDT0L system is the deterministic version of the ET0L system (see [ERS80] for the relationship of these systems to top-down tree transducers). In [Eng82] properness of the ET0L-hierarchy is proved, but it is mentioned as open whether the EDT0L-hierarchy is proper.

(2) There is an $(n+1)$-level IO macro language which cannot be generated by the composition of n MTTs. Here, L' is obtained from L by inserting, before each symbol of a string w in L, a string in $\{l,r\}^*$ that represents (in Dewey notation) the corresponding leaf of some binary tree with yield w. Since every n-level IO macro language can also be generated by the composition of n attributed tree transducers [Fül81,FV98] (ATTs) we obtain that composition of ATTs yields a proper hierarchy of generated string languages. We also obtain properness of the IO-hierarchy of string languages, which was left open in [Dam82].

2 Preliminaries

For $k \in \mathbb{N}$ let $[k] = \{1,\ldots,k\}$. For strings $v, w_1, \ldots, w_n \in A^*$ and distinct $a_1, \ldots, a_n \in A$, we denote by $v[a_1 \leftarrow w_1, \ldots, a_n \leftarrow w_n]$ the result of (simultaneously) substituting w_i for every occurrence of a_i in v. For $A \cap B = \varnothing$ and $w \in (A \cup B)^*$ we denote by $\operatorname{res}_A(w)$ the restriction of w to symbols in A; for $L \subseteq A^*$, $\operatorname{res}_A^{-1}(L) \subseteq (A \cup B)^*$ is also denoted by $\operatorname{rub}_B(L)$, where rub stands for the insertion of 'rubbish'. For functions $f: A \to B$ and $g: B \to C$ their composition is $(f \circ g)(x) = g(f(x))$; note that the order is nonstandard. For sets of functions F and G their composition is $F \circ G = \{f \circ g \mid f \in F, g \in G\}$, and $F^n = F \circ \cdots \circ F$ (n times, $n \geq 1$). For a class $\mathcal{L}$ of languages, $F(\mathcal{L}) = \{f(L) \mid f \in F, L \in \mathcal{L}\}$.

We assume the reader to be familiar with trees, tree automata, and tree translations (see, e.g., [GS84]). A set Σ together with a mapping rank: $\Sigma \to \mathbb{N}$ is called a *ranked set*. For $k \geq 0$, $\Sigma^{(k)}$ is the set $\{\sigma \in \Sigma \mid \operatorname{rank}(\sigma) = k\}$; we also write $\sigma^{(k)}$ to denote that $\operatorname{rank}(\sigma) = k$. For a set A, $\langle \Sigma, A \rangle$ is the ranked set $\{\langle \sigma, a \rangle \mid \sigma \in \Sigma, a \in A\}$ with $\operatorname{rank}(\langle \sigma, a \rangle) = \operatorname{rank}(\sigma)$. The set of all trees over Σ is denoted T_Σ. For a set A, $T_\Sigma(A)$

is the set of all trees over $\Sigma \cup \Delta$, where all elements in Δ have rank zero. We fix the *set of input variables* as $X = \{x_1, x_2, \dots\}$ and the *set of parameters* as $Y = \{y_1, y_2, \dots\}$. For $k \geq 0$, $X_k = \{x_1, \dots, x_k\}$ and $Y_k = \{y_1, \dots, y_k\}$. For a tree $t \in T_\Sigma$, yt denotes the *yield of* t, i.e., the string in $(\Sigma^{(0)})^*$ obtained by reading the leaves of t in pre-order omitting leaves labeled by the special symbol e (thus, $y(\sigma(a, \sigma(e, b))) = ab$). For a class $\mathcal{L}$ of tree languages, $y\mathcal{L}$ denotes $\{yL \mid L \in \mathcal{L}\}$ with $yL = \{yt \mid t \in L\}$. The class of *regular tree languages* is denoted by *REGT*.

Definition 1. A (total, deterministic) *macro tree transducer* (MTT) is a tuple $M = (Q, \Sigma, \Delta, q_0, R)$, where Q is a ranked alphabet of *states*, Σ and Δ are ranked alphabets of *input* and *output symbols*, respectively, $q_0 \in Q^{(0)}$ is the *initial state*, and R is a finite set of *rules*; for every $q \in Q^{(m)}$ and $\sigma \in \Sigma^{(k)}$ with $m, k \geq 0$ there is exactly one rule of the form $\langle q, \sigma(x_1, \dots, x_k)\rangle(y_1, \dots, y_m) \to \zeta$ in R, where $\zeta \in T_{\langle Q, X_k\rangle \cup \Delta}(Y_m)$. □

The rules of M are used as term rewriting rules in the usual way. The derivation relation of M (on $T_{\langle Q, T_\Sigma\rangle \cup \Delta}$) is denoted by $\Rightarrow_M$ and the *translation realized by M*, denoted τ_M, is the total function $\{(s, t) \in T_\Sigma \times T_\Delta \mid \langle q_0, s\rangle \Rightarrow_M^* t\}$. The class of all translations which can be realized by MTTs is denoted by *MTT*. A *top-down tree transducer* is an MTT all states of which are of rank zero. The class of all translations which can be realized by top-down tree transducers is denoted by T. If a top-down tree transducer has only one state, then it is a *tree homomorphism*.

3 Bridge Theorem

Let A and B be disjoint alphabets. Consider a string of the form

$$w_1 a_1 w_2 a_2 \cdots a_{l-1} w_l a_l w_{l+1}$$

with $l \geq 0$, $a_1, \dots, a_l \in A$, $w_1, \dots, w_l, w_{l+1} \in B^*$, and all $w_2, \dots, w_l$ pairwise different. We call such a string a δ-*string for* $a_1 \cdots a_l$. Now let $L \subseteq A^*$ and $L' \subseteq (A \cup B)^*$ such that $\mathrm{res}_A(L') = L$. If L' contains, for every $w \in L$, a δ-string for w, then L' is called δ-*complete for* L. The following theorem shows that if an MTT M generates L', then, due to the structure of δ-strings, M cannot make use of its parameter-copying facility as far as L is concerned. This is, essentially, Theorem 6 of [Man99]. There, the theorem is stated for a particular $L' \in y\mathit{MTT}(\mathcal{L})$, viz., $L' = \mathrm{res}_A^{-1}(L) = \mathrm{rub}_B(L)$ for $B = \{0\}$: the nondeterministic insertion of a new symbol 0. However, the proof only depends on the two facts that (1) $\mathrm{rub}_B(L)$ is δ-complete for L and that (2) L is obtained from L' by deleting all occurrences of 0. Moreover, the conclusion of that theorem is that $L \in y\mathit{MTT}_{\mathrm{sp}}(\mathcal{L})$, the class generated by MTTs which do not copy their parameters; it equals $yT(\mathcal{L})$ by Theorem 5 of [Man99].

Theorem 1. *Let $\mathcal{L}$ be a class of tree languages which is closed under finite state relabelings and under intersection with regular tree languages. Let A, B be disjoint alphabets and let $L \subseteq A^*$, $L' \subseteq (A \cup B)^*$ such that*

(1) *L' is δ-complete for L and*
(2) $\mathrm{res}_A(L') = L$.

If $L' \in y\mathit{MTT}(\mathcal{L})$ then $L \in yT(\mathcal{L})$.

Roughly speaking, the idea of the proof of Theorem 1 is that copying a parameter that contains at least two symbols of $\text{res}_A(w)$, $w \in L'$, yields copies of the same string in B^*, which implies that w is not a δ-string. We now show that Theorem 1 can be applied to $\mathcal{L} = MTT^n(REGT)$. Note that in the applications of Theorem 2, the language L' is often of the form $\varphi(L)$, where φ is an operation on languages (such as rub_B).

Theorem 2. Let A, B be disjoint alphabets and $L \subseteq A^*$, $L' \in (A \cup B)^*$ such that L' is δ-complete for L and $\text{res}_A(L') = L$.

(a) For $n \geq 1$, if $L' \in yMTT^{n+1}(REGT)$, then $L \in yMTT^n(REGT)$.
(b) If $L' \in yMTT(REGT)$, then $L \in yT(REGT)$.

Proof. The class $REGT$ has the required closure properties (cf. [GS84]). Closure of $\mathcal{L} = MTT^n(REGT)$ under finite state relabelings holds by Lemma 7 of [Man99], and closure under intersection with regular tree languages follows from the closure of $REGT$ under inverse macro tree transductions (Theorem 7.4 of [EV85]). Application of Theorem 1 gives (a) and (b), because $MTT \circ T \subseteq MTT$ (Theorem 4.12 of [EV85]). □

4 The yMTT-Hierarchy and the EDT0L-Hierarchy

In this and the next section we apply the bridge theorem of Section 3. The three main results obtained are: (1) The composition of MTTs yields a proper hierarchy of generated string languages, i.e., the *yMTT-hierarchy* $yMTT^n(REGT)$ is proper. (2) The EDT0L-hierarchy, see below, is proper. (3) There is an $(n+1)$-level IO macro language which cannot be generated by the composition of n MTTs; so, the IO-hierarchy is proper.

We now move to the proof of properness of the yMTT-hierarchy. The witnesses for its properness will be generated by controlled EDT0L systems, which are viewed here as string transducers. Essentially, an EDT0L system is a top-down tree transducer M with monadic input alphabet (cf. [ERS80]). However, instead of a tree translation it realizes a string translation as follows: first, the input string $w = a_1 \cdots a_n$ is turned into the monadic tree

$$\text{sm}(w) = a_1(a_2(\cdots a_n(e)\cdots));$$

then it is translated into the string $y\tau_M(\text{sm}(w))$. The *EDT0L translation realized by* M, denoted by τ_M^{EDT0L}, is defined as $\text{sm} \circ \tau_M \circ y$. Hence, the class *EDT0L* of EDT0L translations is $\text{sm} \circ T \circ y$. The *EDT0L-hierarchy* consists of all $EDT0L^n(REG)$, obtained by iterating *EDT0L* on the class *REG* of regular languages. It starts with the class $EDT0L(REG)$ of EDT0L languages. Let us now show that the EDT0L-hierarchy is contained in the yMTT-hierarchy.

Theorem 3. For $n \geq 1$, $EDT0L^{n+1}(REG) \subseteq yMTT^n(REGT)$.

Proof. By definition, $EDT0L^{n+1} = (\text{sm} \circ T \circ y)^{n+1} = \text{sm} \circ (T \circ y \circ \text{sm})^n \circ T \circ y$. Clearly, there is an MTT M which realizes $y \circ \text{sm}$: For a ranked alphabet Σ define

$M_\Sigma = (\{q_0^{(0)}, q^{(1)}\}, \Sigma, \Sigma', q_0, R)$ with $\Sigma' = \{a^{(1)} \mid a \in \Sigma^{(0)}, a \neq e\} \cup \{e^{(0)}\}$. For every $\sigma \in \Sigma^{(k)}, k \geq 1$, let the rules

$$\langle q_0, \sigma(x_1, \ldots, x_k)\rangle \to \langle q, x_1\rangle(\langle q, x_2\rangle(\ldots(\langle q, x_k\rangle(e))\ldots))$$
$$\langle q, \sigma(x_1, \ldots, x_k)\rangle(y_1) \to \langle q, x_1\rangle(\langle q, x_2\rangle(\ldots(\langle q, x_k\rangle(y_1))\ldots))$$

be in R, for every $a \in \Sigma^{(0)} - \{e\}$ let $\langle q_0, a\rangle \to a(e)$ and $\langle q, a\rangle(y_1) \to a(y_1)$ be in R, and let $\langle q_0, e\rangle \to e$ and $\langle q, e\rangle(y_1) \to y_1$ be in R. Then, for $s \in T_\Sigma$ with $ys = a_1 \cdots a_n$, $\langle q_0, s\rangle \Rightarrow^*_{M_\Sigma} \langle q, a_1\rangle(\langle q, a_2\rangle(\ldots(\langle q, a_n\rangle(e))\ldots)) \Rightarrow^*_{M_\Sigma} a_1(a_2(\ldots a_n(e)\ldots)) = \text{sm}(ys)$.

Now $\text{sm} \circ (T \circ y \circ \text{sm})^n \circ T \circ y \subseteq \text{sm} \circ (T \circ MTT)^n \circ T \circ y$ which is included in $\text{sm} \circ MTT^n \circ y$, because $T \circ MTT \circ T \subseteq MTT$ by Theorem 7.6(1) of [EV85]. Applying this to REG gives $yMTT^n(\text{sm}(REG)) \subseteq yMTT^n(REGT)$. $\square$

Based on Theorem 2 it will be proved that there is a language that cannot be generated by the n-fold composition of MTTs, i.e., which is not in $yMTT^n(REGT)$, but which can be generated by the composition of $n+2$ EDT0L translations. This time the language L' in Theorem 2 will be of the form $\text{count}_b(L)$. When applied to a string w, count_b inserts $b^{|w|-i}$ after the i-th symbol of the string w, for $1 \leq i < |w|$. Formally, let A be an alphabet and let $B = \{b\}$ with $b \notin A$. Define the operation $\text{count}_b : A^* \to (A \cup B)^*$ as follows:

$$\text{count}_b(a_1 a_2 \cdots a_l) = \prod_{i=1}^{l} a_i b^{l-i} = a_1 b^{l-1} a_2 b^{l-2} \cdots a_{l-1} b a_l.$$

Clearly, $\text{count}_b(w)$ is a δ-string for w. So, for a language L,

$$\text{count}_b(L) = \{\text{count}_b(w) \mid w \in L\}$$

is δ-complete for L. Since, moreover, $\text{res}_A(\text{count}_b(L)) = L$, we can apply Theorem 2 to L and $\text{count}_b(L)$. For distinct symbols $b_1, \ldots, b_n$, we abbreviate $\text{count}_{b_1} \circ \text{count}_{b_2} \circ \cdots \circ \text{count}_{b_n}$ by $\text{count}_{b_1,\ldots,b_n}$.

Theorem 4. For $n \geq 1$, $EDT0L^{n+2}(REG) - yMTT^n(REGT) \neq \emptyset$.

Proof. Let $b_1, \ldots, b_n$ be distinct symbols not in $\{a, c\}$. We will show that the language $\text{count}_{b_1,\ldots,b_n}(L_{ac})$ with $L_{ac} = \{(a^m c)^{2^m} \mid m \geq 1\}$ is in $EDT0L^{n+2}(REG) - yMTT^n(REGT)$.

(1) $\text{count}_{b_1,\ldots,b_n}(L_{ac}) \in EDT0L^{n+2}(REG)$. First, let us see that the language L_{ac} can be generated by the composition of two EDT0L translations, i.e., that $L_{ac} \in EDT0L^2(REG)$. The first EDT0L system translates a^m into $c^m a^m$. The second EDT0L system recursively generates two new copies of translations for each c in the input. Finally it translates a^m into $a^m c$. Hence, it translates $c^m a^m$ into $(a^m c)^{2^m}$. Next, we define an EDT0L system which realizes count_b. Define $M_b = (\{q_0^{(0)}, q^{(0)}\}, \Sigma, \Delta, q_0, R)$, $\Sigma = \{a^{(1)} \mid a \in A\} \cup \{e^{(0)}\}$, $\Delta = \{a^{(0)} \mid a \in A\} \cup \{\sigma^{(2)}, e^{(0)}, b^{(0)}\}$, A is an arbitrary alphabet not containing b, and R consists of the following rules.

$$\begin{array}{ll}
\langle q_0, a(x_1)\rangle \to \sigma(a, \sigma(\langle q, x_1\rangle, \langle q_0, x_1\rangle)) & \text{for every } a \in A \\
\langle q_0, e\rangle \to e & \\
\langle q, a(x_1)\rangle \to \sigma(b, \langle q, x_1\rangle) & \text{for every } a \in A \\
\langle q, e\rangle \to e &
\end{array}$$

Clearly, for every $w \in A^*$, $\tau_{M_b}^{\text{EDT0L}}(w) = \text{count}_b(w)$. Hence, $\text{count}_b \in \text{EDT0L}$, and so $\text{count}_{b_1,\ldots,b_n}(L_{ac}) \in \text{EDT0L}^{n+2}(\text{REG})$.

(2) $\text{count}_{b_1,\ldots,b_n}(L_{ac}) \notin y\text{MTT}^n(\text{REGT})$. Application of Theorem 2(a) gives: if $\text{count}_{b_1,\ldots,b_n}(L_{ac}) \in y\text{MTT}^n(\text{REGT})$, then $\text{count}_{b_1,\ldots,b_{n-1}}(L_{ac}) \in y\text{MTT}^{n-1}(\text{REGT})$. Hence, by induction, $\text{count}_{b_1}(L_{ac}) \in y\text{MTT}(\text{REGT})$ and by Theorem 2(b) $L_{ac} \in yT(\text{REGT})$. But, by Theorem 3.16 of [Eng82], $L_{ac} \notin yT(\text{REGT})$. $\square$

From Theorems 3 and 4 we obtain the properness of the yMTT-hierarchy.

Theorem 5. For $n \geq 1$, $y\text{MTT}^n(\text{REGT}) \subsetneq y\text{MTT}^{n+1}(\text{REGT})$.

In the proof of Theorem 4 it is shown that $L_{ac} \in \text{EDT0L}^2(\text{REG}) - yT(\text{REGT})$ and thus $L_{ac} \in \text{EDT0L}^2(\text{REG}) - \text{EDT0L}(\text{REG})$, because $\text{EDT0L}(\text{REG}) = yT(\text{sm}(\text{REG})) \subseteq yT(\text{REGT})$. Hence, by Theorems 3 and 4, the EDT0L-hierarchy is proper. This was mentioned as an open problem after Theorem 4.3 of [Eng82].

Theorem 6. For $n \geq 1$, $\text{EDT0L}^n(\text{REG}) \subsetneq \text{EDT0L}^{n+1}(\text{REG})$.

5 The IO-Hierarchy

In this section the relationship between the IO- and the yMTT-hierarchy is investigated. By Theorem 7.5 of [ES78], the *IO-hierarchy* of string languages can be defined in terms of tree translations as follows:

$$\text{for } n \geq 1, \quad IO(n) = y\text{YIELD}^n(\text{REGT}),$$

where *YIELD* denotes the class of YIELD mappings defined below. Since every YIELD mapping can be realized by an MTT, i.e., $\text{YIELD} \subseteq \text{MTT}$ (cf. Theorem 4.6 of [EV85]), it follows that $IO(n) \subseteq y\text{MTT}^n(\text{REGT})$, that is, the IO-hierarchy is inside the yMTT-hierarchy. In fact, the yMTT-hierarchy differs from the IO-hierarchy only by a single application of a top-down tree transducer, because $y\text{MTT}^n(\text{REGT}) = y\text{YIELD}^n(T(\text{REGT}))$ by Corollary 4.13 of [EV85]. It is shown in [Dam82] that the IO-hierarchy is infinite, and that the IO-hierarchy of tree languages $\text{YIELD}^n(\text{REGT})$ is proper.

A YIELD mapping Y_f is a mapping from T_Σ to $T_\Delta(Y)$ defined by a mapping f from $\Sigma^{(0)}$ to $T_\Delta(Y)$, for ranked alphabets Σ and Δ. It realizes the semantics of first-order tree substitution in the following way.

(i) for $a \in \Sigma^{(0)}$, $Y_f(a) = f(a)$ and
(ii) for $\sigma \in \Sigma^{k+1}$, $s_0, s_1, \ldots, s_k \in T_\Sigma$, and $k \geq 0$,
$Y_f(\sigma(s_0, s_1, \ldots, s_k)) = Y_f(s_0)[y_i \leftarrow Y_f(s_i) \mid i \in [k]]$.

Example 1. We show that the language L_{ac} of Theorem 4 is in IO(1). Let K be the regular tree language defined by the regular tree grammar with productions $S \to \sigma(c, \sigma(A, a'))$, $A \to \sigma(c, \sigma(A, a))$, and $A \to e$, and let Y_g be the YIELD mapping defined by $g(a) = \sigma(a, y_1), g(a') = \sigma(a, c), g(c) = \sigma(y_1, y_1)$, and $g(e) = y_1$. Then, for $s = \sigma(c, \sigma(e, a'))$, $Y_g(s) = Y_g(c)[y_1 \leftarrow Y_g(\sigma(e, a'))] = \sigma(y_1, y_1)[y_1 \leftarrow \sigma(a, c)] = \sigma(\sigma(a, c), \sigma(a, c))$. It should be clear that $yY_g(K) = L_{ac}$.

In order to prove that $IO(n+1) - yMTT^n(REGT) \neq \emptyset$ (Theorem 7), we look for an operation φ that can be realized by a YIELD mapping, such that Theorem 2 can be applied to $L' = \varphi(L)$ for a language L. Unlike the operations rub and count of before, we now use a tree translation φ, i.e., $L' = y(\varphi(K))$, where $yK = L$.

Let $\Sigma = \{\sigma^{(2)}, \text{root}^{(1)}\} \cup \Sigma^{(0)}$ be a ranked alphabet and let l, r be symbols not in Σ. A node of a tree $s \in T_\Sigma$ is denoted by its Dewey notation in $\{l, r\}^*$ (e.g., the e-labeled leaf of s in Fig. 1 is denoted by rll); the label of the node ρ of s is denoted by $s[\rho]$ (e.g., $s[rlr] = b$ for the s in Fig. 1). Consider a tree translation τ from T_Σ to T_Δ with $\Delta = \Sigma \cup \{l^{(0)}, r^{(0)}, e^{(0)}\}$. Then τ is an (l, r)-leaf insertion for Σ, if, for every $s' = \text{root}(s)$ and $s \in T_{\Sigma - \{\text{root}\}}$,

(i) $\tau(s') = \text{root}(t)$ for some $t \in T_{\Delta - \{\text{root}\}}$ and
(ii) $y\tau(s') = \rho_1 s[\rho_1] \rho_2 s[\rho_2] \cdots \rho_m s[\rho_m]$, where $\rho_1, \ldots, \rho_m \in \{l, r\}^*$ are those leaves of s in pre-order, that are not labeled by e.

Figure 1 shows an example of trees s' and $\tau(s')$ for some (l, r)-leaf insertion τ (obviously, $y\tau(s') = larlrbrra$ is a δ-string for $ys' = aba$).

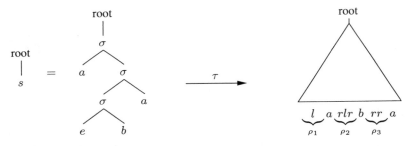

Fig. 1. The trees s' and $\tau(s')$ for an (l, r)-leaf insertion τ

Let $A = \Sigma^{(0)} - \{e\}$ and $B = \{l, r\}$. It should be clear that, for a 'rooted' tree language $K \subseteq \text{root}(T_{\Sigma - \{\text{root}\}})$, the language $L' = y\tau(K)$ is δ-complete for $L = yK$. Moreover, $\text{res}_A(L') = L$ because $\text{res}_A(y\tau(s')) = s[\rho_1]s[\rho_2] \cdots s[\rho_m] = ys$. This means that Theorem 2 can be applied to L and L'.

We now define a tree homomorphism M and a YIELD mapping Y_f such that their composition $\tau_M \circ YIELD$ is an (l, r)-leaf insertion.

Lemma 1. Let $\Sigma = \{\sigma^{(2)}, \text{root}^{(1)}\} \cup \Sigma^{(0)}$ be a ranked alphabet and let l, r be symbols not in Σ. There is a tree homomorphism M and a YIELD mapping Y_f such that $\tau_M \circ Y_f$ is an (l, r)-leaf insertion for Σ.

Proof. Define $M = (\{q^{(0)}\}, \Sigma, \Gamma, q, R)$ with $\Gamma = \{\delta^{(3)}, l^{(0)}, r^{(0)}, c^{(0)}, d^{(0)}, e^{(0)}\} \cup \Sigma$, and R consisting of the following rules.

$$\begin{aligned}
\langle q, \text{root}(x_1) \rangle &\to \sigma(d, \sigma(\langle q, x_1 \rangle, e)) \\
\langle q, \sigma(x_1, x_2) \rangle &\to \delta(c, \sigma(\langle q, x_1 \rangle, l), \sigma(\langle q, x_2 \rangle, r)) \\
\langle q, a \rangle &\to a \qquad\qquad \text{for every } a \in \Sigma^{(0)}
\end{aligned}$$

The mapping f is defined as $f(d) = \text{root}(y_1)$, $f(c) = \sigma(y_1, y_2)$, $f(e) = e$, and, for every $a \in \Sigma^{(0)} \cup \{l, r\}$ with $a \neq e$, $f(a) = \sigma(y_1, a)$.

Let us now prove that $\tau_M \circ Y_f$ is an (l, r)-leaf insertion. For $s \in T_{\Sigma-\{\text{root}\}}$, $Y_f(\tau_M(\text{root}(s))) = Y_f(\sigma(d, \sigma(\tau_M(s), e)))$ by the definition of M. This is

$$= f(d)[y_1 \leftarrow Y_f(\sigma(\tau_M(s), e))]$$
$$= \text{root}(y_1)[y_1 \leftarrow Y_f(\tau_M(s))[y_1 \leftarrow Y_f(e)]]$$
$$= \text{root}(Y_f(\tau_M(s))[y_1 \leftarrow e]).$$

The yield of this tree is equal to $yY_f(\tau_M(s))[y_1 \leftarrow \varepsilon]$ where ε is the empty string, which equals $\rho_1 s[\rho_1]\rho_2 s[\rho_2] \cdots \rho_m s[\rho_m]$ by the following claim (with the ρ_i as in the claim). This proves part (ii) of the definition of (l, r)-leaf insertion.

Claim: For every $s \in T_{\Sigma-\{\text{root}\}}$, $yY_f(\tau_M(s)) = y_1\rho_1 a_1 y_1\rho_2 a_2 \cdots y_1\rho_m a_m$, where $ys = a_1 \cdots a_m$, and $\rho_1, \ldots, \rho_m$ are all leaves of s in pre-order that are not labeled e.

The claim can easily be proved by induction on the structure of s. By the rules of M, $\tau_M(s)$ does not contain occurrences of the symbol d, and thus $Y_f(\tau_M(s))[y_1 \leftarrow e] \in T_{\Delta-\{\text{root}\}}$ with $\Delta = \Sigma \cup \{l^{(0)}, r^{(0)}, e^{(0)}\}$. This proves part (i) of the definition of (l, r)-leaf insertion. □

Theorem 7. For $n \geq 1$, $IO(n+1) - yMTT^n(REGT) \neq \emptyset$.

Proof. We first define a language $L_n \in IO(n+1) - yMTT^n(REGT)$. Let the ranked alphabet Σ be $\{\sigma^{(2)}, \text{root}^{(1)}\} \cup \Sigma^{(0)}$ with $\Sigma^{(0)} = \{a, c, e\}$. Let K and g be as in Example 1. Define the regular tree language K_{ac} as $\{\sigma(d, s) \mid s \in K\}$ and extend g by $g(d) = \text{root}(y_1)$. Then $Y_g(K_{ac}) \subseteq \{\text{root}(s) \mid s \in T_{\Sigma-\{\text{root}\}}\}$ and $yY_g(K_{ac}) = L_{ac}$. Let $l_1, \ldots, l_n, r_1, \ldots, r_n$ be distinct symbols of rank zero, not in Σ. By Lemma 1 there is, for every $i \in [n]$, a tree homomorphism M_i and a YIELD mapping Y_{f_i} such that $\tau_i = \tau_{M_i} \circ Y_{f_i}$ is an (l_i, r_i)-leaf insertion for $\Sigma \cup \{l_1, \ldots, l_{i-1}, r_1, \ldots, r_{i-1}\}$. For $n \geq 0$, define $L_n = yK_n$ where

$$K_n = (Y_g \circ \tau_{M_1} \circ Y_{f_1} \circ \tau_{M_2} \circ Y_{f_2} \circ \cdots \circ \tau_{M_n} \circ Y_{f_n})(K_{ac}).$$

(1) $L_n \in IO(n+1)$. For a YIELD mapping Y_f and a tree homomorphism M it is easy to construct a YIELD mapping $Y_{f'}$ with $Y_{f'} = Y_f \circ \tau_M$; to obtain $f'(a)$, just run M on the tree $f(a)$, leaving the parameters unchanged (cf. Lemma 6.3 of [ES78]). Thus, for every $i \in [n]$, $Y_{f_{i-1}} \circ \tau_{M_i} \in YIELD$ (with $f_0 = g$), and therefore $K_n \in (YIELD^n \circ YIELD)(REGT)$ and $L_n = yK_n \in IO(n+1)$.

(2) $L_n \notin yMTT^n(REGT)$. As discussed before Lemma 1, Theorem 2 can be applied to $L = L_{n-1} = yK_{n-1}$ and $L' = L_n = y\tau_n(K_{n-1})$, for the rooted tree language K_{n-1} and the (l_n, r_n)-leaf insertion $\tau_n = \tau_{M_n} \circ Y_{f_n}$. Thus, by Theorem 2(a), $L_n \in yMTT^n(REGT)$ implies that L_{n-1} is in $yMTT^{n-1}(REGT)$, and by induction, $L_1 = y\tau_1(Y_g(K_{ac})) \in yMTT(REGT)$. By Theorem 2(b) this means that $L_0 = yY_g(K_{ac}) = L_{ac} \in yT(REGT)$ which is a contradiction as observed in the proof of Theorem 4. □

At last, let us consider another type of tree transducer: the attributed tree transducer (ATT) [Fül81,FV98], which is a formal model for attribute grammars. It is well known that $YIELD \subseteq ATT \subseteq MTT$ (cf. Corollary 6.24 and Lemma 6.1 of [FV98]), where ATT denotes the class of all translations realized by ATTs. Thus, $IO(n) \subseteq yATT^n(REGT) \subseteq yMTT^n(REGT)$. By Theorem 7 we obtain the following corollary.

Theorem 8. For $n \geq 1$,

(a) $yATT^{n+1}(REGT) - yMTT^n(REGT) \neq \varnothing$.
(b) $yATT^n(REGT) \subsetneq yATT^{n+1}(REGT)$.

6 Conclusions and Open Problems

We have proved the properness of the yMTT-hierarchy. Moreover, we have shown (in Theorems 4, 7, and 8) that the EDT0L-, IO-, and yATT-hierarchy are subhierarchies of the yMTT-hierarchy. Let $X(n)$ and $Y(n)$ be hierarchies. We call $X(n)$ a *subhierarchy* of $Y(n)$, if there is an $m \in \mathbb{N}$ such that for every $n \geq 1$: $X(m+n) \subseteq Y(n)$ and $X(m+n+1) - Y(n) \neq \varnothing$. Note that for Y being the yMTT-hierarchy, $m = 1$ if X is the EDT0L-hierarchy, and $m = 0$ if X is the IO- or yATT-hierarchy.

Let $X(*)$ denote the union $\bigcup_{n\geq 1} X(n)$. Then $X(*) \subseteq Y(*)$, if $X(n)$ is a subhierarchy of $Y(n)$. We call $X(n)$ *small in* $Y(n)$, if $Y(1) - X(*) \neq \varnothing$. Thus, if $X(n)$ is a small subhierarchy of $Y(n)$, then $X(*) \subsetneq Y(*)$ and, more precisely, the infinite inclusion diagram in Fig. 2 is a Hasse diagram. We now want to show that the EDT0L-hierarchy

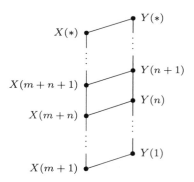

Fig. 2. The Hasse diagram for: "$X(n)$ is a small subhierarchy of $Y(n)$"

is a small subhierarchy of the yMTT-hierarchy. Let CF denote the class of context-free languages.

Theorem 9. $CF - EDT0L^*(REG) \neq \varnothing$.

The proof of Theorem 9 is based on the well-known fact that there are context-free languages L which cannot be generated by EDT0L systems, i.e., which are not in $EDT0L(REG)$ (cf., e.g., Corollary 3.2.18(i) of [ERS80]). Clearly, Theorem 2 also holds if we replace $REGT$ by its restriction $REGT_{\text{mon}}$ to monadic trees. Note that $yT(REGT_{\text{mon}}) = yT(\text{sm}(REG)) = EDT0L(REG)$. Then, for the operation rub_B with $B = \{0, 1\}$, it can be shown, analogous to the proof of Theorem 9 in [Man99], that the context-free language $\text{rub}_B(L)$ is not in $\bigcup_{n\geq 1} yMTT^n(REGT_{\text{mon}})$, which includes $EDT0L^*(REG)$ by the proof of Theorem 3.

Since $CF = yREGT \subseteq yMTT(REGT)$, we get the following theorem.

Theorem 10.
The EDT0L-hierarchy is a small subhierarchy of the yMTT-hierarchy.

The class $ET0L$ of ET0L translations is defined as $EDT0L$, but with nondeterministic top-down tree transducers in place of deterministic ones. In view of Theorem 3, Theorem 9 of [Man99] shows that the EDT0L-hierarchy is small in the ET0L-hierarchy $ET0L^n(REG)$. Obviously, this also holds by Theorem 9, because $CF \subseteq ET0L(REG)$. But, is the EDT0L-hierarchy a small subhierarchy of the ET0L-hierarchy? We can now prove this, using results of [Eng82] and closure properties of $EDT0L^n(REG)$.

Theorem 11.
The EDT0L-hierarchy is a small subhierarchy of the ET0L-hierarchy.

In order to prove Theorem 11 it suffices to show that $EDT0L^{n+1}(REG) - ET0L^n(REG) \neq \emptyset$. For a language L define the copy operations c_2 and c_* as $c_2(L) = \{w\$w \mid w \in L\}$ and $c_*(L) = \{(w\$)^n \mid w \in L, n \geq 1\}$. Let $L_2 = L_{ac}$ and, for $n \geq 2$, let $L_{n+1} = c_2(\text{count}_b(c_*(L_n)))$ for a symbol b not in L_n. (1) $L_n \in EDT0L^n(REG)$. As shown in the proof of Theorem 4, $L_{ac} \in EDT0L^2(REG)$ and $\text{count}_b \in EDT0L$. Hence $L_n \in EDT0L^n(REG)$, because it is easy to show that $EDT0L^n(REG)$ is closed under c_2 and c_*. (2) $L_n \notin ET0L^{n-1}(REG)$. For $n = 2$ this follows from Theorem 3.16 of [Eng82]. For $n > 2$ it follows by the proof of Theorem 4.2 of [Eng82] and the fact that Theorem 3.1 of [Eng82], which is the bridge theorem (Theorem 3.2.14) of [ERS80], is also valid for the operation count in place of the operation rub (with the same proof). Thus, properness of the ET0L-hierarchy is not caused by the alternation of copying and nondeterminism (as stated in [Eng82]), but rather by the alternation of copying and insertion.

At last we mention some open problems: Is the IO-hierarchy small in the yMTT-hierarchy? Is the EDT0L-hierarchy a subhierarchy of the IO-hierarchy? Is $yMTT^*(REGT) \subseteq ET0L^*(REG)$? And for attributed tree transducers: is $yATT^n(REGT) \subsetneq yMTT^n(REGT)$? Note that $yATT^*(REGT)$ is equal to $yMTT^*(REGT)$, see, e.g., Section 6 of [FV98].

References

[BCN81] L. Boasson, B. Courcelle, and M. Nivat. The rational index: a complexity measure for languages. *SIAM Journal on Computing*, 10(2):284–296, 1981.

[Dam82] W. Damm. The IO- and OI-hierarchies. *Theoret. Comput. Sci.*, 20:95–207, 1982.

[DE98]   F. Drewes and J. Engelfriet. Decidability of finiteness of ranges of tree transductions. *Inform. and Comput.*, 145:1–50, 1998.

[Eng82] J. Engelfriet. Three hierarchies of transducers. *Math. Systems Theory*, 15:95–125, 1982.

[ERS80] J. Engelfriet, G. Rozenberg, and G. Slutzki. Tree transducers, L systems, and two-way machines. *J. of Comp. Syst. Sci.*, 20:150–202, 1980.

[ES78]   J. Engelfriet and E.M. Schmidt. IO and OI, Part II. *J. of Comp. Syst. Sci.*, 16:67–99, 1978.
[EV85]   J. Engelfriet and H. Vogler. Macro tree transducers. *J. of Comp. Syst. Sci.*, 31:71–146, 1985.
[Fül81] Z. Fülöp. On attributed tree transducers. *Acta Cybernetica*, 5:261–279, 1981.
[FV98]   Z. Fülöp and H. Vogler. *Syntax-Directed Semantics – Formal Models based on Tree Transducers*. EATCS Monographs on Theoretical Computer Science (W. Brauer, G. Rozenberg, A. Salomaa, eds.). Springer-Verlag, 1998.
[Gre81] S. A. Greibach. Formal languages: origins and directions. *Ann. of the Hist. of Comput.*, 3(1):14–41, 1981.
[GS84]   F. Gécseg and M. Steinby. *Tree Automata*. Akadémiai Kiadó, Budapest, 1984.
[Man99] S. Maneth. String languages generated by total deterministic macro tree transducers. In W. Thomas, editor, *Proc. FOSSACS'99*, volume 1578 of *LNCS*, pages 258–272. Springer-Verlag, 1999.
[Rou70] W.C. Rounds. Mappings and grammars on trees. *Math. Systems Theory*, 4:257–287, 1970.

Partially-Ordered Two-Way Automata: A New Characterization of DA

Thomas Schwentick[1], Denis Thérien[2]*, and Heribert Vollmer[3]

[1] Fachbereich Mathematik und Informatik, Philipps-Universität Marburg,
Hans-Meerwein-Straße, 35032 Marburg, Germany
[2] School of Computer Science, McGill University,
3480 University Street, Montréal (Québec), H3A 2A7 Canada
[3] Theoretische Informatik, Universität Würzburg,
Am Hubland, 97074 Würzburg, Germany

Abstract. In this paper, we consider finite automata with the restriction that whenever the automaton leaves a state it never returns to it. Equivalently we may assume that the states set is partially ordered and the automaton may never move "backwards" to a smaller state.

We show that different types of partially-ordered automata characterize different language classes between level 1 and $\frac{3}{2}$ of the Straubing-Thérien-Hierarchy.

In particular, we prove that partially-ordered 2-way DFAs recognize exactly the class **UL** of *unambiguous languages* introduced by Schützenberger in 1976. As shown by Schützenberger, this class coincides with the class of those languages whose syntactic monoid is in the variety **DA**, a specific subclass of all "group-free" (or "aperiodic") semigroups. **DA** has turned out to possess a lot of appealing characterizations. Our result adds one more to these: partially-ordered two-way automata recognize exactly those languages whose syntactic monoid is in **DA**.

1 Introduction

Finite automata that may move their input head back and forth in both directions are well-known to recognize exactly the regular languages, see, e.g., [HU79], i.e., they are of the same power as usual one-way finite automata.

In spite of the naturalness of the model, there has been comparatively little work done on two-way automata (2DFAs). Some papers compare the conciseness of two-way automata with other models (e.g., [Mic81,Bir92,GH96,HS99]). Other work considered them as a means to compute translations on strings [EH99,MSTV00]. The present authors together with Pierre McKenzie developed the groundwork for an algebraic study of 2DFAs in [MSTV00] by presenting a sensible definition of the so called behavior monoid of such a machine, and by showing how restricted classes of 2-way automata with output correspond in a nice way to algebraically restricted NFAs and first-order logic with monoidal quantifiers. (A concept similar to the behaviour functions from [MSTV00] was examined from a different perspective much previously in [Bir89].)

---

* Supported by NSERC of Canada, by FCAR du Québec, and by the Alexander-von-Humboldt-Gesellschaft. Work done while on leave at the Universität Tübingen, Germany.

Here, we take up the study of subclasses of 2DFAs. In particular, we look at the following natural restriction: Say that an automaton, one-way or two-way, deterministic or not, is partially-ordered (p.-o., for short) if, whenever M leaves a state, it will never return to it. This means that there is a partial order on the states of M such that, whenever M moves from state p to state q, then p is less than or equal to q in that order.

Partially-ordered one-way automata are well studied. It is known that, in the deterministic case, they recognize exactly those languages whose syntactic monoid is R-trivial, where we say that a monoid M is R-trivial if, whenever two monoid elements s, t satisfy $sM = tM$, then $s = t$. (Analogously one can define L-trivial as above but using $Ms = Mt$. Then clearly, p.-o. reverse automata, i.e., FAs moving their head from right to left in a one-way manner, recognize exactly those languages with an L-trivial syntactic monoid.) The intersection of both is known to equal level 1 of the Straubing-Thérien-Hierarchy (STH) of star-free regular languages, known to correspond to so called J-trivial monoids. For background on these classes, we refer the reader to [Pin97], see also [Pin86].

These classes have interesting combinatorial characterizations: Consider an alphabet A. A *left-deterministic product* over A is a concatenation of the form $A_0^* a_1 A_1^* \ldots a_k A_k^*$, where $a_i \in A$, $A_i \subseteq A$, and $a_i \notin A_{i-1}$. Now a language is recognized by a p.-o. FA iff it is a finite disjoint union of left-deterministic products. (Equivalently, a language is recognized by a p.-o. reverse FA iff it is a finite disjoint union of right-deterministic products, where *right-deterministic* is defined replacing in the above the condition $a_i \notin A_{i-1}$ by $a_i \notin A_i$.)

Schützenberger in 1976 [Sch76] studied the following superclass of these: Say that an *unambiguous product* over alphabet A is a concatenation of the form $A_0^* a_1 A_1^* \ldots a_k A_k^*$, where $a_i \in A$, $A_i \subseteq A$, if every word w has at most one factorization $w = w_0 a_1 w_1 \ldots a_k w_k$ with $w_i \in A_i^*$. A language is *unambiguous* if it is a finite disjoint union of unambiguous products. The class **UL** is the class of all unambiguous languages. Schützenberger also obtained an algebraic characterization of **UL**. A monoid M belongs to the variety **DA** if it satisfies for every idempotent element $s \in M$ the condition that if $MsM = MtM$ then t is also idempotent. Schützenberger proved that a language is in **UL** iff its syntactic monoid belongs to **DA**.

Variety **DA** is known to have an impressive number of combinatorial, algebraic and logical characterizations, some of which will be used later in the present paper. It can be shown that **UL** is the intersection of level $\frac{3}{2}$ of the STH and its complement, i.e. languages in **UL** are exactly those that can be defined both by a first-order sentence in prenex normal-form with a Σ_2-prefix and by a sentence with a Π_2-prefix [PW97]. In this paper we develop a new machine characterization of **DA**: We prove that partially-ordered two-way automata recognize exactly those languages in **UL**. Hence a language L has a syntactic monoid in **DA** iff there is a p.-o. 2DFA recognizing L.

An interesting intermediate step in our proof of equivalence of p.-o. 2DFAs and **DA** are so called *turtle languages*. Given an input word w and a position of w, a turtle move is an instruction of the form $(\leftarrow, a)$ or $(\rightarrow, a)$, meaning that the next position is determined by moving to the right (or left), until a letter a occurs. If no a is found, the move fails. A turtle language is a language that can be recognized by a sequence of such turtle moves, in the sense that a word w belongs to the language if and only if none of the moves in

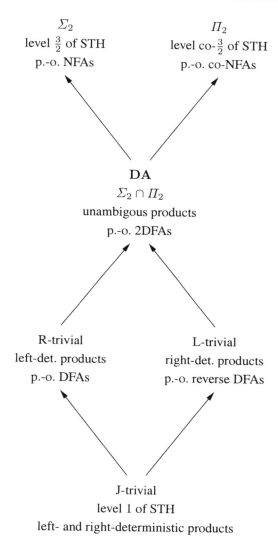

Fig. 1. Inclusions among considered language classes

the sequence fails. We will show that a language has its syntactic monoid in **DA** iff it is a Boolean combination of turtle languages. Particular types of turtle languages can be related to further monoid varieties as we point out in Sect. 4.

Concerning partially-ordered NFAs, we will see that both in the one-way and in the two-way case these characterize level $\frac{3}{2}$ of the STH. Hence, a language can be accepted by a p.-o. 2DFA if and only if it and its complement can be recognized by a p.-o. NFA, either one-way or two-way.

Our results are summarized in Fig. 1. We see that partially-ordered finite automata can be used to characterize a number of classes of current focal interest in the study of concatenation hierarchies.

2 Preliminaries

2.1 Monoids and Congruences

An associative binary operation on a set containing an identity for this operation defines a *monoid*. By a monoid *variety*, we mean a pseudovariety in the sense of Straubing [Str94, pp. 72ff] or Eilenberg [Eil76, pp. 109ff]: it is a set of monoids closed under finite direct product, the taking of submonoids and of homomorphic images. Well-known examples of varieties include the commutative monoids, the groups $\mathbf{G}$, the aperiodics $\mathbf{A}$ (i.e., the monoids containing only trivial groups), the solvable groups $\mathbf{G}_{sol}$.

It is known that any set of monoids satisfying (or ultimately satisfying) a set of identities (in the sense described, e.g., in [Str94, Chapter V.6]) forms a variety. Of particular interest for this paper will be the following case: The variety $\mathbf{DA}$ consists of all finite semigroups satisfying $(xyz)^\omega y(xyz)^\omega = (xyz)^\omega$, where for any semigroup element s, s^ω denotes the least power of s that is idempotent. This variety was originally introduced by Schützenberger [Sch76] using the definition presented in Sect. 1, and is known to recognize the unambiguous languages, i.e., the finite unions of products $A_0^* a_1 A_1^* a_2 \cdots a_k A_k^*$ where every string is unambiguously factorizable w.r.t. this product; here, every $a_i \in A$ and every $A_i \subseteq A$ where A denotes the input alphabet.

Deterministic finite automata are, as usual [HU79, p. 17], given by $M = (S, A, \delta, s_0, F)$, where the different components denote the state set, the input alphabet, the transition function, the initial state, and the set of final states, respectively. The extended transition function [HU79, p. 17] of M is denoted by $\hat{\delta}$. The *transformation monoid* of M is the set $\{\hat{\delta}(\cdot, w) \colon S \to S \mid w \in \Sigma^*\}$ with the operation of composition of functions. Certainly, this set of transformations will always be finite. For a variety $\mathbf{V}$, we say that M is a $\mathbf{V}$-DFA if its transformation monoid is in $\mathbf{V}$. A $\mathbf{V}$-language is a language accepted by a $\mathbf{V}$-DFA. Non-deterministic automata are defined similarly, except that δ is now a function from $Q \times A \to 2^Q$ and we allow a set of possible initial states.

An equivalence relation $\sim$ on A^* is a *congruence* if $u \sim v$ implies $(\forall x, y \in A^*)[xuy \sim xvy]$. A congruence $\sim$ induces a monoid $A^*/\sim$ isomorphic to the transformation monoid of the pre-automaton $(A^*/\sim, A, ([u]_\sim, a) \mapsto [ua]_\sim)$. An example of a congruence is the syntactic congruence of a language L: $u \sim_L v$ iff $(\forall x, y \in A^*)[xuy \in L$ iff $xvy \in L]$. Let a DFA $M = (S, A, \delta, s, F)$ be given. Another example of a congruence is: $u \sim_M v$ iff $\hat{\delta}(\cdot, u) = \hat{\delta}(\cdot, v)$. Then $A^*/\sim_M$ is isomorphic to the transformation monoid of M. Writing $\sim_{pq}$ for the syntactic congruence of the language $L(p, q) = \{w \mid \hat{\delta}(p, w) = q\}$, there is an injective morphism $A^*/\sim_M \longrightarrow \Pi_{p,q \in S}(A^*/\sim_{pq})$ and a surjective morphism $A^*/\sim_M \longrightarrow A^*/\sim_{pq}$. These facts can be shown to imply that M is a $\mathbf{V}$-DFA iff $L(p, q)$ is a $\mathbf{V}$-language for each $p, q \in S$.

2.2 Two-Way Automata

As mentioned in the introduction, the ground for an algebraic study of two-way automata was laid in [MSTV00]. The following definitions are mainly taken from that paper. We remark that our results also remain valid for the standard definition, see [HU79].

A two-way automaton is a tuple $M = (\mathcal{L} \uplus \mathcal{R}, A, \delta, l_0, F)$, where the set S of states is the disjoint union $\mathcal{L} \uplus \mathcal{R}$ of a set $\mathcal{L}$ (the states "entered from the left") and a set $\mathcal{R}$ (the states "entered from the right"), $l_0 \in \mathcal{L}$ is the *initial state*, $F \subseteq S$ is the set of *final states*, A is the *input alphabet*, $\delta \colon (S \times A) \cup (\mathcal{L} \times \{\triangleleft\}) \cup (\mathcal{R} \times \{\triangleright\}) \to S$ is a total *transition function*, where $\triangleright \notin A$ and $\triangleleft \notin A$ are the *leftmarker* and the *rightmarker* respectively,

The meaning of $\delta(s, a) \in \mathcal{L}$ is that M in state s scanning a moves its head to the *right* upon entering state $\delta(s, a)$; M moves its head to the *left* when $\delta(s, a) \in \mathcal{R}$.

The *initial configuration* of M on input $w = w_1 w_2 \cdots w_n \in A^*$ is the situation in which the state of M is l_0 and M scans w_1 within the string $\triangleright w_1 w_2 \cdots w_n \triangleleft$ (M scans $\triangleleft$ when $|w| = 0$). We say that M *eventually exits* $\triangleright w \triangleleft$ if it eventually encounters a transition $\delta(r, \triangleright) \in \mathcal{R}$ or a transition $\delta(l, \triangleleft) \in \mathcal{L}$ (of course M will generally bounce off the end markers several times before exiting). We require that, for any $w = w_1 w_2 \cdots w_n \in A^*$,

- M from its initial configuration on w eventually leaves every w_i to the right, $1 \leq i \leq n$, and eventually exits $\triangleright w \triangleleft$; this is analogous to the (unspoken) requirement that a one-way automaton must traverse its input and halt,
- from *any* state $l \in \mathcal{L}$ scanning w_1, M eventually exits $\triangleright w \triangleleft$; this requirement is analogous to the (unspoken) fact that a one-way automaton eventually runs out of input regardless of its initial configuration,
- from *any* state $r \in \mathcal{R}$ scanning w_n, M also eventually exits $\triangleright w \triangleleft$; this is justified by the natural desire to maintain symmetry between left and right in a two-way machine. The input word is *accepted* by M if M exits $\triangleright w \triangleleft$ in a final state.

Each $w = w_1 w_2 \cdots w_n \in A^*$ coerces M into a behavior described by a *behavior function* $\delta_w \colon \mathcal{L} \uplus \mathcal{R} \to \mathcal{L} \uplus \mathcal{R}$, where $\delta_w(s)$ is the state in which M exits $w_1 w_2 \cdots w_n$ when started in state s scanning w_1, if $s \in \mathcal{L}$; and $\delta_w(s)$ is the state in which M exits $w_1 w_2 \cdots w_n$ when started in state s scanning w_n, if $s \in \mathcal{R}$. Finally we define the *behavior monoid* $\mathcal{B}(M)$ of M to be the monoid $\{\delta_w \mid w \in A^*\}$ under the operation $\delta_u \circ \delta_v =_{\text{def}} \delta_{uv}$. (Note that by this definition, $\circ$ is *not* composition of functions.) M is a **V**-*machine* iff $\mathcal{B}(M) \in \mathbf{V}$.

As in the one-way case, the non-deterministic version of the model is obtained by allowing the transition relation to be multiple-valued and allowing several initial states.

2.3 Partially-Ordered Automata

Central for this paper will be the following definition: An automaton M (one-way or two-way, deterministic or not) is *partially-ordered* (p.-o., for short) if there is a partial order $\prec$ on the state set S of M such that whenever $\delta(s, a) = s'$ then $s = s'$ or $s \prec s'$. This means that whenever M leaves a particular state, it can never return to it.

Let M be a partially ordered automaton with input alphabet A. With the computation of M on an input w we can associate a *characteristic sequence* $S = s_1, \ldots, s_l$, consisting of tuples (p_i, A_i, a_i) where

- $p_1, \ldots, p_l$ is the sequence of distinct states that M assumes on input w in that order (and, of course, p_1 is the initial state of M),
- M stays in p_i while reading symbols from $A_i \subseteq A$, and
- M goes from state p_i into state $p_{i+1} \neq p_i$ by reading the symbol a_i (in w).

Note that each state of M occurs at most once in S, thus, the number of different state sequences is finite and depends only on M. If M is a 2DFA, recall that each state of M is entered either only from the left or only from the right. Therefore, all movements of M that let it stay in a certain state are in the same direction (for that state).

For a characteristic sequence S, we denote by $L(S)$ the set of those strings accepted by M when obeying sequence S.

The following is known about the power of deterministic p.-o. (one-way) automata. Say that a monoid M is R-trivial (L-trivial, resp.) if, whenever two monoid elements s, t satisfy $sM = tM$ ($Ms = Mt$, resp.) then $s = t$. A language has a syntactic monoid that is R-trivial iff it is recognized by a p.-o. DFA. A language has a syntactic monoid that is L-trivial iff it is recognized by a p.-o. DFA that moves its head from right to left in a one-way manner.

To our best knowledge, nondeterministic p.-o. FAs have not been looked at so far. However, it can be seen that both in the one-way and two-way case they recognize exactly level 3/2 of the Straubing-Thérien-Hierarchy.

Proposition 1. *A language L can be recognized by a non-deterministic one-way p.-o. automaton iff it can be recognized by a non-deterministic two-way p.-o. automaton iff it belongs to level $\frac{3}{2}$ of the Straubing-Thérien-Hierarchy (i.e. Σ_2).*

Proof. It is known [Arf91] that a language is in Σ_2 iff it is a union of languages of the form $A_0^* a_1 A_1^* \ldots a_s A_s^*$, where a_i are alphabet symbols and A_i are subsets of the alphabet. The direct construction of a one-way non-deterministic automaton from the expression will yield a p.-o. machine. Of course, every one-way automaton is also a two-way automaton. Finally, consider a non-determinsitic two-way p.-o. automaton M. Take any word w that is accepted by M. Let S be a characteristic sequence of M encountered during the computation, that ends in an accepting state. Let $(q_0, q_1, \ldots, q_s)$ be the sequence of states in S. Consider the induced decomposition of w as $w = w_0 a_1 w_1 \ldots a_s w_s$, where the a_i's correspond to positions of w where the computation of M has changed its state (and possibly its direction). Consider the ith segment of this decomposition. It may have been traversed from left to right a number of times, e.g., while in the states $q_{i_1}, \ldots, q_{i_p}$, and from right to left, a number of times, e.g. while in states $q_{j_1}, \ldots, q_{j_r}$. Let A_i be the set of letters a with the property that for each t between 1 and p, q_{i_t} is in $\delta(q_{i_t}, a)$ and for each t between 1 and r, q_{j_t} is in $\delta(q_{j_t}, a)$. Then all words that are in $A_0^* a_1 A_1^* \ldots a_s A_s^*$ will also be accepted by M, via a computation that traverses exactly the same states as the one on w. Taking the union over the finitely many possible monotone sequences of states that end up in an accepting state will yield a Σ_2 expression for the language recognized by M. □

3   Main Result

Before we state and prove the main result of this paper we need some further definitions. Let A be an alphabet. A *turtle instruction* is a pair $I = (d, \sigma)$ consisting of a *direction* $d \in \{\leftarrow, \rightarrow\}$ and a *target symbol* $\sigma \in A$. If $d = \leftarrow$ we say that I is a $\leftarrow$-instruction, otherwise a $\rightarrow$-instruction. A *turtle program* P is a non-empty sequence $I_1 \cdots I_k$ of turtle instructions.

The semantics of turtle programs is defined as follows. Let $w = w_1 \cdots w_n$ be a string of some length n and let $i \in \{0, \ldots, n+1\}$, i.e., either i is a position in w or $i = 0$ or $i = n+1$. Let $I = (\leftarrow, \sigma)$ be a $\leftarrow$-instruction. Let $j_0 = \max\{j \in \{1, \ldots, i-1\} \mid w_j = \sigma\}$. We define $I(w, i) = (w, j_0)$ if j_0 exists, i.e., if $\{j \in \{1, \ldots, i-1\} \mid w_j = \sigma\}$ is non-empty. Otherwise, $I(w, i)$ is undefined, denoted as $I(w, i) = \bot$. Correspondingly, we define $I(w, i)$ for $\rightarrow$-instructions but in this case j_0 is defined as $\min\{j \in \{i+1, \ldots, n\} \mid w_j = \sigma\}$. For a turtle program $P = I_1 \cdots I_k$ and a string w of length n we define the *final position of P on w* as follows.

- If I_1 is a $\rightarrow$-instruction and $(w, j) = I_k(\cdots (I_1(w, 0)) \cdots)$ is defined then we set $P(w) = j$;
- If I_1 is a $\leftarrow$-instruction and $(w, j) = I_k(\cdots (I_1(w, n+1)) \cdots)$ is defined then, again, we set $P(w) = j$;
- Otherwise, we say that $P(w)$ is undefined and write $P(w) = \bot$.

Note that the interpretation of a turtle program starts at the left end of a string if its first instruction is a $\rightarrow$-instruction and at the right end otherwise. It should also be noted that if $P(w)$ is not undefined it is always a position of w, i.e., $P(w) \in \{1, \ldots, n\}$.

We write $L(P)$ for the set of strings w for which $P(w)$ is defined. Languages of the form $L(P)$ are called *turtle languages*.

It is not hard to see that turtle languages can be accepted by partially ordered two-way automata. On the other hand, we will show that, if a language has a syntactic monoid in **DA**, it can be expressed as a Boolean combination of turtle languages. Therefore, turtle languages are, in a sense, a link between **DA** and partially ordered two-way automata. The above relations and their convereses are the contents of our main result, stated formally as follows:

Theorem 1. *For a language L, the following are equivalent.*

*(a) The syntactic monoid of L is in **DA**.*
(b) L is a Boolean combination of turtle languages.
(c) L is accepted by a partially ordered two-way automaton.

The proof is given by a series of lemmas. The first one is of a technical nature. Given two turtle programs, P_1, P_2, we write $L(P_1, P_2)$ for the set of strings $w \in L(P_1) \cap L(P_2)$ for which $P_1(w) < P_2(w)$. In the following, we do not distinguish in notation between a turtle instruction I and program with the single instruction I.

Lemma 1. *For every turtle program P and every turtle instruction I, the languages $L(I, P)$ and $L(P, I)$ can be expressed as Boolean combinations of turtle languages.*

Proof. Let $I = (d, \sigma)$. We write I^{-1} for the *inverse* of I, i.e., the instruction $(\rightarrow, \sigma)$ if $d = \leftarrow$, and $(\leftarrow, \sigma)$ if $d = \rightarrow$. If $d = \rightarrow$ then the following hold, for every string w.

$$w \in L(I, P) \iff w \in L(PI^{-1}).$$
$$w \in L(P, I) \iff w \in \overline{L(PII^{-1})} \cap L(PI).$$

In the first equivalence, the term $L(PI^{-1})$ intuitively says that there is a σ in w to the left of $P(w)$, hence $I(w) < P(w)$. In the second equivalence, $L(PI)$ contains all w for which there is a σ to the right of $P(w)$. The term $\overline{L(PII^{-1})}$ rules out all strings for which $I(w) \leq P(w)$.

If $d = \leftarrow$ then, similarly, the following hold, for every string w.

$$w \in L(I, P) \iff w \in \overline{L(PII^{-1})} \cap L(PI),$$
$$w \in L(P, I) \iff w \in L(PI^{-1}).$$

$\square$

Lemma 2. *If the syntactic monoid of a language L is in* **DA** *then L is a Boolean combination of turtle languages.*

Proof. Let the syntactic monoid of L be in **DA**.

Recall the following characterization of **DA** from [TW98]: For a word w, let $\alpha(w)$ denote the set of letters occurring in w. If $a \in \alpha(w)$ we say that $w = uav$ is an a-left decomposition of w if $a \notin \alpha(u)$. Analogously, $w = uav$ is an a-right decomposition of w if $a \notin \alpha(v)$. We define a relation $\sim_n$ on words (shown to be a congruence in [TW98]) by induction on n:

First, $x \sim_0 y$ for all strings x, y. Next, for $n > 0$, $x \sim_n y$ if $\alpha(x) = \alpha(y)$ and

(1) for all $a \in \alpha(x)$, if $x = x_0 a x_1$ and $y = y_0 a y_1$ are the a-left decompositions of x and y, then $x_0 \sim_n y_0$ and $x_1 \sim_{n-1} y_1$, and
(2) for all $a \in \alpha(x)$, if $x = x_0 a x_1$ and $y = y_0 a y_1$ are the a-right decompositions of x and y, then $x_0 \sim_{n-1} y_0$ and $x_1 \sim_n y_1$.

[TW98] prove that a language has its syntactic monoid in **DA** iff it is a finite union of equivalence classes of $\sim_n$ for some n. As in [TW98] it should be noted that this characterization is well-defined because $|\alpha(x_0)| < |\alpha(x)|$ in (1) and $|\alpha(x_1)| < |\alpha(x)|$ in (2).

We show in the following that each class C of $\sim_n$ is a Boolean combination of turtle languages. The proof is by induction on n and the size of the alphabet of C. Note that by the above definition, all strings in C use the same letters, if $n > 0$.

For $n = 0$ the proof is straightforward, as A^* is $L(I) \cup \overline{L(I)}$ for some turtle move I.

Now let $n > 0$, C be an equivalence class of $\sim_n$ with alphabet $B \subseteq A$, y a representant of C and $a \in B$. Let $y_0 a y_1$ be the a-left-decomposition of y. We show next how the language of strings $w = w_0 a w_1$ with $w_0 \sim_n y_0$ and $w_1 \sim_{n-1} y_1$ can be expressed as a Boolean combination of turtle languages. Given the above characterization of **DA**-languages, and as the case of right-decompositions can be handled completely analogously, the statement of the lemma follows directly. L is simply the intersection of all the Boolean combination obtained.

First, we show how the language L_0 of strings $w = w_0 a w_1$ with $w_0 \sim_n y_0$ can be expressed. By induction, as $|\alpha(y_0)| < |B|$, the equivalence class C_0 of y_0 (over the alphabet $B - \{a\}$) can be expressed as a Boolean combination of turtle languages. Let $L(P)$ be a language occuring in this characterization. Let $I = (\rightarrow, a)$. If the first instruction of P is a $\rightarrow$-instruction then

$$w_0 \in L(P) \quad \text{if and only if} \quad w \in L(P) \cap \bigcap_{P' \sqsubseteq P} L(P', I).$$

Here, $P' \sqsubseteq P$ denotes that P' is a non-empty prefix of P. If the first instruction of P is a $\leftarrow$-instruction then

$$w_0 \in L(P) \quad \text{if and only if} \quad w \in L(IP) \cap \bigcap_{P' \sqsubseteq P} L(IP', I).$$

It is straightforward to check that these equivalences indeed hold. Note that the big intersections on the right ensure that the application of P on w only uses w_0. By Lemma 1 it follows that the languages of the form $L(P', I)$ and $L(IP', I)$ can be expressed as Boolean combinations of turtle languages.

By replacing each $L(P)$ in the characterization of C_0 with the corresponding right hand expression we obtain a characterization for L_0.

Finally, we show how the language L_1 of strings $w = w_0 a w_1$ with $w_1 \sim_{n-1} y_1$ can be expressed. Again, by induction, the equivalence class C_1 of y_1 can be expressed as a Boolean combination of turtle languages. Let $L(P)$ be a language occuring in this characterization. If P starts with a $\rightarrow$-instruction then

$$w_1 \in L(P) \quad \text{if and only if} \quad w \in L(IP) \cap \bigcap_{P' \sqsubseteq P} L(I, IP').$$

If P starts with a $\leftarrow$-instruction then

$$w_1 \in L(P) \quad \text{if and only if} \quad w \in L(P) \cap \bigcap_{P' \sqsubseteq P} L(I, P').$$

Again, the languages on the right hand side can be expressed as Boolean combinations of turtle languages and we obtain the desired characterization of L_1. This completes the proof of the lemma. $\square$

Lemma 3. *If a language L is accepted by a p.-o. 2DFA, then its syntactic monoid is in* **DA**.

Proof. Let M be a p.-o. 2DFA accepting L. Let S be a characteristic sequence of M (see Sect. 2.3), $S = (p_1, A_1, a_1)(p_2, A_2, a_2) \cdots (p_n, A_n, a_n)$, where the p_i are states of M, A_i are alphabet subsets, and a_i are alphabet symbols. We will show that the syntactic monoid of $L(S)$ is in **DA** by proving that it satisfies the equation $(xyz)^n y (xyz)^n = (xyz)^n$. For this, it is sufficient to show that, for all words u, v, x, y, z, we have

$u(xyz)^n y(xyz)^n v \in L(S)$ iff $u(xyz)^n v \in L(S)$. The statement of the lemma then follows as L is a union of languages $L(S)$ and **UL** is closed under Boolean combinations.

Let w_l be a tape content of M of the form $\triangleright u(xyz)^n y(xyz)^n v \triangleleft$, and let w_r be of the form $\triangleright u(xyz)^n v \triangleleft$. Suppose that M switches from p_1 to p_2 while reading the prefix $\triangleright uxyz$ of w_l. Then, certainly, the same switch will occur at the same position when reading w_r. If M, reading w_l, does not switch from p_1 to p_2 while within $\triangleright uxyz$ then this switch can only occur in the suffix $v\triangleleft$. Again, the same switch must occur at exactly the same position within $v\triangleleft$ when M reads w_r. Note that this means that in the latter case, M in state p_1 skips every letter in u, x, y, and z.

In general, consider the switch from p_i to p_{i+1} in M working on w_l. Then one of the following cases holds:

- The switch occurs in the prefix $\triangleright u(xyz)^i$ of w_l, and the switch occurs at exactly the same position in w_r.
- The switch occurs in the suffix $(xyz)^i v \triangleleft$ of w_l, and the switch will occur at the same position of the suffix of w_r.

This means that the middle y of w_l is *always* skipped entirely, hence we conclude $u(xyz)^n y(xyz)^n v \in L(S) \iff u(xyz)^n v \in L(S)$, what we had to show. □

Proof. (*of Theorem 1.*) (a) $\implies$ (b) is the contents of Lemma 2.

(b) $\implies$ (c) follows from the trivial fact that every turtle language is accepted by some partially ordered two-way automaton, and the fact that the class of languages accepted by partially-ordered two-way automata is closed under Boolean operations. For complement this is trivial, since we are dealing with deterministic machines. An automaton M recognizing the union of the languages accepted by the partially-ordered two-way automata M_1 and M_2, given an input word w, first simulates M_1 on w. If M_1 accepts then M exits the input accepting. If not, M is transferred into a new state s; then it uses s to move its head back to the left end-marker and then, moving into the initial state of M_2, simulates M_2 on w. The order of states of M is the ordinal addition of the state set of M_1, the set $\{s\}$, and the state set of M_2.

Finally, (c) $\implies$ (a) is the contents of Lemma 3. □

4 Conclusion

In the present paper we gave a characterization of the class **UL**, or, alternatively, the class of languages whose syntactic monoid is in **DA**, using finite two-way automata. A question that arises immediately is if our characterization leads to new decidability results. A first step would be to determine the complexity of deciding if a 2DFA is p.-o. It is relatively easy to see that this problem is in coNP; we conjecture that it is actually complete for that class.

Coming back to Fig. 1, our results suggest a new proof of one of the results of [PW97], namely that languages with a syntactic monoid in **DA** are exactly those definable in $\Sigma_2 \cap \Pi_2$: For this, one would have to show independently, what follows from our paper, namely that a language and its complement can be recognized by p.-o. NFAs if and only if it can be recognized by a p.-o. 2DFA.

Turtle languages, which were introduced in the present paper as a technical vehicle, also turn out to capture other nice classes of varieties: It can be shown that a language has an R-trivial syntactic monoid iff it is a Boolean combination of turtle languages of the form $(\rightarrow, a_1)(\rightarrow, a_2) \ldots (\rightarrow, a_s)(\leftarrow, b_1) \ldots (\leftarrow, b_t)$, i.e., they start from the left and are allowed only one turn in direction. Starting from right, one obtains L-trivial; allowing one turn, starting from either direction gives the join of the varieties. No turn turtle programs gives J-trivial. These results hint in the direction that turtle languages might turn out to be a helpful tool for further studies in algebraic language theory. It would be nice to add more evidence to this thesis.

References

[Arf91]   M. Arfi. Opération polynomiales et hiérarchies de concaténation. *Theoretical Computer Science*, 91:71–84, 1991.

[Bir89]   J.-C. Birget. Concatenation of inputs in a two-way automaton. *Theoretical Computer Science*, 63:141–156, 1989.

[Bir92]   J.-C. Birget. Positional simulation of two-way automata: Proof of a conjecture of R. Kannan and generalizations. *Journal of Computer and System Sciences*, 45(2):154–179, 1992.

[EH99]   J. Engelfriet and H. Hoogeboom. Two-way finite state transducers and monadic second-order logic. In *Proceedings 26th International Colloqium on Automata, Languages and Programming*, volume 1644 of *Lecture Notes in Computer Science*, pages 311–320, Berlin Heidelberg, 1999. Springer Verlag.

[Eil76]   S. Eilenberg. *Automata, Languages, and Machines*, volume B. Academic Press, New York, 1976.

[GH96]   N. Globerman and D. Harel. Complexity results for two-way and multi-pebble automata and their logics. *Theoretical Computer Science*, 169:161–184, 1996.

[HS99]   J. Hromkovic and G. Schnitger. On the power of Las Vegas II. Two-way finite automata. In *Proceedings 26th International Colloqium on Automata, Languages and Programming*, volume 1644 of *Lecture Notes in Computer Science*, pages 433–442, Berlin Heidelberg, 1999. Springer Verlag.

[HU79]   J. E. Hopcroft and J. D. Ullman. *Introduction to Automata Theory, Languages, and Computation*. Addison-Wesley Series in Computer Science. Addison-Wesley, Reading, MA, 1979.

[Mic81]   S. Micali. Two-way deterministic finite automata are exponentially more succinct than sweeping automata. *Information Processing Letters*, 12:103–105, 1981.

[MSTV00]   P. McKenzie, T. Schwentick, D. Thérien, and H. Vollmer. The many faces of a translation. In *Proceedings 27th International Colloqium on Automata, Languages and Programming*, volume 1853 of *Lecture Notes in Computer Science*, pages 890–901, Berlin Heidelberg, 2000. Springer Verlag. By a mistake of the organizers while reformatting this paper, alas, all equality signs disappeared in the proceedings version. Refer to ftp://ftp-info4.informatik.uni-wuerzburg.de/pub/TRs/mc-sc-th-vo00.ps.gz for a full version *with* equality signs.

[Pin86]   J. E. Pin. *Varieties of Formal Languages*. Plenum Press, New York, 1986.

[Pin97]   J. E. Pin. Syntactic semigroups. In G. Rozenberg and A. Salomaa, editors, *Handbook of Formal Languages*, volume I, pages 679–746. Springer Verlag, Berlin Heidelberg, 1997.

[PW97] J.-E. Pin and P. Weil. Polynomial closure and unambiguous product. *Theory of Computing Systems*, 30:383–422, 1997.

[Sch76] M. Schützenberger. Sur le produit de concatenation non ambigu. *Semigroup Forum*, 13:47–75, 1976.

[Str94] H. Straubing. *Finite Automata, Formal Logic, and Circuit Complexity*. Birkhäuser, Boston, 1994.

[TW98] D. Thérien and T. Wilke. Over words, two variables are as powerful as one quantifier alternation: $FO^2 = \Sigma_2 \cap \Pi_2$. In *Proceedings 30th Symposium on Theory of Computing*, pages 234–240, New York, 1998. ACM Press.

Level 5/2 of the Straubing-Thérien Hierarchy for Two-Letter Alphabets

Christian Glaßer[1,*] and Heinz Schmitz[2,**]

[1] Theoretische Informatik, Universität Würzburg, Am Hubland,
97074 Würzburg, Germany, glasser@informatik.uni-wuerzburg.de
[2] sd&m AG, software design & management, Thomas-Dehler-Str. 27,
81737 München, Germany, heinz.schmitz@sdm.de

Abstract. We prove an effective characterization of level 5/2 of the Straubing-Thérien hierarchy for the restricted case of languages defined over a two-letter alphabet.

1 Introduction

We provide in this paper a new decidability result concerning the Straubing-Thérien hierarchy (STH, for short) [15,18,16] and show that its level 5/2 in the restricted case of a two-letter alphabet is decidable. The STH, as well as the closely related dot-depth hierarchy (DDH, for short) [4], puts a natural parameterization on the class of starfree regular languages by counting the alternating use of Boolean operations versus concatenation. For background on starfree languages and on the history of the pending decidability questions we refer to [10].

We state one possibility to define the STH and the DDH. Let A be some finite alphabet with $|A| \geq 2$. For a class $\mathcal{C}$ of languages let $\text{Pol}(\mathcal{C})$ be its polynomial closure, i.e., the closure under finite union and concatenation, and denote by $\text{BC}(\mathcal{C})$ its Boolean closure (taking complements w.r.t. A^+ since we consider ε-free languages). The classes $\mathcal{L}_{n/2}(A)$ of the STH and the classes $\mathcal{B}_{n/2}(A)$ of the DDH can be defined as follows.

$$\mathcal{L}_{1/2}(A) := \text{Pol}(\{A^*aA^* \,|\, a \in A\}) \quad \mathcal{B}_{1/2}(A) := \text{Pol}(\{\{a\} \,|\, a \in A\} \cup \{A^+\})$$
$$\mathcal{L}_{n+1}(A) := \text{BC}(\mathcal{L}_{n+1/2}(A)) \quad \mathcal{B}_{n+1}(A) := \text{BC}(\mathcal{B}_{n+1/2}(A)) \quad \text{for } n \geq 0$$
$$\mathcal{L}_{n+3/2}(A) := \text{Pol}(\mathcal{L}_{n+1}(A)) \quad \mathcal{B}_{n+3/2}(A) := \text{Pol}(\mathcal{B}_{n+1}(A)) \quad \text{for } n \geq 0$$

By definition, all these classes are closed under union and it is known, that they are also closed under intersection and under taking residuals [2,11]. Both hierarchies are strict and closely related [3,16,11,12]. The question whether there exists an algorithm that exactly locates a given language in these hierarchies is commonly referred to as the dot-depth problem. A lot of effort has been invested in the past to cope with the levelwise membership problems. So far the best known results for the STH and DDH are the decidability of levels 1/2, 1 and 3/2 of both hierarchies [14,7,1,11,5] while the question is open for any other level. A step beyond this was achieved in [17] where it was

---

[*] Supported by the Studienstiftung des Deutschen Volkes.
[**] Partially supported by the Deutsche Forschungsgemeinschaft, grant Wa 847/4-1.

shown that level 2 of the STH is also decidable—in the restricted case that A contains only two letters.

Among the mentioned decidability results, the present result involves for the first time *two* quantifier alternations, as becomes clear when we recall the natural connection of these concatenation hierarchies to the quantifier alternation hierarchy of first-order logic over finite words. It is shown in the seminal work of McNaughton and Papert [8] that the starfree languages over alphabet A are exactly those definable by sentences of the logic FO[<] having unary relations for the letters in A and the binary relation < (for an introduction see, e.g., [20]). Let Σ_n be the subclass of this logic that is defined by at most $n-1$ quantifier alternations, starting with an existential quantifier. It has been proved in [19,9] that Σ_n-formulas describe just the $\mathcal{L}_{n-1/2}(A)$ languages and that the Boolean combinations of Σ_n-formulas describe just the $\mathcal{L}_n(A)$ languages. Due to this correspondence our result implies that the class of languages over $\{0,1\}$ definable by Σ_3-formulas of the logic FO[<] is decidable.

On the technical side we work with the so-called forbidden-pattern approach: The decidability of $\mathcal{L}_{5/2}(B)$ for $B = \{a,b\}$ is a consequence of an effective characterization of the type "*a language belongs to $\mathcal{L}_{5/2}(B)$ if and only if the accepting automaton does not have a certain subgraph in its transition graph*". In [6] a theory of forbidden-patterns in the context of concatenation hierarchies is developed. In order to preview the main theorem of the present paper we need to recall some of this theory. Based on an iteration rule working on patterns, in [6] strict hierarchies $\{\mathbb{L}_n^B(A)\}_{n \geq 0}$ and $\{\mathbb{L}_n^C(A)\}_{n \geq 0}$ of language classes defined via forbidden-patterns are introduced. It is shown there that for all $n \geq 0$ the classes $\mathbb{L}_n^B(A)$ and $\mathbb{L}_n^C(A)$ have decidable membership problems and that the following inclusions hold (for an arbitrary alphabet A):

$$\mathcal{B}_{1/2}(A) = \mathbb{L}_0^B(A) \qquad \mathcal{L}_{1/2}(A) = \mathbb{L}_0^C(A)$$
$$\mathcal{B}_{3/2}(A) = \mathbb{L}_1^B(A) \qquad \mathcal{L}_{3/2}(A) = \mathbb{L}_1^C(A)$$
$$\mathcal{B}_{n+1/2}(A) \subseteq \mathbb{L}_n^B(A) \qquad \mathcal{L}_{n+1/2}(A) \subseteq \mathbb{L}_n^C(A)$$

Now we can state our main technical result (Theorem 4) which gives a general relationship. Let $n \geq 1$ and let B be a two-letter alphabet.

If $\mathcal{B}_{n+1/2}(A) = \mathbb{L}_n^B(A)$ for every finite A then $\mathcal{L}_{n+3/2}(B) = \mathbb{L}_{n+1}^C(B)$.

So whenever level $n+1/2$ of the DDH coincides with the respective forbidden-pattern class, then this holds also for level $n + 3/2$ of the STH in the two-letter case. From [5,6] it follows that the prerequisite of the above implication holds for $n = 1$. So we obtain the forbidden-pattern characterization $\mathcal{L}_{5/2}(B) = \mathbb{L}_2^C(B)$ which implies the decidability of $\mathcal{L}_{5/2}(B)$.

The proof of Theorem 4 rests on a reduction that makes use of the following observation. Let $\mathcal{M}$ be a deterministic finite automaton with the input alphabet $B = \{a,b\}$ and which has the property that it is permutation-free. Then $\mathcal{M}$ cannot distinguish the words $a^r, a^{r+1}, \ldots$ with $r := |\mathcal{M}|$ and $a \in B$. So we consider in the reduction every arbitrary long block of a's (resp., block of b's) in the input only up to threshold r. This finite number of possibilities is then encoded into a larger alphabet.

2 Preliminaries

We consider finite alphabets having at least two letters. Let $B := \{a, b\}$, and let $A_r^a := \{a_1, \ldots, a_r\}$, $A_r^b := \{b_1, \ldots, b_r\}$ and $A_r := A_r^a \cup A_r^b$ for $r \geq 1$. The empty word is denoted by ε and for a fixed alphabet A the set of all words (resp., non-empty words) is denoted by A^* (resp., A^+). We consider languages as subsets of A^+. For a class $\mathcal{C}$ of languages the set of complements is denoted by $\mathrm{co}\mathcal{C} := \{ A^+ \setminus L \mid L \in \mathcal{C} \}$. For a word w denote by $|w|$ its length, and let $\alpha(w)$ be the set of letters occurring in w.

A deterministic finite automaton (DFA) $\mathcal{M}$ is given by (A, S, δ, s_0, S'), where A is its input alphabet, S the set of states, $\delta : A \times S \to S$ the total transition function, $s_0 \in S$ the starting state and $S' \subseteq S \setminus \{s_0\}$ the set of accepting states (we restrict to DFAs accepting subsets from A^+). The language accepted by $\mathcal{M}$ is denoted as $L(\mathcal{M})$. We extend transition functions to input words and denote by $|\mathcal{M}|$ the number of states of $\mathcal{M}$. Say that a state $s \in S$ has a v-loop if and only if $\delta(s, v) = s$. If the DFA $\mathcal{M}$ is fixed we write $s_1 \xrightarrow{w} s_2$ instead of '$\delta(s_1, w) = s_2$'. Moreover, we use $s_1 \xrightarrow{w} +$ and $s_1 \xrightarrow{w} -$ instead of '$\delta(s_1, w) \in S'$' and '$\delta(s_1, w) \notin S'$'. We say that a word w leads to a certain structure in a DFA (e.g. a v-loop) if and only if for all $s \in S$ the state $\delta(s, w)$ has this structure (has a v-loop).

Theorem 1 ([13,8]). *Let $\mathcal{M} = (A, S, \delta, s_0, S')$ be a minimal DFA. $L(\mathcal{M})$ is starfree if and only if there is some $m \geq 0$ such that for all $w \in A^+$ and all $p \in S$ it holds that $\delta(p, w^m) = \delta(p, w^{m+1})$.*

Minimal DFAs of this type are called permutation-free. The next theorem connects the quantifier alternation hierarchy of $\mathrm{FO}[<]$ and the STH.

Theorem 2 ([19,9]). *Let A be an alphabet, $n \geq 1$ and $L \subseteq A^+$. Then $L \in \mathcal{L}_{n-1/2}(A)$ if and only if L is $\mathrm{FO}[<]$-definable by a Σ_n-formula.*

3 Forbidden-Pattern Classes

In [6] a method for a uniform definition of hierarchies via forbidden-patterns is developed. We recall some results needed for this paper and start with the definition of hierarchies that consist of classes of iterated patterns. A so-called class of initial patterns determines the first level of such hierarchies, and using an iteration rule we obtain more complicated classes of patterns which define the higher levels.

Definition 1. *Let A be some alphabet. We define a class of initial patterns to be a subset $\mathcal{I} \subseteq A^* \times A^*$ such that for all $r \geq 1$ and $v, w \in A^*$ it holds that $(v, w) \in \mathcal{I} \implies (v, v), (v^r, w \cdot v^r) \in \mathcal{I}$. For $p = (v, w) \in \mathcal{I}$ and given states s, s_1, s_2 of some DFA $\mathcal{M}$ we say:*

- *p appears at s $\overset{\mathrm{def}}{\iff}$ s has a v-loop*
- *s_1, s_2 are connected via p (in symbols $s_1 \overset{p}{\underset{\circ\circ\circ}{\longrightarrow}} s_2$) $\overset{\mathrm{def}}{\iff}$ p appears at s_1 and at s_2, and $s_1 \xrightarrow{w} s_2$*

We come to the definition of the iteration rule that transforms a class of initial patterns into a class of more complicated (i.e., iterated) patterns.

Definition 2. *Let A be some alphabet. For every set $\mathbb{P}$ we define*

$$\mathrm{IT}(\mathbb{P}) := \{ (w_0, p_0, \ldots, w_m, p_m) \mid p_i \in \mathbb{P}, w_i \in A^+ \}.$$

Now we say what it means that we find an iterated pattern in some DFA.

Definition 3. *Let A be some alphabet. For a class of initial patterns $\mathcal{I}$ let $\mathbb{P}_0^{\mathcal{I}} := \mathcal{I}$ and $\mathbb{P}_{n+1}^{\mathcal{I}} := \mathrm{IT}(\mathbb{P}_n^{\mathcal{I}})$ for $n \geq 0$. For $p = (w_0, p_0, \ldots, w_m, p_m) \in \mathrm{IT}(\mathbb{P}_n^{\mathcal{I}})$ and given states s, s_1, s_2 of some DFA $\mathcal{M}$ we say:*

- *p appears at $s \stackrel{\text{def}}{\iff}$ there exist states $q_0, r_0, \ldots, q_m, r_m$ such that $s \stackrel{w_0}{\longrightarrow} q_0 \stackrel{p_0}{\dashrightarrow} r_0 \stackrel{w_1}{\longrightarrow} q_1 \stackrel{p_1}{\dashrightarrow} r_1 \stackrel{w_2}{\longrightarrow} \cdots \stackrel{w_m}{\longrightarrow} q_m \stackrel{p_m}{\dashrightarrow} r_m = s$*
- *s_1, s_2 are connected via p (i.e., $s_1 \stackrel{p}{\dashrightarrow} s_2$) $\stackrel{\text{def}}{\iff}$ p appears at s_1 and at s_2, there exist states $q_0, \ldots, q_m$ with $s_1 \stackrel{w_0}{\longrightarrow} q_0 \stackrel{w_1}{\longrightarrow} q_1 \stackrel{w_2}{\longrightarrow} \cdots \stackrel{w_m}{\longrightarrow} q_m = s_2$ and p_i appears at the state q_i for $i = 0, \ldots, m$*

Each of the classes of patterns $\mathbb{P}_n^{\mathcal{I}}$ induces a certain class of regular languages if we forbid all elements of $\mathbb{P}_n^{\mathcal{I}}$ in the transition graph of DFAs.

Definition 4. *Let A be some alphabet. For a DFA $\mathcal{M} = (A, S, \delta, s_0, S')$, a class of initial patterns $\mathcal{I}$ and $n \geq 0$ we say that $\mathcal{M}$ has a pattern from $\mathbb{P}_n^{\mathcal{I}}$ if and only if there exist $s_1, s_2 \in S$, $x, z \in A^*$, $p \in \mathbb{P}_n^{\mathcal{I}}$ such that $s_0 \stackrel{x}{\longrightarrow} s_1 \stackrel{z}{\longrightarrow} +$, $s_2 \stackrel{z}{\longrightarrow} -$ and $s_1 \stackrel{p}{\dashrightarrow} s_2$.*

If $\mathcal{M}$ is minimal then we do not need to require the word x above.

Definition 5. *Let A be some alphabet and $\mathcal{I}$ be a class of initial patterns. For $n \geq 0$ we define the class of languages corresponding to $\mathbb{P}_n^{\mathcal{I}}$ as*

$$\mathbb{L}_n^{\mathcal{I}}(A) := \{ \mathrm{L}(\mathcal{M}) \mid \mathcal{M} \text{ is a DFA that does not have a pattern from } \mathbb{P}_n^{\mathcal{I}} \}.$$

We call the classes $\mathbb{L}_n^{\mathcal{I}}(A)$ also *forbidden-pattern classes*. For all DFAs $\mathcal{M}_1, \mathcal{M}_2$ with $\mathrm{L}(\mathcal{M}_1) = \mathrm{L}(\mathcal{M}_2)$ it holds that $\mathcal{M}_1$ has a pattern from $\mathbb{P}_n^{\mathcal{I}}$ if and only if $\mathcal{M}_2$ has a pattern from $\mathbb{P}_n^{\mathcal{I}}$. So we can restrict ourselves to minimal DFAs in Definition 5. From now on we consider for any fixed alphabet A two particular classes of initial patterns $\mathcal{L} := \{\varepsilon\} \times A^*$ and $\mathcal{B} := A^+ \times A^+$. For their forbidden-pattern classes the following is known.

Theorem 3 ([5,6]). *Let A be an alphabet and $n \geq 0$.*

1. *$\mathbb{L}_n^{\mathcal{L}}(A)$ and $\mathbb{L}_n^{\mathcal{B}}(A)$ are starfree and decidable in nondet. logspace.*
2. *$\mathcal{L}_{n+1/2}(A) \subseteq \mathbb{L}_n^{\mathcal{L}}(A)$*
3. *$\mathcal{B}_{n+1/2}(A) \subseteq \mathbb{L}_n^{\mathcal{B}}(A)$*
4. *$\mathcal{B}_{3/2}(A) = \mathbb{L}_1^{\mathcal{B}}(A)$*

4 $\mathcal{L}_{5/2}(B)$ Is Decidable for $B = \{a, b\}$

In this section we show that $\mathbb{L}_2^{\mathcal{L}}(B) = \mathcal{L}_{5/2}(B)$ for the alphabet $B = \{a, b\}$. This implies in particular the decidability of $\mathcal{L}_{5/2}(B)$ and has the announced consequences in first-order logic. In fact, the main theorem shows something more general: If $\mathcal{B}_{n+1/2}(A) =$

$\mathbb{L}_n^B(A)$ for some $n \geq 1$ and all alphabets A then $\mathcal{L}_{n+3/2}(B) = \mathbb{L}_{n+1}^{\mathcal{C}}(B)$ for $B = \{a,b\}$ (cf. Theorem 4).

Assume that $\mathcal{B}_{n+1/2}(A) = \mathbb{L}_n^B(A)$ for some $n \geq 1$. Then our approach is as follows. First of all, by Theorem 3, it suffices to show $\mathbb{L}_{n+1}^{\mathcal{C}}(B) \subseteq \mathcal{L}_{n+3/2}(B)$. So we start with some $L \in \mathbb{L}_{n+1}^{\mathcal{C}}(B)$ and introduce for a certain $r \geq 1$ an operation f_r that transforms $L \subseteq B^+$ to a language $f_r(L)$ over the larger alphabet A_r. Both f_r and its inverse f_r^{-1} can be carried out effectively on a given regular language. Using structural arguments of DFAs we show the benefit of this transformation: On one hand we achieve a decrease of the pattern complexity, i.e., $f_r(L) \in \mathbb{L}_n^B(A_r)$ (cf. Lemma 1). Together with our assumption it follows that $f_r(L) \in \mathcal{B}_{n+1/2}(A_r)$. On the other hand we deal with regular expressions (in terms of words, Boolean operations and concatenation). The expression witnessing $f_r(L) \in \mathcal{B}_{n+1/2}(A_r)$ allows the construction of an expression for $f_r^{-1}(f_r(L)) = L$ showing that $L \in \mathcal{L}_{n+3/2}(B)$ (cf. Lemma 2). This shows $\mathcal{L}_{n+3/2}(B) = \mathbb{L}_{n+1}^{\mathcal{C}}(B)$ under the assumption that $\mathcal{B}_{n+1/2}(A) = \mathbb{L}_n^B(A)$ for all alphabets A. In particular this can be applied to the case $n = 1$. There we have $\mathcal{B}_{3/2}(A_r) = \mathbb{L}_1^B(A_r)$ by Theorem 3.4 and it follows that $\mathcal{L}_{5/2}(B) = \mathbb{L}_2^{\mathcal{C}}(B)$.

The idea of the transformation f_r is straightforward: If a permutation-free DFA $\mathcal{M}$ with $r := |\mathcal{M}|$ reads more than r times the same letter from the input, it remains in the same state. Hence, for blocks that consist of the same letters, the automaton can only determine their lengths up to the threshold r. We prune large blocks in arbitrary words from B^+ to length r and encode these pruned blocks using the larger alphabet A_r.

Note that every word $w \in B^+$ can be decomposed into maximal blocks of equal letters (i.e., $w = w_1 w_2 \cdots w_k$ for some $k \geq 1$ and factors w_i of maximal length such that $w_i \in \{a\}^+ \cup \{b\}^+$). Call this the B-factorization of w and observe that it is unique due to the maximality condition. We say that $c \in A_r$ has type a if $c \in A_r^a$, and it has type b if $c \in A_r^b$.

The B-factorization induces for $r \geq 1$ the mapping $f_r : B^+ \longrightarrow (A_r)^+$ which connects both alphabets, and which allows the transformation of languages over B into languages over A_r. It translates every block of the B-factorization of a given word to a single letter from A_r, where the index of the letter corresponds to the block size up to threshold r.

Definition 6. *Let $r \geq 1$, $w \in B^+$ and let $w = w_1 w_2 \cdots w_k$ for some $k \geq 1$ be the B-factorization of w. Then $f_r(w) := c_1 c_2 \cdots c_k \in (A_r)^+$ with*

$$c_i := \begin{cases} a_{\min\{l,r\}} & : \quad \text{if } w_i = a^l \text{ for some } l \geq 1 \\ b_{\min\{l,r\}} & : \quad \text{if } w_i = b^l \text{ for some } l \geq 1. \end{cases}$$

Moreover, for $L \subseteq B^+$ and $L' \subseteq (A_r)^+$ we define $f_r(L) := \bigcup_{w \in L} \{f_r(w)\}$ and $f_r^{-1}(L') := \{w \in B^+ \mid f_r(w) \in L'\}$.

However, not all words from $(A_r)^+$ can appear in the range of f_r. The maximality condition in B-factorizations ensures that the types of letters in $f_r(w)$ alternate between a and b. We call these words well-formed, and let $WF_r := f_r(B^+)$ be the set of well-formed words of $(A_r)^+$.

Hence f_r is a surjective function from B^+ onto WF_r. Note that $\mu \in WF_r$ if and only if the letters in μ alternate between type a and b. Moreover, every non-empty factor of a well-formed word is well-formed. The following proposition contains some basic facts of f_r.

Proposition 1. *Let* $\mathcal{M} = (B, S, \delta, s_0, S')$ *be a permutation-free DFA,* $r := |\mathcal{M}|$ *and* $v, w \in B^+$ *such that* $w = w_1 \cdots w_k$ *is the B-factorization of w.*

1. $f_r(w)f_r(v) \in WF_r \iff f_r(w)f_r(v) = f_r(wv)$
2. $f_r(w) = f_r(w_1)f_r(w_2) \cdots f_r(w_k)$ *and* $f_r(w_i) \in A_r$ *for all* i.
3. *If* $f_r(w) = f_r(v)$ *then* $\delta(s, w) = \delta(s, v)$ *for all* $s \in S$.
4. *If* $f_r(w) = c_1 \cdots c_k$ *with* $c_i \in A_r$ *then* $f_r(w_j \cdots w_{j'}) = c_j \cdots c_{j'}$ *for* $j \leq j'$.

The nice thing about f_r is that it forges links between $\mathbb{L}^{\mathcal{L}}_{n+1}(B)$ and $\mathbb{L}^{\mathcal{B}}_n(A_r)$ on one hand, and between $\mathcal{B}_{n+1/2}(A_r)$ and $\mathcal{L}_{n+3/2}(B)$ on the other hand. We make this precise in the following lemmas.

Lemma 1. *Let* $\mathcal{M} = (B, S, \delta, s_0, S')$ *be a permutation-free DFA, let* $n \geq 1$ *and* $r := |\mathcal{M}|$. *If* $L(\mathcal{M}) \in \mathbb{L}^{\mathcal{L}}_{n+1}(B)$ *then* $f_r(L(\mathcal{M})) \in \mathbb{L}^{\mathcal{B}}_n(A_r)$.

Lemma 2. *Let* $B = \{a, b\}$ *and* $n, r \geq 1$. *For all* $L \in \mathcal{B}_{n+1/2}(A_r)$ *with* $L \subseteq WF_r$ *it holds that* $f_r^{-1}(L) \in \mathcal{L}_{n+3/2}(B)$.

The proofs of both lemmas are sketched in the subsections 4.1 and 4.2. With these lemmas we are able to prove the main theorem.

Theorem 4. *Let* $n \geq 1$ *and* $B = \{a, b\}$. *If* $\mathcal{B}_{n+1/2}(A) = \mathbb{L}^{\mathcal{B}}_n(A)$ *for every finite alphabet* A, *then* $\mathcal{L}_{n+3/2}(B) = \mathbb{L}^{\mathcal{L}}_{n+1}(B)$.

Proof. By Theorem 3 it suffices to show the inclusion $\mathbb{L}^{\mathcal{L}}_{n+1}(B) \subseteq \mathcal{L}_{n+3/2}(B)$ under the assumption that $\mathcal{B}_{n+1/2}(A) = \mathbb{L}^{\mathcal{B}}_n(A)$ for arbitrary alphabets A. So let $L \in \mathbb{L}^{\mathcal{L}}_{n+1}(B)$. By Theorem 3 and Theorem 1 there is some permutation-free DFA $\mathcal{M}$ with $L = L(\mathcal{M}) \subseteq B^+$. With $r := |\mathcal{M}|$ it follows from Lemma 1 and our assumption that $f_r(L(\mathcal{M})) \in \mathbb{L}^{\mathcal{B}}_n(A_r) = \mathcal{B}_{n+1/2}(A_r)$. Since $f_r(L(\mathcal{M})) \subseteq WF_r$ we obtain from Lemma 2 that $f_r^{-1}(f_r(L(\mathcal{M}))) \in \mathcal{L}_{n+3/2}(B)$.

It holds that $L(\mathcal{M}) \subseteq f_r^{-1}(f_r(L(\mathcal{M})))$ and we want to argue that also the reverse inclusion holds. So let $w \in f_r^{-1}(f_r(L(\mathcal{M})))$ and hence $f_r(w) \in f_r(L(\mathcal{M}))$. It follows that there is some $v \in L(\mathcal{M})$ with $f_r(v) = f_r(w)$. Proposition 1.3 shows that also $w \in L(\mathcal{M})$. This shows $L = L(\mathcal{M}) = f_r^{-1}(f_r(L(\mathcal{M}))) \in \mathcal{L}_{n+3/2}(B)$. □

With Theorem 3.4 this implies the following.

Corollary 1. *For* $B = \{a, b\}$ *it holds that* $\mathcal{L}_{5/2}(B) = \mathbb{L}^{\mathcal{L}}_2(B)$.

An easy rewriting of the patterns from $\mathbb{P}^{\mathcal{L}}_2$ shows that a DFA $\mathcal{M}$ has a pattern from $\mathbb{P}^{\mathcal{L}}_2$ if and only if we find the pattern given in Figure 1 in its transition graph. Therefore, the pattern in Figure 1 characterizes $\mathcal{L}_{5/2}(B)$. Note that this pattern is similar to that of the forbidden-pattern characterization for $\mathcal{B}_{3/2}(A)$ but here we have the additional conditions $\alpha(b_i) = \alpha(l_i)$. With Theorem 3.1 we get the following.

Corollary 2. *The membership problem for* $\mathcal{L}_{5/2}(B)$ *with* $B = \{a, b\}$ *is decidable in nondeterministic logspace.*

Due to Theorem 2 we draw the connection to the first-order logic FO[<].

Corollary 3. *Given a regular language* $L \subseteq \{0, 1\}^+$ *it is decidable whether* L *is definable by a* Σ_3-*formula of the logic* FO[<].

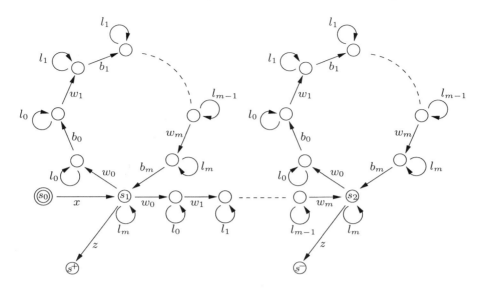

Fig. 1. Forbidden-pattern for $\mathcal{L}_{5/2}(B)$ with initial state s_0, accepting state s^+, rejecting state s^- and words $x, z \in A^*$, $b_i, l_i, w_i \in A^+$ with $\alpha(b_i) = \alpha(l_i)$.

4.1 Proof of Lemma 1: A Transformation of Patterns

The proof of Lemma 1 links together the transformation f_r and the non-existence of certain patterns in DFAs. More precisely, if $\mathcal{M}$ does not have a pattern from $\mathbb{P}_{n+1}^{\mathcal{L}}$ then no DFA accepting $f_r(L(\mathcal{M})) \subseteq (A_r)^+$ has a pattern from $\mathbb{P}_n^{\mathcal{B}}$. Note that we reduce here the index of the pattern from $n+1$ to n at the cost of an increased alphabet size. To put it another way, f_r shifts some of the structural complexity of $\mathcal{M}$ to the increased alphabet while we maintain the essential properties of the accepted language.

In the proof we start with the given permutation-free DFA $\mathcal{M}$ and we define another DFA $\tilde{\mathcal{M}}$ with $L(\tilde{\mathcal{M}}) = f_r(L(\mathcal{M}))$. Then we show that if $\tilde{\mathcal{M}}$ had a pattern from $\mathbb{P}_n^{\mathcal{B}}$ then $\mathcal{M}$ would have even a pattern from $\mathbb{P}_{n+1}^{\mathcal{L}}$ (which is a contradiction to $L(\mathcal{M}) \in \mathbb{L}_{n+1}^{\mathcal{L}}(B)$). The fact that $|B| = 2$ and that only well-formed words can appear in the pattern in $\tilde{\mathcal{M}}$ are the crucial points in our argumentation. We carry out three steps where we use the previous notations.

Step 1: Construction of $\tilde{\mathcal{M}}$. The automaton $\tilde{\mathcal{M}}$ we have in mind has the input alphabet A_r, the set of states $(S \times \{a,b\}) \cup \{(s_0, \varepsilon)\} \cup \{\bot\}$ and simulates $\mathcal{M}$ in a straightforward way: If $\tilde{\mathcal{M}}$ reads for instance a_j for $1 \leq j \leq r$ from the input then it behaves like $\mathcal{M}$ on input a^j. Since we want that $\tilde{\mathcal{M}}$ rejects whenever the input is not well-formed, we store in the second component of the states the letter type a or b of the most recent input letter. Since at the beginning of the input there is no previous input letter we introduce a new starting state (s_0, ε) which has only outgoing edges and which behaves like s_0. With help of the second component of the states we can detect non well-formed inputs (i.e., inputs where the letter types do not alternate) and we can lead them to the rejecting sink state $\bot$. We refrain from giving a formal definition of $\tilde{\mathcal{M}}$, here is an example instead. Suppose $\tilde{\mathcal{M}}$ is in state (s, a) where $s \in S$ is a state of $\mathcal{M}$. This means that the last input letter from A_r was of type a. If the next input letter is for example b_j then $\tilde{\mathcal{M}}$ moves to the state $(\delta(s, b^j), b)$. On the other hand, if the next input letter is of type a then $\tilde{\mathcal{M}}$ moves to its sink state $\bot$. Accepting states of $\tilde{\mathcal{M}}$ are those which have an accepting state of $\mathcal{M}$ in their first component. In particular, $\bot$ is rejecting.

The following two claims state that $\mathcal{M}$ and $\tilde{\mathcal{M}}$ have a similar behavior, i.e., we find a path in the transition graph of $\mathcal{M}$ also in the transition graph of $\tilde{\mathcal{M}}$ and vice versa.

Claim 1. Let $w \in B^+$ and $s \xrightarrow{w} s'$ in $\mathcal{M}$ for some $s, s' \in S$. If $t \in \{a, b, \varepsilon\}$ such that (s, t) is a state of $\tilde{\mathcal{M}}$ and t differs from the first letter of w then it holds that $(s, t) \xrightarrow{f_r(w)} (s', t_l)$ in $\tilde{\mathcal{M}}$ where t_l is the last letter of w.

Claim 2. Suppose $(s, t) \xrightarrow{\mu} (s', t')$ in $\tilde{\mathcal{M}}$ for some $\mu \in (A_r)^+$ and states $(s, t), (s', t')$ of $\tilde{\mathcal{M}}$. Then $\mu \in WF_r$ and $s \xrightarrow{w} s'$ in $\mathcal{M}$ for all $w \in f_r^{-1}(\{\mu\})$.

The proof of the first claim is an easy induction on the length of the B-factorization of w. For the second claim note that μ has to be well-formed since it does not lead to $\bot$ in $\tilde{\mathcal{M}}$. The second part follows from the construction of $\tilde{\mathcal{M}}$ together with Proposition 1.3. Once we have this structural correspondence between $\mathcal{M}$ and $\tilde{\mathcal{M}}$ it is easy to see that $L(\tilde{\mathcal{M}}) = f_r(L(\mathcal{M}))$.

Step 2: Pattern transformation. We assume in this second step that two states $(s, t), (s', t')$ of $\tilde{\mathcal{M}}$ are connected via some pattern $\tilde{p} \in \mathbb{P}_n^B$ for $n \geq 0$ such that all states involved in this pattern are different from $\bot$. We show that this implies the existence of some pattern $p \in \mathbb{P}_{n+1}^C$ such that s, s' are connected in $\mathcal{M}$ via p.

Claim 3. Let $n \geq 0$, $\tilde{p} \in \mathbb{P}_n^B$ and let $(s, t), (s', t')$ be states in $\tilde{\mathcal{M}}$ such that $(s, t) \xrightarrow{\tilde{p}} (s', t')$ and all states involved in this pattern are different from $\bot$. Then there is some $p \in \mathbb{P}_{n+1}^C$ such that $s \xrightarrow{p} s'$ in $\mathcal{M}$.

This claim is shown as follows. With help of Claim 2 we immediately find a pattern $p' \in \mathbb{P}_n^B$ such that $s \xrightarrow{p'} s'$ in $\mathcal{M}$: If we apply f_r^{-1} to the labels of the paths involved in $\tilde{p}$ in $\tilde{\mathcal{M}}$ we obtain a subgraph of exactly the same structure between s and s' (since all states $\neq \bot$). This gives rise to p' in $\mathcal{M}$. However, we still have to argue that this subgraph is in fact a pattern from $\mathbb{P}_{n+1}^C$ in $\mathcal{M}$. Due to the inductive definition of patterns it suffices to show that the inner-most patterns from $\mathbb{P}_0^B$ that are part of p' are in fact patterns from $\mathbb{P}_1^C$. This is the crucial point where it becomes important that B has only two letters.

The argument is as follows. Fix some $\tilde{p}_0 = (\mu, \nu) \in \mathbb{P}_0^B$ involved in $\tilde{p}$ in $\tilde{\mathcal{M}}$ and its counterpart $p_0' = (v, w) \in \mathbb{P}_0^B$ that we find in $\mathcal{M}$ with help of f_r^{-1} and Claim 2. Since μ is a non-empty loop at some state $\neq \bot$ in $\tilde{\mathcal{M}}$ it must contain letters of *both* types. It follows from the definition of f_r^{-1} that both letters $a, b \in B$ appear somewhere in v. An easy observation makes clear that any $p_0' = (v, w) \in \mathbb{P}_0^B$ with $\alpha(w) \subseteq \alpha(v)$ can be rewritten as some pattern from $\mathbb{P}_1^{\mathcal{L}}$. This applies in particular to our case since there are *all* letters from B somewhere in v.

Step 3: From $\tilde{\mathcal{M}}$ to $\mathcal{M}$. Finally we prove the following claim.

Claim 4. *Let $n \geq 1$. If $\tilde{\mathcal{M}}$ has a pattern from $\mathbb{P}_n^{\mathcal{B}}$ then $\mathcal{M}$ has a pattern from $\mathbb{P}_{n+1}^{\mathcal{L}}$.*

In order to apply Claim 3 we need to observe that if $\tilde{\mathcal{M}}$ has a pattern from $\mathbb{P}_n^{\mathcal{B}}$ then no state involved in this pattern can be $\bot$.

Let $\tilde{s}_1, \tilde{s}_2$ be states in $\tilde{\mathcal{M}}$ and let $\tilde{p} \in \mathbb{P}_n^{\mathcal{B}}$ for $n \geq 1$ such that $\tilde{s}_1 \xrightarrow{\tilde{p}} \tilde{s}_2$. An induction over n shows the following fact (by comparing the alternating letter types in the paths in $\tilde{p}$): If $\tilde{s}_1 \neq \bot$ then $\tilde{s}_1 = (s_1, t)$ and $\tilde{s}_2 = (s_2, t)$ for suitable states s_1, s_2 from $\mathcal{M}$ and $t \in \{a, b\}$.

Assume that $\tilde{\mathcal{M}}$ has the pattern $\tilde{p} \in \mathbb{P}_n^{\mathcal{B}}$ witnessed by the states $\tilde{s}_1$ and $\tilde{s}_2$. Since $\tilde{s}_1 \longrightarrow +$ by definition and because $\bot$ is a rejecting sink in $\tilde{\mathcal{M}}$ we already know that $\tilde{s}_1 \neq \bot$. From the fact above it follows that also $\tilde{s}_2 \neq \bot$. Since the state $\tilde{s}_2$ is reachable from all states involved in $\tilde{p}$ we obtain that all these states are different from $\bot$. Finally, Claim 4 follows from Definition 4, Claim 2 and Claim 3.

We complete the proof of Lemma 1 as follows. Assume $f_r(L(\mathcal{M})) \notin \mathbb{L}_n^{\mathcal{B}}(A_r)$. Since $L(\tilde{\mathcal{M}}) = f_r(L(\mathcal{M}))$ by Step 1 this implies that $\tilde{\mathcal{M}}$ has a pattern from $\mathbb{P}_n^{\mathcal{B}}$. From Claim 4 we obtain that $\mathcal{M}$ has a pattern from $\mathbb{P}_{n+1}^{\mathcal{L}}$ and therefore $L(\mathcal{M}) \notin \mathbb{L}_{n+1}^{\mathcal{L}}(B)$. This contradicts the assumption of Lemma 1.

4.2 Proof of Lemma 2: A Transformation of Expressions

To prepare the proof of Lemma 2 we observe the following proposition.

Proposition 2. *For $r \geq 1$ and $B = \{a, b\}$ the following holds.*

1. *If $\mu \in WF_r$ then $f_r^{-1}(\mu) \in \mathcal{L}_{3/2}(B)$.*
2. *$f_r^{-1}(L_1 L_2) = f_r^{-1}(L_1) f_r^{-1}(L_2)$ for all $L_1, L_2 \subseteq (A_r)^+$ with $L_1 L_2 \subseteq WF_r$.*
3. *$WF_r \in \text{co}\mathcal{B}_{1/2}(A_r)$.*

We turn to the proof of Lemma 2. Let $B = \{a, b\}$ and $r \geq 1$.

Claim 5. *$f_r^{-1}(L \cap WF_r) \in \mathcal{L}_{3/2}(B)$ for every $L \in \mathcal{B}_{1/2}(A_r)$.*

We prove this as follows. W.l.o.g. we may assume that $L \cap WF_r$ is a finite union of languages $L' \cap WF_r$ with $L' = u_0 \cdot c_1(A_r)^+ d_1 \cdot u_1 \cdots c_m(A_r)^+ d_m \cdot u_m$ such that $c_i \in A_r^a$, $d_i \in A_r^b$, $u_i \in (A_r)^+$ and $u_0 c_1 d_1 u_1 \cdots c_m d_m u_m \in WF_r$. To see this we use the facts that on one hand $(A_r)^+$ can be written as the union of $c \cup c(A_r)^+$ for all $c \in A_r$, and on the other hand that $L' \cap WF_r = \emptyset$ is implied if $u_{i-1} c_i \notin WF_r$ or $d_i u_i \notin WF_r$ for some $i \geq 1$. So it suffices to show $f_r^{-1}(L' \cap WF_r) \in \mathcal{L}_{3/2}(B)$. Note that $L' \cap WF_r = u_0 c_1 D d_1 u_1 \cdots c_m D d_m u_m$ with $D := A_r^b(A_r)^* \cap (A_r)^* A_r^a \cap WF_r$.

Now by Proposition 2, it suffices to show $f_r^{-1}(D) \in \mathcal{L}_{3/2}(B)$. This is easy to see, since $f_r^{-1}(D) = \{b\}^+\{a,b\}^*\{a\}^+$. This proves Claim 5.

Let $L \in \text{co}\mathcal{B}_{1/2}(A_r)$ with $L \subseteq WF_r$. Then $L = WF_r \setminus L' = WF_r \setminus (L' \cap WF_r)$ for some $L' \in \mathcal{B}_{1/2}(A_r)$. Note that $f_r^{-1}(WF_r \setminus L'') = B^+ \setminus f_r^{-1}(L'')$ for every $L'' \subseteq WF_r$. So with $L'' := L' \cap WF_r$ and Claim 5 we obtain:

Claim 6. $f_r^{-1}(L) \in \text{co}\mathcal{L}_{3/2}(B)$ for every $L \in \text{co}\mathcal{B}_{1/2}(A_r)$ with $L \subseteq WF_r$.

With Claim 6 we can complete the proof of Lemma 2 by induction on n. For the induction base let $n = 1$ and $L \in \mathcal{B}_{3/2}(A_r)$ with $L \subseteq WF_r$. Since $\mathcal{B}_{m+1/2}(A) = \text{Pol}(\text{co}\mathcal{B}_{m-1/2}(A))$ for $m \geq 1$ it suffices to show $f_r^{-1}(L_0 \cdots L_m) \in \mathcal{L}_{5/2}(B)$ for all $L_i \in \text{co}\mathcal{B}_{1/2}(A_r)$ with $L_0 \cdots L_m \subseteq WF_r$. By Proposition 2.2, we have $f_r^{-1}(L_0 \cdots L_m) = f_r^{-1}(L_0) \cdots f_r^{-1}(L_m)$. >From Claim 6 we get $f_r^{-1}(L_i) \in \text{co}\mathcal{L}_{3/2}(B)$. This shows the induction base.

For the induction step we assume that the lemma has been shown for some $n \geq 1$ and we want to show it for $n+1$. For this let $L \in \mathcal{B}_{n+3/2}(A_r)$ with $L \subseteq WF_r$. We proceed analogously to the induction base and see that it suffices to show $f_r^{-1}(L_i) \in \text{co}\mathcal{L}_{n+3/2}(B)$ for all $L_i \in \text{co}\mathcal{B}_{n+1/2}(A_r)$ with $L_i \subseteq WF_r$. By Proposition 2.3, we may assume that $L_i = WF_r \setminus L_i'$ for some $L_i' \in \mathcal{B}_{n+1/2}(A_r)$ with $L_i' \subseteq WF_r$ (note that $\mathcal{B}_{n+1/2}(A_r)$ is closed under intersection [11]). It follows that $f_r^{-1}(L_i) = f_r^{-1}(WF_r \setminus L_i') = B^+ \setminus f_r^{-1}(L_i')$. Together with the induction hypothesis we obtain $f_r^{-1}(L_i) \in \text{co}\mathcal{L}_{n+3/2}(B)$. This completes the induction step.

References

1. M. Arfi. Polynomial operations on rational languages. In *Proceedings 4th Symposium on Theoretical Aspects of Computer Science*, volume 247 of *Lecture Notes in Computer Science*, pages 198–206. Springer-Verlag, 1987.
2. M. Arfi. Opérations polynomiales et hiérarchies de concaténation. *Theoretical Computer Science*, 91:71–84, 1991.
3. J. A. Brzozowski and R. Knast. The dot-depth hierarchy of star-free languages is infinite. *Journal of Computer and System Sciences*, 16:37–55, 1978.
4. R. S. Cohen and J. A. Brzozowski. Dot-depth of star-free events. *Journal of Computer and System Sciences*, 5:1–16, 1971.
5. C. Glaßer and H. Schmitz. Languages of dot-depth 3/2. In *Proceedings 17th Symposium on Theoretical Aspects of Computer Science*, volume 1770 of *Lecture Notes in Computer Science*, pages 555–566. Springer Verlag, 2000.
6. C. Glaßer and H. Schmitz. Decidable hierarchies of starfree languages. In *Proceedings 20th Conference on the Foundations of Software Technology and Theoretical Computer Science*, volume 1974 of *Lecture Notes in Computer Science*, pages 503–515. Springer Verlag, 2000.
7. R. Knast. A semigroup characterization of dot-depth one languages. *RAIRO Inform. Théor.*, 17:321–330, 1983.
8. R. McNaughton and S. Papert. *Counterfree Automata*. MIT Press, Cambridge, 1971.
9. D. Perrin and J. E. Pin. First-order logic and star-free sets. *Journal of Computer and System Sciences*, 32:393–406, 1986.

10. J. E. Pin. Syntactic semigroups. In G. Rozenberg and A. Salomaa, editors, *Handbook of formal languages*, volume I, pages 679–746. Springer, 1996.
11. J. E. Pin and P. Weil. Polynomial closure and unambiguous product. *Theory of computing systems*, 30:383–422, 1997.
12. H. Schmitz. *The Forbidden-Pattern Approach to Concatenation Hierarchies*. PhD thesis, Fakultät für Mathematik und Informatik, Universität Würzburg, 2001.
13. M. P. Schützenberger. On finite monoids having only trivial subgroups. *Information and Control*, 8:190–194, 1965.
14. I. Simon. Piecewise testable events. In *Proceedings 2nd GI Conference*, volume 33 of *Lecture Notes in Computer Science*, pages 214–222. Springer-Verlag, 1975.
15. H. Straubing. A generalization of the Schützenberger product of finite monoids. *Theoretical Computer Science*, 13:137–150, 1981.
16. H. Straubing. Finite semigroups varieties of the form V * D. *J. Pure Appl. Algebra*, 36:53–94, 1985.
17. H. Straubing. Semigroups and languages of dot-depth two. *Theoretical Computer Science*, 58:361–378, 1988.
18. D. Thérien. Classification of finite monoids: the language approach. *Theoretical Computer Science*, 14:195–208, 1981.
19. W. Thomas. Classifying regular events in symbolic logic. *Journal of Computer and System Sciences*, 25:360–376, 1982.
20. W. Thomas. Languages, automata, and logic. Technical Report 9607, Institut für Informatik und praktische Mathematik, Universität Kiel, 1996.

On the Power of Randomized Pushdown Automata

Juraj Hromkovič[1] and Georg Schnitger[2]

[1] Lehrstuhl für Informatik I, RWTH Aachen,
Ahornstraße 55, 52074 Aachen, Germany
[2] Institut für Informatik, Johann Wolfgang Goethe-Universität,
60054 Frankfurt am Main, Germany

Abstract. Although randomization is now a standard tool for the design of efficient algorithms or for building simpler systems, we are far from fully understanding the power of randomized computing. Hence it is advisable to study randomization for restricted models of computation. We follow this approach by investigating the power of randomization for pushdown automata.
Our main results are as follows. First we show that deterministic pushdown automata are weaker than Las Vegas pushdown automata, which in turn are weaker than one-sided-error pushdown automata. Finally one-sided-error pushdown automata are weaker than (nondeterministic) pushdown automata.
In contrast to many other fundamental models of computing there are no known methods of decreasing error probabilities. We show that such methods do not exist by constructing languages which are recognizable by one-sided-error pushdown automata with error probability $\frac{1}{2}$, but not by one-sided-error pushdown automata with error probability $p < \frac{1}{2}$. On the other hand we construct languages which are not deterministic context-free (resp. not context-free) but are recognizable with arbitrarily small error by one-sided-error (resp. bounded-error) pushdown automata.

1 Introduction

A computing problems is tractable in practice, if it can be solved by a randomized polynomial-time algorithm. Thus a comparative study of the computational power of deterministic, randomized and nondeterministic computations is one of the central tasks of complexity and algorithm theory. Recent research has focused on a comparison of different modes of randomization with determinism and nondeterminism for restricted models of computation.

Whereas a Las Vegas algorithms never errs but is allowed to output the answer "I don't know" (resp. ?) with some fixed probability $p < 1$, a one-sided-error algorithm may err with some fixed probability $p < 1$ when rejecting; finally a bounded-error algorithm may err with error probablities bounded away from $\frac{1}{2}$ when rejecting as well as when accepting.

There are only few results comparing Las Vegas and determinism and in many cases determinism and Las Vegas are polynomially related: this remark applies to the combinational complexity of Boolean circuits, to communication complexity [MS82] and the size of finite automata [HS01] as well as to the size of OBDDs [ĎHRS97,HS01]). For the time complexity of CREW PRAMs [DKR94] and for one-way communication complexity [ĎHRS97] there is even a linear relation between Las Vegas and determinism. On the other hand there is an exponential gap between Las Vegas and determinism for the size of read-once branching programs [Sa99] and there is an at least superpolynomial gap for the size of k-OBDDs [HS00]. Moreover a strong separation holds for two-dimensional finite automata [ĎHI00], since two-dimensional Las Vegas automata recognize languages which are not recognizable by two-dimensional deterministic automata.

An exponential gap between Las Vegas and one-sided error as well as between one-sided error and nondeterminism seems to be typical (see [ĎHRS97] for one-way finite automata, [Hr97,Hr00,KN97,ĎHRS97] for two-way and one-way communication complexity and [Sa97] for OBDDs.) One conjectures that there is an exponential gap between Las Vegas and nondeterminism for the size of two-way finite automata too, but up till now only a quadratic gap was proved [HS99]. On the other hand for space complexity for Turing machines, there is a linear relation between nondeterminism and Las Vegas for space classes above $\log_2 n$ [Gi77,MS99].

In this paper we focus on the study of randomization for pushdown automata. We assume that the reader is familiar with the model of pushdown automata and the fundamentals of context-free languages. We denote by CF the family of context-free languages and by DCF the family of deterministic context-free languages (i.e., DCF is the class of languages recognizable by deterministic pushdown automata). A transition of a pushdown automaton is called an ϵ-transition if no symbol of the input tape is read in this transition.

Our main contributions are as follows.

(i) A strong separation between determinism, Las Vegas randomization, one-sided-error randomization and nondeterminism for pushdown automata and counter automata.
(ii) We show that in general error probabilities cannot be reduced arbitrarily. However we also show that one-sided-error pushdown automata can recognize languages with arbitrarily small error which are not deterministic context-free and bounded-error pushdown automata can recognize non-context-free languages with arbitrarily small error.

In contrast to the intensive investigation of different versions of randomized finite automata, there is little known about randomized pushdown automata. In [AFK97] it is shown that there is no difference between determinism, nondeterminism and bounded-error randomness for pushdown automata recognizing tally languages. Further results are known for unbounded-error randomization [MO94], but these results are not applicable to our bounded-error setting.

A **Las Vegas pushdown automaton** A can be defined as follows. The states of A are divided into three pairwise disjoint sets of accepting, rejecting and neutral states. A can be viewed as a nondeterministic pushdown automaton with a probability distribution assigned to every nondeterministic branching. A computation of A is a sequence $C_1, \ldots, C_m$ of configurations of A such that A can move from C_i to C_{i+1} in one step (transition) for all $i = 1, \ldots, m-1$. The probability of the computation $C = C_1, \ldots, C_m$ is the product of the probabilities of transitions executed in C. An **accepting (resp. rejecting) computation of A on an input** w is a computation in which w

is completely read and the state of the last configuration is an accepting (resp. rejecting) state. A **neutral computation** C of A on an input w is a computation of A in which w is completely read, all states occurring in C after reading the last symbol of w are neutral states and there is no possibility to reach any accepting or rejecting state by ϵ-transitions only.

We demand for any input w, that both accepting and rejecting computations on w may not exist. Thus we may assume that if A reaches an accepting (resp. rejecting) state p, then all states reachable from p by ϵ-transitions are also accepting (resp. rejecting) states.

When calculating the probability of A to accept (resp. reject) an input w, we consider only the set of all accepting (resp. rejecting) computations $C_1, \ldots, C_m$, where C_m is the first accepting (resp. rejecting) configuration after w was read (i.e., either C_{m-1} contains a neutral state or in the move from C_{m-1} to C_m the last symbol of w was read). As usual, the probability to accept (resp. reject) is the sum of the probabilities of all accepting (resp. rejecting) computations of the above described set. To determine the probability of the "I do not know" answer for an input w, one considers the following subset S_{neutral} of neutral computations. If $C \in S_{\text{neutral}}(w)$, then using ϵ-transitions from the last configuration of C one can reach neutral states only. Moreover no prefix of C is a neutral computation on w.

Note, that A may have infinite computations on an input. However we construct only "nice" Las Vegas pushdown automata without any infinite computations.

We say that a Las Vegas pushdown automaton A accepts a language $L(A)$ with probability at least $1 - \epsilon$, $0 \leq \epsilon < 1$ if A never gives a wrong answer and if the probability of the "I do not know" answer is bounded by ϵ for every input.

$LVCF_\epsilon$ denotes the set of languages accepted by Las Vegas pushdown automata with probability at least $1 - \epsilon$. We set

$$LVCF = \bigcup_{0 < \epsilon < 1} LVCF_\epsilon \quad \text{and} \quad LVCF^* = \bigcap_{0 < \epsilon < 1} LVCF_\epsilon,$$

where we collect in LVCF (resp. LVCF*) all languages recognizable by Las Vegas pda's where the "I don't know" answer has arbitrarily large (resp. small) probability.

When considering one-sided error randomization we again assume that there is no ϵ-move from an accepting state to a rejecting state. In contrast to Las Vegas we do not have neutral states. We say that a randomized pda A is a one-sided-error (Monte Carlo) pushdown automaton that accepts a language $L(A)$ with error probability at most ϵ, if

(i) for every $w \in L(A)$, $\Pr(A \text{ accepts } w) \geq 1 - \epsilon$, and
(ii) for every $w \notin L(A)$, $\Pr(A \text{ rejects } w) = 1$.

We define RandomCF$_\epsilon$ to be the set of languages accepted by Las Vegas pushdown automata with error probability at most ϵ and introduce the classes

$$\text{RandomCF} = \bigcup_{0 < \epsilon < 1} \text{RandomCF}_\epsilon, \quad \text{RandomCF}^* = \bigcap_{0 < \epsilon < 1} \text{RandomCF}_\epsilon.$$

We say that a randomized pda A is a bounded-error pda that accepts $L(A)$ with error probability at most ϵ, if

(i) for every $w \in L(A)$, $\Pr(A \text{ rejects } w) \leq \epsilon$, and
(ii) for every $w \notin L(A)$, $\Pr(A \text{ accepts } w) \leq \epsilon$.

The set of languages accepted by bounded-error pda's with error probability at most ϵ will be denoted by BPCF_ϵ and we also introduce

$$\text{BPCF} = \bigcup_{0<\epsilon<1/2} \text{BPCF}_\epsilon, \quad \text{BPCF}^* = \bigcap_{0<\epsilon<1/2} \text{BPCF}_\epsilon.$$

Bounded-error pda's are very powerful. It's not hard to show that BPCF is closed under complementation, under finite union and thus under finite intersection. Therefore BPCF contains languages outside of CF, since DCF is contained in BPCF. We will show that also BPCF* contains languages outside of CF and thus there are languages outside of CF which are recognizable with arbitrarily small error. Moreover, we obviously have

$$\text{DCF} \subseteq \text{LVCF} \subseteq \text{RandomCF} \subseteq \text{CF}.$$

Our first result shows in particular, that all above inclusions are proper.

Theorem 1.

(a) *DCF $\subseteq$ LVCF $\subseteq$ RandomCF $\subseteq$ CF and all inclusions are proper.*
(b) *LVCF* is a proper subset of LVCF and RandomCF* is a proper subset of RandomCF. Moreover RandomCF* and LVCF are incomparable: in particular, there is a language $L \in$ RandomCF* which is not deterministic context-free.*
(c) *LVCF* $\subseteq$ RandomCF* $\subseteq$ BPCF* and all inclusions are proper.*
(d) *There is a language $L \in$ BPCF* which is not context-free.*

The incomparability of RandomCF* and LVCF builds on two further results. First we show

Theorem 2. *Let $L = \{\, a^n \cdot b^n \mid n \in I\!N \,\} \cup \{\, a^n \cdot b^{2n} \mid n \in I\!N \,\}$. Then*

$$L \in LVCF_{1/2} - \bigcup_{p<1/2} RandomCF_p.$$

Thus L can be accepted by a Las Vegas pda which commits with probability at least $\frac{1}{2}$; however error probability $p < \frac{1}{2}$ is unreachable even when the more powerful mode of one-sided error is used. As a consequence, error probability $\frac{1}{2}$ is a sharp threshold and thus there is no procedure to decrease error probabilities.

The second result shows that arbitrarily small error is indeed non-trivially obtainable.

Theorem 3. *Let $L = \{a^i \cdot b^j \cdot c^k \mid i \neq k \text{ or } j \neq k\,\}$. Then*

$$L \in RandomCF^* - LVCF.$$

Moreover, the random pda's accepting L with arbitrarily small error can even be chosen to be one-counter automata.

Hence L can be recognized with arbitrarily small error by a one-sided error pda, but not by a Las Vegas pda even when the "I don't know" answer has arbitrarily large probability.

This paper is organized as follows. Closure properties for the above probabilistic language classes are studied in section 2. We utilize these results in section 3 to obtain the separation results.

2 Closure Properties

We say that a union $L_1 \cup \cdots \cup L_k$ of languages over Σ is a *marked union*, if the alphabet Σ can be partitioned into $\Sigma = \Sigma_1 \cup \cdots \cup \Sigma_k$ and words in L_i end in letters from Σ_i. We say that a Kleene closure L^* is marked, if the words in L end in letters that only appear at the end.

We first consider closure properties of Las Vegas languages.

Lemma 1.

(a) Let L_1, L_2 be deterministic context-free languages over an alphabet not containing the symbol \$. Then $L_1 \cup L_2\$ \in LVCF_{1/2}$.

(b) Assume $0 < p < 1$. If $L \in LVCF_p$, then $\overline{L} \in LVCF_p$.

Proof. (a) We describe a Las Vegas pushdown automata P that recognizes $K = L_1 \cup L_2\$$. For input w, P first decides randomly by tossing a fair coin whether to bet on $w \in L_1$ or to bet on $w \in L_2\$$.

Case 1: P bets on $w \in L_1$. P simulates a deterministic pda D_1 for L_1 and accepts, if D_1 accepts. Moreover, P rejects w if and only if D_1 rejects w and the last letter of w is different from \$. Finally, if D_1 rejects and the last letter of w is equal to \$, then P answers with a question mark.

Case 2: P bets on $w \in L_2\$$. P simulates a deterministic pda D_2 for L_2 and accepts (resp. rejects), if the last letter is equal to \$ and D_2 has accepted (resp. rejected) in the previous step. Finally, if the last letter is not equal to \$, then P answers with a question mark.

Observe that P does not make any error and outputs a question mark with probability $\frac{1}{2}$.

(b) The argument is analogous to the case of deterministic pda's. □

We show that the probability of $\frac{1}{2}$ for a commitment of a Las Vegas pda, resp. for a correct answer of a random pda cannot be improved for a rather large class of language pairs.

Lemma 2. *Assume that languages L_1, L_2 are given. Let \$ and # be symbols that do not occur in the alphabet of L_1 or L_2.*

(a) *Assume that $L_1 \cup L_2\$$ can be recognized by a pda with one-sided error smaller than $\frac{1}{2}$. Then the languages*

$$L_1 \cap L_2, \text{ and } (L_1, L_2) = \{\, u \# v \mid u \in L_1 \text{ and } u \cdot v \in L_2 \,\}$$

are context-free.

(b) *If $L_1 \cap L_2$ or if (L_1, L_2) is not context-free and if both L_1 and L_2 are deterministic context-free, then*

$$L_1 \cup L_2\$ \notin LVCF_{1/2} - \bigcup_{p<1/2} RandomCF_p.$$

Thus $\frac{1}{2}$ is a sharp error threshold for a large class of languages. For specific examples set $L_1 = \{a^n \cdot b^n \cdot c^m \mid n, m \in \mathbb{N}\}$ and $L_2 = \{a^m \cdot b^n \cdot c^n \mid n, m \in \mathbb{N}\}$, resp. $K_1 = \{\, a^n \cdot b^n \mid n \in \mathbb{N}\,\}$ and $K_2 = \{\, a^n \cdot b^{2n} \mid n \in \mathbb{N}\,\}$.

Proof. (a) Let Q be a random pda which accepts $L_1 \cup L_2\$$ with probability greater than $\frac{1}{2}$. Let u be arbitrary and assume that $u \cdot v_1 \in L_1$ as well as $u \cdot v_2 \in L_2\$$. Then there must be a Q-computation on u which is simultaneously extendable to an accepting computation on $u \cdot v_1$ and to an accepting computation on $u \cdot v_2\$$, since otherwise the error probability on one of $u \cdot v_1$ or $u \cdot v_2\$$ is at least $\frac{1}{2}$.

Thus $L_1 \cap L_2$ can be accepted by a pda Q_1 which simulates Q on input u and accepts u if and only if Q accepts u and then subsequently $u \cdot \$$. Thus $L_1 \cap L_2$ is context-free.

We accept (L_1, L_2) by a pda Q_2 as follows. Q_2 simulates Q until Q reads the symbol $\#$. Then Q_2 checks whether Q is in an accepting state and if no $\$$ was previously read. If this is the case, then Q_2 continues its simulation and otherwise enters a terminally rejecting state. From now on however Q_2 accepts only if Q would have accepted *after* reading the dollar symbol.

(b) The claim for L_1 and L_2 is obvious, since $L_1 \cap L_2$ is not context-free. Finally consider K_1 and K_2 and observe that $(K_1, K_2) = \{\, a^n \cdot b^n \cdot \# \cdot b^n \mid n \in I\!N \,\}$ is not context-free. □

Neither LVCF nor RandomCF turn out to be closed under the marked Kleene closure.

Lemma 3. *Let L be a language over an alphabet not containing the letter $\#$.*
(a) If $(L\#)^* \in$ LVCF, then $L \in$ LVCF*.
(b) If $(L\#)^* \in$ RandomCF, then $L \in$ RandomCF*.
(c) LVCF as well as RandomCF are not closed under the marked Kleene closure.

Proof. (a): Let P be a Las-Vegas pda which recognizes $(L\#)^*$ with acceptance probability $\delta > 0$. We assume that P is equipped with a prediction mechanism. Hence, if P accepts $u\#v\#$ in some computation, then it accepts $u\#$ in that computation as well. For a given $\varepsilon > 0$ choose words $w_1, \ldots, w_k \in L$ such that

$$\text{prob}[\, P \text{ accepts } w_1\# \cdots \#w_k\# \,] \leq \frac{\delta}{1 - \varepsilon/2}.$$

Let $w \in L$ be arbitrary and set $x = w_1\# \cdots \#w_k\#$. We get

$$\text{prob}[\, P \text{ accepts } x \cdot w\# \mid P \text{ accepts } x \,] = \frac{\text{prob}[\, P \text{ accepts } x \text{ and } x \cdot w\# \,]}{\text{prob}[\, P \text{ accepts } x \,]}$$

$$= \frac{\text{prob}[\, P \text{ accepts } x \cdot w\# \,]}{\text{prob}[\, P \text{ accepts } x \,]}$$

$$\geq \frac{\delta}{\delta/(1-\varepsilon/2)} = 1 - \varepsilon/2.$$

This observation suggests the following Las Vegas pda P'. P' simulates P on the "virtual" input $w_1\# \cdots \#w_k\#$ until an accepting state is reached. (In order not to get caught in an infinite computation P' will count the number of steps per try in its states and stop the try, if a predetermined threshold is reached.) If an accepting state is eventually reached, then P' continues the simulation of P by reading the "real" input w. By supplying a sufficiently large threshold we can guarantee that P' accepts any $w \in L$ with acceptance probability at least $1 - \varepsilon$. (b) follows analogously.

(c) is a consequence of (a) and (b), since LVCF* (resp. RandomCF*) is a proper subset of LVCF (resp. RandomCF) as a consequence of Lemma 2 (c). □

We summarize the closure properties for Las Vegas-CFL and Random-CFL.

Lemma 4.

(a) *LVCF is closed under complementation and under finite marked union of languages from DCF, but not closed under finite union of languages from DCF.*

(b) *LVCF is not closed under concatenation of languages from DCF. Moreover LVCF is not closed under marked Kleene closure.*

Proof. (a) Closure under complementation respectively under finite marked union follows from Lemma 1. If LVCF would be closed under finite union of languages from DCF, then it would also be closed under finite intersection of deterministic context-free languages. This is obviously false as for instance the language $\{\, a^n \cdot b^n \cdot a^m \mid n, m \in \mathbb{N} \,\} \cap \{\, a^m \cdot b^n \cdot a^n \mid n, m \in \mathbb{N} \,\}$ is not context-free.

(b) We use the standard construction to show non-closure under concatenation. Let $L_1, L_2 \in$ DCF be languages over the alphabet Σ. Observe that $K = \$L_1 \cup L_2$ is a deterministic context-free language. If the concatenation $\$^* \cdot K$ belongs to LVCF, then so does $\$L_1 \cup \$L_2$. But then obviously $L_1 \cup L_2 \in$ LVCF and we obtain a contradiction since, by closure under complementation, the intersection of deterministic context-free languages belongs to LVCF and is thus context-free.

LVCF is not closed under marked Kleene closure as a consequence of Lemma 3 (c). □

Lemma 5.

(a) *RandomCF is closed under finite (unmarked) union of languages from DCF.*

(b) *RandomCF is not closed under marked Kleene closure and hence RandomCF is a proper subset of CF. Moreover RandomCF is not closed under complementation.*

Proof. (a) Closure under finite (unmarked) union is obvious, since a union of k deterministic context-free languages can be accepted with probability at least $\frac{1}{k}$. (b) We obtain RandomCF $\subseteq$ CF, since a random pda does not err when accepting. But RandomCF is not closed under the marked Kleene closure as a consequence of Lemma 3 (c) and hence RandomCF is a proper subset of CF.

RandomCF is closed under finite union of deterministic context-free languages and hence closure under complementation implies that any finite intersection of deterministic context-free languages belongs to RandomCF and thus to CF. Therefore RandomCF is not closed under complementation. □

3 Separation Results

Next we show Theorem 3, that is we show that

$$L \in \text{RandomCF}^* - \text{LVCF},$$

for $L = \{a^i \cdot b^j \cdot c^k \mid i \neq k \text{ or } j \neq k \}$. Thus we have to show in particular that L can be accepted with arbitrarily small error probabilities.

Proof of Theorem 3: $\{a^i \cdot b^i \cdot c^i \mid i \in I\!\!N \}$ is the complement of L. Since LVCF is closed under complementation and since LVCF $\subseteq$ CF, we obtain $L \notin$ LVCF.

We construct a random pda P_N for L which randomly decides to simulate one of a collection of deterministic one-counter automata $(Q_x \mid 1 \leq x \leq N)$. For an input word $w = a^i \cdot b^j \cdot c^k$ the automaton Q_x determines $\alpha_{i,j,k}(x) = i + j \cdot x - k \cdot (x+1)$ thru appropriate counter movements and accepts w iff $\alpha_{i,j,k}(x) \neq 0$. Any input w which does not belong to $a^* \cdot b^* \cdot c^*$ is rejected.

The random pda P_N picks $x \in \{1, \ldots, N\}$ uniformly at random and simulates Q_x. If Q_x accepts, then P_N accepts as well and acceptance is error-free, since $\alpha_{i,j,k}(x) = 0$ whenever $i = j = k$.

Now assume that $w = a^i \cdot b^j \cdot c^k$ belongs to L. Observe that

$$\alpha_{i,j,k}(x) = i + j \cdot x - k \cdot (x+1) = (i-k) + x \cdot (j-k)$$

and the condition $\alpha_{i,j,k}(x) = 0$ is equivalent to $(i-k) = x \cdot (k-j)$. Thus there is at most one choice for x with $\alpha_{i,j,k}(x) = 0$ and P_N accepts L with error probability at most $\frac{1}{N}$. □

We are now able to verify the separation results claimed in Theorem 1.

Proof of Theorem 1 (a): For $L = \{ a^n \cdot b^n \mid n \in I\!\!N \} \cup \{ a^n \cdot b^{2n} \cdot \$ \mid n \in I\!\!N \}$ we get

$$L \in \text{LVCF}_{1/2} - \bigcup_{p<1/2} \text{Random-CFL}_p \tag{1}$$

with Lemma 2 (c). Thus DCF is a proper subset of LVCF, since

$$L \in \text{LVCF}_{1/2} - \bigcup_{p<1/2} \text{RandomCF}_p \subseteq L \in \text{LVCF} - \text{DCF}.$$

Obviously LVCF $\subseteq$ RandomCF. If however LVCF = RandomCF, then RandomCF is closed under intersection. This contradicts Lemma 5 (b) and LVCF is a proper subset of RandomCF.

Finally observe that RandomCF is a proper subset of CFL as a consequence of Lemma 5 (b).

(b) As a consequence of (1) LVC* (resp. RandomCF*) is a proper subset of LVCF (resp. RandomCF). Moreover, also as a consequence of (1), LVCF is not a subset of RandomCF*. Finally, with Theorem 3, RandomCF* is not a subset of LVCF and hence RandomCF* and LVCF are incomparable.

(d,c) Let $L = \{a^n \cdot b^n \cdot c^n \mid n \in I\!\!N \}$. Then obviously $L \notin$ CF and it suffices to show that $L \in$ BPCF*. Observe that $\overline{L} \in$ RandomCF* with Theorem 3. But $\overline{L} \in$ RandomCF* $\subseteq$ BPCF* and hence $\overline{L} \in$ BPCF*. Hence $L \in$ BPCF*, since BPCF* is closed under complementation.

Since RandomCF* $\subseteq$ CF, we get $L \notin$ RandomCF* and RandomCF* is a proper subset of BPCF*. Observe that LVCF* is a proper subset of RandomCF* as a consequence of Theorem 3. □

4 Conclusion and Open Problems

In this paper we started the investigation of randomized pushdown automata. The following research problems offer further potential progress in the investigation of the nature of randomization within the framework of context-free languages.

Is DCF a proper subset of LVCF* or does the requirement of arbitrarily large commitment probability collapse Las Vegas to determinism?

Is it possible to bound the number of random decisions (random bits) by a moderately growing function of the input size without decreasing the power of one-sided pushdown automata? Note, that we have proved that even one random bit is sufficient for Las Vegas pushdown automata to be more powerful than deterministic pushdown automata. On the other hand, we conjecture that the number of random bits of two-way finite automata cannot be bounded by any polynomial function of the input length without essentially decreasing their power [HS99].

If the number of random bits can be bounded by a constant, then RandomCF is nothing else than the class of unions of deterministic context-free languages. Moreover, in this case, it is possible to move all random decisions to the very beginning of the computation. Observe that our constructions in Lemma 1 (a) as well as in Theorem 3 use random decisions only at the beginning.

Acknowledgement. We thank Rainer Kemp for lots of very helpful discussions.

References

[AFK97] J. Kaneps, D. Geidmanis, R. Freivalds: Tally languages accepted by Monte Carlo pushdown automata. In: *RANDOM '97, Lexture Notes in Computer Science* 1269, pp. 187-195.

[AHY83] A.V. Aho, J.E. Hopcroft, and M. Yannakakis, "On notions of information transfer in VLSI circuits", Proc. *15th Annual ACM STOCS*, ACM 1983, pp. 133–139.

[Ba79] L. Babai, "Monte Carlo algorithms in graph isomorphism techniques", Research Report no. 79-10, Département de mathématiques et statistique, Université de Montréal, 1979.

[ĎHI00] Ďuriš, P., Hromkovič, J., Inone, K.: A separation of determinism, Las Vegas and nondeterminism for picture recognition. In: *Proc. IEEE Conference on Computational Complexity*, IEEE 2000, pp. 214–228.
Full Version: *Electronic Colloqium on Computational Complexity*, Report No. 27 (2000).

[ĎHRS97] P. Ďuriš, J. Hromkovič, J.D.P. Rolim, and G. Schnitger, "Las Vegas versus determinism for one-way communication complexity, finite automata and polynomial-time computations", Proc. *STACS'97, Lecture Notes in Computer Science* 1200, Springer, 1997, pp. 117–128.

[DKR94] M. Dietzfelbinger, M. Kutylowski, and R. Reischuk, "Exact lower bounds for computing Boolean functions on CREW PRAMs", *J. Computer System Sciences* 48, 1994, pp. 231-254.

[Fr81] R. Freivalds: Projections of languages recognizable by probabilistic and alternating multitape automata. *Information Processing Letters* 13 (1981), 195-198.

[Gi77] J. Gill, "Computational complexity of probabilistic Turing machines", *SIAM J. Computing* 6, 1977, pp. 675–695.

[Hr97] J. Hromkovič, *Communication Complexity and Parallel Computing*, Springer 1997.

[Hr00] J. Hromkovič, "Communication Protocols - An Exemplary Study of the Power of Randomness", *Handbook on Randomized Computing*, (P. Pardalos, S. Kajasekaran, J. Reif, J. Rolim, Eds.), Kluwer Publisher 2001, to appear.

[HS00] J. Hromkovič, and M. Sauerhoff, "Tradeoffs between nondeterminism and complexity for communication protocols and branching programs", Proc. *STACS 2000, Lecture Notes in Computer Science* 1770, Springer 2000, pp. 145–156.

[HS01] J. Hromkovič, G. Schnitger, "On the power of Las Vegas for one-way communication complexity, OBDD's and finite automata". *Information and Computation*, to appear.

[HS99] J. Hromkovič, and G. Schnitger, "On the power of Las Vegas II, Two-way finite automata", Proc. *ICALP'99, Lecture Notes in Computer Science* 1644, Springer 1999, pp. 433–443. (extended version: to appear in *Theoretical Computer Science*)

[KN97] E. Kushilevitz, and N. Nisan, *Communication Complexity*, Cambridge University Press 1997.

[Mac94] Macarie, I, "On the structure of log-space probabilistic complexity classes." *Technical Report* TR-506, Dept. of Computer Science, University of Rochester 1994.

[MO94] Macarie, I., Ogihara, M., "Properties of probabilistic pushdown automata." *Technical Report* TR-554, Dept. of Computer Science, University of Rochester 1994.

[MS82] K. Mehlhorn, and E. Schmidt, "Las Vegas is better than determinism in VLSI and distributed computing", Proc. *14th ACM STOC'82*, ACM 1982, pp. 330–337.

[MS99] I.I. Macarie, and J.I. Seiferas, "Amplification of slight probabilistic advantage at absolutely no cost in space", *Information Processing Letters* 72, 1999, pp. 113–118.

[PS84] Ch. Papadimitriou, and M. Sipser, "Communication complexity", Proc. *14th ACM STOC*, ACM 1982, pp. 196–200, also in
J. Computer System Sciences 28, 1984, pp. 260–269.

[Sa97] M. Sauerhoff, "On nondeterminism versus randomness for read-once branching programs", *Electronic Colloquium on Computational Complexity*, TR 97 - 030, 1997.

[Sa99] M. Sauerhoff, "On the size of randomized OBDDs and read-once branching programs for k-stable functions", Proc. *STACS '99, Lecture Notes in Computer Science* 1563, Springer 1999, pp. 488–499.

[Ya79] A.C. Yao, "Some complexity questions related to distributed computing", Proc. *11th ACM STOC*, ACM 1979, pp. 209–213.

The Root of a Language and Its Complexity*

Gerhard Lischke

Institute of Informatics
Faculty of Mathematics and Informatics
Friedrich Schiller University Jena
D-07743 Jena, Germany
lischke@minet.uni-jena.de

Abstract. The root of a language L is the set of all primitive words p such that p^n belongs to L for some $n \geq 1$. We show that the gap between the time complexity and space complexity, respectively, of a language and that of its root can be arbitrarily great. From this we conclude that there exist regular languages the roots of which are not even context-sensitive. Also we show that the quadratic time complexity for deciding the set of all primitive words by an 1-tape Turing machine is optimal.

1 Introduction

The concept of primitive words and primitive roots came up with the paper [11] of Lyndon and Schützenberger. A word p over some alphabet X is said to be primitive if $p = q^n$ for $q \in X^*$ implies $n = 1$. The set of all primitive words over X is denoted by $Q(X)$, or simply by Q if X is understood. This set Q has received special interest in the algebraic theory of codes and formal languages (see, for instance, [16] and [10]). Regarding relationships to the Chomsky hierarchy it is easy to see that Q is deterministic context-sensitive and it was strongly conjectured that it is not context-free. This conjecture was formulated in 1991 by Dömösi, Horváth and Ito [3,4]. Even though all attempts failed up to now to prove or disprove this conjecture it is mostly assumed that it is true. In 1994 Petersen [12,13] has shown that Q is not generated by any unambiguous context-free grammar.

Strongly connected with the notion of primitive words is that of the root of a word. For a nonempty word x, there exists a unique primitive word p and $n \geq 1$ such that $x = p^n$. This primitive word p is called the root of x, denoted by $\sqrt{x}$. The set of all roots of words from a language L create the root of L, denoted by $\sqrt{L}$. Of course $\sqrt{L} \subseteq Q$ and $\sqrt{X^*} = Q$ hold. Horváth and Kudlek [9] have shown that Q is decidable in quadratic time. We show that this result is optimal, that means there does not exist any 1-tape

---

* This work was done in part during a visit of the Department of Computer Science at the Eötvös Loránd University Budapest within the scope of a partnership agreement between Friedrich Schiller University and Eötvös Loránd University to the collaboration project "Complexity theory and combinatorics on words and languages".

Turing machine deciding Q within time bound smaller than n^2. Further, we investigate general relationships between the computational complexity of a language L and that of its root. We show that the complexity of the root can be arbitrarily great.

More exactly: For almost arbitrary functions t and f we can construct a language L which is decidable within complexity bound t but whose root is not decidable within complexity bound $f \circ t$.[1]

The functions t and f have to fulfill some quite natural conditions: $t \geq id$ or $t \in \omega(id)$,[2] and both t and $f \circ t$ must be monotone nondecreasing, unbounded, and computable within their own complexity bound. More formally we shall give these conditions in Sections 2.2 and 4. Our computational model is the usual more-tape Turing machine, and by complexity we mean the time or the space complexity of Turing machines or some similar measure.

In Section 2 we will give the fundamental definitions from formal language theory and from complexity theory. In Section 3 we prove the optimality of the quadratic time bound for 1-tape Turing machines deciding Q. Our main theorem we formulate and prove in Section 4. Finally, in Section 5 we discuss some consequences from our result.

2 Preliminaries and Definitions

2.1 Languages

Even though the following is standard in the literature (see, e.g., the textbooks [15,16, 8]) we briefly recall the most important notions. For our whole paper, let X be a fixed finite nonempty alphabet. X^* is the free monoid generated by X or the set of all words over X. The empty word we denote by e, and $X^+ =_{Df} X^* \setminus \{e\}$. A (formal) language (over X) is a subset L of X^*, $L \subseteq X^*$. $|p|$ denotes the length of a word $p \in X^*$. For a natural number n ($n \in \mathbb{N}$), p^n denotes the concatenation of n copies of the word p.

Definition 1. A word p is *primitive* iff it is not of the form q^n for any word q and $n \geq 2$. $Q =_{Df} \{p : p \in X^* \wedge \neg \exists q \exists n (q \in X^* \wedge n \in \mathbb{N} \wedge n > 1 \wedge p = q^n)\}$ is the set of all primitive words (over X).

Remark. e is not primitive because of $e^n = e$ for each $n \in \mathbb{N}$.

Definition 2. The *root* of a word $p \in X^+$ is the unique primitive word q such that $p = q^n$ for some $n \in \mathbb{N}$. $\sqrt{p}$ denotes the root of p.
For a language L, $\sqrt{L} =_{Df} \{\sqrt{p} : p \in L \wedge p \neq e\}$ is the root of L.

It is easy to see that $\sqrt{X^*} = Q$, $\sqrt{L} \subseteq Q$ for each language L, and $\sqrt{L} = L$ if and only if $L \subseteq Q$.

---
[1] $f \circ t$ denotes the function defined by $(f \circ t)(x) = f(t(x))$.
[2] id denotes the identity function over the set $\mathbb{N}$ of natural numbers. $t \geq id$ which means $t(n) \geq n$ for all $n \in \mathbb{N}$, is need in the case of space complexity. $t \in \omega(id)$ also written as $t \in \omega(n)$ or $n \in o(t)$ means $\lim_{n \to \infty} \frac{n}{t(n)} = 0$, and this is need in the case of time complexity.

2.2 Complexity

Our basic model for complexity investigations is the model of deterministic more-tape off-line Turing machines used as acceptors as it is standard in the literature (see, e.g., the textbooks [1,8,17]). If $\mathcal{M}$ is such a machine over the alphabet X then we denote by $Acc(\mathcal{M})$ (which is a subset of X^*) the set of all words accepted by $\mathcal{M}$, also called the *language of* $\mathcal{M}$. The time and space complexity of $\mathcal{M}$ are defined in the following way:

If $p \in X^*$ and $\mathcal{M}$ on input p reaches a final state (we also say $\mathcal{M}$ *halts* on p) then $t_{\mathcal{M}}(p)$ is the number of computation steps required by $\mathcal{M}$ to halt, and $s_{\mathcal{M}}(p)$ is the maximum number of tape cells scanned during the computation of $\mathcal{M}$ on p. If $\mathcal{M}$ doesn't halt on p then both $t_{\mathcal{M}}(p)$ and $s_{\mathcal{M}}(p)$ are not defined. We transfer this to functions over $\mathbb{N}$ in the usual way:

for $n \in \mathbb{N}$, $t_{\mathcal{M}}(n) =_{Df} \max\{t_{\mathcal{M}}(p) : p \in X^* \wedge |p| = n\}$ and
$s_{\mathcal{M}}(n) =_{Df} \max\{s_{\mathcal{M}}(p) : p \in X^* \wedge |p| = n\}$ if $\mathcal{M}$ halts on each word of length n, otherwise both $t_{\mathcal{M}}(n)$ and $s_{\mathcal{M}}(n)$ are not defined.

These partial functions from $\mathbb{N}$ to $\mathbb{N}$ we call the *time complexity* and the *space complexity of* $\mathcal{M}$, respectively.

Definition 3. Let f be a total function from $\mathbb{N}$ to $\mathbb{N}$.
TIME$(f) =_{Df} \{Acc(\mathcal{M}) : \mathcal{M}$ is a Turing machine $\wedge$
$\wedge\ t_{\mathcal{M}}(n) \leq f(n)$ for all $n \in \mathbb{N}\}$,
SPACE$(f) =_{Df} \{Acc(\mathcal{M}) : \mathcal{M}$ is a Turing machine $\wedge$
$\wedge\ s_{\mathcal{M}}(n) \leq f(n)$ for all $n \in \mathbb{N}\}$.
Restricting to 1-tape Turing machines these classes are denoted by 1-TIME(f) and 1-SPACE(f), respectively.

Remark that 1-SPACE(f) = SPACE(f) but not so for the time complexity (see, e.g., [1,8,17]).

As is well known (from [5] and [6], see also [1,8,17]) the time measure and the space measure have a *linear speed-up* in the following sense:

Theorem 1. *If* $s' \in O(s)$, $t' \in O(t)$ *and* $t \in \omega(n)$ [3] *then*
SPACE$(s') \subseteq$ *SPACE*(s) *and TIME*$(t') \subseteq$ *TIME*(t).
If $t(n) \geq n^2$ *for all* n *then also 1-TIME*$(t') \subseteq$ *1-TIME*(t).

Of some importance are such functions which are computable within their own complexity bound and which we will define in the following way (cf. [1]):

Definition 4. A total function f from $\mathbb{N}$ to $\mathbb{N}$ is said to be *time constructible* (*space constructible*) if and only if there is a Turing machine which, for each $n \in \mathbb{N}$, on every input of length n halts in exactly $f(n)$ steps (halts after touching exactly $f(n)$ tape cells).

---

[3] $f' \in O(f)$ means: $\exists c \exists n_0 (c > 0 \wedge n_0 \in \mathbb{N} \wedge \forall n (n \geq n_0 \to f'(n) \leq c \cdot f(n)))$.
For *omega* remember to footnote [2].

The sets of time constructible and space constructible functions, respectively, are very rich and contain the most common functions (see, e.g., [8]). One can set such functions as "clocks" to any machine to measure or to bound its complexity. Because of the linear speed-up (Theorem 1 which is also true for machines which compute functions), a function f is space constructible if and only if it can be computed in space $O(f)$, and a function f with $f \in \omega(n)$ is time constructible if and only if it can be computed within time $O(f)$.

Finally, let us refer to the fact that in the usual way (cf. [1]) by arithmetization or gödelization we can get an effective numbering $\mathcal{M}_0, \mathcal{M}_1, \mathcal{M}_2, \ldots$ of all deterministic more-tape Turing machines used as acceptors over X, and that there exists an algorithm realized by an 1-tape Turing machine itself which for a given number n produces the (program of the) Turing machine $\mathcal{M}_n$. The latter we will call the *decoding process*, and this will be extensively used in the proof of our main theorem.

3 Deciding Primitiveness in Quadratic Time Is Optimal

If the basic alphabet X contains only one symbol then we get $Q = X$ and the root of a language is either $\emptyset$ or Q. Therefore all will become trivial, and we want to assume for the rest of this paper that X has at least two symbols. We denote, without loss of generality, two fixed different symbols from X by a and b.

As remarked in the introduction the exact relation of Q to the Chomsky language classes is not known. On the other side, its relation to complexity classes is clear:

Theorem 2. $Q \in 1\text{-}TIME(n^2) \cap SPACE(n)$. [4]

That Q is decidable in quadratic time was shown by Horváth and Kudlek [9]. $Q \in SPACE(n)$ is trivial and obviously it cannot be done with less space.

We show that the time bound of Horváth and Kudlek is optimal for 1-tape Turing machines.

Theorem 3. *For each 1-tape Turing machine $\mathcal{M}$ deciding Q, $t_\mathcal{M} \in \Omega(n^2)$* [5] *must hold.*

For the proof we use the method of counting of crossing sequences which goes back to Rabin [14] and was used by Hennie [7] and Barzdin [2] to prove the quadratic lower bound for deciding symmetric words (cf. [17]). First we give an informal definition of crossing sequences.

Let us assume that the boundaries between any two tape cells of the unique tape of an 1-tape Turing machine $\mathcal{M}$ are numbered in a natural way, and that $\mathcal{M}$ starts with a configuration where the input word is given in such a way that its first letter appears in the cell immediately before the boundary 1. A sequence $cs_\mathcal{M}(p, j) = (z_1, z_2, \ldots, z_k)$ where $p \in X^*, j \in \mathbb{N}$, and $z_1, \ldots, z_k$ are states of $\mathcal{M}$, is called the *crossing sequence of $\mathcal{M}$ on*

---
[4] This notion is vividly clear but not formally correct. A more correct but fussy notation would be 1-TIME(f) ∩ SPACE(id) where $f(n) = n^2$ for each $n \in \mathbb{N}$ or $f = \lambda n[n^2]$.
[5] This means, more exactly, $\exists c \exists n_0 (c > 0 \wedge n_0 \in \mathbb{N} \wedge \forall n(n \geq n_0 \rightarrow t_\mathcal{M}(n) \geq c \cdot n^2))$.

p *in* j if during the computation of $\mathcal{M}$ on input p the head crosses the boundary j exactly k times, being in state z_i when crossing j for the ith time $(i = 1, \ldots, k)$. $|cs_{\mathcal{M}}(p,j)| = k$ is the *length* of this crossing sequence. Obviously, we have $t_{\mathcal{M}}(p) \geq \sum_{j \in \mathbb{N}} |cs_{\mathcal{M}}(p,j)|$.

For a word $p = x_1 \cdots x_n$ and $0 \leq i \leq j \leq n$, by p_i^j we denote the subword $x_{i+1} \cdots x_j$ of p if $i < j$ and $p_i^j = e$ if $i = j$.

Further define: $p \sim_{\mathcal{M}} q =_{Df} p \in Acc(\mathcal{M}) \leftrightarrow q \in Acc(\mathcal{M})$.

The following basic lemma can be proved straightforwardly.

Lemma 1. *If* $p \sim_{\mathcal{M}} q$, $0 \leq j \leq |p|$, $0 \leq i \leq |q|$, and $cs_{\mathcal{M}}(p,j) = cs_{\mathcal{M}}(q,i)$, then the words p, $p_0^j q_i^{|q|}$, $q_0^i p_j^{|p|}$ and q are pairwise equivalent with respect to $\sim_{\mathcal{M}}$.*

Now we prove our theorem. Let $\mathcal{M}$ be an 1-tape Turing machine deciding Q which has exactly k states. Let h be the cardinality of X. Remember that by agreement $h \geq 2$ and $a, b \in X$, $a \neq b$. For $n \in \mathbb{N}$ we define:
$S_n =_{Df} \{wa^m bwa^m b : w \in X^* \wedge |w| = [\frac{n}{2}] \wedge m = n - [\frac{n}{2}]\}$.
For each n, S_n is a subset of $\overline{Q} = X^* \setminus Q$, containing exactly $h^{[\frac{n}{2}]}$ words of length $2n + 2$.

Lemma 2. *For any two different words $p, q \in S_n$ and $i, j \in \{[\frac{n}{2}], [\frac{n}{2}]+1, \ldots, n\}$ there holds $cs_{\mathcal{M}}(p,j) \neq cs_{\mathcal{M}}(q,i)$.*

Proof of Lemma 2. Otherwise and because of $p \sim_{\mathcal{M}} q$ and Lemma 1, also $p_0^j q_i^{2n+2}$ must be nonprimitive which means $p_0^j q_i^{2n+2} = r^k$ for some word r and $k \geq 2$. Because of $q_{n+1}^{n+1+[\frac{n}{2}]} = q_0^{[\frac{n}{2}]}$ and therefore $p_0^j q_i^{2n+2} =$
$= p_0^{[\frac{n}{2}]} a^{j-[\frac{n}{2}]} a^{n-i} b q_0^{[\frac{n}{2}]} a^{n-[\frac{n}{2}]} b$ it follows that $k = 2$, $r = p_0^{[\frac{n}{2}]} a^{n-[\frac{n}{2}]+j-i} b$ and $p_0^{[\frac{n}{2}]} = q_0^{[\frac{n}{2}]}$. On the other hand, because of $p \neq q$ and $p, q \in S_n$ we must have $p_0^{[\frac{n}{2}]} \neq q_0^{[\frac{n}{2}]}$, which proves the lemma. $\square$

Let us set $N = \frac{[\frac{n}{2}]}{\log_h k}$. We count the number of crossing sequences with length smaller than N. It is bounded by $1 + k + k^2 + \cdots + k^{N-1} = \frac{k^N - 1}{k - 1} < k^N = h^{[\frac{n}{2}]}$. If there would be for each $p \in S_n$ some $j \in \{[\frac{n}{2}], \ldots, n\}$ with $|cs_{\mathcal{M}}(p,j)| < N$ then because of Lemma 2 we would have at least $h^{[\frac{n}{2}]}$ different crossing sequences with length smaller than N contradicting our observation. Therefore some $p \in S_n$ must exist with $|cs_{\mathcal{M}}(p,j)| \geq N$ for each $j \in \{[\frac{n}{2}], \ldots, n\}$. Now we can conclude

$t_{\mathcal{M}}(2n+2) = \max\{t_{\mathcal{M}}(p) : |p| = 2n+2\} \geq \max\{\sum_{j=[\frac{n}{2}]}^{n} |cs_{\mathcal{M}}(p,j)| : p \in S_n\} \geq$

$\frac{n}{2} \cdot N \geq c \cdot (2n+2)^2$ for some suitable constant $c > 0$ and sufficiently great n, which means $t_{\mathcal{M}} \in \Omega(n^2)$. $\square$

4 The Complexity of the Root of L with Respect to the Complexity of L

We have seen that the time complexity of deciding the root of X^* is quadratic and cannot be better in general. But the complexity of deciding the root of a restricted language can be much larger. Even if the language L has a known (possibly, small) c omplexity, the complexity of its root can be arbitrarily much greater. Thereby it doesn't matter whether we consider time complexity or space complexity or some other similar measure. Only the necessary presuppositions must differ to allow the linear speed-up from Theorem 1. We shall give a detailed proof for the case of time complexity and a sketched one for space complexity.

Theorem 4. *Let t and f be arbitrary total functions over $\mathbb{N}$ such that $t \in \omega(n)$ and both t and $f \circ t$ are monotone nondecreasing, unbounded, and time constructible. Then there exists a language L such that $L \in \text{TIME}(t)$ but $\sqrt{L} \notin \text{TIME}(f \circ t)$. Using only 1-tape machines it holds that $L \in \text{1-TIME}(t)$ if $t(n) \geq n^2$ for all n, and $L \in \text{1-TIME}(O(t))$ if $t \in \omega(n)$.*

Proof. Assume t and f have the properties formulated as hypothesis in the theorem. We shall construct in stages an appropriate language L making use of the effective numbering $\mathcal{M}_0, \mathcal{M}_1, \mathcal{M}_2, \ldots$ of all deterministic more-tape Turing machines used as acceptors over X which was mentioned in 2.2. We consider only special primitive words which we denote by p_l and define $p_l =_{Df} a^{l-1}b$ for $l \geq 1$. Remark that for each l, $|p_l| = l$. During our construction we maintain and sometimes modify a finite list I of natural numbers. By L_l and I_l we denote that part of L and I, respectively, which was constructed up to the stage l. Initially, we start with $L_0 =_{Df} \emptyset$ and $I_0 =_{Df} \emptyset$.
Stage $l \geq 1$ of the construction is as follows.

Look for the smallest number i fulfilling the following conditions:
(a) $i < 2^l$ and $i \notin I_{l-1}$,
(b) the decoding process to determine $\mathcal{M}_i$ doesn't need more than $f(t(l))$ steps,
(c) $\mathcal{M}_i$ on p_l halts in no more than $f(t(l))$ steps.

If no such i can be found then let L and I unchanged, that means $L_l =_{Df} L_{l-1}$ and $I_l =_{Df} I_{l-1}$, and go to the next stage $l+1$.
If i is found then if $p_l \in Acc(\mathcal{M}_i)$ then put $L_l =_{Df} L_{l-1}$, $I_l =_{Df} I_{l-1} \cup \{i\}$, and go to the next stage $l+1$.
Finally, if $p_l \notin Acc(\mathcal{M}_i)$ then let k be the smallest number such that
$(*)$ $l^3 \cdot 2^l \cdot f(t(l)) \leq \frac{t(k \cdot l)}{k}$,
and put $L_l =_{Df} L_{l-1} \cup \{p_l^{k'} : k' \geq k\}$, $I_l =_{Df} I_{l-1} \cup \{i\}$,
and go to the next stage $l+1$.

Completing the construction by $L =_{Df} \bigcup_{l=0}^{\infty} L_l$.

Now we have to prove that $L \in \text{TIME}(t)$ (or 1-TIME(t) or 1-TIME($O(t)$), respectively) and $\sqrt{L} \notin \text{TIME}(f \circ t)$. Let us start with the latter.

We have $\sqrt{L} \subseteq \{p_l : l \geq 1\}$ and $p_l \in \sqrt{L}$ if and only if during our construction it was found that $p_l \notin Acc(\mathcal{M}_i)$ for an appropriate i. Assume $\sqrt{L} \in \text{TIME}(f \circ t)$. Then there must exist a Turing machine working within time bound $f \circ t$ and accepting $\sqrt{L}$. This machine must occur in our numbering, let us say as the machine $\mathcal{M}_{i_0}$, and we have $Acc(\mathcal{M}_{i_0}) = \sqrt{L}$. In our construction there must exist a stage l_0 where i_0 is the smallest number i fulfilling (a), (b), (c). If $\mathcal{M}_{i_0}$ accepts p_{l_0} then no word $p_{l_0}^k$ will ever be added to L and therefore $p_{l_0} \notin \sqrt{L}$. Otherwise $p_{l_0} \notin Acc(\mathcal{M}_{i_0})$ and we get $p_{l_0} \in \sqrt{L}$. Both contradicts $Acc(\mathcal{M}_{i_0}) = \sqrt{L}$.

Now we describe a process for deciding L. Given a word $p \in X^*$ we have to decide whether $p \in L$.
(I) Determine $l \geq 1$ such that $a^{l-1}b$ is a prefix of p. If no such l exists then $p \notin L$. Otherwise, simultaneously with the following steps check whether p has the form p_l^k for some $k \geq 1$. If this is not the case then $p \notin L$.
(II) Proceed the stages 0 up to $l-1$ of our construction, thereby taking care only of the lists $I_1, \ldots, I_{l-1}$, which altogether contain at most $l-1$ numbers $< 2^l$.
(III) Look for the smallest i fulfilling (a), (b), (c) of stage l.
Simultaneously, as soon as k from (I) is found, check whether
$l^3 \cdot 2^l \cdot f(t(l)) \leq \frac{t(k \cdot l)}{k}$ (remember that $k \cdot l = |p|$).
If this inequality is not true then break off with answer $p \notin L$. Otherwise $p \in L$ if and only if i was found under (III) and $p_l \notin Acc(\mathcal{M}_i)$.

Now we count the number of steps needed for this process on a Turing machine. (I) can be done in linear time (in the length of p) on a more-tape machine, and in $k \cdot l^2$ steps on an 1-tape machine. Bookkeeping of the lists up to I_{l-1} can be done, for instance, by handling $l-1$ blocks of length l, each containing one or no number in binary, and using several traces of the Turing tape. For updating I_j the first $j-1$ blocks have to be inspected to determine the smallest i, $j = 2, \ldots, l-1$. Because of time constructibility of $f \circ t$, this can be done, for each of these j, in less than $2^j \cdot (j-1) \cdot l \cdot c \cdot f(t(j))$ steps for some constant c. Therefore, and because of time constructibility and monotonicity of t and $f \circ t$, the whole process including checking of the unequality can be done in less than $k \cdot l^2 + c \cdot l \cdot \sum_{j=1}^{l}(2^j \cdot j \cdot f(t(j))) < c' \cdot k \cdot l^3 \cdot 2^l \cdot f(t(l)) \leq c' \cdot t(|p|)$ steps for some constant c'. This means, $L \in \text{1-TIME}(O(t))$ and, if the linear speed-up takes effect, $L \in \text{1-TIME}(t)$ or $L \in \text{TIME}(t)$, respectively. $\square$

Theorem 5. *Let s and f be arbitrary total functions over $\mathbb{N}$ such that $s \geq id$ and both s and $f \circ s$ are monotone nondecreasing, unbounded, and space constructible. Then there exists a language L such that $L \in \text{SPACE}(s)$ but $\sqrt{L} \notin \text{SPACE}(f \circ s)$.*

Sketch of proof. The construction of L can be done in the same - in fact, in a little simpler - way as in the former proof replacing t by s, (b) and (c) by

(b') the decoding process to determine $\mathcal{M}_i$ can be done within space $f(s(l))$,

(c') $\mathcal{M}_i$ on p_l halts after touching no more than $f(s(l))$ tape cells,
and (∗) by: $\quad l^2 \cdot f(s(l)) \leq s(k \cdot l)$.

Because of space constructibility of $f \circ s$ it is easy to mark a space amount of $f(s(l))$ cells at the beginning of each stage l. In the same way as before we see that $\sqrt{L} \notin \text{SPACE}(f \circ s)$, and also it is easy to see that $L \; in\text{SPACE}(s)$. □

By the same method we can also get the appropriate result for any other computational complexity measure for Turing machines which allows a linear speed-up and the notion of constructibility. This is true, for instance, for the reversal measure and the crossing measure and each other Blum measure (see [17]).

5 Some Consequences

We have seen, by our theorems 4 and 5, that there can be an arbitrarily great gap between the complexity of a language and that of its root. This has some consequences for languages which are characterizable by complexity bounds. For instance, we can conclude that a language and its root need not lie in the same language class of the Chomsky hierarchy. The greatest gap of this kind is given by the following theorem.

Theorem 6. *There exist regular languages the roots of which are not even context-sensitive.*

Proof. Let us define $t(n) =_{Df} n \cdot \log \log n$ and $f(n) =_{Df} 2^n$. Then the hypotheses of Theorem 4 are true, and there exist languages L such that $L \in 1\text{-TIME}(O(t))$ and $\sqrt{L} \notin \text{TIME}(f \circ t)$. But in this case, by [17, Theorem 12.1], $1\text{-TIME}(O(t))$ is the class of regular languages. $f \circ t$ is approximately greater than 2^g for each linear function g and therefore $\text{TIME}(f \circ t)$ includes $\text{TIME}(2^{Lin})$ which for its part includes the class of context-sensitive languages.[6] □

If we consider the relationships in the opposite direction, that means from the root to the language, we can establish an almost arbitrary speed-up from the complexity of deciding a set of primitive words, namely the root, to that of deciding a language having this root. Unfortunately, our results seem not to help for proving the conjecture that the set Q of primitive words is not context-free. But we think that they may be interesting in themselves.

Acknowledgments. I am grateful to Sándor Horváth for commending to me studying properties of sets of primitive words and for helpful discussions and suggestions, and to him and other colleagues from the Eötvös Loránd University for their hospitality during my visits in Budapest. I also thank H.J. Shyr for letting to me one copy of his book [16].

References

[1] J.L.BALCÁZAR, J.DIAZ, J.GABARRÓ, *Structural complexity I*, Springer-Verlag, Berlin-Heidelberg, 1988.

---

[6] Without any explanation, let us refer to the inclusion $\text{CS} \subseteq \text{NSPACE}(id) \subseteq \text{TIME}(2^{Lin})$ which is shown in complexity theory (see, e.g., [17]).

[2] J.M.BARZDIN, *Complexity of the recognition of the symmetry predicate on Turing machines* (in Russian), Problemy Kibernetiki 15 (1965), 245–248.
[3] P.DÖMÖSI, S.HORVÁTH, M.ITO, *On the connection between formal languages and primitive words*, in Proc. First Session on Scientific Communication, Univ. of Oradea, Oradea, Romania, June 1991, 59–67.
[4] P.DÖMÖSI, S.HORVÁTH, M.ITO, *Formal languages and primitive words*, Publ. Math., Debrecen, 42 (1993), 315–321.
[5] J.HARTMANIS, P.M.LEWIS II, R.E.STEARNS, *Hierarchies of memory limited computations*, in Proc. Sixth Annual IEEE Symp. on Switching Circuit Theory and Logical Design, 1965, 179–190.
[6] J.HARTMANIS, R.E.STEARNS, *On the computational complexity of algorithms*, Trans. AMS 117 (1965), 285–306.
[7] F.C.HENNIE, *One-tape off-line Turing machine computations*, Information and Control 8 (1965), 553–578.
[8] J.E.HOPCROFT, J.D.ULLMAN, *Introduction to automata theory, languages, and computation*, Addison-Wesley, Reading (Mass.), 1979.
[9] S.HORVÁTH, M.KUDLEK, *On classification and decidability problems of primitive words*, PU.M.A. 6 (1995), 171–189.
[10] H.JÜRGENSEN, S.KONSTANTINIDIS, *Codes*, in [15], 511–607.
[11] R.C.LYNDON, M.P.SCHÜTZENBERGER, *On the equation $a^M = b^N c^P$ in a free group*, Michigan Math. Journ. 9 (1962), 289–298.
[12] H.PETERSEN, *The ambiguity of primitive words*, in Proc. STACS 94, Caen, Lecture Notes in Computer Science 775, Springer-Verlag, Berlin-Heidelberg, 1994, 679–690.
[13] H.PETERSEN, *On the language of primitive words*, Theoretical Computer Science 161 (1996), 141–156.
[14] M.O.RABIN, *Real-time computation*, Israel J. Math. 1 (1963), 203–211.
[15] G.ROZENBERG, A.SALOMAA, *Handbook of formal languages, Vol. 1*, Springer-Verlag, Berlin-Heidelberg, 1997.
[16] H.J.SHYR, *Free monoids and languages*, Hon Min Book Company, Taichung, 1991.
[17] K.WAGNER, G.WECHSUNG, *Computational complexity*, Deutscher Verlag der Wissenschaften, Berlin / D. Reidel Publ. Comp., Dordrecht-Boston-Lancaster-Tokyo, 1986.

Valuated and Valence Grammars: An Algebraic View

Henning Fernau[1]* and Ralf Stiebe[2]

[1] University of Newcastle
Department of Computer Science and Software Engineering
University Drive, NSW 2308 Callaghan, Australia
fernau@cs.newcastle.edu.au
[2] Fakultät für Informatik
Otto-von-Guericke-Universität Magdeburg
PF-4120, D-39016 Magdeburg, Germany
stiebe@iws.cs.uni-magdeburg.de

Abstract. Valence grammars were introduced by Gh. Păun in [8] as a grammatical model of chemical processes. Here, we focus on discussing a simpler variant which we call *valuated grammars*. We give some algebraic characterizations of the corresponding language classes. Similarly, we obtain an algebraic characterization of the linear languages. We also give some Nivat-like representations of valence transductions.

1 Introduction

Algebraic techniques often lead to elegant and mathematically appealing characterizations of language families. For example, the regular languages are characterized as preimages of subsets of finite monoids under morphisms [4]. A similar characterization—using groupoids instead of monoids—is known for the context-free languages [2]. The first contribution of this paper is to give an algebraic characterization of the linear languages in this spirit.

Then, we introduce valuated languages, a concept which appears to be a rather natural variant of valence languages. We can derive algebraic characterizations of these language classes, as well. To our knowledge, these are, besides [10], the first results of this type of algebraic flavour ever obtained in the area of regulated rewriting. In passing, we also derive Nivat-type (and other) theorems for valence transductions.

The now classical algebraic characterizations of regular and context-free languages initiated vivid research, finding deep connections to algebra and complexity theory, see [1,9,11]. We hope that similar insights can be obtained for some of the language families characterized in this paper. As a first application of the algebraic characterizations, we derive several closure properties of valuated language families.

---

* Work was done while the author was with Wilhelm-Schickard-Institut für Informatik, Universität Tübingen, Sand 13, D-72076 Tübingen, Germany.

2 Definitions

Conventions: pr_i denotes the projection to the ith component of a given tuple. The empty word is denoted by λ. The number of letters a occurring in a word x is denoted by $|x|_a$. If G is a grammar with nonterminal alphabet N and terminal alphabet T, we denote the total alphabet by $V := N \cup T$. We consider two languages L_1, L_2 to be equal iff $L_1 \setminus \{\lambda\} = L_2 \setminus \{\lambda\}$. We term two devices describing languages equivalent if the two described languages are equal. If $\tau \subseteq I^* \times O^*$ and $L \subset I^*$, then $\tau(L) := \{x \in O^* : \exists w (w \in L \wedge (w,x) \in \tau)\}$.

We expect some knowledge of formal language theory on side of the reader. In particular, the (extended) Chomsky hierarchy

$$\mathcal{L}(\mathrm{REG}) \subset \mathcal{L}(\mathrm{LIN}) \subset \mathcal{L}(\mathrm{CF}) \subset \mathcal{L}(\mathrm{CS}) \subset \mathcal{L}(\mathrm{RE})$$

should be known. For details, see [3,6].

We are interested in algebraic characterizations of language classes. Therefore, we need the following notions.

If $\mathfrak{G}$ is a *groupoid*, i.e., a set with a binary (not necessarily associative) multiplicatively written operation, and $\theta : T \to \mathfrak{G}$ is extended to a monoid morphism $\theta : T^* \to \mathfrak{G}^*$, then $\mathfrak{G}(g)$, $g \in \mathfrak{G}^*$, denotes the set of all elements of $\mathfrak{G}$ obtained by multiplying the elements of g in the given order following an arbitrary bracketization. The empty word is mapped onto the neutral element $1 \in \mathfrak{G}$ which we always assume to lie in any considered groupoid. Observe that in this setting, the (not necessarily finite) set $\mathfrak{G}$ is considered as an alphabet generating the free word monoid $\mathfrak{G}^*$. We have to distinguish the groupoid operation within $\mathfrak{G}$ from the catenation defined on $\mathfrak{G}^*$. The difference should be clear from the context. An associative groupoid with neutral element is also termed *monoid*.

We now introduce the central notions of this paper:

Definition 1. *A (context-free) valence grammar (over a monoid $\mathfrak{M}$) is given as $G = (N, T, P, S, \mathfrak{M})$, where P is a set of productions labelled by elements of the multiplicatively written monoid $\mathfrak{M}$ with neutral element 1: $(A \to w, \mathrm{m})$. (More formally, there is a labelling mapping $P \to \mathfrak{M}$.) Let $V = N \cup T$. On $V^* \times \mathfrak{M}$, we define a derivation relation $\Rightarrow$ as follows: $(u_1, \mathrm{m}_1) \Rightarrow (u_2, \mathrm{m}_2)$ iff there is a production $(A \to w, \mathrm{m})$ in P such that $u_1 = u'Au''$, $u_2 = u'wu''$, and $\mathrm{m}_2 = \mathrm{m}_1 \cdot \mathrm{m}$. Let*

$$L(G) = \{w \in T^* : (S, 1) \stackrel{*}{\Rightarrow} (w, 1)\}.$$

The corresponding language classes are denoted by $\mathcal{L}(\mathrm{Val}, X, \mathfrak{M})$, where $X \in \{\mathrm{REG}, \mathrm{LIN}, \mathrm{CF}\}$ refers to the permitted core rules. If we want to exclude erasing core rules, we use the notation $\mathcal{L}(\mathrm{Val}, X - \lambda, \mathfrak{M})$.

We consider the following variant:

Definition 2. *A (context-free) valuated grammar (over a monoid $\mathfrak{M}$) is given by a tuple $G = (N, T, P, S, \mathfrak{M}, \varphi)$, where P is a set of productions $A \to w$, $\mathfrak{M}$ is a multiplicatively written monoid with neutral element 1, and $\varphi : T \to \mathfrak{M}$ induces a monoid morphism. Let $\Rightarrow$ refer to the derivation relation of the underlying grammar (N, T, P, S). Then, let*

$$L(G) = \{w \in T^* : S \stackrel{*}{\Rightarrow} w \wedge 1 = \varphi(w)\}.$$

The corresponding language classes are denoted by $\mathcal{L}(\text{Val}', X, \mathfrak{M})$, *where* $X \in \{\text{REG}, \text{LIN}, \text{CF}\}$ *refers to the permitted core rules.*

Definition 3. *A language* $L \subseteq T^*$ *is recognized by the groupoid* $\mathfrak{G}$ *if there exists a finite subset* $F \subseteq \mathfrak{G}$ *and a "translation function"* $\theta : T \to \mathfrak{G}$ *such that* $L = \{w \in T^* : \mathfrak{G}(\theta(w)) \cap F \neq \emptyset\}$. *More precisely, we write* $L = L(\mathfrak{G}, T, \theta, F)$, *and we call the quadruple* $(\mathfrak{G}, T, \theta, F)$ *the recognizer of* L. *A recognizer is called*

- regular *if* $\mathfrak{G}$ *is a finite monoid;*
- linear *if it satisfies the following properties:*
  1. $\mathfrak{G}$ *is a finite groupoid;*
  2. $\mathfrak{G}$ *has a zero element* $0 \notin \theta(T)$ *and a neutral element* $1 \notin \theta(T)$;
  3. $\forall x, y \in \mathfrak{G}(x, y \notin (\theta(T) \cup \{1\}) \Rightarrow x \cdot y = 0)$; *and*
  4. $\forall x, y \in \mathfrak{G}(x \cdot y \notin \theta(T))$;
- context-free *if* $\mathfrak{G}$ *is a finite groupoid.*

More specifically, in the following we treat groupoids which are in fact product groupoids of a **finite** groupoid $\mathfrak{G}$ and a monoid $\mathfrak{M}$. In this case, a language $L \subseteq T^*$ is normally recognized by the groupoid $\mathfrak{G} \times \mathfrak{M}$ if there exists a finite subset $F \subseteq \mathfrak{G}$ and a "translation function" $\theta : T \to \mathfrak{G} \times \mathfrak{M}$ such that

$$L = \{w \in T^* : ((\mathfrak{G} \times \mathfrak{M})(\theta(w))) \cap (F \times \{1\}) \neq \emptyset\}.$$

Let $\mathcal{L}(\mathbb{G}, \mathfrak{M})$ denote the family of languages normally recognized by some $\mathfrak{G} \times \mathfrak{M}$, where $\mathfrak{G}$ is some finite groupoid and $\mathfrak{M}$ is some monoid. If, in addition, $(\mathfrak{G}, T, \theta, F)$ is a linear (or regular, resp.) recognizer, we come to the language class $\mathcal{L}(\mathbb{L}, \mathfrak{M})$ (or $\mathcal{L}(\mathbb{M}, \mathfrak{M})$, resp.).

Definition 4. *A* valence transducer (over a monoid $\mathfrak{M}$), *is a tuple*

$$g = (Q, I, O, q_0, F, P, \mathfrak{M}),$$

where (Q, I, O, q_0, F, P) *is a finite transducer whose moves are encoded by rewriting rules of the form* $qw \to xq'$, *where* $q, q' \in Q$, $w \in I^*$ *and* $x \in O^*$. *The rules are labelled by elements of* $\mathfrak{M}$.

For a derivation sequence

$$p_0 w_1 \cdots w_n \Rightarrow x_1 p_1 w_2 \cdots w_n \overset{*}{\Rightarrow} x_1 \cdots x_{n-1} p_{n-1} w_n \Rightarrow x_1 \cdots x_n p_n$$

with $w_i \in I^*$, $x_i \in O^*$, $p_i \in Q$, $(p_{i-1} w_i \to x_i p_i, \mathfrak{m}_i) \in P$, $1 \leq i \leq n$, $\mathfrak{m}_1 \cdots \mathfrak{m}_n = \mathfrak{m} \in \mathfrak{M}$, *we write*

$$(qw_1 \cdots w_n \overset{*}{\Rightarrow} x_1 \cdots x_n p_n, \mathfrak{m})$$

as a shorthand. Only rewritings $(q_0 w \overset{*}{\Rightarrow} xq, 1)$, *where* $w \in I^*$, $x \in O^*$, *and* $q \in F$ *are accepted as completely correct. In that case, we also write* $x \in g(w)$.

We call a relation $\tau \subseteq I^* \times O^*$ *a* valence relation (over a monoid $\mathfrak{M}$) *if there exists a valence transducer (over a monoid $\mathfrak{M}$)*

$$g = (Q, I, O, q_0, F, P, \mathfrak{M})$$

such that $\tau = \{(w,x) : x \in g(w) \wedge w \in I^* \wedge x \in O^*\}$. We write τ_g in order to denote the valence relation realized by the valence transducer g.

For a language family $\mathcal{L}$ and a monoid $\mathfrak{M}$, let

$$TD(\mathcal{L}, \mathfrak{M}) = \{\tau_g(L) : g \text{ is a valence transducer over } \mathfrak{M}, \ L \in \mathcal{L}\}.$$

A valence transducer $g = (Q, I, O, q_0, F, P, \mathfrak{M})$ is called a valence generalized sequential machine, or valence gsm for short, if the underlying finite transducer is a gsm, i.e., has rewriting rules of the form $qa \to xq'$, where $q, q' \in Q$, $a \in I$ and $x \in O^*$. In analogy, we use $GSM(\mathcal{L}, \mathfrak{M})$ for collecting all the $g(L)$ for some valence gsm g over $\mathfrak{M}$ and some $L \in \mathcal{L}$.

A valence transducer g is called λ-free if there are no productions of the form $qw \to q'$. A valence gsm g is called deterministic if $qa \to xp \in P$ and $qa \to x'p' \in P$ imply $x = x'$ and $p = p'$.

The following fact is known from [7,12]. It connects valence languages and valence transducers.

Proposition 1. *For* $\mathfrak{M} \in \{(\mathbb{Z}^n, +, 0), (\mathbb{Q}_+, \cdot, 1) : n \geq 1\}$ *and* $X \in \{\text{REG}, \text{CF}\}$, *we have* $\mathcal{L}(\text{Val}, X, \mathfrak{M}) = GSM(\mathcal{L}(X), \mathfrak{M})$. □

Below, we will strengthen the previous proposition.

3 Simple Properties and Auxiliary Results

We start with observing an obvious inclusion relation between valuated and valence language families.

Lemma 1. *Let* $\mathfrak{M}$ *be some commutative monoid. Let* $X \in \{\text{REG}, \text{LIN}, \text{CF}\}$. *Then,* $\mathcal{L}(\text{Val}', X, \mathfrak{M}) \subseteq \mathcal{L}(\text{Val}, X, \mathfrak{M})$.

Proof. Firstly, we extend φ to $V = T \cup N$ by $A \mapsto \{1\}$ for $A \in N$. For every rule $A \to w$, we introduce a rule $(A \to w, \varphi(w))$ in the simulating valence grammar. Since $\mathfrak{M}$ is commutative, the order in which the productions are applied in the valence grammar does not matter. □

In the case of non-commutative monoids, a grammar constructed as in the proof above would contain "additional words" due to malicious derivations, as explained by the following example.

Example 1. Consider the non-commutative monoid $(\mathfrak{M}, \cdot, 1)$ given as the factor monoid of the free word monoid $(\{a,b\}^*, \cdot, \lambda)$ by the congruence relation $ab \equiv \lambda$. It is well-known that the language corresponding to the congruence class $[\lambda]$ is just the one-sided Dyck language $\mathbb{D}_1$ with left parenthesis a and right parenthesis b. Let $\varphi : \{a,b\}^* \to \mathfrak{M}$ be the canonical (factor) morphism. If we consider the valuated grammar

$$G = (\{S\}, \{a,b\}, \{S \to SS, S \to a, S \to b, S \to \lambda\}, S, \varphi),$$

then $L(G) = \mathbb{D}_1$. If we take the proof above, we obtain the following valence grammar

$$G' = (\{S\}, \{a,b\}, \{(S \to SS, 1), (S \to a, a), (S \to b, b), (S \to \lambda, 1)\}, S, \mathfrak{M}).$$

Obviously, $L(G') = \{ w \in \{a,b\}^* : |w|_a = |w|_b \}$ is the two-sided Dyck language. $\mathbb{D}_1$ can be generated by a (Val′, REG, $\mathfrak{M}$) grammar, namely,

$$G = (\{S\}, \{a,b\}, \{S \to aS, S \to bS, S \to \lambda\}, S, \varphi),$$

and, hence, by a (Val, REG, $\mathfrak{M}$) grammar, namely

$$G' = (\{S\}, \{a,b\}, \{(S \to aS, a), (S \to bS, b), (S \to \lambda, 1)\}, S, \mathfrak{M}).$$

□

We do not know any concrete example of a non-commutative monoid $\mathfrak{M}$ which leads to a valuated context-free language not lying in $\mathcal{L}(\text{Val}, \text{CF}, \mathfrak{M})$. On the other hand, the proof of Lemma 1 yields:

Lemma 2. *Let $\mathfrak{M}$ be some arbitrary monoid. Then, we have:*

$$\mathcal{L}(\text{Val}', \text{REG}, \mathfrak{M}) \subseteq \mathcal{L}(\text{Val}, \text{REG}, \mathfrak{M}).$$

□

It is well-known that not admitting erasing productions in regular, linear or context-free grammars only influences the (non-)derivability of the empty word. Therefore, we can observe:

Proposition 2. *Let $\mathfrak{M}$ be some monoid. For $X \in \{\text{REG}, \text{LIN}, \text{CF}\}$, we have: $\mathcal{L}(\text{Val}', X, \mathfrak{M}) = \mathcal{L}(\text{Val}', X - \lambda, \mathfrak{M})$.* □

A classical topic of formal languages is to find normal forms. The following simple normal forms are needed as prerequisites to our algebraic approach. Since, in valuated grammars, the derivation is only determined by the underlying "classical" grammar, these normal form results are immediate corollaries from classical language theory, see, e.g., [6]. Recall that a grammar is termed *invertible* if, for each right-hand side, there is at most one left-hand side.

Lemma 3. *For any (Val′, LIN, $\mathfrak{M}$) grammar, there is an equivalent (Val′, LIN, $\mathfrak{M}$) grammar $G = (N, T, P, S, \mathfrak{M}, \varphi)$ satisfying:*

- *the only rules involving the start symbol S are of the form $S \to \lambda$ or $S \to A$ for some $A \in N \setminus \{S\}$ and*
- *every rule $A \to w \in P$, $A \in N \setminus \{S\}$, satisfies either $w \in T$, $w \in TN$ or $w \in NT$.*

□

Lemma 4. *Let $\mathfrak{M}$ be an arbitrary monoid. For every (Val′, CF, $\mathfrak{M}$) grammar, there exists an equivalent invertible (Val′, CF, $\mathfrak{M}$) grammar $G = (N, T, P, S, \mathfrak{M}, \varphi)$ in Chomsky normal form such that the only rules involving the start symbol S are of the form $S \to \lambda$ or $S \to A$ for some $A \in N \setminus \{S\}$.* □

From [5, Theorem 5.1], we may conclude the non-trivial normal form theorem for valence grammars:

Lemma 5. *Let $\mathfrak{M}$ be some commutative monoid. For every (Val, CF, $\mathfrak{M}$) grammar, there exists an equivalent (Val, CF, $\mathfrak{M}$) grammar $G = (N, T, P, S, \mathfrak{M}, \varphi)$ in Greibach normal form such that the only rules involving the start symbol S are of the form $S \to \lambda$ or $S \to A$ for some $A \in N \setminus \{S\}$.* □

We conclude by listing some remarks concerning the algebraic notions.

Lemma 6. *The monoid $\mathbb{Q}_+ \times \mathbb{Q}_+$ is isomorphic to $(\mathbb{Q}_+, \cdot, 1)$.* □

Lemma 7. *Let $\mathfrak{M}$ be a finite monoid and $F \subseteq \mathfrak{M}$. Consider a morphism $\theta : T^* \to \mathfrak{M}^*$, where T is some alphabet. If $R = (\mathfrak{G}, T, \theta', F')$ is a regular (or, resp., linear or context-free) recognizer, then $\bar{R} = (\mathfrak{G} \times \mathfrak{M}, T, \bar{\theta}, \bar{F})$ is a regular (or, resp., linear or context-free) recognizer, as well, where $\bar{\theta}(w) = (\theta'(w), \theta(w))$ and $\bar{F} = F' \times F$.*

Proof. Observe that

$$(\mathfrak{G} \times \mathfrak{M})(\bar{\theta}(w)) \cap \bar{F} = (\mathfrak{G} \times \mathfrak{M})((\theta'(w), \theta(w))) \cap (F' \times F) \neq \emptyset$$

iff $\mathfrak{G}(\theta'(w)) \cap F' \neq \emptyset$ and $\mathfrak{M}(\theta(w)) \cap F \neq \emptyset$, since the bracketization in $\mathfrak{M}$ is arbitrary.
□

Example 2. Consider the groupoid $\mathfrak{G} = \{a, b, A, B, 0, 1\}$ defined by $aB = ab = A$ and $Ab = B$; all other multiplications of elements from $\{a, b, A, B\}$ yield 0. Taking θ as the inclusion mapping from $T = \{a, b\}$ into $\mathfrak{G}$ and setting $F = \{A\}$, we obtain a linear recognizer $(\mathfrak{G}, T, \theta, F)$ for $\{a^n b^n : n > 0\}$. For example, $\mathfrak{G}(\theta(aabb))$ contains A due to the bracketization $\theta(a)((\theta(a)\theta(b))\theta(b)) = \theta(a)(A\theta(b)) = \theta(a)B = A$. □

4 Characterizations

This section contains two new algebraic characterizations: one of linear languages[1] and one of valuated languages.

Theorem 1. *A language is regular (or, resp., linear or context-free) if it is describable by a regular (or, resp., linear or context-free) recognizer.*

Proof. For the regular and context-free cases, refer to [1,2,4].[2] We show the construction in the linear case. For a detailed proof, refer to the long version of the paper. Let $\mathfrak{G} = (\{g_1, \ldots, g_k\}, \cdot, g_1 = 1)$ be a finite groupoid. Let $\theta : T^* \to \mathfrak{G}^*$ be a monoid morphism such that $(\mathfrak{G}, T, \theta, F)$ is a linear recognizer of L. We construct a linear grammar $G' = (N, T, P, A_0)$, $N = \{A_i : 0 \leq i \leq k\}$, as follows:

$$P = \{A_i \to a : \theta(a) = g_i\}$$
$$\cup \{A_0 \to A_i : g_i \in F\} \cup \{A_1 \to \lambda : g_1 \in F\}$$
$$\cup \{A_i \to A_j a : g_j \theta(a) = g_i\}$$
$$\cup \{A_i \to aA_j : \theta(a)g_j = g_i\}.$$

On the other hand, let $G = (N, T, P, S)$ be a linear grammar in the normal form according to Lemma 3, and let $V = N \cup T$. Define the groupoid $\mathfrak{G} = 2^V \cup \{1\}$ by

$$A_1 \circ A_2 := \{X \in N : X \to X_1 X_2 \in P, X_i \in A_i\}.$$

---
[1] A. Muscholl (1992) and F. Lémieux (2001) both wrote unpublished manuscripts which contain slightly different groupoid characterizations of the linear languages.
[2] Observe that in the regular case, the assertion is basically Kleene's Theorem in algebraic disguise.

The zero element is just the empty set, and the neutral element is 1. The translation function is given by: $\theta(a) = \{a\} \cup \{A \in N : A \to a \in P\}$. Define

$$F = \{N' \subseteq N \setminus \{S\} : \exists X \in N' : S \to X \in P\} \cup \{1 : S \to \lambda \in P\}.$$

Since $a \notin A_1 \circ A_2$ for all sets A_1, A_2 and terminals a, $(\mathfrak{G}, T, \theta, F)$ is a linear recognizer. □

Theorem 2. *For each monoid* $\mathfrak{M}$, $\mathcal{L}(\mathbb{G}, \mathfrak{M}) = \mathcal{L}(\text{Val}', \text{CF}, \mathfrak{M})$.
Similarly, $\mathcal{L}(\mathbb{M}, \mathfrak{M}) = \mathcal{L}(\text{Val}', \text{REG}, \mathfrak{M})$ *and* $\mathcal{L}(\mathbb{L}, \mathfrak{M}) = \mathcal{L}(\text{Val}', \text{LIN}, \mathfrak{M})$.

Proof. We only give the details of the context-free case, which is similar to [2, Lemma 3.1]. As to the (right-)linear cases, the constructions of [4] and Theorem 1 can be analogously modified. Let $\mathfrak{G} = (\{g_1, \ldots, g_k\}, \cdot, g_1 = 1)$ be a finite groupoid and $(\mathfrak{M}, \cdot, 1)$ be a monoid. Let $\theta : T^* \to (\mathfrak{G} \times \mathfrak{M})^*$ be a monoid morphism, such that $L = \{w \in T^* : ((\mathfrak{G} \times \mathfrak{M})(\theta(w))) \cap (F \times \{1\}) \neq \emptyset\}$. We construct a context-free valuated grammar $G' = (N, T, P, A_0, \mathfrak{M}, \varphi)$, $N = \{A_i : 0 \leq i \leq k\}$, as follows:

$$P = \{A_i \to a : \mathrm{pr}_1(\theta(a)) = g_i\}$$
$$\cup \{A_0 \to A_i : g_i \in F\} \cup \{A_1 \to \lambda : g_1 \in F\}$$
$$\cup \{A_i \to A_j A_{j'} : g_j g_{j'} = g_i\}.$$
$$\varphi(a) = \mathrm{pr}_2(\theta(a)).$$

By induction on the length of words, $L \in \mathcal{L}(\text{Val}', \text{CF}, \mathfrak{M})$ follows.

Now, let $L \in \mathcal{L}(\text{Val}', \text{CF}, \mathfrak{M})$ be given by a context-free valuated grammar $G = (N, T, P, S, \mathfrak{M}, \varphi)$ in the normal form described in Lemma 4. Let $N = \{S, g_1, \ldots, g_k\}$. We define the groupoid $G' = (\{g_1, \ldots, g_k, 1, 0\}, \cdot, 1)$, where 1 is the neutral and 0 the zero element. Define $A \cdot B = C$ iff $C \to AB \in P$ for every $A, B, C \in N$. Let

$$F = \{A \in N \setminus \{S\} : S \to A \in P\} \cup \{1 : \lambda \in L(G)\}.$$

Define θ by $\theta(a) = (A, \varphi(a))$ iff $A \to a \in P$. By induction on the length of words, the claim follows. □

5 Valence Transducers

For rational transducers, there exists the famous theorem of Nivat, see also [3]. We prove a similar statement for valence transducers in the following.

Theorem 3. $\tau \subseteq I^* \times O^*$ *is a valence relation over a monoid* $(\mathfrak{M}, \cdot, 1)$ *iff there exists an alphabet* Y, *a valuated language* $R \in \mathcal{L}(\mathbb{M}, \mathfrak{M})$ *with* $R \subseteq Y^*$, *a morphism* $f : Y^* \to I^*$, *and a morphism* $g : Y^* \to O^*$ *such that* $\tau = \{(f(y), g(y)) : y \in R\}$.

Observe that this theorem automatically implies Nivat's theorem for classical transducers, viewed as valence transducers over the trivial monoid.

Proof. Let τ be given by the valence transducer $g = (Q, I, O, q_0, F, P, \mathfrak{M})$. Formally, we set $Y := P$. Consider the set

$$R := \{(q_0 w_1 \to x_1 q_1) \cdots (q_{n-1} w_n \to x_n q_n) :$$
$$q_n \in F \wedge (\forall 1 \leq i \leq n (q_{i-1} w_i \to x_i q_i, \mathfrak{m}_i) \in P) \wedge \mathfrak{m}_1 \cdots \mathfrak{m}_n = 1\}.$$

R is generated by the valuated regular grammar $G = (N, T, P', S, \mathfrak{M}, \varphi)$ with $N = Q$, $T = Y$, $S = q_0$, φ is just the labelling function of g, and

$$P' = \{q \to (qw \to xq')q' : q, q' \in Q \wedge (qw \to xq') \in Y = P\} \cup$$
$$\{q \to \lambda : q \in F\}$$

Now, the two homomorphisms $f : Y \to I$ and $g : Y \to O$ are mainly projections defined by $(qw \to xq') \mapsto w$ and $(qw \to xq') \mapsto x$, respectively. It is easily seen that $\tau = \{(f(y), g(y)) : y \in R\}$.

On the other hand, let $G = (N, Y, P, S, \mathfrak{M}, \varphi)$ be a valuated regular grammar generating R, and let f and g the required morphisms. Without loss of generality, the productions of P are of the form $A \to aB$ or $A \to \lambda$ for $A, B \in N$, $a \in Y$. Consider the valence transducer $g = (Q, I, O, q_0, F, P', \mathfrak{M})$, where $Q = N$, $F = \{A \in N : A \to \lambda \in P\}$, and $P' = \{(Af(a) \to g(a)B, \varphi(a)) : A \to aB \in P\}$. It is easily shown by induction that $(Aw \overset{*}{\Rightarrow} xB, \mathfrak{m})$ iff $A \overset{*}{\Rightarrow} yB$ for some $y \in Y^*$ with $f(y) = w$, $g(y) = x$ and $\varphi(y) = \mathfrak{m}$. Hence, $\tau_g = \{(f(y), g(y)) : y \in R\}$. □

Similar characterizations can be given for valence gsm mappings (where f is required to be non-erasing) and λ-free valence transductions (where g has to be non-erasing). A formal proof of the gsm case appeared in the preproceedings' preliminary version of this paper.

Analogously to finite transducers, there is a *normal form for valence transducers*.

Theorem 4. *For any valence transduction τ, there is valence transducer $g = (Q, X, Y, q_0, F, P, \mathfrak{M})$ such that $\tau = \tau_g$, and all rules of P have the form $(qa \to q', \mathfrak{m})$ or $(q \to bq', \mathfrak{m})$ or $(q \to q', \mathfrak{m})$ with $q, q' \in Q$, $a \in I$, $b \in O$, $\mathfrak{m} \in \mathfrak{M}$.*

Proof. Let $\tau = \tau_{g'}$, for some valence transducer $h = (Q', I, O, q_0, F, P', \mathfrak{M})$. First, we construct the transducer g'', where any rule $p = (qw \to xq', \mathfrak{m})$ is replaced by the pair of rules $(qw \to q_p, \mathfrak{m})$ and $(q_p \to xq', 1)$. Obviously, g'' generates the same relation as g'.

In a second step, we obtain g from g'' by replacing any rule

$$p = (qa_1 \cdots a_\ell \to q', \mathfrak{m}), \quad \ell \geq 2,$$

in g'' by

$$(qa_1 \to q_{p,1}, \mathfrak{m}), \quad (q_{p,1} a_2 \to q_{p,2}, 1), \quad \ldots, \quad (q_{p,\ell-1} a_\ell \to q_{p,2}, 1),$$

and any rule

$$p = (q \to b_1 \cdots b_\ell q', \mathfrak{m}), \quad \ell \geq 2,$$

in g'' by

$$(q \to b_1 q_{p,1}, \mathfrak{m}), \quad (q_{p,1} \to b_2 q_{p,2}, 1), \quad \ldots, \quad (q_{p,\ell-1} \to b_\ell q_{p,2}, 1). \quad \square$$

By a similar construction, we can prove the following analogue for the λ-free case:

Corollary 1. *For any λ-free valence transduction τ, there is valence transducer $g = (Q, X, Y, q_0, F, P, \mathfrak{M})$ such that $\tau = \tau_g$, and all rules of P have the form $(qw \to bq', \mathfrak{m})$ with $q, q' \in Q$, $w \in I^*$, $b \in O$, and $\mathfrak{m} \in \mathfrak{M}$.* □

Marcus and Păun proved in [7, Theorem 2] an assertion on the composition of two valence gsm's. In the following, we generalize this statement within our framework.

Theorem 5. *Let $g = (Q, X, Y, q_0, F, P, \mathfrak{M})$ be a valence transducer over the monoid $(\mathfrak{M}, \cdot, 1)$ and $g' = (Q', Y, Z, q'_0, F', P', \mathfrak{M}')$ be a valence transducer over the monoid $(\mathfrak{M}', \cdot, 1')$. Then, the relation $\tau_g \circ \tau_{g'}$ is realized by a valence transducer g'' over the monoid $\mathfrak{M} \times \mathfrak{M}'$. g'' is a gsm if g and g' are gsm's. g'' is λ-free if g and g' are λ-free. g'' is a deterministic gsm if g and g' are deterministic gsm's.*

Proof. In the general case, we can assume that g and g' are in normal form. Consider $g'' = (Q \times Q', X, Z, (q_0, q'_0), F \times F', P'', \mathfrak{M} \times \mathfrak{M}')$, where

$$P'' = \{\,((s, s') \to (t, t'), (\mathfrak{m}, \mathfrak{m}'))\,:$$
$$(s \to bt, \mathfrak{m}) \in P \wedge (s'b \to t', \mathfrak{m}') \in P' \wedge b \in Y\,\}$$
$$\cup\,\{\,((s, s') \to c(s, t'), (1, \mathfrak{m}'))\,:\,(s' \to ct', \mathfrak{m}') \in P' \wedge c \in Z \cup \{\lambda\}\,\}$$
$$\cup\,\{((s, s')a \to (t, s'), (\mathfrak{m}, 1'))\,:\,(sa \to t, \mathfrak{m}) \in P \wedge a \in X \cup \{\lambda\}\,\}.$$

If g and g' are λ-free, then we can assume that g obeys the normal form of Cor. 1. Consider $g'' = (Q \times Q', X, Z, (q_0, q'_0), F \times F', P'', \mathfrak{M} \times \mathfrak{M}')$, where

$$P'' = \{\,((s, s')a \to z(s, t'), (1, \mathfrak{m}'))\,:\,(s' \to zt', \mathfrak{m}') \in P'\,\}$$
$$\cup\,\{((s, s')w \to z(t, t'), (\mathfrak{m}, \mathfrak{m}'))\,:$$
$$(sw \overset{*}{\Rightarrow} ys', \mathfrak{m})\text{ by }g \wedge (ty \to zt', \mathfrak{m}') \in P'\,\}.$$

Since g is in normal form, every $y \in Y^+$ occurring in the left-hand side of a rule $(ty \to zt', \mathfrak{m}') \in P'$ is trivially spelled in a unique way as $y = a_1 \ldots a_\ell$, $a_i \in Y$, where each a_i occurs in the right-hand side of a rule $(s_i w_i \to a_i s_{i+1}) \in P$.
If g and g' are gsm's, consider $g'' = (Q \times Q', X, Z, (q_0, q'_0), F \times F', P'', \mathfrak{M} \times \mathfrak{M}')$,

$$P'' = \{\,((s, s')a \to y(t, t'), (\mathfrak{m}, \mathfrak{m}'))\,:$$
$$(sa \to xt, \mathfrak{m}) \in P \wedge x \neq \lambda \wedge (s'x \overset{*}{\Rightarrow} yt', \mathfrak{m}')\text{ by }g'\,\}$$
$$\cup\,\{((s, s')a \to (t, s'), (\mathfrak{m}, 1'))\,:\,(sa \to t, \mathfrak{m}) \in P\,\}.$$

Obviously, if g and g' are deterministic gsm's, then the just constructed gsm g'' is also deterministic. □

By Lemma 6, [7, Theorem 2] becomes an easy corollary. More generally, we can state:

Corollary 2. *The class of valence relations over the monoid $(\mathbb{Q}_+, \cdot, 1)$ is closed under composition.* □

We further the analysis of results from [7,12].

Theorem 6. *Let $\mathfrak{M}$ be a commutative monoid, and let $X \in \{\text{REG}, \text{LIN}, \text{CF}\}$. Then, each language from $\mathcal{L}(\text{Val}, X, \mathfrak{M})$ can be represented as the image of a language from $\mathcal{L}(X)$ under a non-erasing one-state deterministic $\mathfrak{M}$-valence gsm.*[3]

Proof. Due to Lemma 5, $L \in \mathcal{L}(\text{Val}, \text{CF}, \mathfrak{M})$ can be generated by a non-erasing valence grammar $G = (N, T, P, S, \mathfrak{M})$ in Greibach normal form. We construct a context-free grammar $G' = (N, T \times P, P', S)$ in order to make the valences explicit. This means that $A \to (a, \mathfrak{m})B_1 \cdots B_m \in P'$ iff $(A \to aB_1 \cdots B_m, \mathfrak{m}) \in P$. Now, the non-erasing one-state deterministic gsm is simply defined by $((a, \mathfrak{m})q_0 \to aq_0, \mathfrak{m})$.

In the (right-)linear case, the only problems are posed by chain rules, which can be eliminated by the techniques exhibited in [5]. □

6 Closure Properties

Lemma 8. *For each monoid $\mathfrak{M}$ and for every $\mathbb{X} \in \{\mathbb{M}, \mathbb{L}, \mathbb{G}\}$, $\mathcal{L}(\mathbb{X}, \mathfrak{M})$ is closed under inverse morphisms.*

Proof. Let $L = \{w \in T^* : ((\mathfrak{G} \times \mathfrak{M})(\theta(w))) \cap (F \times \{1\}) \neq \emptyset\}$. Then, given some monoid morphism $h : Y^* \to X^*$,

$$h^{-1}(L) = \{v \in Y^* : ((\mathfrak{G} \times \mathfrak{M})(\theta(h(v)))) \cap (F \times \{1\}) \neq \emptyset\},$$

where $\theta \circ h : Y^* \to (\mathfrak{G} \times \mathfrak{M})^*$ is the new translation function. □

Let $\mathcal{H}(\mathcal{L})$ denote the closure of family $\mathcal{L}$ under homomorphisms, and let $\mathcal{C}(\mathcal{L})$ denote the closure of $\mathcal{L}$ under non-erasing letter-to-letter homomorphisms (codings). Due to Lemmas 1 and 8 and the trivial fact that $\mathcal{L}(\text{Val}, X, \mathfrak{M})$ is closed under morphisms for $X \in \{\text{REG}, \text{LIN}, \text{CF}\}$, we may deduce:

Lemma 9. *Let $\mathfrak{M}$ be a commutative monoid. Then,*
$\mathcal{L}(\text{Val}', \text{CF}, \mathfrak{M}) = \mathcal{L}(\mathbb{G}, \mathfrak{M}) \subseteq \mathcal{H}(\mathcal{L}(\text{Val}', \text{CF}, \mathfrak{M})) \subseteq \mathcal{L}(\text{Val}, \text{CF}, \mathfrak{M}).$
Similar statements are true for the (right-)linear cases.[4] □

Theorem 7. *Let $\mathfrak{M}$ be a commutative monoid, and let $X \in \{\text{REG}, \text{LIN}, \text{CF}\}$. Then, we have $\mathcal{C}(\mathcal{L}(\text{Val}', X, \mathfrak{M})) = \mathcal{L}(\text{Val}, X, \mathfrak{M})$.*

Proof. (Sketch) By Lemma 9, we have to prove $\mathcal{C}(\mathcal{L}(\text{Val}', X, \mathfrak{M})) \supseteq \mathcal{L}(\text{Val}, X, \mathfrak{M})$. By the proof of Theorem 6, this claim easily follows. □

As regards closure properties, we may state the following (using Berstel's terminology [3]):

Theorem 8.
1. *Let $\mathfrak{M}$ be some monoid. Then, for each $\mathbb{X} \in \{\mathbb{M}, \mathbb{L}, \mathbb{G}\}$, $\mathcal{L}(\mathbb{X}, \mathfrak{M})$ is a cylinder, i.e., they are closed under inverse morphisms and intersection with regular sets.*
2. *If the monoid $\mathfrak{M}$ is isomorphic to the product monoid $\mathfrak{M} \times \mathfrak{M}$, then $\mathcal{L}(\mathbb{M}, \mathfrak{M})$ is closed under intersection.*[5]

---
[3] These valence gsm might be viewed as "valence morphisms."
[4] In the regular case, we may also admit non-commutative monoids.
[5] One example of such a monoid is $(\mathbb{Q}_+, \cdot, 1)$, see Lemma 6.

3. Let $\mathfrak{M}$ be from $\{(\mathbb{Z}^n, +, 0), (\mathbb{Q}_+, \cdot, 1) : n \geq 1\}$, $X \in \{\text{REG}, \text{LIN}, \text{CF}\}$. Then, $\mathcal{L}(\text{Val}, X, \mathfrak{M})$ is the cone generated by $\mathcal{L}(\text{Val}', X, \mathfrak{M})$.[6]

Proof. (Sketch)

1. As regards inverse morphisms, we refer to Lemma 8. As to intersection with regular sets, the techniques exhibited in Lemma 7 can be used to prove the claim.
2. If $\mathfrak{M} \times \mathfrak{M}$ is isomorphic to $\mathfrak{M}$, then modify the hinted proof of the closure under intersection with regular sets.
3. This claim follows from the fact that blind counter languages form cones, as proved, e.g., in [12], together with Theorem 7.

□

7 Conclusions

We exhibited algebraic characterizations of linear and valuated languages, thereby extending now classical works on algebraic characterizations of the regular and of the context-free languages. As to valuated languages, we introduced the technical notion of *normal* recognizability. In the future, we plan to study also the language classes which result from the classical notion of recognizability. Moreover, it would be interesting to see whether algebraically defined subclasses of linear and valuated languages lead to useful language families.

References

1. D. A. M. Barrington and D. Thérien. Finite monoids and the fine structure of NC^1. *Journal of the Association for Computing Machinery*, 35(4):941–952, October 1988.
2. F. Bédard, F. Lémieux, and P. McKenzie. Extensions to Barrington's M-program model. *Theoretical Computer Science*, 107:31–61, 1993.
3. J. Berstel. *Transductions and Context-Free Languages*, volume 38 of *LAMM*. Stuttgart: Teubner, 1979.
4. S. Eilenberg. *Automata, Languages, and Machines, Volume B*. Pure and Applied Mathematics. New York: Academic Press, 1976.
5. H. Fernau and R. Stiebe. Sequential grammars and automata with valences. Technical Report WSI–2000–25, Universität Tübingen (Germany), Wilhelm-Schickard-Institut für Informatik, 2000. Revised version to appear in *Theoretical Computer Science*. Up to publication, it can be retrieved from the webpage of Elsevier.
6. M. A. Harrison. *Introduction to Formal Language Theory*. Addison-Wesley series in computer science. Reading (MA): Addison-Wesley, 1978.
7. M. Marcus and Gh. Păun. Valence gsm-mappings. *Bull. Math. Soc. Sci. Math. Roumanie*, 31(3):219–229, 1987.
8. Gh. Păun. A new generative device: valence grammars. *Rev. Roumaine Math. Pures Appl.*, XXV(6):911–924, 1980.
9. J.-E. Pin. *Varieties of formal languages*. New York: Plenum Press, 1986.
10. V. Red'ko and L. Lisovik. Regular events in semigroups (in Russian). *Problems of Cybernetics*, 37:155–184, 1980.

---

[6] Due to [5], these are the only interesting commutative monoids for valence grammars.

11. D. Thérien. Classification of finite monoids: the language approach. *Theoretical Computer Science*, 14:195–208, 1981.
12. S. Vicolov-Dumitrescu. Grammars, grammar systems, and gsm mappings with valences. In Gh. Păun, editor, *Mathematical Aspects of Natural and Formal Languages*, volume 43 of *World Scientific Series in Computer Science*, pages 473–491. Singapore: World Scientific, 1994.

Context-Free Valence Grammars – Revisited

Hendrik Jan Hoogeboom

Institute for Advanced Computer Science
Universiteit Leiden, The Netherlands
http://www.liacs.nl/~hoogeboo/

Abstract. Context-free valence languages (over $\mathbb{Z}^k$) are shown to be codings of the intersection of a context-free language and a blind k-counter language. This AFL-style characterization allows one to infer some of the properties of the family of valence languages, in particular the λ-free normal form proved by Fernau and Stiebe.

1 Introduction

Valence grammars were introduced by Păun as a new way of regulated rewriting ([Pau80], or [DP89, p. 104]): associate with each production of a Chomsky grammar an integer value, the valence of the production, and consider only those derivations for which the valences of the productions used add to zero. The formalism is surprisingly simple, and adding valences to context-free grammars one obtains a family of languages strictly in between context-free and context-sensitive. More importantly, that family has convenient closure properties.

The formalism has been extended to valence monoids other than $\mathbb{Z}$, like positive rational numbers ($\mathbb{Q}^+$ under multiplication), or vectors of integers ($\mathbb{Z}^k$ under componentwise addition).

Fernau and Stiebe study the context-free valence grammars over arbitrary monoids in a uniform presentation, and give additional results for the context-free valence grammars over $\mathbb{Z}^k$ ([FS97], full paper [FS00]). One of their important contributions is that the Chomsky and Greibach normal forms hold in the valence case. This is somewhat unexpected, as chain productions and λ-productions do not contribute to the derived string, and yet their valences have to be taken into account – which makes their removal a nontrivial task.

This paper reconsiders those basic properties of context-free valence grammars over $\mathbb{Z}^k$. It has been observed in the cited papers that the valence mechanism is very similar to the storage type used by the blind counter automata of Greibach [Gre78], in fact, regular valence grammars are easily seen to be equivalent to blind counter automata. Here we make this connection explicit in another way. In the style of AFL theory [Gin75] we characterize context-free valence languages (over $\mathbb{Z}^k$) as homomorphic images of the intersection of a context-free language and a blind k-counter language, see Lemma 1. In this fashion we have explicitly separated the context-free derivation process from the phase where the valences are checked. As the context-free languages and the blind counter language share specific closure properties, generic results from AFL theory allow us to conclude those properties for context-free valence languages (Theorem 2).

Then the characterization is improved by replacing the (arbitrary) homomorphisms by codings (Theorem 1). Technically this is much more complicated, involving the prediction of the valences of derivations that yield the empty string (ending in λ-productions). Apart from its elegance, this characterization has some immediate use. A classic result of Latteux shows that the blind k-counter languages are accepted by *real-time* automata, i.e., automata reading a symbol in each step. The construction of a valence grammar starting with a context-free grammar and such a real-time blind counter automaton does not change the shape of the grammar. From this we immediately conclude the existence of Chomsky and Greibach normal forms for context-free valence grammars, see Theorems 4 and 5, i.e., directly from the existence of those normal forms for context-free grammars. This means that we have avoided the lengthy and involved technical analysis of the derivations in valence grammars [FS00, Sections 5.3–5.5], at the "cost" of applying an abstract result from AFL theory (which thus hides the actual complexity of the construction).

Our motivation to study valence language comes from the formal language theory of splicing, see the Appendix.

2 Preliminaries

The reader is assumed to be familiar with standard formal language notions, see [Sal73] for a broad introduction. After some generalities, this section presents basic facts on language families, valence grammars, and blind counter automata.

We use λ to denote the empty string, and $\|$ to denote the shuffle operation on languages.

For a vector $v \in \mathbb{Z}^k$, $\|v\|$ denotes the 1-norm $\sum_{i=1}^{k} |v_i|$, where $v = (v_1, \ldots, v_k)$. Vectors in $\mathbb{Z}^k$ are added componentwise; the vector all of which elements are 0 is denoted by **0**.

Language Families. We use REG and CF to denote the families of regular and of context-free languages. The family of homorphisms is denoted by HOM; the family of codings (letter-to-letter homomorphisms) by COD. Thus, e.g., HOM($\mathcal{F}$) denotes the family of homomorphic images of languages from $\mathcal{F}$.

For language families $\mathcal{F}$ and $\mathcal{G}$, $\mathcal{F} \wedge \mathcal{G}$ denotes $\{F \cap G \mid F \in \mathcal{F}, G \in \mathcal{G}\}$.

The family $\mathcal{F}$ is a *trio* (or *faithful cone*) if it is closed under λ-free homomorphisms, inverse homomorphisms, and intersection with regular languages. A *full trio* (or *cone*) is additionally closed under (arbitrary) homomorphisms. The smallest trio and full trio containing the language L are denoted by $\mathcal{M}(L)$, and $\widehat{\mathcal{M}}(L)$, respectively. It can be shown that

$$\widehat{\mathcal{M}}(L) = \{\, g(h^{-1}(L) \cap R) \mid h, g \in \mathsf{HOM}, R \in \mathsf{REG} \,\}.$$

For $\mathcal{M}(L)$ there is an analogous characterization where g is λ-free (Corollary 1 to Theorem 3.2.1 in [Gin75]). A (full) trio of this form is called (full) *principal*, i.e., it is generated by a single language L. It is automatically a (full) *semi*-AFL, i.e., additionally closed under union [Gin75, Proposition 5.1.1].

As explained in [Gin75, Example 5.1.1], it follows from the Chomsky-Schutzenberger Theorem that CF equals $\widehat{\mathcal{M}}(D_2)$; the full trio generated by D_2, the Dyck set on two symbols, a language generated by the context-free grammar with productions $S \to SS \mid a_1 S b_1 \mid a_2 S b_2 \mid \lambda$.

Valence Grammars. A *valence grammar* over $\mathbb{Z}^k$ is a context-free grammar in which every production has an associated value (the valence) from $\mathbb{Z}^k$. A string is in the language of the valence grammar if it can be derived in the usual way, under the additional constraint that the valences of the productions used add up to $\mathbf{0}$.

Formally the grammar is specified as four-tuple $G = (N, T, R, S)$, where N, T are alphabets of *nonterminals* and *terminals*, $S \in N$ is the *axiom*, and R is a finite subset of $N \times (N \cup T)^* \times \mathbb{Z}^k$, the *rules*. When $(A, w, \boldsymbol{r})$ is an element of R, then we write as usual $(A \to w, \boldsymbol{r})$, and $A \to w$ is the (underlying) *production* and $\boldsymbol{r}$ the *valence* of the rule.

Omitting the valences from the rules of G, we obtain the underlying context-free grammar of G. The derivation relation of G is an obvious extension of the one for context-free grammars; for $x, y \in (N \cup T)^*$ and $\boldsymbol{v} \in \mathbb{Z}^k$ we write $(x, \boldsymbol{v}) \Rightarrow (y, \boldsymbol{v} + \boldsymbol{r})$ if there is a rule $(A \to w, \boldsymbol{r})$, such that $x = x_1 A x_2$, and $y = x_1 w x_2$. Then, the *language* of G equals $L(G) = \{ w \mid w \in T^*, (S, \mathbf{0}) \Rightarrow^* (w, \mathbf{0}) \}$. The family of all *valence languages* over $\mathbb{Z}^k$ is denoted here by $\mathsf{VAL}(\mathbb{Z}^k)$.

Blind Counter Automata. Let $k \geq 0$. A k *blind counter automaton*, or kBCA, is a finite state automaton with k integers $\boldsymbol{v} = (v_1, v_2, \ldots, v_k)$ as additional external storage. In each step the automaton may increment and decrement each of its counters. The storage is called *blind* as the automaton is not allowed to test the contents of the counters during the computation. The only, implicit, test is at the end of the computation, as we accept only computations from initial state to final state, that start and end with empty counters.

Formally, an instruction of a kBCA is of the form $(p, a, q, \boldsymbol{r})$, where p, q are states, a is an input symbol or λ, and $\boldsymbol{r} \in \mathbb{Z}^k$. Executing this instruction, the automaton changes from state p into state q, reads a from its input tape, and changes the counter values from $\boldsymbol{v}$ into $\boldsymbol{v} + \boldsymbol{r}$.

The automaton is called *real-time* if none of the instructions read λ.

We use $k\mathsf{BC}$ to denote the family of languages accepted by kBCA.

According to standard AFA interpretation (cf. [Gin75, Lemma 5.2.3]) an automaton can be seen as a finite state transducer translating input symbols into sequences of storage instructions; the input string is accepted only if the resulting sequence of storage instructions is "legal", i.e., it can be executed by the storage, leading from initial to final storage. This implies that the languages accepted are precisely the inverses under finite state transductions of the language of legal storage instructions.

Let Σ_k be the alphabet $\{a_1, b_1, \ldots, a_k, b_k\}$. Define $B_k = \{x \in \Sigma_k^* \mid \#_{a_i}(x) = \#_{b_i}(x)$ for each $1 \leq i \leq k\}$. Observe that B_k models the possible legal instruction sequences to the blind counter storage, interpreting a_i and b_i as increments and decrements of the i-th counter.

As every transducer (and its inverse) can be decomposed into homomorphisms and intersection with regular languages (known as Nivat's Theorem, cf. equation 2.4 in

[Sal73, Chapter IV]), the language accepted by a kBCA can be written as $g(h^{-1}(B_k) \cap R)$, where h, g are homomorphisms, and R is a regular language.

This being the characterization of $\widehat{\mathcal{M}}(B_k)$ that was recalled on the previous page, it then follows that kBC equals $\widehat{\mathcal{M}}(B_k)$; the full trio generated by B_k. For real-time kBCA h is a λ-free homomorphism, and we obtain along the same lines the trio $\mathcal{M}(B_k)$ generated by B_k.

By a result of Greibach, [Gre78, Theorem 2], blind counter automata are equivalent to their real-time restriction. Unfortunately, the proof of this result does not keep the number of counters constant, see the proof of Lemma 1 in [Gre78], which states "[..] this can be done [..] with delay 0, *by adding another counter*".

The equivalence of the real-time restriction for k blind counter automata, for every fixed k, is a special case of a general language theoretic result of Latteux [Lat79].

Proposition 1. *Real-time kBCA are as powerful as (unrestricted) kBCA.*

Proof. By [Lat79, Theorem II.11] $\widehat{\mathcal{M}}(L) = \mathcal{M}(L)$ for every permutation closure L of a regular language. Hence the proposition follows, as B_k is the permutation closure of the regular language $(a_1 b_1)^* \cdots (a_k b_k)^*$. □

For kBCA we have a normal form in which the automaton never changes more than one of its counter values in each step.

Proposition 2. *For each (real-time) kBCA there exists an equivalent (real-time) kBCA such that $\|\boldsymbol{r}\| \leq 1$ for each instruction $(p, a, q, \boldsymbol{r})$.*

Proof. We use a variant of the normalisation procedure for counter automata, see [FMR68] – a rather flexible method, cf. [HvV00, Lemma 11].

Let m be the maximal amount by which any of the counters can change in one step; hence in k consecutive steps each of the counters is changed by a value at most $\pm km$. The normal form is obtained by accessing the counters in turn, such that in k consecutive steps each of the counters is changed (at most) once. We store the original counter value divided by $2km$ on the counter, while leaving the remainder (the original counter value modulo $2km$) in the finite state.

To be more precise, when a particular counter has just been accessed, the finite state keeps a remainder for that counter in the interval $[-km, \ldots, km - 1]$. In the next $k - 1$ steps we do not update that counter, but store the changes in the finite memory. Thus, the remainder grows to $v \in [-2km + m, \ldots, 2km - m - 1]$. After the k steps we may again access that particular counter. If the original automaton at that moment wants to add $r \in [-m, \ldots, m]$ to the counter, the new automaton now computes $v' = v + r$ in its finite state. If $v' \geq km$ it increments the counter (by one) and stores $v' - 2km$ in the state; if $v' < -km$ it decrements the counter and stores $v' + 2km$ in the state.

Acceptance as before, by final state and empty counter; a state is final if it is final in the original automaton and if it stores 0. □

3 Characterization of VAL($\mathbb{Z}^k$)

Let G be a valence grammar. We say that G is in *2-normal form* if G has only productions of the form $A \to BC$, $A \to B$, $A \to a$, and $A \to \lambda$, where $A, B, C \in N$, and $a \in T$. Indeed this is a normal form, and can be obtained easily similar to Chomsky normal form for context-free grammars, by splitting longer productions, and by introducing "shadow"-nonterminals that introduce terminals ($N_a \to a$) whenever necessary [FS00, Theorem 4.2].

We start by showing that valence languages are the homomorphic images of the intersection of a context-free language and a blind k-counter language. This reminds us of the characterization in the proof of Theorem 6 in [Pau80], in our notation $L(G) = h(L(G') \cap (B_1 || T^*))$, where G is a *regular* valence grammar over $\mathbb{Z}^1$ and G' is a regular grammar constructed from G.

Lemma 1. VAL($\mathbb{Z}^k$) = HOM(CF $\wedge$ kBC).

Proof. Let G be a valence grammar over $\mathbb{Z}^k$ with terminal alphabet T. We obtain a context-free grammar G' from G by replacing each rule $(A \to w, \boldsymbol{r})$ by the production $A \to \boldsymbol{r}w$, and by adding the valence vectors that occur in the rules of G to the terminal alphabet T.

Let $\mathcal{A}$ be the blind counter automaton that operates on the terminal alphabet of G', and which interprets the vectors $\boldsymbol{r}$ by adding their values to its counters. The automaton ignores the terminal symbols from T. It is clear that $L(G)$ is equal to $h(L(G') \cap L(\mathcal{A}))$, where h is the homomorphism that acts as identity on T, while it erases the valence vectors.

The converse implication can be obtained using a variant of the classical triple construction for the intersection of a context-free language with a regular language, as used in [FS00, Theorem 4.1] to show closure properties of VAL($\mathbb{Z}^k$). Starting with a context-free grammar G in Chomsky normalform, a real-time blind counter automaton $\mathcal{A}$, and a homomorphism h, we introduce rules $([p, A, r] \to [p, B, q][q, C, r], \boldsymbol{0})$ for each production $A \to BC$ of G and states p, q, r of $\mathcal{A}$. Moreover we add terminal rules $([p, A, q] \to h(a), \boldsymbol{r})$ whenever there is an instruction $(p, a, q, \boldsymbol{r})$ of $\mathcal{A}$, and a production $A \to a$ of G.

The axiom S of the valence grammar gets rules $(S \to [p, S, q], \boldsymbol{0})$ such that p is the initial state of $\mathcal{A}$, and q a final state of $\mathcal{A}$. If we want, we can get rid of those initial chain-rules by combining them with rules for the nonterminals $[p, S, q]$. □

As an immediate consequence of this basic characterization we can infer the main closure properties of VAL($\mathbb{Z}^k$), see Theorem 2 below.

The main difficulty in improving the characterization to codings instead of arbitrary homomorphisms, is the removal of λ-rules from the grammar, while taking into account their valences. The main tool is the observation that the *Szilard language* of a context-free grammar, the production sequences used in derivations, has a semilinear Parikh image (see [DP89, Lemma 2.1.9], [FS00, Proposition 5.6]). This implies that we can count the corresponding valences using a sequential device, in our case a blind counter automaton.

For a sequence p of productions, we write $x \Rightarrow^p y$ if y is derived from x using the sequence p.

Lemma 2. Let $G = (N, T, P, S)$ be a context-free grammar, and let A and B be nonterminals of G. The set $\{ p \in P^* \mid A \Rightarrow^p B \}$ has a semilinear Parikh image.

Proof. Replace each production $\pi : C \to \alpha$ by the production $C \to \pi\alpha$, where π is used as a label to represent the production. Consider the sentential forms of the new grammar, using A as axiom. This is a context-free language, and it remains context-free after intersection with P^*BP^* to select the derivations matching $A \Rightarrow^* B$. By Parikh's theorem, the resulting language has a semilinear image (from which we may drop the occurrence of B). □

We now show that it is possible to replace the homomorphisms in Lemma 1 by codings. For this we need the notion of the *binary structure* of a derivation tree of a valence grammar in 2-normal form (or of its underlying context-free grammar). It suffices to introduce that notion in an informal way. This binary structure is created from the derivation tree by deleting each node that has no terminal symbols among its descendants. The remaining tree can be decomposed into "linear" derivations $(A, \mathbf{0}) \Rightarrow^*_G (A', \mathbf{v})$ leading into a rule which is either of the type $(A' \to BC, \mathbf{r})$ or of the type $(A' \to a, \mathbf{r})$. In fact, this decomposition by the binary structure induces a decomposition of the original derivation itself (but we need to find a way to include the valences of the rules that were cut from the tree).

Theorem 1. $\mathsf{VAL}(\mathbb{Z}^k) = \mathsf{COD}(\mathsf{CF} \wedge k\mathsf{BC})$.

Proof. By Lemma 1 it suffices to prove the inclusion from left to right. Let $G = (N, T, R, S)$ be a valence grammar over $\mathbb{Z}^k$. We may assume that G is in 2-normal form. In Lemma 1 the new grammar G' was constructed such that the (valence of) every production used in the derivation according to G was present in the language. As a coding cannot remove symbols, we cannot insert information in the string other than by associating it to terminals of $L(G)$.

Thus, we will code in our new grammar G' only the binary structure of the derivation tree. Actually, this structure is a presentation of the full derivation, up to derivations of the form $A \Rightarrow^* B$. These "linear" derivations may not only involve chain productions, but also large subtrees with empty yield. The blind counter language used in the intersection will not only take care of the computation of the valences, but it will also verify that those "linear" derivations exist in G.

The context-free grammar G' contains the productions

- $A \to B[A.BC]C$, for $A, B, C \in N$, where $[A.BC]$ is a new terminal symbol making the applied production visible in the language;
- $A \to [A.a]$, for $A \in N$ and $a \in T$, where again $[A.a]$ is a new terminal symbol;
- $S' \to [.S]S$, $S' \to \lambda$.

The terminal alphabet of G' consists the new symbols $[A.BC]$, $[A.a]$, and $[.S]$. The axiom of G' is the new nonterminal S'; the other nonterminals of G' are copied from G. Note G' does not depend on R.

Observe that $L(G')$ consists of strings of the form $\pi_1\rho_1\pi_2\rho_2\ldots\pi_n\rho_n$ recording the binary structure of a possible derivation tree in the in-order of the nodes in the tree.

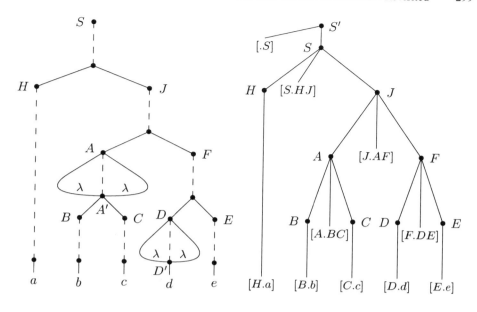

Fig. 1. Left: the binary structure of a derivation tree for $(S, \mathbf{0}) \Rightarrow^* (abcde, \mathbf{u})$ in valence grammar G. Right: the corresponding context-free derivation tree in G' for $S \Rightarrow^* [.S][H.a][S.HJ][B.b][A.BC]\ldots[E.e]$. Here the terminal symbol $[A.BC]$ records the production $A \to B[A.BC]C$ which represents a derivation $(A, \mathbf{0}) \Rightarrow^* (A', \mathbf{v})$ and a rule $(A' \to BC, \mathbf{r})$ for some nonterminal A' of G.

Each π_i is of type $[A.BC]$, each ρ_i is of type $[A.a]$ (with the exception of $\pi_1 = [.S]$, which was introduced to obtain an even number of symbols). If $\rho_i = [A_i.a_i]$, then $a_1 a_2 \ldots a_n$ is a string over the terminal alphabet of G, which *may* have a derivation with the given binary structure. The intersection with the blind counter language will get rid of unwanted binary structures.

We now consider the blind counter automaton $\mathcal{A}$ that should accept input of the form $\pi_1 \rho_1 \pi_2 \rho_2 \ldots \pi_n \rho_n$ only if the productions coded in that string correspond to "linear" derivations in G, with valences adding to $\mathbf{0}$.

Reading a symbol π, with $\pi = [A.BC]$, $\mathcal{A}$ tries to accomplish the following. It guesses a rule $(A' \to BC, \mathbf{r})$ of G, and a derivation $(A, \mathbf{0}) \Rightarrow^*_G (A', \mathbf{v})$ to match a part of a derivation in G. The resulting vectors $\mathbf{r}$ and $\mathbf{v}$ are added using the counters of $\mathcal{A}$. Once the production $A' \to BC$ is guessed, the nondeterministic simulation of the "linear" derivation $A \Rightarrow^*_G A'$ can be done by $\mathcal{A}$ using Lemma 2. As the sequence of productions for such a derivation is an element of a semilinear set, $\mathcal{A}$ may simulate a finite state automaton that accepts a regular language with the same semilinear image.

For a symbol $\rho = [A.a]$, $\mathcal{A}$ guesses a rule $(A' \to a, \mathbf{r})$ and a matching derivation $(A, \mathbf{0}) \Rightarrow^*_G (A', \mathbf{v})$ following the same principle. The symbol $\pi_1 = [.S]$ can be ignored.

When $\mathcal{A}$ receives the empty string as input, it guesses a derivation $(S, \mathbf{0}) \Rightarrow^*_G (A, \mathbf{v})$ and a rule $(A \to \lambda, \mathbf{r})$.

It is now straightforward to verify that the intersection $L(G') \cap L(\mathcal{A})$ represents $L(G)$ in the above manner, and that $L(G)$ can be obtained from this intersection in CF $\wedge$ kBC by projecting the $[A.a]$ symbols to their a-components.

Of course, we cannot hide the π_i using a coding. The languages we need are obtained from $L(G')$ and $L(\mathcal{A})$ by an inverse homomorphism g^{-1} which combines $\pi_i \rho_i$ into a single symbol $[\pi_i \rho_i]$. Recall that both CF and kBC are closed under inverse homomorphisms.

Again, $g^{-1}(L(G')) \cap g^{-1}(L(\mathcal{A})) = g^{-1}(L(G') \cap L(\mathcal{A}))$ represents $L(G)$, which can be obtained from this new language using a coding that maps $[\pi_i \rho_i]$ into a_i. □

4 Implications

In this section we apply our basic characterizations of VAL($\mathbb{Z}^k$) to obtain some of the (known) properties of this family.

Theorem 2. VAL($\mathbb{Z}^k$) *is a (full) principal semi-*AFL.

Proof. According to Theorem 5.5.1(e) in [Gin75], if $\mathcal{F}$ and $\mathcal{G}$ are full principal semi-AFLs, then so is HOM($\mathcal{F} \wedge \mathcal{G}$). Both CF and kBC are full principal semi-AFLs, as discussed, hence so is VAL($\mathbb{Z}^k$) = HOM(CF $\wedge$ kBC). Similarly we find that VAL($\mathbb{Z}^k$) is a principal semi-AFL, i.e., it is of the form $\mathcal{M}(L)$ rather than $\widehat{\mathcal{M}}(L)$, by [Gin75, Theorem 5.5.1(d)] and the characterization with codings, Theorem 1. Here we additionally use the fact that both CF and kBC are principal semi-AFLs, by the real-time results (of Greibach and Latteux). □

In fact, as is shown in [Gin75], the generator of the full principal semi-AFL VAL($\mathbb{Z}^k$) is the shuffle $D_2 \| B_k$ of the respective generators for CF and kBC, after renaming one of the alphabets to make them disjoint.

This also suggests a machine interpretation as (obviously) $D_2 \| B_k$ describes the "legal" instruction sequences of the storage type push-down combined with k blind counters cf. [FS00, Remark. 5.3]. The connection between AFL and automata discussed before Proposition 1 for kBCA now yields the following characterization (cf. [Gin75, Section 4.6]).

Theorem 3. VAL($\mathbb{Z}^k$) *is the family of languages accepted by (real-time) automata equipped with a push-down and k blind counters.*

As a corollary to our characterizations we now obtain the "Chomsky II" normal form, Proposition 5.16 from [FS00].

Theorem 4. *For each valence grammar over $\mathbb{Z}^k$ there is an equivalent valence grammar $G = (N, T, R, S)$ over $\mathbb{Z}^k$, such that each rule is either of the form $(A \to BC, \mathbf{0})$, or of the form $(A \to a, \mathbf{r})$, with $A, B, C \in N$, $a \in T$, and $\|\mathbf{r}\| \leq 1$.*

Proof. Let G be an arbitrary valence grammar over $\mathbb{Z}^k$. By Theorem 1, we may write $L(G) = h(L(G') \cap L(\mathcal{A}))$, where h is a coding, G' a context-free grammar, and $\mathcal{A}$ is a kBCA. By classic formal language theory we may assume that G' is in Chomsky normal form, and according to Latteux we may assume that $\mathcal{A}$ is real-time (Proposition 1). Additionally we may assume that $\|r\| \leq 1$ for each of the counter instructions of $\mathcal{A}$ (Proposition 2).

Now reassemble G' and $\mathcal{A}$ into a valence grammar, using the triple construction, cf. the proof of Lemma 1. We obtain a valence grammar in Chomsky II normal form. □

Of course, an important consequence of this normal form is the equivalence of context-free valence grammars with their λ-free restriction, one of the important contributions of [FS00]; in their notation $\mathcal{L}(\text{Val}, \text{CF}, \mathbb{Z}^k) = \mathcal{L}(\text{Val}, \text{CF-}\lambda, \mathbb{Z}^k)$.

In the same way as Chomsky II above, we obtain Greibach normal form, [FS00, Proposition 5.17]. Note that we do not have to extend the classical construction (to obtain Greibach starting from Chomsky) from CF to $\mathsf{VAL}(\mathbb{Z}^k)$ as we directly use the classical result for the context-free languages.

Theorem 5. *For each valence grammar over $\mathbb{Z}^k$ there is an equivalent valence grammar $G = (N, T, R, S)$ over $\mathbb{Z}^k$, such that each rule is of the form $(A \to a\alpha, r)$, with $A \in N$, $\alpha \in N^*$, $|\alpha| \leq 2$, $a \in T$, and $\|r\| \leq 1$.*

Proof. Similar to the previous proof. Reassemble a context-free grammar G' in Greibach normal form with a real-time kBCA with the $\|r\| \leq 1$ restriction: productions of the form $A \to a$, $A \to aB_1$, and $A \to aB_1B_2$ can be combined with the instruction (p, a, q, r) to form the rules $([p, A, q] \to h(a), r)$, $([p, A, s] \to h(a)[q, B_1, s], r)$, and $([p, A, s] \to h(a)\,[q, B_1, q_1][q_1, B_2, s], r)$ respectively, for all states q_1, s. □

5 Final Words

The generality of the result of Latteux, $\mathcal{M}(L) = \widehat{\mathcal{M}}(L)$ for the permutation closure L of any regular language, suggests other types of valence grammars to apply the AFL techniques. Consider the permutation closure A_k of $(a_1 \ldots a_k)^*$, i.e., the strings in which each letter from $\{a_1, \ldots, a_k\}$ occurs an equal number of times. Then A_k describes a set of legal storage operations, here interpreted as increments on a vector of k natural numbers, with acceptance if all these "counters" are equal.

Then we have, in the style of Theorem 1, $\mathsf{VAL}(\mathbb{N}^k) = \mathrm{COD}(\mathsf{CF} \wedge \mathcal{M}(A_k))$, where $\mathsf{VAL}(\mathbb{N}^k)$ denotes the family accepted by the corresponding context-free valence grammars, i.e., valences in $\mathbb{N}^k$ and the condition that all components in the total valence of the derivation are equal. At first glance this type of valence grammar does not seem to fit the framework of [FS00] as the additional valence condition is not on the identity $\mathbf{0}$ but on equality of the components.

Unfortunately, this does not lead to an extension of the model, as it can be shown that $\mathsf{VAL}(\mathbb{N}^{k+1}) = \mathsf{VAL}(\mathbb{Z}^k)$, either directly, or by observing that $\mathcal{M}(A_{k+1}) = \mathcal{M}(B_k)$. For example, $\mathsf{VAL}(\mathbb{N}^{k+1}) \supseteq \mathsf{VAL}(\mathbb{Z}^k)$, as $B_k = h^{-1}(A_{k+1})$ for the homomorphism that maps b_i to $a_1 \ldots a_{i-1}a_{i+1} \ldots a_{k+1}$ (the string of all a's, except a_i) and leaves each a_i unchanged. This relation can be better understood if we interpret the symbols $a_1, \ldots, a_k$, a_{k+1} in another way as storage instructions: $a_1, \ldots, a_k$ increment a counter, as before, but a_{k+1} acts as a synchronous decrement on all k counters.

Independently, Fernau and Stiebe [FS01, Theorem 7] show that the context-free valence languages over $\mathbb{Z}^k$ are precisely the codings of the *valuated* context-free languages, in their notation $\mathcal{L}(\text{Val}, X, \mathfrak{M}) = \mathcal{C}(\mathcal{L}(\text{Val}', X, \mathfrak{M}))$ with $X = \mathsf{CF}$ and $\mathfrak{M} = \mathbb{Z}^k$. This result is very close to our Theorem 1, relating internal valences to an external valuation through codings. We stress however that we have derived our Theorem 1 as a tool for obtaining the normal forms for context-free valence grammars in Section 4, whereas in [FS01] this is reversed.

Acknowledgements. The author is most obliged to Joost Engelfriet: for lending his copy of [Gin75] and teaching how to read it. The referees of DLT'01 kindly commented on a previous version of this paper.

References

[DP89]   J. Dassow, G. Păun. *Regulated Rewriting in Formal Language Theory.* EATCS Monographs in Theoretical Computer Science, vol. 18. Springer-Verlag, 1989.

[FS97]   H. Fernau, R. Stiebe. Regulation by Valences. In: B. Rovan (ed.) Proceedings of MFCS'97, Lecture Notes in Computer Science, vol. 1295, pages 239-248. Springer-Verlag, 1997.

[FS00]   H. Fernau, R. Stiebe. Sequential Grammars and Automata with Valences. Technical Report WSI-2000-25, Wilhelm- Schickard-Institut für Informatik, Universität Tübingen, 2000. Revised version to appear in *Theoretical Computer Science*.

[FS01]   H. Fernau, R. Stiebe. Valuated and Valence Grammars: An Algebraic View. Proceedings of Developments in Language Theory 2001, *this volume*.

[FMR68]   P.C. Fischer, A.R. Meyer, A.R. Rosenberg. Counter Machines and Counter Languages. *Mathematical Systems Theory* 2 (1968) 265-283.

[Gin75]   S. Ginsburg. *Algebraic and Automata-theoretic Properties of Formal Languages*, Fundamental Studies in Computer Science, vol. 2, North-Holland, 1975.

[Gre78]   S.A. Greibach. Remarks on Blind and Partially Blind One-Way Multicounter Machines. *Theoretical Computer Science* 7 (1978) 311- 324.

[HvV00]   H.J. Hoogeboom, N. van Vugt. Fair Sticker Languages. *Acta Informatica* 37 (2000) 213-225.

[Lat79]   M. Latteux. Cônes Rationnels Commutatifs. *Journal of Computer and Systems Sciences* 18 (1979) 307-333.

[Pau80]   G. Păun. A New Generative Device: Valence Grammars. *Revue Roumaine de Mathématiques Pures et Appliquées* 6 (1980) 911-924.

[Sal73]   A. Salomaa. *Formal Languages*. ACM Monograph Series, Academic Press, 1973.

A Motivation

This appendix describes our motivation for considering valence grammars, [HvV01].

In the language theory of splicing systems the cutting and recombination of DNA molecules with the help of restriction enzymes is modelled by an operation on languages: two strings $x = x_1 x_2$ and $y = y_1 y_2$ are recombined to give the new string $z = x_1 y_2$. The place where the two original strings are cut (in between x_1 and x_2, and y_1 and y_2, respectively) is specified by a set of rules. We refer to [HPP97] for an overview.

The power of this operation is studied by looking at its effect on languages in the Chomsky hierarchy under different types of rules. If the initial language L is regular and the rules are given as context-free language R (or reversely, L is context-free and R regular) then the operation can be modelled as follows. Both L and R are coded in a context-free language L_R in which the strings are of the form $x_1 \# x_2 \$ y_1 \# y_2$, the splicing operation σ can be performed by a finite state transduction, mapping the above string into the result $x_1 y_2$. Consequently, the resulting splicing language $\sigma(L, R)$ is context-free.

Additional constraints involving the lengths of the two strings that participate in the splicing (increasing splicing, same length splicing [KPS96]) require that $|x_1 x_2| = |y_1 y_2|$ or that $|x_1| \geq |y_1|$. As a consequence, L_R is no longer context-free, but context-sensitive, a family not closed under finite state transductions.

In order to get a reasonable upper bound on the family to which $\sigma(L, R)$ belongs, we discovered the valence grammars (more precisely, additive context-free valence grammars) to be a suitable formalism. The family of valence languages is powerful enough to model initial language L and rules R with additional constraints, it is strictly within the family of context sensitive languages, and finally, is closed under finite state transductions.

References

[HPP97]   T. Head, G. Păun, D. Pixton. Language Theory and Molecular Genetics: Generative Mechanisms Suggested by DNA Recombination. In: G. Rozenberg and A. Salomaa (eds.) Handbook of Formal Languages, volume 2. Springer-Verlag, 1997.

[HvV01]   H.J. Hoogeboom, N. van Vugt. Upper Bounds for Restricted Splicing. LIACS Technical Report 01-05, 2001. Submitted.

[KPS96]   L. Kari, G. Păun, A. Salomaa. The Power of Restricted Splicing with Rules from a Regular Language. *Journal of Universal Computer Science*, 2 (1996) 224-240.

An Undecidability Result Concerning Periodic Morphisms

Vesa Halava and Tero Harju

Department of Mathematics
and TUCS - Turku Centre for Computer Science
University of Turku
FIN-20014 Turku, Finland
vesa.halava@utu.fi, harju@utu.fi

Abstract. The following universe problem for the equality sets is shown to be undecidable: given a weak coding h, and two morphisms g_1, g_2, where g_2 is periodic, determine whether or not $h(E_G(g_1, g_2)) = \Sigma^+$, where $E_G(g_1, g_2)$ consists of the solutions w to the equation $g_1(w) = \#g_2(w)$ for a fixed letter $\#$. The problem is trivially decidable, if instead of $E_G(g_1, g_2)$ the equality set $E(g_1, g_2)$ (without a marker symbol $\#$) is chosen.

1 Introduction

In the *Post Correspondence Problem*, PCP for short, we are given two morphisms g_1 and g_2 from Γ^* into Δ^* and it is asked whether there exists a nonempty word $w \in \Gamma^+$ such that $g_1(w) = g_2(w)$. Here the pair (g_1, g_2) is called an *instance* of the PCP and the word w is called a *solution*.

The PCP was proved to be undecidable in the general form by E. Post [11]. By restricting the instances of the PCP by posing further assumptions to the morphisms or the form of the solutions, may result in a problem that is decidable. For example, if the domain alphabet Γ is binary, then the PCP is decidable, see [2], or [6] for a somewhat simpler proof. On the other hand, if the size of the domain alphabet is $n \geq 7$, then the PCP is undecidable, see [10]. The decidability status of the PCP is open for the instances of sizes $3 \leq n \leq 6$.

The PCP states that the existence of a solution is undecidable for morphisms. In this paper we shall prove a simple undecidability result for the corresponding *universality* problem. We show that it is undecidable for a weak coding (a letter-to-letter morphism) h, a morphism g_1 and a periodic morphism g_2 whether or not $h(E_G(g_1, g_2)) = \Sigma^+$, where $E_G(g_1, g_2)$ consists of the solutions w to the equation $g_1(w) = \#g_2(w)$ for a fixed letter $\#$. In this special universe problem the distance to the border that lies between decidability and undecidability is very short. Indeed, the problem is trivially decidable, if instead of $E_G(g_1, g_2)$ we choose the equality set $E(g_1, g_2)$ of all solutions w to the equation $g_1(w) = g_2(w)$ without the marker symbol $\#$.

2   Preliminaries

Let Γ be an alphabet, and denote by Γ^* the monoid of all finite words under the operation of catenation of words over the alphabet Γ. Note that the *empty word* ε is in Γ^*. The set $\Gamma^* \setminus \{\varepsilon\}$ is denoted by Γ^+.

Let Δ be another alphabet. A mapping $g\colon \Gamma^* \to \Delta^*$ is a *morphism* if, for all $u, v \in \Gamma^*, g(uv) = g(u)g(v)$. Furthermore, a morphism g is a *weak coding*, if $g(a) \in \Delta$ for all $a \in \Gamma$. Also, a morphism g is said to be *periodic*, if there exists a word $w \in \Delta^*$ such that $g(x) \in w^*$ for all $x \in \Gamma^*$.

We shall denote $g^{-1}(x) = \{w \in \Gamma^* \mid g(w) = x\}$ for $x \in \Delta^*$.

We shall mainly consider a modification of the PCP called the *generalized* Post Correspondence Problem, GPCP for short. In the GPCP we are given two morphisms $g_1, g_2\colon \Gamma^* \to \Delta^*$ and words $p_1, p_2, s_1, s_2 \in \Delta^*$, and it is asked whether there exists a nonempty word $w \in \Gamma^*$ such that

$$p_1 g_1(w) s_1 = p_2 g_2(w) s_2. \tag{1}$$

The 6-tuple $J = (p_1, p_2, s_1, s_2, g_1, g_2)$ is called an *instance* of the GPCP and a word w satisfying the equation (1) is called a *solution* of J. It is known that the GPCP is decidable when $|\Gamma| \leq 2$, see [2] (or [6]), and it is undecidable when $|\Gamma| \geq 7$, see [8]. For the alphabets of sizes between these limits the decidability status of the GPCP is still open.

For an instance $I = (g_1, g_2)$ of the PCP, let

$$E(I) = \{w \in \Gamma^* \mid g_1(w) = g_2(w)\}$$

be its *equality set*. Similarly, for an instance $J = (p_1, p_2, s_1, s_2, g_1, g_2)$ of the GPCP, we shall define

$$E_G(J) = \{w \in \Gamma^* \mid p_1 g_1(w) s_1 = p_2 g_2(w) s_2\}$$

as the *generalized equality set* of J. We shall consider a special decision problem concerning the equality sets of the instances of the GPCP by restricting ourselves to the instances, where the second morphism g_2 is periodic. We shall prove that the universe problem, whether

$$h(E_G(J)) = \Sigma^+,$$

where h is a weak coding, is undecidable for the generalized instances J. Note that the same problem for the equality sets $E(I)$ (of the PCP) is trivially decidable. Actually, we prove that the universe problem is undecidable for the instances of the simple form

$$J_a = (\varepsilon, a, \varepsilon, \varepsilon, g_1, g_2), \tag{2}$$

where $a \in \Delta$, and g_2 is a periodic morphism. Recall also that the PCP is decidable for the instances (g_1, g_2), where g_2 is periodic, see [2] or [7].

3 Integer Weighted Finite Automata

In the proof of our main result we shall use a result from [5], where it is shown that the universe problem is undecidable for the integer weighted finite automata.

Consider first a nondeterministic finite automaton $\mathcal{A} = (Q, \Sigma, T, \sigma, q_0)$ *without final states*, where Q is the set of the *states*, Σ is the *input alphabet*, T is the alphabet of *edges*, $\sigma \colon T \to Q \times \Sigma \times Q$ is the *transition mapping*, and q_0 is the *initial state*.

The values $\sigma(t)$, for $t \in T$, are the *transitions* of the automaton $\mathcal{A}$. An application of a transition $\sigma(t) = (q, a, p)$ (or of the edge t) of $\mathcal{A}$ will change the state $q \in Q$ to $p \in Q$ after reading the input symbol $a \in \Sigma$. We use a set of edges T rather than a transition relation ($\delta \subseteq Q \times \Sigma \times Q$) in order to allow multiple transitions with the same value, that is, we allow edges $t, t' \in T$ with $t \neq t'$ such that $\sigma(t) = \sigma(t')$. Clearly, this definition of transitions does not affect the language accepted by the finite automata.

Let $\mathbb{Z}$ be the additive group of the integers. An *integer weighted finite automaton* $\mathcal{A}^\gamma$ consists of a finite automaton $\mathcal{A} = (Q, \Sigma, T, \sigma, q_0)$, and a *weight function* $\gamma \colon T \to \mathbb{Z}$ of the edges. To simplify the notation, we shall write the edges in the form

$$t = \langle q, a, p, z \rangle,$$

if $\sigma(t) = (q, a, p)$ and $\gamma(t) = z$. Similarly, we shall write the transition mapping σ as a set, $\sigma \subseteq Q \times \Sigma \times Q \times \mathbb{Z}$, where

$$\sigma = \{\langle q, a, p, z \rangle \mid \exists t \in T \colon \sigma(t) = (q, a, p) \text{ and } \gamma(t) = z\}.$$

Note that the weight function γ does not affect the computations of the original finite automaton $\mathcal{A}$.

The family of all finite automata with integer weights is denoted by $\mathrm{FA}(\mathbb{Z})$.

Let $\mathcal{A}^\gamma \in \mathrm{FA}(\mathbb{Z})$, and let $\pi = t_0 t_1 \ldots t_n$ be a *path* of $\mathcal{A}$, where $\sigma(t_j) = (q_j, a_j, q_{j+1})$ for $0 \leq j \leq n$. The *weight of the path* π is the integer

$$\gamma(\pi) = \sum_{i=0}^{n} \gamma(t_i).$$

Define a morphism $\|\cdot\| \colon T^* \to \Sigma^*$ by setting $\|t\| = a$, if $\sigma(t) = (q, a, p)$, and let $L(\mathcal{A}^\gamma) = \|\gamma^{-1}(0)\|$, that is,

$$L(\mathcal{A}^\gamma) = \{w \in \Sigma^* \mid w = \|\pi\|, \gamma(\pi) = 0\},$$

be the language *accepted by* $\mathcal{A}^\gamma$.

The following lemma is a special case of a result in [5].

Lemma 1. *It is undecidable for automata* $\mathcal{A}^\gamma \in \mathrm{FA}(\mathbb{Z})$ *whether or not* $L(\mathcal{A}^\gamma) = \Sigma^*$.

Actually, it was proved in [5] that the universe problem $L(\mathcal{A}^\gamma) = \Sigma^*$ is undecidable for 4-state unimodal integer weighted finite automata. An $\mathcal{A}^\gamma \in \mathrm{FA}(\mathbb{Z})$ is said to be *unimodal*, if for all paths in $\mathcal{A}^\gamma$ (not only for the accepting paths of weight 0) the weights of the transitions are first strictly positive, then 0, then strictly negative and finally 0.

We note that the model of finite automaton defined above is similar to the blind counter machine introduced by Greibach [4] and it is also closely related to the 1-turn counter automata as considered by Baker and Book [1], Greibach [3], and especially by Ibarra [9]. However, in the present model the counter is replaced by a weight function of the transitions, and while doing so, the finite automaton becomes independent of the counter.

We have defined the integer weighted finite automata without final states. The existence of a final state simplifies the proof in the next section, and therefore we introduce an integer weighted finite automaton *with a final state* as $\mathcal{A}_f^\gamma = (Q, \Sigma, T, \sigma, q_0, q_f)$, where $\mathcal{A}^\gamma = (Q, \Sigma, T, \sigma, q_0) \in \mathrm{FA}(\mathbb{Z})$, and q_f is the (unique) *final state*. A word w is accepted by $\mathcal{A}_f^\gamma$ if and only if it is accepted in the state q_f.

The family of all integer weighted finite automata with a final state is denoted by $\mathrm{FA}_f(\mathbb{Z})$.

Note that each automaton in $\mathrm{FA}(\mathbb{Z})$ accepts the empty word, but an automaton in $\mathrm{FA}_f(\mathbb{Z})$ accepts the empty word if and only if $q_0 = q_f$.

The following lemma is useful in the next section.

Lemma 2. *Let $\mathcal{A}^\gamma \in \mathrm{FA}(\mathbb{Z})$. There exists an $\mathcal{A}_f^\gamma \in \mathrm{FA}_f(\mathbb{Z})$ such that $L(\mathcal{A}_f^\gamma) = L(\mathcal{A}^\gamma) \setminus \{\varepsilon\}$.*

Proof. Let $\mathcal{A}^\gamma = (Q, \Sigma, T, \sigma, q_0)$, and let $q_f \notin Q$ be a new state. Define $\mathcal{A}_f^\gamma = (Q \cup \{q_f\}, \Sigma, T', \sigma', q_0, q_f)$ as follows. For each edge $t \in T$ with $t = \langle q, a, p, z \rangle$ in $\mathcal{A}^\gamma$, add a new edge $t' \in T'$ such that $t' = \langle q, a, q_f, z \rangle$ in $\mathcal{A}_f^\gamma$. A word $w \in L(\mathcal{A}^\gamma)$, $w \neq \varepsilon$, is accepted in q_f with a path, which differs from an accepting path in $\mathcal{A}^\gamma$ only by the last transition, which now enters in q_f. The claim easily follows from this construction. □

The next theorem follows from Lemmas 1 and 2.

Theorem 1. *It is undecidable for the automata $\mathcal{A}_f^\gamma \in \mathrm{FA}_f(\mathbb{Z})$ whether or not $L(\mathcal{A}_f^\gamma) = \Sigma^+$.*

Actually, by the proof of Lemma 2, the previous theorem holds even in the case, where the automata $\mathcal{A}_f^\gamma \in \mathrm{FA}_f(\mathbb{Z})$ has no edges leaving the final state. Using the same argument to the initial state, we may assume that in the automata of $\mathrm{FA}_f(\mathbb{Z})$ there are no transitions to the initial state. In fact, we simply introduce a new initial state for which we copy all transitions of the original initial state. Therefore we have

Corollary 1. *It is undecidable for automata $\mathcal{A}_f^\gamma \in \mathrm{FA}_f(\mathbb{Z})$, where there are no transitions to the initial state and no transitions from the final state, whether or not $L(\mathcal{A}_f^\gamma) = \Sigma^+$.*

4 Undecidability of the Universe Problem

In this section we shall give a proof for the undecidability of the universe problem mentioned in the introduction. We begin with a simple decidability property.

Theorem 2. *Let $I = (g_1, g_2)$ with $g_1, g_2 \colon \Gamma^* \to \Delta^*$ be an instance of the PCP, and let $h \colon \Gamma^* \to \Sigma^*$ be a weak coding. It is decidable whether or not $h(E(I)) = \Sigma^+$.*

Proof. We have $h(E(I)) = \Sigma^+$ if and only if for all letters $a \in \Sigma$, the set $h^{-1}(a) \cap E(I)$ is nonempty. Since the sets $h^{-1}(a)$ are finite, and the membership problem (given a word w, determine whether $w \in E(I)$) is decidable for all instances, the claim follows. Indeed, if $u, v \in E(I)$, also $uv \in E(I)$. □

We shall now prove our main theorem.

Theorem 3. *It is undecidable whether or not $h(E_G(J)) = \Sigma^+$ for weak codings h, and the instances $J = (\varepsilon, \#, \varepsilon, \varepsilon, g_1, g_2)$, where g_2 is periodic and $\#$ is a letter.*

Proof. Let $\mathcal{A}_f^\gamma = (Q, \Sigma, T, \sigma, q_0, q_n) \in \mathrm{FA}_f(\mathbb{Z})$ be as in Corollary 1, where $Q = \{q_0, q_1, \ldots, q_n\}$. Especially, we require that there are no transitions to q_0 and from q_n.

Define the alphabet

$$\Gamma = \{[i, a, j, z] \mid \langle q_i, a, q_j, z \rangle \in \sigma\}$$

and let $\Delta = \{c, d, \#\}$ consist of new symbols. Furthermore, define two mappings $+$ and $-$ for $z \in \mathbb{Z}$ by

$$z_+ = \begin{cases} z & \text{if } z > 0, \\ 0 & \text{otherwise,} \end{cases} \quad \text{and} \quad z_- = \begin{cases} 0 & \text{if } z > 0, \\ |z| & \text{otherwise.} \end{cases}$$

The morphisms g_1, g_2 and h are defined as follows for each $[i, a, j, z] \in \Gamma$,

$$g_1([i, a, j, z]) = \begin{cases} \#d^n c(d^n c)^{z_+} d^{j+1} & \text{if } i = 0, \\ d^{n-i-1} c(d^n c)^{z_+} d^{j+1} & \text{if } j \neq n, \\ d^{n-i-1} c(d^n c)^{z_+} & \text{if } j = n, \end{cases}$$

and

$$g_2([i, a, j, z]) = (d^n c)^{z_- + 1} \quad \text{and} \quad h([i, a, j, z]) = a.$$

Denote $J = (\varepsilon, \#, \varepsilon, \varepsilon, g_1, g_2)$. We shall prove that $h(E_G(J)) = L(\mathcal{A}_f^\gamma)$.

Assume first that $x = a_1 \ldots a_k \in h(E_G(J))$, where $a_\ell \in \Sigma$ for all $1 \leq \ell \leq k$. Then there exists a word $w \in h^{-1}(x) \subseteq \Gamma^*$,

$$w = [i_1, a_1, j_1, z_1][i_2, a_2, j_2, z_2] \ldots [i_k, a_k, j_k, z_k],$$

such that $w \in E_G(J)$. Consequently, $g_1(w) = \# g_2(w)$, where

$$g_2(w) = (d^n c)^{\sum_{\ell=1}^k (z_\ell)_- + k}. \tag{3}$$

Since $g_1(w)$ begins with the letter $\#$, necessarily $i_1 = 0$. Similarly, since $g_2(w)$ ends with $d^n c$, necessarily $j_k = n$.

Consider next the images of g_1 of two consecutive letters in w. The image is of the form

$$d^{n-i_\ell-1} c(d^n c)^{(z_\ell)_+} d^{j_\ell+1} d^{n-i_{\ell+1}-1} c(d^n c)^{(z_{\ell+1})_+} d^{j_{\ell+1}+1},$$

for some $1 \leq \ell \leq n-1$. In order for $d^{j_{\ell}+1}d^{n-i_{\ell+1}-1}c = d^n c$ to hold, necessarily $j_\ell = i_{\ell+1}$ and $i_{\ell+1}, j_\ell \notin \{0, n\}$ for $\ell = 1, \ldots, n-1$. This shows that

$$w = [0, a_1, j_1, z_1][j_1, a_2, j_2, z_2]\ldots[j_{k-1}, a_k, j_k, z_k],$$

and, therefore, $g_1(w) = \#(d^n c)^{\sum_{\ell=1}^k (z_\ell)_+ + k}$. On the other hand, the equality $g_1(w) = \#g_2(w)$ together with (3) implies

$$\sum_{\ell=1}^k (z_\ell)_+ = \sum_{\ell=1}^k (z_\ell)_-,$$

and, therefore

$$\sum_{\ell=1}^k z_\ell = \sum_{\ell=1}^k (z_\ell)_+ - \sum_{\ell=1}^k (z_\ell)_- = 0.$$

By the definition of Γ, we obtain that there exists an accepting path for the word $a_1 a_2 \ldots a_n = h(w) = x$ in $\mathcal{A}^\gamma$. It follows that $h(E_G(J)) \subseteq L(\mathcal{A}_f^\gamma)$.

Assume next that $x = a_1 \ldots a_k \in L(\mathcal{A}_f^\gamma)$, where $a_i \in \Sigma$ for $1 \leq i \leq k$. Then there exists a sequence

$$\langle q_{j_{i-1}}, a_i, q_{j_i}, z_i \rangle \in \sigma,$$

of transitions, where $1 \leq i \leq k$, $j_i \in \{0, \ldots, n\}$, and $q_{j_0} = 0$ such that $\sum_{i=1}^k z_i = 0$. Clearly,

$$w = [0, a_1, j_1, z_1][j_1, a_2, j_2, z_2]\ldots[j_{k-1}, a_k, j_k, z_k] \in h^{-1}(x).$$

By the definitions of g_1 and g_2,

$$g_1(w) = \#(d^n c)^{\sum_{\ell=1}^k (z_\ell)_+ + k} \text{ and } g_2(w) = (d^n c)^{\sum_{\ell=1}^k (z_\ell)_- + k}.$$

Since $\sum_{i=1}^k z_i = 0$, we have

$$\sum_{i=1}^k (z_i)_+ = \sum_{i=1}^k (z_i)_-,$$

which implies $g_1(w) = \#g_2(w)$, i.e., $w \in E_G(J)$. Moreover, since $h(w) = x$, we obtain that $x \in h(E_G(J))$. It follows that $L(\mathcal{A}_f^\gamma) \subseteq h(E_G(J))$.

Because $h(E_G(J)) = L(\mathcal{A}_f^\gamma)$, we have that $h(E_G(J)) = \Sigma^+$ if and only if $L(\mathcal{A}_f^\gamma) = \Sigma^+$, and the claim follows from the undecidability result in Corollary 1. □

Note that in the above proof the special symbol $\#$ is used only to force the transition sequences in $\mathcal{A}_f^\gamma$ simulated by morphisms g_1 to start from the initial state. It also guarantees that concatenations of such sequences are not allowed in $E_G(J)$.

Note also that in Theorem 2 we did not restrict to the periodic instances of the PCP as in Theorem 3. In the case of periodic instances the decidability result in Theorem 2 is true for all morphisms h, not only for weak codings. This follows, since the PCP is decidable in the periodic cases.

References

1. B. Baker and R. Book, *Reversal-bounded multipushdown machines*, J. Comput. System Sci. **8** (1974), 315–332.
2. A. Ehrenfeucht, J. Karhumäki, and G. Rozenberg, *The (generalized) Post Correspondence Problem with lists consisting of two words is decidable*, Theoret. Comput. Sci. **21** (1982), 119–144.
3. S. A. Greibach, *An infinite hierarchy of context-free languages*, J. Assoc. Comput. Mach. **16** (1969), 91–106.
4. S. A. Greibach, *Remarks on blind and partially blind one-way multicounter machines*, Theoret. Comput. Sci. **7** (1978), 311–324.
5. V. Halava and T. Harju, *Undecidability in integer weighted finite automata*, Fund. Inform. **38** (1999), 189 – 200.
6. V. Halava, T. Harju, and M. Hirvensalo, *Binary (generalized) Post Correspondence Problem*, Tech. Report 357, Turku Centre for Computer Science, August 2000, to appear in Theoret. Comput. Sci.
7. T. Harju and J. Karhumäki, *Morphisms*, Handbook of Formal Languages (G. Rozenberg and A. Salomaa, eds.), vol. 1, Springer-Verlag, 1997.
8. T. Harju, J. Karhumäki, and D. Krob, *Remarks on generalized Post Correspondence Problem*, STACS'96, Lecture Notes in Comput. Sci., vol. 1046, Springer-Verlag, 1996, pp. 39–48.
9. O. H. Ibarra, *Restricted one-counter machines with undecidable universe problems*, Math. Systems Theory **13** (1979), 181–186.
10. Y. Matiyasevich and G. Sénizergues, *Decision problems for semi-Thue systems with a few rules*, Proceedings, 11th Annual IEEE Symposium on Logic in Computer Science (New Brunswick, New Jersey), IEEE Computer Society Press, 27–30 July 1996, pp. 523–531.
11. E. Post, *A variant of a recursively unsolvable problem*, Bulletin of Amer. Math. Soc. **52** (1946), 264–268.

A Universal Turing Machine with 3 States and 9 Symbols

Manfred Kudlek[1] and Yurii Rogozhin[2]

[1] Fachbereich Informatik, Universität Hamburg, Vogt-Kölln-Straße 30, D-22527 Hamburg,
Germany
kudlek@informatik.uni-hamburg.de
[2] Institute of Mathematics and Computer Science, Academy of Sciences of Moldova, Str.
Academiei 5, MD-2028 Chişinău, Republica Moldova
rogozhin@math.md

Abstract. With an **UTM(3,9)** we present a new small universal Turing machine with 3 states and 9 symbols, improving a former result of an **UTM(3,10)**.

1 Introduction

After introducing the problem of the construction of simple universal Turing machines by Shannon [14], a number of small universal Turing machines have been found [5, 9]. A presentation of the records until 1996 was presented in [12], stating that there are universal Turing machines of the types **UTM(24,2)**, **UTM(10,3)**, **UTM(7,4)**, **UTM(5,5)**, **UTM(4,6)**, **UTM(3,10)**, and **UTM(2,18)**. Here **UTM(m,n)** denotes the class of universal Turing machines with m states and n symbols, the halting state not counted. In [13] a universal Turing machine in **UTM(22,2)** was presented. In this paper we present a machine of type **UTM(3,9)**, improving the result from [12].

The result in [13] on **UTM(22,2)**, recent results in [1] on **UTM(19,2)**, and the present result reduce the number of classes **UTM(m,n)** with an unsettled emptiness problem to 45, i.e if **UTM(m,n)** is empty.

For the basic definitions of universal Turing machines we refer to [12], and for the historical development of the subject to the bibliography.

2 Construction of Universal Turing Machines

We deal with the ordinary notion of a *deterministic Turing machine* (TM) with a one-dimensional tape and one head, the notion of a *configuration* of TM, and of a *tag-system* (cf. [2]). In what follows, denote via α, β (with or without subscripts) the configurations of a TM (tag-system or other algorithm model). Let $\beta_1 \overset{M}{\to} \beta_2$ mean that TM M moves from a configuration β_1 to a configuration β_2 by one step, and we write $\beta_1 \overset{M}{\Rightarrow} \beta_j$ for $\beta_1 \overset{M}{\to} \beta_2 \overset{M}{\to} \ldots \overset{M}{\to} \beta_j$.

Definition : A μ-*tag-system* is a triple $T = (\mu, A, \Pi)$, where μ is a positive integer, the *tag-number*, $A = \{a_1, \cdots, a_n, a_{n+1}\}$ a finite alphabet, and Π a mapping $\Pi : A \to A^* \cup \{STOP\}$ with $\Pi(a_{n+1}) = STOP$.

The words $\alpha_i = \Pi(a_i) \in A^* \cup \{STOP\}$ are called the productions of T. The symbol a_{n+1} is called the *halting symbol*.

A computation of a tag-system $T = (\mu, A, \Pi)$ on a word $\beta \in A^*$ is a sequence $\beta_0, \beta_1, \cdots$ such that for all $k \geq 0$ β_k is transformed into β_{k+1} by deleting the first μ symbols of β_k and appending α_i to the result if the first symbol of β_k is a_i. The computation halts if the length $|\beta_k| < \mu$ or if the first symbol of β_k is a_{n+1}. We will also use the notation $a_i \to \alpha_i$ $1 \leq i \leq n$, $a_{n+1} \to STOP$, respectively.

Minsky [5] has proved the existence of universal 2-tag-systems. Such systems have been used for the construction of all small universal Turing machines so far, and will be used here too. Furthermore, he has shown that
1. The computation of a tag-system stops only if the first symbol is the halting symbol a_{n+1}.
2. The productions α_i, $1 \leq i \leq n$, are not empty.

A universal Turing machine U simulates a tag-system T in the following way. Let $T = (2, A, \Pi)$ with $A = \{a_1, \cdots, a_n, a_{n+1}\}$ and $a_i \to \alpha_i$. Associated to each symbol $a_i \in A$ there is a $N_i > 0$, and codes A_i and $\tilde{A}_i$ (possibly $A_i = \tilde{A}_i$ as for the construction of our machine), of the form u^{N_i} with u a string over the alphabet of the Turing machine U.

The codes A_i ($\tilde{A}_i$) are separated by markers on the tape of U.

For $1 \leq i \leq n$, the production $a_i \to \alpha_i = a_{i1} \cdots a_{im_i}$ of the tag-system T is encoded by $\quad P_i = A_{im_i} \cdots A_{i2} A_{i1}$.

The initial word $\beta = a_r a_s a_t \cdots a_w$ to be transformed by T, is encoded by $S = A_r A_s A_t \cdots A_w$ (or $S = \tilde{A}_r \tilde{A}_s \tilde{A}_t \cdots \tilde{A}_w$).

The initial inscription on the tape of the *UTM* U is :
$Q_L PSQ_R$ with $P = P_{n+1} P_n \cdots P_1 P_0$ and $S = A_r A_s A_t \cdots A_w$,

where Q_L and Q_R are the infinite parts to the left and to the right of the relevant inscription on the tape of U, consisting only of blanks, P_{n+1} is the encoding of the halting symbol a_{n+1}, P_0 an additional code, generally consisting of several markers, and the head of U is located on the left of the code S in state q_1, the initial state of U.

For a tag-system T let S_1 and S_2 be the encodings of words β_1 and β_2, respectively, and $\beta_1 \xrightarrow{T} \beta_2$. Then U transforms as :
$$Q_L P R_1 S_1 Q_R \xRightarrow{U} Q_L P R_2 S_2 Q_R$$
where the rightmost two symbols of R_2 correspond to the cells having contained the first two symbols of β_1. For the initial tape inscription of U we have $R_1 = \lambda$.

U simulates the tag-system in 3 stages :

(i) In the first stage, U searches the code P_r corresponding to the code A_r, and after that U deletes the codes A_r and A_s.

(ii) In the second stage, U writes the code P_r at the left end of Q_R as $A_{r1} \cdots A_{rm_r}$ (note that is done in the reversed order of P_r).

(iii) In the third stage, U restores its tape (essentially the part where $P_{r-1} \cdots P_1 P_0$ is located), and a new simulation cycle can start.

The number N_r corresponding to the symbol a_r ($1 \leq r \leq n+1$) of the tag-system T, has the property that there are exactly N_r markers between the codes P_r and A_r at each simulation cycle.

In the first simulation stage, the head of U passes a number of markers in part P, equal to the number of symbols x in code A_r. After this first stage of simulating, the tape of U is
$$Q_L P_{n+1} P_n \cdots P_{r+1} P_r P'_{r-1} \cdots P'_1 P'_0 R'_1 A'_r A_s A_t \cdots A_w Q_R,$$
the head of U is located between A'_r and A_s. Then U deletes A_s, and the second stage starts.

After the second simulation stage, the tape of U is
$$Q_L P_{n+1} P_n \cdots P_{r+1} P''_r P''_{r-1} \cdots P''_1 P''_0 R''_1 A_t \cdots A_w A_{r1} \cdots A_{rm_r} Q_R,$$
the head of U is located on the left side of P''_r, and the third stage starts.

After the third simulation stage, the tape of U is
$$Q_L P_{n+1} P_n \cdots P_1 P_0 R_2 A_t \cdots A_w A_{r1} \cdots A_{rm_r} Q_R,$$
and the head of U is located on the right side of R_2. Now a new simulation cycle can start.

Let $a_1, a_2, \cdots, a_k, b_1, b_2, \cdots, b_k$ be symbols of U. $a_1 a_2 \cdots a_k R b_1 b_2 \cdots b_k$ means the change of the group $a_1 a_2 \cdots a_k$ into the group $b_1 b_2 \cdots b_k$ and the move of the head of U from symbol a_1 k positions to the right (i.e. $q a_1 a_2 \cdots a_k \overset{U}{\Rightarrow} b_1 b_2 \cdots b_k q'$ where q, q' are states). Similarly, the notation $a_1 a_2 \cdots a_k L b_1 b_2 \cdots b_k$ means such a change and move to the left (i.e. $x a_1 a_2 \cdots q a_k \overset{U}{\Rightarrow} q' x b_1 b_2 \cdots b_k$ where x is an arbitrary symbol).

group
$L a_1 a_2 \cdots a_k (b_1 b_2 \cdots b_k) R$ means the change of the group $a_1 a_2 \cdots a_k$ into the group $b_1 b_2 \cdots b_k$ and the move of the head of U from a_k one position to the right (i.e. $a_1 a_2 \cdots q a_k \overset{U}{\Rightarrow} b_1 b_2 \cdots b_k q'$).

Similarly, $R a_1 a_2 \cdots a_k (b_1 b_2 \cdots b_k) L$ means such a change in the opposite direction (i.e. $x q a_1 a_2 \cdots a_k \overset{U}{\Rightarrow} q' x b_1 b_2 \cdots b_k$).

3 The UTM with 3 States and 9 Symbols

The symbols of the machine in **UTM(3,9)** are $\overleftarrow{1}$ (blank symbol), 1, $\overrightarrow{1}$, b, $\overleftarrow{b}$, $\overrightarrow{b}$, c, $\overleftarrow{c}$, $\overrightarrow{c}$; and the states are $- q_1, q_2, q_3$.
$$N_1 = 1, \quad N_{k+1} = N_k + 2m_k + 2 \quad (k \in \{1, \ldots, n\}).$$
The code of the production $\alpha_i = a_{i1} a_{i2} \ldots a_{im_i}$ ($i \in \{1, \ldots, n\}$) of the tag-system is:
$$P_i = b1bbb1^{N_{im_i}} bb1^{N_{im_i-1}} \ldots bb1^{N_{i2}} bb1^{N_{i1}},$$
where $A_j = 1^{N_j}$, $j \in \{1, \ldots, n+1\}$ and the symbol b is a marker.
$$P_0 = b, \quad P_{n+1} = \overrightarrow{1} b.$$
The code S of the initial word $\beta = a_r a_s a_t \ldots a_w$, to be transformed by the tag-system, is:
$$S = c1^{N_r} c1^{N_s} c1^{N_t} \ldots c1^{N_w},$$
where the symbol c is a marker.

The program of the machine in UTM(3,9):

$q_1\ b\overleftarrow{b}\ Rq_1$	$q_2\ b\overrightarrow{b}\ Lq_3$	$q_3\ b\overrightarrow{c}\ Rq_1$
$q_1\ \overleftarrow{b}b\ Lq_1$	$q_2\ \overleftarrow{b}\overrightarrow{b}\ Lq_2$	$q_3\ b\ b\ Lq_3$
$q_1\ \overrightarrow{b}\overleftarrow{b}\ Rq_1$	$q_2\ \overrightarrow{b}\overleftarrow{b}\ Rq_2$	$q_3\ \overrightarrow{b}b\ Rq_3$
$q_1\ 1\overrightarrow{c}\ Lq_1$	$q_2\ 1\overleftarrow{1}\ Rq_2$	$q_3\ 11\ Rq_3$
$q_1\ \overrightarrow{1}\overleftarrow{c}\ Rq_1$	$q_2\ \overrightarrow{1}\overleftarrow{1}\ Rq_2$	$q_3\ \overrightarrow{1}\ -$
$q_1\ \overleftarrow{1}\overrightarrow{c}\ Lq_1$	$q_2\ \overleftarrow{1}\overrightarrow{1}\ Lq_2$	$q_3\ \overleftarrow{1}c\ Lq_3$
$q_1\ c\overrightarrow{1}\ Lq_2$	$q_2\ c\overleftarrow{c}\ Rq_2$	$q_3\ c1\ Rq_1$
$q_1\ \overleftarrow{c}1\ Lq_3$	$q_2\ \overleftarrow{c}\overrightarrow{c}\ Lq_2$	$q_3\ \overleftarrow{c}1\ Lq_3$
$q_1\ \overrightarrow{c}\overleftarrow{1}\ Rq_1$	$q_2\ \overrightarrow{c}\overleftarrow{c}\ Rq_2$	$q_3\ \overrightarrow{c}\overrightarrow{b}\ Lq_2$

(i) On the first stage of modelling (rules used in parentheses):

$$1L\ \overrightarrow{c} \quad (q_1\ 1\overrightarrow{c}\ Lq_1),$$
$$\overleftarrow{1}\ L\ \overrightarrow{c} \quad (q_1\ \overleftarrow{1}\overrightarrow{c}\ Lq_1),$$
$$b\ L\ b \quad (q_1\ \overleftarrow{b}b\ Lq_1),$$
$$L\ b\ (\overleftarrow{b})R \quad (q_1\ b\overleftarrow{b}\ Rq_1),$$
$$\overrightarrow{c}\ R\ \overleftarrow{1} \quad (q_1\ \overrightarrow{c}\overleftarrow{1}\ Rq_1),$$
$$b\ R\ \overleftarrow{b} \quad (q_1\ b\overleftarrow{b}\ Rq_1),$$
$$R\ 1\ (\overrightarrow{c})L \quad (q_1\ 1\overrightarrow{c}\ Lq_1).$$

If the head of the UTM moves to the right and meets the marker c, the first stage of modelling is over. The UTM deletes this marker and the second stage of modelling begins ($q_1\ c\overrightarrow{1}\ Lq_2$).

(ii) On the second stage of modelling the UTM writes the markers $\overrightarrow{c}$ and the symbols $\overrightarrow{1}$ in Q_R; moreover the UTM writes the marker $\overrightarrow{c}$ only after the symbol $\overrightarrow{1}$.

If the UTM writes the symbol $\overrightarrow{1}$ in Q_R either after writing the marker $\overrightarrow{c}$ or after the first stage of modelling, then:

$$\overrightarrow{b}\ R\ \overleftarrow{b} \quad (q_2\ \overrightarrow{b}\overleftarrow{b}\ Rq_2),$$
$$1\ R\ \overleftarrow{1} \quad (q_2\ 1\overleftarrow{1}\ Rq_2),$$
$$c\ R\ \overleftarrow{c} \quad (q_2\ c\overleftarrow{c}\ Rq_2),$$
$$R\ \overleftarrow{1}\ (\overrightarrow{1})L \quad (q_2\ \overleftarrow{1}\overrightarrow{1}\ Lq_2).$$

If the UTM writes the symbol $\overrightarrow{1}$ in Q_R after the same symbol $\overrightarrow{1}$, then:

$$\overrightarrow{b}\ R\ \overleftarrow{b} \quad (q_2\ \overrightarrow{b}\overleftarrow{b}\ Rq_2),$$
$$\overrightarrow{1}\ R\ \overleftarrow{1} \quad (q_2\ \overrightarrow{1}\overleftarrow{1}\ Rq_2),$$
$$\overrightarrow{c}\ R\ \overleftarrow{c} \quad (q_2\ \overrightarrow{c}\overleftarrow{c}\ Rq_2),$$
$$R\ \overleftarrow{1}\ (\overrightarrow{1})L \quad (q_2\ \overleftarrow{1}\overrightarrow{1}\ Lq_2).$$

A Universal Turing Machine with 3 States and 9 Symbols

The UTM goes to the left after writing the symbol $\vec{1}$ in Q_R:

$$\begin{array}{ll} \overleftarrow{1}\, L\, \vec{1} & (q_2\, \overleftarrow{1}\, \vec{1}\, Lq_2), \\ \overleftarrow{c}\, L\, \vec{c} & (q_2\, \overleftarrow{c}\, \vec{c}\, Lq_2), \\ \overleftarrow{b}\, L\, \vec{b} & (q_2\, \overleftarrow{b}\, \vec{b}\, Lq_2). \end{array}$$

Then, if the head of the UTM meets the symbol 1 in P_r, the head will change the direction of its motion, the UTM changes the symbol 1 to the symbol $\overleftarrow{1}$ and writes the symbol $\vec{1}$ in Q_R:

$$L\, 1\, (\overleftarrow{1})R \quad (q_2\, 1\overleftarrow{1}\, Rq_2).$$

If the head of the UTM meets the markers bb in P_r, then the UTM writes the marker $\vec{c}$ in Q_R:

$$L\, bb\, (\vec{c}\,\overleftarrow{b})R \quad (q_2\, b\vec{b}\, Lq_3,\; q_3\, b\vec{c}\, Rq_1,\; q_1\, \vec{b}\,\overleftarrow{b}\, Rq_1),\text{ then}$$

$$\begin{array}{ll} \vec{b}\, R\, \overleftarrow{b} & (q_1\, \vec{b}\,\overleftarrow{b}\, Rq_1), \\ \vec{1}\, R\, \overleftarrow{c} & (q_1\, \vec{1}\,\overleftarrow{c}\, Rq_1), \\ \vec{c}\, R\, \overleftarrow{1} & (q_1\, \vec{c}\,\overleftarrow{1}\, Rq_1), \\ R\, \overleftarrow{1}\, (\vec{c})L & (q_1\, \overleftarrow{1}\,\vec{c}\, Lq_1). \end{array}$$

When the head of the UTM moves to the left after writing the marker $\vec{c}$ in Q_R, then:

$$\begin{array}{ll} \overleftarrow{c}\, L\, 1 & (q_1\, \overleftarrow{c}1\, Lq_3,\; q_3\, \overleftarrow{c}1\, Lq_3), \\ \overleftarrow{1}\, L\, c & (q_3\, \overleftarrow{1}c\, Lq_3), \end{array}$$

and the UTM restores the part S of the tape. Then the UTM meets the marker $\overleftarrow{b}$ in P:

$$\begin{array}{ll} \overleftarrow{b}\, L\, \vec{b} & (q_3\, \overleftarrow{b}\,\vec{b}\, Lq_3), \\ \overleftarrow{c}\, L\, 1 & (q_3\, \overleftarrow{c}1\, Lq_3), \\ \vec{c}\, L\, \vec{b} & (q_3\, \vec{c}\,\vec{b}\, Lq_2). \end{array}$$

The UTM halts when it meets the pair $\vec{1}b$ $(q_2\, b\vec{L}\, q_3,\; q_3\, \vec{1}\, -)$.
If the UTM meets the pair $1b$, then the second stage of modelling is over. Then

$$L1b(1b)R \quad (q_2\, b\vec{b}\, Lq_3,\; q_3 11Rq_3,\; q_3\, \vec{b}b\, Rq_3)$$

and the third stage of modelling begins.

(iii) On the third stage of modelling the UTM restores the tape in P (the tape is restored in S after writing the marker $\vec{c}$ in Q_R):

$$\begin{array}{ll} 1R1 & (q_3 11Rq_3), \\ \vec{b}\, Rb & (q_3\, \vec{b}\, bRq_3). \end{array}$$

When the head of the UTM moves to the right and meets the marker c, both the third stage and the whole cycle of modelling is over. The UTM deletes the marker c and a new cycle of modelling begins ($q_3 c1Rq_1$).

4 Example

The following example illustrates the simulation of a 2-tag-system with alphabet $A = \{a_1, a_2, a_3, a_4\}$ and tag productions $\alpha_1 = a_2 a_1 a_3$, $\alpha_2 = a_3$, $\alpha_3 = a_1 a_2$ by the universal Turing machine.

With $N_1 = 1, N_2 = 9, N_3 = 13, N_4 = 19$ the symbols are encoded by $c1, c1^9, c1^{13}, c1^{19}$, and the tag productions by $P_0 = b$, $P_1 = b1bbb1^{13}bb1bb1^9$, $P_2 = b1bbb1^{13}$, $P_3 = b1bbb1^9bb1$, $P_4 = \overrightarrow{1}\,b$.

The initial word $a_3 a_2 a_1 a_3 a_2$, transformed into $a_1 a_3 a_2 a_1 a_2$, gives the initial configuration

$\cdots \overset{\leftarrow\rightarrow}{1\,1}\, bb1bbb1^9bb1b1bbb1^{13}b1bbb1^{13}bb1bb1^9bq_3c1^{13}c1^9c1c1^{13}c1^9\,\overset{\rightarrow\leftarrow}{c\,1}\cdots$

or

$\cdots \overset{\leftarrow\rightarrow}{1\,1}\, bb1bbb1^9bb1b1bbb1^{13}b1bbb1^{13}bb1bb1^9b1q_11^{13}c1^9c1c1^{13}c1^9\overset{\leftarrow}{c\,1}\cdots$

One step of the tag-system is simulated by the following sequence of the Turing machine (not all steps are given) :

First stage

$\cdots bq_3c1^{13}c\cdots$

$\cdots b1q_11^{13}c\cdots$

$\cdots q_1b\,\overset{\rightarrow\rightarrow}{c\,c}\,1^{12}c\cdots$

$\cdots \overset{\leftarrow\leftarrow\leftarrow}{b\,1\,1}\,q_11^{12}c\cdots$

$\cdots$

$\cdots bq_1b\,\overset{\rightarrow 9}{c}\,b\,\overset{\rightarrow\rightarrow\rightarrow}{c\,c\,c}\,1^{11}c\cdots$

$\cdots b\,\overset{\leftarrow\leftarrow 9}{b\,1}\,\overset{\leftarrow\leftarrow\leftarrow\leftarrow}{b\,1\,1\,1}\,q_11^{11}c\cdots$

$\cdots bb1\,\overset{\leftarrow\leftarrow\leftarrow\leftarrow\leftarrow 13}{b\,1\,b\,b\,b\,1}\,\cdots\,\overset{\leftarrow 13}{1}\,q_1c1^9c\cdots$

$\cdots bb1\,\overset{\leftarrow\leftarrow\leftarrow\leftarrow\leftarrow 13}{b\,1\,b\,b\,b\,1}\,\cdots\,\overset{\leftarrow 12}{1}\,q_2\,\overset{\leftarrow\rightarrow}{1\,1}\,1^9c\cdots$

Second stage

$\cdots bbq_21\,\overset{\rightarrow\rightarrow\rightarrow\rightarrow\rightarrow 13}{b\,1\,b\,b\,b\,1}\,\cdots\,\overset{\rightarrow 12}{1}\,\overset{\rightarrow\rightarrow}{1\,1}\,1^9c\cdots$

$\cdots bb\,\overset{\leftarrow\leftarrow\leftarrow\leftarrow\leftarrow\leftarrow 13}{1\,b\,1\,b\,b\,b\,1}\,\cdots\,\overset{\leftarrow 14}{1}\,q_21^9c\cdots$

$\cdots bb\,\overset{\leftarrow\leftarrow\leftarrow\leftarrow\leftarrow\leftarrow 13}{1\,b\,1\,b\,b\,b\,1}\,\cdots\,\overset{\leftarrow 23}{1}\,q_2c\cdots$

$\cdots bb\,\overset{\leftarrow\leftarrow\leftarrow\leftarrow\leftarrow\leftarrow 13}{1\,b\,1\,b\,b\,b\,1}\,\cdots\,\overset{\leftarrow 23}{1}\,\overset{\leftarrow}{c}\,q_2\cdots$

$\cdots bb\,\overset{\leftarrow\leftarrow\leftarrow\leftarrow\leftarrow\leftarrow 13}{1\,b\,1\,b\,b\,b\,1}\,\cdots\,\overset{\leftarrow 23}{1}\,\overset{\leftarrow}{c}\,\cdots\,\overset{\leftarrow 9}{1}\,\overset{\leftarrow}{c}\,q_2\,\overset{\leftarrow\leftarrow}{1\,1}\,\cdots$

$\cdots bb\,\overset{\leftarrow\leftarrow\leftarrow\leftarrow\leftarrow\leftarrow 13}{1\,b\,1\,b\,b\,b\,1}\,\cdots\,\overset{\leftarrow 23}{1}\,\overset{\leftarrow}{c}\,\cdots\,\overset{\leftarrow 9}{1}\,q_2\,\overset{\leftarrow\rightarrow\leftarrow}{c\,1\,1}\,\cdots$

$\cdots bq_2b\,\overset{\rightarrow\rightarrow\rightarrow\rightarrow\rightarrow\rightarrow 13}{1\,b\,1\,b\,b\,b\,1}\,\cdots\,\overset{\rightarrow 23\rightarrow}{1\;c}\,\cdots\,\overset{\rightarrow 9\rightarrow\leftarrow}{1\;c\,1\,1}\,\cdots$

$\cdots q_3b\,\overset{\rightarrow\rightarrow\rightarrow\rightarrow\rightarrow\rightarrow\rightarrow 13}{b\,1\,b\,1\,b\,b\,b\,1}\,\cdots\,\overset{\rightarrow 23\rightarrow}{1\;c}\,\cdots\,\overset{\rightarrow 9\rightarrow\leftarrow}{1\;c\,1\,1}\,\cdots$

$\cdots \vec{c}\,q_1\,\overset{\rightarrow\rightarrow\rightarrow\rightarrow\rightarrow\rightarrow 13}{b\,1\,b\,1\,b\,b\,b\,1}\,\cdots\,\overset{\rightarrow 23\rightarrow}{1\;c}\,\cdots\,\overset{\rightarrow 9\rightarrow\leftarrow}{1\;c\,1\,1}\,\cdots$

$\cdots \vec{c} \overset{\leftarrow}{b} \overset{\leftarrow}{c} \overset{\leftarrow}{b} \overset{\leftarrow}{c} \overset{\leftarrow}{b} \overset{\leftarrow}{b} \overset{\leftarrow}{b} \overset{\leftarrow 13}{c} \cdots \overset{\leftarrow 23 \leftarrow}{c} 1 \cdots \overset{\leftarrow 9 \leftarrow}{c} 1 \overset{\leftarrow \leftarrow}{c} q_1 1 1 \cdots$

$\cdots \vec{c} \overset{\leftarrow}{b} \overset{\leftarrow}{c} \overset{\leftarrow}{b} \overset{\leftarrow}{c} \overset{\leftarrow}{b} \overset{\leftarrow}{b} \overset{\leftarrow}{b} \overset{\leftarrow 13}{c} \cdots \overset{\leftarrow 23 \leftarrow}{c} 1 \cdots \overset{\leftarrow 9}{c} 1 q_1 \overset{\leftarrow \rightarrow \leftarrow}{c\ c} 1 \cdots$

$\cdots \vec{c} \overset{\leftarrow}{b} \overset{\leftarrow}{c} \overset{\leftarrow}{b} \overset{\leftarrow}{c} \overset{\leftarrow}{b} \overset{\leftarrow}{b} \overset{\leftarrow}{b} \overset{\leftarrow 13}{c} \cdots \overset{\leftarrow 23}{c} 1 \cdots \overset{\leftarrow 9}{c} q_3 1 1 \overset{\rightarrow \leftarrow}{c} 1 \cdots$

$\cdots 1^9 q_3 \overset{\rightarrow \rightarrow}{c\ b} 1 \vec{b} 1 \overset{\rightarrow \rightarrow \rightarrow}{b\ b\ b} 1^{13} \cdots 1^{23} c \cdots 1^9 c 1 \overset{\rightarrow \leftarrow}{c} 1 \cdots$

$\cdots 1^8 q_2 1 \overset{\rightarrow \rightarrow}{b\ b} 1 \vec{b} 1 \overset{\rightarrow \rightarrow \rightarrow}{b\ b\ b} 1^{13} \cdots 1^{23} c \cdots 1^9 c 1 \overset{\rightarrow \leftarrow}{c} 1 \cdots$

$\cdots 1^8 \overset{\leftarrow \leftarrow \leftarrow \leftarrow \leftarrow \leftarrow \leftarrow \leftarrow 13}{1\ b\ b\ 1\ b\ 1\ b\ b\ b\ 1} \cdots 1 \overset{\leftarrow 23 \leftarrow}{c} \cdots 1 \overset{\leftarrow 9}{c} 1 c q_2 \overset{\leftarrow}{1} \cdots$

$\cdots$

$\cdots b1bbq_2 b \overset{\rightarrow 9}{1} \overset{\rightarrow \rightarrow \rightarrow \rightarrow \rightarrow \rightarrow \rightarrow 13}{b\ b\ 1\ b\ 1\ b\ b\ b\ 1} \cdots \overset{\rightarrow 23 \rightarrow}{1\ c} \cdots \overset{\rightarrow 9 \rightarrow \rightarrow \rightarrow 9 \leftarrow}{1\ c\ 1\ c\ 1} 1 \cdots$

$\cdots b1bq_3 b \overset{\rightarrow \rightarrow 9}{b\ 1} \overset{\rightarrow \rightarrow \rightarrow \rightarrow \rightarrow \rightarrow \rightarrow 13}{b\ b\ 1\ b\ 1\ b\ b\ b\ 1} \cdots \overset{\rightarrow 23 \rightarrow}{1\ c} \cdots \overset{\rightarrow 9 \rightarrow \rightarrow \rightarrow 9 \leftarrow}{1\ c\ 1\ c\ 1} 1 \cdots$

$\cdots b1b \vec{c} q_1 \overset{\rightarrow \rightarrow 9}{b\ 1} \overset{\rightarrow \rightarrow \rightarrow \rightarrow \rightarrow \rightarrow \rightarrow \rightarrow 13}{b\ b\ 1\ b\ 1\ b\ b\ b\ 1} \cdots \overset{\rightarrow 23 \rightarrow}{1\ c} \cdots \overset{\rightarrow 9 \rightarrow \rightarrow \rightarrow 9 \leftarrow}{1\ c\ 1\ c\ 1} 1 \cdots$

$\cdots b1b \overset{\rightarrow \leftarrow 9 \leftarrow \leftarrow \leftarrow \leftarrow \leftarrow \leftarrow \leftarrow 13}{c\ b\ c\ b\ b\ c\ b\ c\ b\ b\ b\ c} \cdots \overset{\leftarrow 23 \leftarrow}{c} 1 \cdots \overset{\leftarrow 9 \leftarrow \leftarrow \leftarrow 9}{c} 1 c 1 c q_1 1 1 \cdots$

$\cdots b1b \overset{\rightarrow \leftarrow 9 \leftarrow \leftarrow \leftarrow \leftarrow \leftarrow \leftarrow \leftarrow 13}{c\ b\ c\ b\ b\ c\ b\ c\ b\ b\ b\ c} \cdots \overset{\leftarrow 23 \leftarrow}{c} 1 \cdots \overset{\leftarrow 9 \leftarrow \leftarrow \leftarrow 8}{c} 1 c 1 c q_1 \overset{\leftarrow \rightarrow \leftarrow}{c\ c} 1 \cdots$

$\cdots b1b \overset{\rightarrow \leftarrow 9 \leftarrow \leftarrow \leftarrow \leftarrow \leftarrow \leftarrow \leftarrow 13}{c\ b\ c\ b\ b\ c\ b\ c\ b\ b\ b\ c} \cdots \overset{\leftarrow 23 \leftarrow}{c} 1 \cdots \overset{\leftarrow 9 \leftarrow \leftarrow \leftarrow 7}{c} 1 c 1 c q_3 \overset{\leftarrow}{c} 1 \overset{\rightarrow \leftarrow}{c} 1 \cdots$

$\cdots b1bq_3 \overset{\rightarrow \rightarrow}{c\ b} 1^9 \overset{\rightarrow \rightarrow}{b\ b} 1 \vec{b} 1 \overset{\rightarrow \rightarrow \rightarrow}{b\ b\ b} 1^{13} \cdots 1^{23} c \cdots 1^9 c 1 c 1^9 \overset{\rightarrow \leftarrow}{c} 1 \cdots$

$\cdots b1q_2 b \overset{\rightarrow \rightarrow}{b\ b} 1^9 \overset{\rightarrow \rightarrow}{b\ b} 1 \vec{b} 1 \overset{\rightarrow \rightarrow \rightarrow}{b\ b\ b} 1^{13} \cdots 1^{23} c \cdots 1^9 c 1 c 1^9 \overset{\rightarrow \leftarrow}{c} 1 \cdots$

$\cdots bq_3 1 \overset{\rightarrow \rightarrow \rightarrow}{b\ b\ b} 1^9 \overset{\rightarrow \rightarrow}{b\ b} 1 \vec{b} 1 \overset{\rightarrow \rightarrow \rightarrow}{b\ b\ b} 1^{13} \cdots 1^{23} c \cdots 1^9 c 1 c 1^9 \overset{\rightarrow \leftarrow}{c} 1 \cdots$

Third stage

$\cdots b1q_3 \overset{\rightarrow \rightarrow \rightarrow}{b\ b\ b} 1^9 \overset{\rightarrow \rightarrow}{b\ b} 1 \vec{b} 1 \overset{\rightarrow \rightarrow \rightarrow}{b\ b\ b} 1^{13} \cdots 1^{23} c \cdots 1^9 c 1 c 1^9 \overset{\rightarrow \leftarrow}{c} 1 \cdots$

$\cdots b1bbb1^9 bb1b1bbb1^{13} \cdots bb1^9 b1^{23} q_3 c 1 c \cdots 1^9 c 1 c 1^9 \overset{\rightarrow \leftarrow}{c} 1 \cdots$

$\cdots b1bbb1^9 bb1b1bbb1^{13} \cdots bb1^9 b1^{24} q_1 1 c \cdots 1^9 c 1 c 1^9 \overset{\rightarrow \leftarrow}{c} 1 \cdots$

Acknowledgement. The authors acknowledge the very helpful contribution of *INTAS project* 97-1259 for enhancing their cooperation, giving the best conditions for producing the present result.

References

1. C. Baiocchi, Three Small Universal Turing Machines. Proc. MCU'2001, ed. M. Margenstern, Yu. Rogozhin, Springer, LNCS 2055, pp. 1-10, 2001.
2. M.D. Davis and E.J. Weyuker, Computability, Complexity, and Languages. Academic Press, Inc., 1983.
3. M. Kudlek, Small deterministic Turing machines. Theoretical Computer Science, Elsevier Science B.V., vol. 168 (2), 1996, pp. 241–255.
4. M. Margenstern, Frontier between decidability and undecidability: a survey. Proc. of 2nd International Colloquium Universal Machines and Computations, vol.1, March 23-27, 1998, Metz, France, pp. 141-177.

5. M.L. Minsky, Size and structure of universal Turing machines using tag systems. Recursive Function Theory, Symp. in pure mathematics, Amer. Math. Soc., 5, 1962, pp. 229–238.
6. Gh. Păun, DNA Computing Based on Splicing: Universality Results. Proc. of 2nd International Colloquium Universal Machines and Computations, vol.1, March 23-27, 1998, Metz, France, pp. 67–91.
7. L.M. Pavlotskaya, Sufficient conditions for halting problem decidability of Turing machines. Avtomati i mashini (Problemi kibernetiki), Moskva, Nauka, 1978, vol. 33, pp. 91–118, (Russian).
8. R.M. Robinson, Minsky's small universal Turing machine. International Journal of Mathematics, vol.2, N.5, 1991, pp. 551–562.
9. Yu. Rogozhin, Seven universal Turing machines. Systems and Theoretical Programming, Mat. Issled. no.69, Academiya Nauk Moldavskoi SSR, Kishinev, 1982, pp. 76–90, (Russian).
10. Yu. Rogozhin, A universal Turing machine with 10 states and 3 symbols. Izvestiya Akademii Nauk Respubliki Moldova, Matematika, 1992, N 4(10), pp. 80–82 (Russian).
11. Yu. Rogozhin, About Shannon's problem for Turing machines. Computer Science Journal of Moldova, vol.1, no 3(3), 1993, pp. 108–111.
12. Yu. Rogozhin, Small universal Turing machines. Theoretical Computer Science, Elsevier Science B.V., vol. 168 (2), 1996, pp. 215–240.
13. Yu. Rogozhin, A Universal Turing Machine with 22 States and 2 Symbols. Romanian Journal of Information Science and Technology, vol. 1, N. 3, 1998, pp. 259–265.
14. C.E. Shannon, A universal Turing machine with two internal states. Automata studies, Ann. of Math. Stud. 34, Princeton, Princeton Univ.Press, 1956, pp. 157–165.

Minimal Covers of Formal Languages

Michael Domaratzki[1], Jeffrey Shallit[2]*, and Sheng Yu[3]

[1] Department of Computing and Information Science, Queen's University
Kingston, Ontario, Canada K7L 3N6
`domaratz@cs.queensu.ca`
[2] Department of Computer Science, University of Waterloo
Waterloo, Ontario, Canada N2L 3G1
`shallit@graceland.uwaterloo.ca`
[3] Department of Computer Science, University of Western Ontario
London, Ontario, Canada N6A 3K7
`syu@csd.uwo.ca`

Abstract. Let L, L' be languages. If $L \subseteq L'$, we say that L' covers L. Let $\mathcal{C}, \mathcal{D}$ be two classes of languages. If $L' \in \mathcal{C}$, we say that L' is a minimal $\mathcal{C}$-cover with respect to $\mathcal{D}$ if whenever $L \subseteq L'' \subseteq L'$ and $L'' \in \mathcal{C}$, we have $L' - L'' \in \mathcal{D}$. In this paper we discuss minimal $\mathcal{C}$-covers with respect to finite languages, when $\mathcal{C}$ is the class of regular languages.

1 Introduction

Let $L, L' \subseteq \Sigma^*$ be languages. If $L \subseteq L'$, then we say L' *covers* L. In this paper, we are interested in studying the case where the covering language L' is

(i) restricted to lie in some language class—in particular, the regular languages (REG)—and
(ii) is *minimal* in some sense.

The motivation for studying minimal covers is that arbitrary languages L may be arbitrarily difficult to recognize. However, a regular cover L' is easy to recognize, and if the regular cover is minimal, then we might hope that the difference between L' and L is not too large. A recognition algorithm based on L' will mistakenly accept some words it shouldn't, but never mistakenly reject a word in L.

One definition of minimal that at first sight seems natural is the following. Let $\mathcal{C}$ be a class of languages. If $L' \in \mathcal{C}$, we might say L' is a minimal $\mathcal{C}$-cover of L if $L'' \in \mathcal{C}$ and $L \subseteq L'' \subseteq L'$ implies $L'' = L'$. If $\mathcal{C}$ is closed under finite modification — which is the case for nearly every interesting class of languages — then under this definition only members of $\mathcal{C}$ would have minimal covers, and every language is minimally covered by itself! Suppose that L has a minimal cover $L' \in \mathcal{C}$. Then either $L = L'$, in which case $L' \in \mathcal{C}$, or $L \neq L'$. In the latter case, choose any $x \in L' - L$, and consider $L'' = L' - \{x\}$. Since $\mathcal{C}$ is closed under finite modification, $L'' \in \mathcal{C}$. Now $L \subseteq L'' \subseteq L'$, so L' was not minimal, a contradiction.

Thus it is clear we need to seek an alternative definition of minimal cover. To do so we introduce a second language class, $\mathcal{D}$.

---
* Research supported in part by a grant from NSERC.

Definition 1. *We say that L' is a minimal $\mathcal{C}$-cover of L with respect to $\mathcal{D}$ if the following conditions hold:*

(i) $L' \in \mathcal{C}$, $L \subseteq L'$ and
(ii) for all languages $L'' \in \mathcal{C}$ with $L \subseteq L'' \subseteq L'$, we have $L' - L'' \in \mathcal{D}$.

The class of languages L having a minimal $\mathcal{C}$-cover with respect to $\mathcal{D}$ is denoted by $MC(\mathcal{C}, \mathcal{D})$.

An alternative characterization of minimal covers is possible in some cases, as follows.

Proposition 1. *Suppose $\mathcal{C}$ is a class of languages that is closed under intersection and complement. Suppose $L \subseteq L'$ and $L' \in \mathcal{C}$. Then L' is a minimal $\mathcal{C}$-cover of L with respect to $\mathcal{D}$ if and only if every subset $S \subseteq L' - L$ with $S \in \mathcal{C}$ satisfies $S \in \mathcal{D}$.*

Proof. Suppose L' is a minimal $\mathcal{C}$-cover with respect to $\mathcal{D}$. Let $S \subseteq L' - L$ and $S \in \mathcal{C}$. Now let $L'' := L' - S$. Since $\mathcal{C}$ is closed under intersection and complement, we have $L'' \in \mathcal{C}$. But then $S = L' - L'' \in \mathcal{D}$.

On the other hand, suppose every subset $S \subseteq L' - L$ with $S \in \mathcal{C}$ satisfies $S \in \mathcal{D}$. Let $L'' \in \mathcal{C}$ be such that $L \subseteq L'' \subseteq L'$. Define $S := L' - L''$; then by the assumed closure properties $S \in \mathcal{C}$. Since $S \subseteq L' - L$, it follows that $L' - L'' = S \in \mathcal{D}$. ∎

If a language L is infinite and no infinite subset of L lies in the class $\mathcal{C}$, it is said to be $\mathcal{C}$-*immune*. The terminology was apparently introduced by Post [11], who proved among other things that if L is an infinite recursively enumerable set, then L is not RECURSIVE-immune. For other works on immunity, see, for example, [1, p. 13] and [13, p. 107].

In this paper, our main focus is when $\mathcal{C} = $ REG and $\mathcal{D} = $ FINITE. In this case, Proposition 1 can be rephrased as follows:

Proposition 2. *Let L be a non-regular language, L' be a regular language, and $L \subseteq L'$. Then L' is a minimal regular cover of L with respect to finite languages iff $L' - L$ is* REG-*immune.*

Proof. We have $T := L' - L$ is infinite, for if T were finite, then $L = L' - T$ would be regular, a contradiction. Now use Proposition 1. ∎

(Flajolet and Steyaert briefly mentioned REG-immunity in a 1974 paper [4].)

We point out that the term "minimal cover" was recently used by Câmpeanu, Sântean, and Yu [2] in a different context.

2 Some Examples

In this section we consider some specific examples of context-free languages and determine if they have minimal regular covers. These examples show that there exist context-free languages that do not have a minimal regular cover, and there exist non-regular context-free languages that do have minimal regular covers.

We recall two theorems of Lyndon and Schützenberger [10]:

Theorem 1. *Let $y \in \Sigma^*$, and $x, z \in \Sigma^+$. Then $xy = yz$ if and only if there exist $u, v \in \Sigma^*$, and an integer $e \geq 0$ such that $x = uv$, $z = vu$, and $y = (uv)^e u = u(vu)^e$.*

Theorem 2. *Let $x, y \in \Sigma^+$. Then the following three conditions are equivalent:*
(1) $xy = yx$;
(2) There exist integers $i, j > 0$ such that $x^i = y^j$;
(3) There exist $z \in \Sigma^+$ and integers $k, l > 0$ such that $x = z^k$ and $y = z^l$.

We also recall a theorem of Fine and Wilf [3]:

Theorem 3. *Let $x, y \in \Sigma^*$. If there exist integers $p, q \geq 0$ such that x^p and y^q have a common prefix (resp., suffix) of length $\geq |x| + |y| - \gcd(|x|, |y|)$, then there exist $z \in \Sigma^*$ and integers $i, j \geq 0$ such that $x = z^i$, $y = z^j$.*

We now prove two useful lemmas. The first is of independent interest.

Lemma 1. *Let $u, v, w, x, y \in \Sigma^*$ be fixed words with $v, x \neq \epsilon$. Define*
$$L = L(u, v, w, x, y) = \{uv^i wx^i y : i \geq 0\}.$$

Then the following are equivalent:
(a) L is regular;
(b) L has an infinite regular subset;
(c) There exist integers $k, l \geq 1$ such that $v^k w = wx^l$;
(d) There exist words $r, s \in \Sigma^*$ and integers $m, n \geq 1$ and $p \geq 0$ such that
$$v = (rs)^m, w = (rs)^p r, \text{ and } x = (sr)^n.$$
(e) There exist integers $a, b \geq 0$ with $a \neq b$ such that $uv^a wx^b y \in L$.

Proof. Note: the implications (e) $\Longrightarrow$ (a) and (e) $\Longleftrightarrow$ (b) were previously proved by Hunt, Rosenkrantz, and Szymanski [8, pp. 239–240]. The implications (a) $\Longleftrightarrow$ (c) were proved by Latteux and Thierrin [9].

We prove (a) $\Longrightarrow$ (b) $\Longrightarrow$ (c) $\Longrightarrow$ (d) $\Longrightarrow$ (e). The reader is referred to Hunt, Rosenkrantz and Symanski [8, pp. 239–240] for the implication (e) $\Longrightarrow$ (a).

(a) $\Longrightarrow$ (b): L is clearly infinite, so if L is regular, then it has an infinite regular subset, namely L itself.

(b) $\Longrightarrow$ (c): Suppose R is an infinite regular language, and $R \subseteq L$. Then by the pumping lemma, there exist words $r, s \neq \epsilon$, and t such that $rs^*t \subseteq R$. Then there exist integers $m, n \geq 0$ such that $rt = uv^m wx^m y$ and $rst = uv^{m+n} wx^{m+n} y$. Since L contains at most one word of each length, it then follows that $rs^i t = uv^{m+ni} wx^{m+ni} y$ for all $i \geq 0$.

By replacing r with rs^a for some a and t with $s^b t$ for some b, if necessary, we may assume without loss of generality that $|r| > |u|$ and $|t| > |y|$. Similarly, by replacing s with s^c for some c, if necessary, we may assume $|r| < |uv^{m+ni}|$ and $|t| < |x^{m+ni} y|$ for all $i \geq 1$.

It now follows that $r = uv^e v'$, $s = v''v^f wx^g x'$, $t = x''x^h y$ for some $e, f, g, h \geq 0$ where $v = v'v''$ and $x = x'x''$. Now choose i sufficiently large such that $m + ni - e - h - 1 > |s| + |v| + |x|$. Then

$$rs^i t = uv^{m+ni} wx^{m+ni} y.$$

By cancelling $r = uv^e v'$ on the left and $t = x''x^h y$ on the right, we get

$$s^i = v'' v^{m+ni-e-1} wx^{m+ni-h-1} x'$$
$$= (v''v')^{m+ni-e-1} v'' wx' (x''x')^{m+ni-h-1}. \qquad (1)$$

We now observe that s^i and $(v''v')^{m+ni-e-1}$ agree on a prefix of size $\geq m + ni - e - 1 > |s| + |v|$, and so by Theorem 3, there exists a word Y such that s and $v''v'$ are both powers of Y. Similarly, s^i and $(x''x')^{m+ni-h-1}$ agree on a suffix of size $\geq m + ni - h - 1 > |s| + |x|$, and so by Theorem 3, there exists a word Z such that s and $x''x'$ are both powers of Z. From Theorem 2 it now follows that there exists a word X such that both Y and Z are powers of X. Hence s, $v''v'$, and $x''x'$ are all powers of X. Then from Eq. (1) it follows that $v''wx'$ is a power of X. Now write $v''v' = X^l$, $x''x' = X^k$, and $v''wx' = X^d$ for some integers l, k, d. Then

$$(v''v')^k v'' wx' = v'' wx' (x''x')^l.$$

Now cancel v'' on the left and x' on the right; we get

$$(v'v'')^k w = w(x'x'')^l,$$

and hence $v^k w = wx^l$, as desired.

(c) $\implies$ (d): If $v^k w = wx^l$, then by Theorem 1 there exist $t, z \in \Sigma^*$ and an integer $e \geq 0$ such that $v^k = tz$, $w = t(zt)^e$, and $x^l = zt$.

Now $v^k = tz$ implies there exists a decomposition $v = v'v''$ such that $t = v^i v'$ and $z = v''v^j$ for some integers $i, j \geq 0$. Then $x^l = zt = v''v^j v^i v' = (v''v')^{i+j+1}$. Set $h = v''v'$. Then $x^l = h^{i+j+1}$. Hence, by Theorem 2, there exists a word $g \in \Sigma^+$ and integers $n, a \geq 1$ such that $x = g^n$ and $h = g^a$.

From this last equality we get $v''v' = g^a$. Thus there exists a decomposition $g = sr$ such that $v'' = g^b s$ and $v' = rg^c$ for some integers $b, c \geq 0$. Then we find $v = v'v'' = (rg^c)(g^b s) = (rs)^{b+c+1}$.

Finally, we have

$$w = t(zt)^e = v^i v'(x^l)^e$$
$$= (rs)^{(b+c+1)i} rg^c (g^n)^{le}$$
$$= (rs)^{(b+c+1)i} r(sr)^{c+nle} = (rs)^{(b+c+1)i+c+nle} r.$$

so (d) holds with $m = b + c + 1$, and $p = (b + c + 1)i + c + nle$.

(d) $\implies$ (e): Choose $a = 0$, $b = m + n$. Then

$$uv^a wx^b y = u(rs)^p r(sr)^{n(m+n)} y$$
$$= u(rs)^{mn} (rs)^p r(sr)^{n^2} y$$
$$= uv^n wx^n y \in L.$$

∎

The languages $L = L(u, v, w, x, y)$ are also important in the study of slender context-free languages (cf. Păun and Salomaa [12])

Theorem 4. *Let $u, v, w, x, y \in \Sigma^*$ be fixed words with $v, x \neq \epsilon$. Let $L = \{uv^i wx^i y : i \geq 0\}$. Then L is context-free, and has a minimal regular cover with respect to finite languages iff L is regular.*

Proof. It is clear that L is context-free, since it is generated by the context-free grammar with the following set of productions: $\{S \to uBy, B \to w, B \to vBx\}$. First, suppose that L is minimally covered by the regular language R. Let $M = (Q, \Sigma, \delta, q_0, F)$ be a deterministic finite automaton (DFA) accepting R. Then there exists a state q such that $\delta(q_0, uv^i w) = q$ for infinitely many i. Let $I = \{i : \delta(q_0, uv^i w) = q\}$ and i_0 be the smallest element of I. Define

$$R' = \{x^j y : \delta(q, x^j y) \in F \text{ and } j > i_0\}.$$

Then R' is infinite since it contains $\{x^i y : i \in I - \{i_0\}\}$. Furthermore, R' is regular since $R' = \{t \in \Sigma^* : \delta(q, t) \in F\} \cap x^{i_0} x^+ y$. Let $R'' = uv^{i_0} wR'$. Then R'' is infinite and regular, and clearly $R'' \subseteq R$. Now if $R'' \subseteq R - L$, then $R''' = R - R''$ is a regular cover of L with $R - R'''$ infinite. Hence R is not minimal, a contradiction. Thus $R'' \not\subseteq R - L$, and hence $uv^{i_0} wx^{j_0} y \in L$ for some $j_0 > i_0$. Then by Lemma 1, L is regular.

On the other hand, if L is regular then L itself is trivially a minimal regular cover. ∎

Let $\overline{L}$ denote the complement of L, that is $\Sigma^* - L$.

Theorem 5. *Let $u, v, w, x, y \in \Sigma^*$ be fixed words with $v, x \neq \epsilon$. Let $L = \{uv^i wx^i y : i \geq 0\}$. Then $\overline{L}$ is context-free. Further, if L is not regular, then Σ^* is a minimal regular cover of $\overline{L}$.*

3 Characterizations of MC(REG, FINITE)

In this section, we obtain some characterization of the class of languages having minimal regular covers.

First we consider the unary case.

Lemma 2. *If $L \subset 0^*$ is a regular language, then there exist integers $m \geq 0$, $n \geq 1$ and sets $A \subseteq \{\epsilon, 0, \ldots, 0^{m-1}\}$ and $B \subseteq \{0^m, 0^{m+1}, \ldots, 0^{m+n-1}\}$ such that $L = A + B(0^n)^*$. Furthermore, L is infinite iff $B \neq \emptyset$.*

By an *arithmetic progression* we mean a set of the form $\{nt + b : t \geq 0\}$ for integers $n \geq 1$, $b \geq 0$.

Theorem 6. *Let $L \subseteq 0^*$. Then L has a minimal regular cover 0^* iff the set $C(L) := \{j : 0^j \in L\}$ has a nonempty intersection with every arithmetic progression.*

Proof. Suppose L has 0^* as a minimal regular cover. Then $0^* - L$ has no infinite regular subset. Let $R = 0^b(0^n)^*$. Then R is infinite and regular, and hence $R \not\subseteq 0^* - L$. It follows that there exists t such that $0^{nt+b} \in L$.

On the other hand, suppose $C(L)$ intersects every arithmetic progression. Then by Lemma 2, any infinite regular set R can be written as $R = A + B(0^n)^*$ for some $B \neq \emptyset$. Let $b \in B$. Then if $R' := 0^b(0^n)^*$ we have $R' \subseteq R$. But $L \cap R' \neq \emptyset$, so $R \not\subseteq 0^* - L$. ∎

As an example of Theorem 6, we have

Proposition 3. *The language $L = \{0^{n!+n} : n \geq 0\}$ has minimal regular cover 0^*.*

Proof. Let $r \geq 1, b \geq 0$. Consider the arithmetic progression $\{rt + b : t \geq 0\}$. Then let $n = r + b$ and $t = n!/r - 1$. Clearly t is a non-negative integer, since $n \geq r$. We have $n! + n = rt + b$. ∎

The next lemma gives a sufficient condition for a unary language to have 0^* as a minimal regular cover.

Lemma 3. *Suppose $a_1 < a_2 < a_3 < \cdots$ is a strictly increasing sequence of non-negative integers such that $a_n/n \to \infty$ as $n \to \infty$. Let $A = \{0^{a_i} : i \geq 0\}$. Then 0^* is a minimal regular cover of $L = 0^* - A$.*

Proof. We have $0^* - L = A$. If A contained an infinite regular set, then by Lemma 2, there would be integers $i \geq 0, j \geq 1$ such that $0^i(0^j)^* \subseteq A$. But then there would be $\epsilon > 0$ such that $a_n/n \leq j + \epsilon$ for all n sufficiently large, a contradiction. ∎

Next, we consider a generalization of Lemma 6. Recall that a language L is *bounded* if there exist finitely many words $w_1, w_2, \ldots, w_n$ such that $L \subseteq w_1^* w_2^* \cdots w_n^*$.

We now completely characterize those bounded languages which possess a minimal regular cover. We begin by recalling a characterization of bounded regular languages due to Ginsburg and Spanier [5].

Theorem 7. *Let $L \subseteq w_1^* w_2^* \cdots w_n^*$ be a bounded language. Then L is regular if and only if there exist an integer N and indices $a_{i,j}, b_{i,j}$ for $1 \leq i \leq N, 1 \leq j \leq n$ such that*

$$L = \bigcup_{i=1}^{N} w_1^{a_{i,1}} (w_1^{b_{i,1}})^* \cdots w_n^{a_{i,n}} (w_n^{b_{i,n}})^*.$$

We now use Theorem 7 as a means of classifying bounded regular languages with minimal regular covers.

Let $L \subseteq w_1^* \cdots w_n^*$ be a bounded language. Given fixed integers $a_{i,j}$ and $b_{i,j}$ ($1 \leq i \leq N, 1 \leq j \leq n$), we define, for each $1 \leq i \leq N$ and $1 \leq k \leq n$, the sets

$I_{i,k}(L, j_1, \ldots, j_{k-1}, j_{k+1}, \ldots, j_n) = \{j :$
$w_1^{a_{i,1}+j_1 b_{i,1}} \cdots w_{k-1}^{a_{i,k-1}+j_{k-1} b_{i,k-1}} w_k^{a_{k,i}+jb_{k,i}} w_{k+1}^{a_{i,k+1}+j_{k+1} b_{i,k+1}} \cdots w_n^{a_{i,n}+j_n b_{i,n}} \in L\}$

for any $j_1, j_2, \ldots, j_{i-1}, j_{i+1}, \ldots, j_n \geq 0$.

Theorem 8. *Let $L \subseteq R \subseteq w_1^* \cdots w_n^*$ be bounded languages. Let R be regular, with representation $R = \bigcup_{i=1}^{N} w_1^{a_{i,1}}(w_1^{b_{i,1}})^* \cdots w_n^{a_{i,n}}(w_n^{b_{i,n}})^*$. Then L has minimal regular cover R if and only if for all $1 \leq i \leq N$, $1 \leq k \leq n$ and for all $j_1, j_2, \ldots, j_{i-1}, j_{i+1}, \ldots, j_n \geq 0$, the set $I_{i,k}(L, j_1, j_2, \ldots, j_{i-1}, j_{i+1}, \ldots, j_n)$ intersects every arithmetic progression.*

As an example of Theorem 8, we have

Theorem 9. *Let $L = \{a^i b^j : \text{there exists } n \text{ such that } i + j = n! + n\}$. Then L has a^*b^* as a minimal regular cover.*

4 Closure Properties of Minimal Regular Covers

In this section we consider the closure properties of the class $MC(\text{REG}, \text{FINITE})$, i.e., the class of languages having minimal regular covers with respect to finite languages. We show that $MC(\text{REG}, \text{FINITE})$ is closed under the operations of union, concatenation with a finite set, intersection with a regular language, homomorphism, quotient, and inverse ϵ-free homomorphism.

The following theorem shows that if a language has at least one minimal regular cover, then it has exactly one (up to finite modification).

Theorem 10. *Let L be a language, and suppose R, R' are minimal regular covers of L. Then $(R - R') \cup (R' - R)$ is finite.*

The class of languages with minimal regular covers is closed under finite union. More precisely we have

Theorem 11. *Suppose L_1, L_2 are languages with minimal regular covers R_1, R_2 respectively. Then $L_1 \cup L_2$ has $R_1 \cup R_2$ as a minimal regular cover.*

Once we have a language possessing a regular cover, we can add to that language while maintaining the property.

Theorem 12. *If L_1 is minimally covered by a regular language R, and $L_2 \subseteq R$, then $L_1 \cup L_2$ is also minimally covered by R.*

We now consider concatenation with a finite set.

Theorem 13. *Suppose R is a minimal regular cover of a language L, and F is finite. Then FR is a minimal regular cover of FL and RF is a minimal regular cover of LF.*

Although we will see in the next section that the class of languages possessing minimal regular covers is not closed under intersection, this class is closed under intersection with regular languages.

Theorem 14. *Let L be a language with R a minimal regular cover of L, and let R' be any regular language. Then $L \cap R'$ is minimally covered by $R \cap R'$.*

We now consider closure under homomorphism. Recall that $\varphi : \Sigma^* \to \Delta^*$ is a homomorphism if $\varphi(xy) = \varphi(x)\varphi(y)$ for all $x, y \in \Sigma^*$. We prove

Theorem 15. *Suppose a language $L \subseteq \Sigma^*$ is minimally covered by the regular language $R \subseteq \Sigma^*$, and suppose $\varphi : \Sigma^* \to \Delta^*$ is a homomorphism. Then $\varphi(L)$ is minimally covered by $\varphi(R)$.*

Proof. First, we have the following set-theoretic lemma:

Lemma 4. *Let L_1, L_2 be languages with $L_1 \subseteq \Sigma^*$, $L_2 \subset \Delta^*$, and let $\varphi : \Sigma^* \to \Delta^*$ be a homomorphism. Then $\varphi(L_1) - L_2 \subseteq \varphi(L_1 - \varphi^{-1}(L_2))$.*

We can now prove Theorem 15. Suppose $\varphi(L)$ is not minimally covered by $\varphi(R)$. Then there exists a regular language R_0 such that $\varphi(L) \subseteq R_0 \subseteq \varphi(R)$ and $\varphi(R) - R_0$ is infinite. Now

$$L \subseteq \varphi^{-1}(\varphi(L)) \subseteq \varphi^{-1}(R_0)$$

and $L \subseteq R$, so $L \subseteq \varphi^{-1}(R_0) \cap R$.

But $R' = \varphi^{-1}(R_0) \cap R$ is regular, and $L \subseteq R' \subseteq R$. By Lemma 4, $\varphi(R) - R_0 \subseteq \varphi(R - \varphi^{-1}(R_0))$. Since $\varphi(R) - R_0$ is infinite, we must have that $\varphi(R - \varphi^{-1}(R_0))$ is infinite. Hence $R - \varphi^{-1}(R_0)$ is infinite. Thus $R - R'$ is infinite, contradicting the minimality of the cover. ∎

Note: Theorem 15 is not true if "homomorphism" is replaced by "substitution". For example, consider the case $L = \{a\}$, $R = \{a, b\}$, and φ maps a to a and b to $(a + b)^*$. The following theorem shows that minimal regular covers are closed under ϵ-free inverse homomorphism.

Theorem 16. *Let $\varphi : \Delta^* \to \Sigma^*$ be an ϵ-free homomorphism. Let $L, R \subseteq \Sigma^*$ be languages with R a minimal regular cover for L. Then $\varphi^{-1}(L)$ has minimal regular cover $\varphi^{-1}(R)$.*

We now examine closure under quotient. Recall that for $L_1, L_2 \subseteq \Sigma^*$, we define the quotient L_1/L_2 by $L_1/L_2 = \{x \in \Sigma^* \ : \ \exists \, y \in L_2 \text{ such that } xy \in L_1\}$.

Theorem 17. *Let L_1 be a language with R_1 a minimal regular cover. If L_2 is any language, then L_1/L_2 has minimal regular cover R_1/L_2.*

Proof. Clearly if $L_1 \subseteq R_1$, then $L_1/L_2 \subseteq R_1/L_2$. Furthermore, it is well known that if R is regular, then R/L is regular for all languages L [7, Thm. 3.6, pp. 62–63]. Hence R_1/L_2 is a regular cover of L_1/L_2.

Suppose, contrary to what we want to prove, that R_1/L_2 is not a minimal regular cover of L_1/L_2. Then there exists a regular language R' such that

$$L_1/L_2 \subseteq R' \subseteq R_1/L_2$$

and $R_1/L_2 - R'$ is infinite. We claim that there exists a word $w \in L_2$ such that $R_1/\{w\} - R'$ is infinite.

To see this, let $M_1 = (Q_1, \Sigma, \delta_1, q_1, F_1)$ be a DFA for R_1. By the usual construction [7, p. 63], let $F'_1 = \{q \in Q \ : \ \text{there exists } y \in L_2 \text{ such that } \delta_1(q, y) \in F\}$, then $M'_1 = (Q_1, \Sigma, \delta_1, q_1, F'_1)$ is a DFA for R_1/L_2. Let $M_2 = (Q_2, \Sigma, \delta_2, q_2, F_2)$ be a DFA for R'. Let $M = (Q_1 \times Q_2, \Sigma, \delta, q_0, F)$ where $q_0 = [q_1, q_2]$, $F = F'_1 \times (Q_2 - F_2)$ and $\delta([p, q], a) = [\delta_1(p, a), \delta_2(q, a)]$ for $p \in Q_1, q \in Q_2$, and $a \in \Sigma$. Then M accepts the language $R_1/L_2 - R'$. Since $R_1/L_2 - R'$ is infinite, by the pigeonhole principle, there must be a state $t \in F$ such that $L = \{x \in \Sigma^* \ : \ \delta(q_0, x) = t\}$ is infinite. Suppose

$t = [r, s]$. Then $r \in F_1'$, and hence there exists $w \in L_2$ such that $\delta_1(r, w) \in L_2$. It follows that $L \subseteq R_1/\{w\} - R'$, and the result follows.

Now let $B = (R_1/L_2 - R')\{w\} \cap R_1$. Clearly B is regular and $B \subseteq R_1$. We claim that

$$L_1 \subseteq R_1 - B \subseteq R_1 \tag{2}$$

and B is infinite.

First, let's prove (2). Since $L_1 \subseteq R_1$, it suffices to show that $L_1 \cap B = \emptyset$. Suppose there exists a word x with $x \in L_1$ and $x \in B$. Since $x \in B$, we can write $x = tw$ where $t \in R_1/L_2 - R'$. Hence $t \notin R'$. But $tw \in L_1$, which implies $t \in L_1/\{w\} \subseteq L_1/L_2 \subseteq R'$, so $t \in R'$, a contradiction.

Now let us prove that B is infinite. We know from above that $R_1/\{w\} - R'$ is infinite, say $R_1/\{w\} - R' = \{x_1, x_2, \ldots, \}$. Since $x_i \in R_1/\{w\}$ for each $i \geq 1$, we have $x_i w \in R_1$ for each $i \geq 1$. Also $x_i w \in (R_1/L_2 - R')\{w\}$. Hence $x_i w \in B$. But all the $x_i w$ are distinct and hence B is infinite.

Since $L_1 \subseteq R_1 - B \subseteq R_1$, and $R_1 - B$ is regular, and $R_1 - (R_1 - B) = B$ is infinite, it now follows that R_1 is not a minimal regular cover of L_1, a contradiction. ∎

Corollary 1. *The following decision problems are unsolvable:*

(a) Given a context-free grammar G, does $L(G)$ have a minimal regular cover?
(b) Given a context-free grammar G, does $\Sigma^* - L(G)$ have a minimal regular cover?

Proof. (a) We use Greibach's theorem [6], which says in particular that if P is a nontrivial property that is true for all regular languages and preserved under quotient with a single symbol, then P is undecidable for the class of context-free languages.

Let P be the property of having a minimal regular cover. Then P is true for all regular languages, and by Theorem 17, P is preserved under quotient. Assertion (b) is proved similarly. ∎

5 Non-closure Properties for Minimal Regular Covers

In this section we prove that the class of languages having minimal regular covers with respect to finite languages is not closed under intersection, concatenation, Kleene star and inverse homomorphism.

First we prove a useful lemma:

Lemma 5. *Let $L_s = \{0^{n^2} : n \geq 1\}$. Then L_s has no minimal regular cover.*

Proof. Suppose R is a minimal regular cover of L_s. Then by Lemma 2 we can write $R = A \cup B(0^n)^*$ with $B \subseteq \{0^m, 0^{m+1}, \ldots, 0^{m+n-1}\}$. Let p be an odd prime not dividing n, and let a be a quadratic non-residue (mod p), i.e., the Legendre symbol $\left(\frac{a}{p}\right) = -1$. Since L_s is infinite, R must be infinite, and so $B \neq \emptyset$. Choose a b such that $0^b \in B$. Let r be an integer, $m \leq r < m + pn$ such that $r \equiv \quad \mod bn$ and $r \equiv \mod ap$; such an r exists by the Chinese remainder theorem.

Now let $T = (0^{pn})^*0^r$. Note that $T \subseteq R$. If $0^c \in T$, then c is not a square, since $\left(\frac{c}{p}\right) = \left(\frac{r}{p}\right) = \left(\frac{a}{p}\right) = -1$. It follows that $L_s \subseteq R - T$, so $R - T$ is a regular cover of L_s. On the other hand, $R - (R - T) = T$ is infinite, so R is not minimal, a contradiction. ∎

First, we give an example proving that the class of languages possessing minimal regular covers is not closed under intersection. We introduce the following notation: $\omega(n)$ denotes the total number of prime factors of n, counted with multiplicity. For example, $\omega(2^3 5^7 11^8) = 18$.

Theorem 18. *Let*

$$L_0 = \{0^r : \omega(r) \equiv \mod 02\};$$
$$L_1 = \{0^r : r \text{ is a square or } \omega(r) \equiv \mod 12\}.$$

Then 0^* is a minimal regular cover of both L_0 and L_1, but $L_0 \cap L_1$ possesses no minimal regular cover.

Proof. First, consider L_0. By Theorem 6, it suffices to show that $C(L_0) = \{n : \omega(n) \equiv \mod 02\}$ intersects every arithmetic progression. Let $A_{n,a} := \{nt + a : t \geq 0\}$ be an arithmetic progression. Let $g = \gcd(a, n)$; note that if $a = 0$ then $g = n$. Then by Dirichlet's theorem there exists infinitely many prime numbers $p \equiv \mod a/g n/g$, so in particular there exists such a p with $p \geq n$.

Now there are two cases to consider. If $\omega(g) \equiv \mod 12$, then $\omega(gp) \equiv \mod 02$. Also $gp \equiv \mod an$ and $gp \geq n$, so $gp \in A_{n,a}$ and hence $C(L_0)$ intersects $A_{n,a}$.

If $\omega(g) \equiv \mod 02$, then by Dirichlet's theorem there exists a prime $q \equiv \mod 1 n/g$. Then $\omega(gpq) \equiv \mod 02$. Also $gpq \equiv \mod an$ and $gpq \geq n$, so $gpq \in A_{n,a}$ and hence $C(L_0)$ intersects $A_{n,a}$. It follows that 0^* is a minimal regular cover of L_0.

A similar argument applies to $\{0^r : \omega r \equiv \mod 12\}$, and hence, by Theorem 12, L_1 has 0^* as a minimal regular cover. However, $L_0 \cap L_1 = \{0^{n^2} : n \geq 1\} = L_s$, and L_s has no minimal regular cover by Lemma 5. ∎

We now prove that the class of languages with minimal regular covers is not closed under concatenation — even under concatenation with regular languages.

Theorem 19. *Let $L_1 = 1^+ - \{1^p : p \text{ prime}\}$, and $L_2 = 0^*$. Then L_1 has a minimal regular cover (namely, 1^+) and L_2 trivially does. However, $L_1 L_2$ has no minimal regular cover.*

Suppose L is a unary language. Then it is well-known that L^* is regular; hence trivially if L is a unary language possessing a minimal regular cover, then so does L^*. However, this result is not true for larger alphabets:

Theorem 20. *Let $D = \{0^{n!+n} : n \geq 0\}$ and let $\overline{D} = 0^* - D$. Then $T := 1\overline{D}$ is minimally covered by 10^*, but T^* has no minimal regular cover.*

We now consider closure under inverse homomorphism. The next lemma shows that $MC(\text{REG}, \text{FINITE})$ is not closed under arbitrary inverse homomorphism. (We have already seen that this closure holds if we restrict ourselves to ϵ-free inverse homomorphism.)

Lemma 6. *Let $L = \{0^{n+n!} : n \geq 1\}$ and $\varphi : \{a, b\} \to \{0\}$ be the homomorphism defined by $\varphi(a) = 0$ and $\varphi(b) = \epsilon$. Then $\overline{L}$ has minimal regular cover 0^*, but $\varphi^{-1}(\overline{L})$ has no minimal regular cover.*

References

1. J. Balcázar, J. Díaz, and J. Gabarró. *Structural Complexity I*, volume 11 of *EATCS Monographs on Theoretical Computer Science*. Springer-Verlag, 1988.
2. C. Câmpeanu, N. Sântean and S. Yu. Minimal cover-automata for finite languages. In J.-M. Champarnaud, D. Maurel and D. Ziadi, editors, *WIA '98*, volume 1660 of *Lecture Notes in Computer Science*, pages 43–56. Springer-Verlag, 1999.
3. N.J. Fine and H.S. Wilf. Uniqueness theorems for periodic functions. *Proc. Amer. Math. Soc.*, 16:109–114, 1965.
4. P. Flajolet and J. M. Steyaert. On sets having only hard subsets. In J. Loeckx, editor, *Proc. 2nd Colloq. on Automata, Languages and Programming (ICALP)*, volume 14 of *Lecture Notes in Computer Science*, pages 446–457. Springer-Verlag, 1974.
5. S. Ginsburg and E. H. Spanier. Bounded regular sets. *Proc. Amer. Math. Soc.*, 17:1043–1049, 1966.
6. S. A. Greibach. A note on undecidable properties of formal languages. *Math. Systems Theory*, 2:1–6, 1968.
7. J. E. Hopcroft and J. D. Ullman. *Introduction to Automata Theory, Languages, and Computation*. Addison-Wesley, 1979.
8. H. B. Hunt III, D. J. Rosenkrantz and T. G. Szymanski. On the equivalence, containment, and covering problems for the regular and context-free languages. *J. Comput. System Sci.*, 12:222-268, 1976.
9. M. Latteux and G. Thierrin. Semidiscrete context-free languages. *Internat. J. Comput. Math.* 14(1):3–18, 1983.
10. R.C. Lyndon and M. P. Schützenberger. The equation $a^M = b^N c^P$ in a free group. *Michigan Math. J.*, 9:289–298, 1962.
11. E. L. Post. Recursively enumerable sets of positive integers and their decision problems. *Bull. Amer. Math. Soc.*, 50:284–316, 1944.
12. G. Păun and A. Salomaa. Thin and slender languages. *Disc. Appl. Math.*, 61:257–270, 1995.
13. H. Rogers, Jr. *Theory of Recursive Functions and Effective Computability*. McGraw-Hill, 1967.

Some Regular Languages That Are Church-Rosser Congruential

Gundula Niemann[1] and Johannes Waldmann[2]

[1] Fachbereich Mathematik/Informatik
Universität Kassel, D–34109 Kassel, Germany
niemann@theory.informatik.uni-kassel.de
[2] Fakultät für Mathematik und Informatik
Universität Leipzig, D–04109 Leipzig, Germany
joe@informatik.uni-leipzig.de

Abstract. In 1988 McNaughton et al introduced the class CRCL of Church-Rosser congruential languages as a way to define formal languages by confluent length-reducing string-rewriting systems. As other congruential language classes CRCL is quite limited, although it contains some languages that are not context-free. In 2000 Niemann has shown that at least each regular language with polynomial density is Church-Rosser congruential. It is still an open question whether the class of regular languages is contained in CRCL. Here we give some families of regular languages of exponential density that are Church-Rosser congruential. More precisely, we show that some shuffle languages, as well as Level 1 of the Straubing-Thérien hierarchy, are in CRCL, using a sufficient condition under which a regular language is Church-Rosser congruential. Last, we give a family of group languages that are Church-Rosser congruential, but do not fulfill this condition.

1 Introduction

For a finite length-reducing string-rewriting system R there exists an algorithm that, given a string w as input, computes an irreducible descendant $\hat{w}$ of w with respect to R in linear time [1]. If R is confluent, then this irreducible descendant is uniquely determined. Accordingly, two strings u and v are congruent with respect to the Thue congruence generated by R if and only if their irreducible descendants $\hat{u}$ and $\hat{v}$ coincide, and hence the *word problem* for R is solvable in linear time.

Motivated by this observation McNaughton et al [2] used the finite, length-reducing, and confluent string-rewriting systems to define two classes of languages: the class CRL of *Church-Rosser languages* and the class CRCL of *Church-Rosser congruential languages* (see Section 2 for the definitions), which is a proper subclass of CRL. The membership problem for these languages is solvable in linear time, and hence, CRL is contained in the class CSL of context-sensitive languages. In addition, it was shown in [2] that CRL contains the class DCFL of deterministic context-free languages, that CRCL and DCFL are incomparable under set inclusion and that CRCL contains languages that are not context-free.

However it is not known whether the class REG of regular languages is contained in CRCL. Niemann gave a partial answer to this question showing that at least regular languages with polynomial density are Church-Rosser congruential [3]. CRCL also contains regular languages of exponential density like A^* or the set of all strings over $\{a,b\}$ of even length.

In this paper we introduce some more regular languages of exponential density that are Church-Rosser congruential. More precisely, in Section 3 we show that if the syntactic congruence of a regular language L partitions A^n over all infinite congruence classes for some natural number n, then L is Church-Rosser congruential. In the next section we conclude that for regular $L \subseteq A^*$, the shuffle language $L \sqcup B^*$ is in CRCL, provided the alphabet B contains at least one letter not in A. In Section 5, we show that Level 1 of the Straubing-Thérien hierarchy is in CRCL. The syntactic monoid of such a language is always group-free. In Section 6 we show that the group languages $(A^2)^*$, for any size of the alphabet A, are Church-Rosser congruential, even though they do not fulfill the condition given in Section 3.

But it still remains open whether CRCL contains all regular languages.

2 Preliminaries

Let A be a finite alphabet. Then A^* denotes the set of strings over A including the empty string ε, and $A^+ := A^* \setminus \{\varepsilon\}$. Further, A_m denotes an alphabet with m symbols.

For a language $L \subseteq A^*$, we define the relation $\sim_L$ on A^* by

$$u \sim_L v \iff \forall x,y \in A^* : (xuy \in L \iff xvy \in L).$$

An equivalence relation $\sim$ on A^* is called *stable* iff

$$\forall u \sim v : \forall x, y \in A^* : xuy \sim xvy.$$

A stable equivalence is called a *congruence*. The relation $\sim_L$ defined above is the coarsest congruence that separates L from $A^* \setminus L$. It is called the *syntactic congruence* of L.

We let $\mathsf{Mon}(L)$ denote the set of congruence classes $A^*/\sim_L$. Indeed $\mathsf{Mon}(L)$ is a monoid under concatenation, called the *syntactic monoid* of L. It is well known that $\mathsf{Mon}(L)$ is finite iff $L \in$ REG.

A *string-rewriting system* R on A is a subset of $A^* \times A^*$. An element $(\ell, r) \in R$ is called a *(rewrite) rule*, and it will be denoted as $(\ell \to r)$. Here we will only be dealing with finite string-rewriting systems.

The string-rewriting system R induces a *reduction relation* $\to_R^*$ on A^*, which is the reflexive, transitive closure of the single-step reduction relation $\to_R = \{(x\ell y, xry) : x, y \in A^*, (\ell \to r) \in R\}$. If $u \to_R^* v$, then u is an *ancestor* of v, and v is a *descendant* of u. If there is no $v \in A^*$ such that $u \to_R v$ holds, then the string u is called *irreducible* (mod R). By IRR(R) we denote the set of all such irreducible strings.

The smallest equivalence relation $\leftrightarrow_R^*$ containing $\to_R$ is called the *Thue congruence* generated by R. For $u \in A^*$, $[u]_R$ denotes the congruence class $\{v \in A^* : v \leftrightarrow_R^* u\}$ of u.

The string-rewriting system R is called

- *length-reducing* if $|\ell| > |r|$ holds for each rule $(\ell \to r) \in R$,
- *confluent* if, for all $u, v, w \in A^*$, $u \to_R^* v$ and $u \to_R^* w$ imply that v and w have a common descendant,
- *locally confluent* if, for all $u, v, w \in A^*$, $u \to_R v$ and $u \to_R w$ imply that v and w have a common descendant.

To determine whether or not the string-rewriting system R is locally confluent, it is sufficient to consider pairs of rewriting rules where the left-hand sides overlap. For each pair of not necessarily distinct rewriting rules (ℓ_1, r_1) and (ℓ_2, r_2) from R, let the set of *critical pairs* corresponding to this pair be $\{\langle xr_1, r_2y\rangle : x, y \in A^* \wedge x\ell_1 = \ell_2 y \wedge |x| < |\ell_2|\} \cup \{\langle r_1, xr_2y\rangle : x, y \in A^* \wedge \ell_1 = x\ell_2 y\}$. We will say that a critical pair $\langle z_1, z_2\rangle$ resolves if z_1 and z_2 have a common descendant. The string-rewriting system R is locally confluent if and only if every critical pair resolves.

Each string-rewriting system R that is length-reducing and locally confluent is also confluent. Using string-rewriting systems of this restricted form the following two language classes have been defined in [2].

Definition 1. *(a) A language $L \subseteq A^*$ is a* Church-Rosser language *if there exist an alphabet $\Gamma \supsetneq A$, a finite, length-reducing, and confluent string-rewriting system R on Γ, two irreducible strings $t_1, t_2 \in (\Gamma \smallsetminus A)^*$, and an irreducible letter $Y \in \Gamma \smallsetminus A$ such that, for all $w \in A^*$, $t_1 w t_2 \to_R^* Y$ if and only if $w \in L$.*
(b) A language $L \subseteq A^*$ is a Church-Rosser congruential language *if there exist a finite, length-reducing, and confluent string-rewriting system R on A and finitely many strings $w_1, \ldots, w_n \in \mathrm{IRR}(R)$ such that $L = \bigcup_{i=1}^{n} [w_i]_R$.*

In other words, a language $L \subseteq A^*$ is a Church-Rosser congruential language if it can be expressed as the union of finitely many congruence classes of a finite, length-reducing, and confluent string-rewriting system. In this case R is called a CRCL-system. By CRL we denote the class of Church-Rosser languages, while CRCL denotes the class of Church-Rosser congruential languages. CRCL is obviously contained in CRL, and already CRCL contains non-regular languages. For example, if $R = \{aabb \to ab\}$ and $L_1 = [ab]_R$, then we see that $L_1 \in$ CRCL is the non-regular language $L_1 = \{a^n b^n : n \geq 1\}$. Further, CRL contains DCFL [2], the class of deterministic context-free languages, while CRCL is incomparable to DCFL, as DCFL contains the language $L_2 = L_1 \cup b^+$, which is not even congruential, while CRCL contains some languages that are not even context-free, as shown by the following example.

Example 2. [2] Consider the language $L_3 = \{a^{2^n} : n \geq 0\}$, which is not context-free, and the rewrite system $R = \{\textcent aaaa \to \textcent aaF, Faa \to aF, F\$ \to \$, \textcent aa\$ \to Y, \textcent a\$ \to Y\}$. Then R is length-reducing and confluent, and it is easily verified that, for all $n \geq 0$, $a^n \in L_3$ iff $\textcent a^n \$ \to_R^* Y$ holds. Thus, $L_3 \in$ CRL.

Now let $A = \{a, F, \textcent, \$, Y\}$, and let $L_4 = [Y]_R$. Then $L_4 \in$ CRCL. On the other hand, $L_4 \notin$ CFL, as $\textcent \cdot L_3 \cdot \$ = L_4 \cap \textcent \cdot a^* \cdot \$$, and CFL is closed under intersection with regular sets and left- and right-derivatives. $\square$

The *density function* p_L for a language $L \subseteq A^*$ tells us how many strings of a given length are in the language. It is defined by $p_L(n) = |L \cap A^n|$. We say that L has a polynomial density if $p_L(n) \in \mathcal{O}(n^k)$ for some integer $k \geq 0$.

The lemma below characterizes regular languages of polynomial density by regular expressions of a specific form (see also [7]).

Lemma 3. [6]: *A regular language $L \subseteq A^*$ has polynomial density, if and only if L can be written as a finite union of languages of the form*

$$u_0 v_1^* u_1 \ldots v_n^* u_n \quad ,$$

where $u_i \in A^*$ and $v_i \in A^+$. □

The following is known.

Theorem 4. *[3] Each regular language of polynomial density is a Church-Rosser congruential language.* □

So the question whether all regular languages are in CRCL is reduced to the question of whether there exists a regular language with exponential density that is not in CRCL.

3 Building a CRCL-System from the Syntactic Congruence

By the definition of $\sim_L$ and CRCL, we get

Proposition 5. *A finite, length-reducing, and confluent string rewriting system R is a CRCL-system for a language L if and only if*

1. $\leftrightarrow_R^*$ *is a refinement of $\sim_L$, and*
2. $\mathrm{IRR}(R) \cap L$ *is finite.* □

We are mainly interested in regular languages L. These induce a syntactic congruence $\sim_L$ of finite index. This suggests to look for CRCL rewrite systems R whose Thue congruence $\leftrightarrow_R^*$ has finite index as well:

Definition 6. *A language $L \subseteq A^*$ is a* strongly Church–Rosser congruential language *iff there exists a finite, length reducing, and confluent string-rewriting system R on A such that*

- $\mathrm{IRR}(R)$ *is finite, and*
- *the Thue congruence of R is a refinement of the syntactic congruence of L.*

The system R is then called a sCRCL-system for L. The set of strongly Church–Rosser congruential languages is denoted by sCRCL.

It is immediate that sCRCL $\subseteq$ CRCL. The inclusion is strict since sCRCL $\subsetneq$ REG, due to the finiteness of the index of $\leftrightarrow_R^*$.

It is an open question whether REG $\subseteq$ CRCL implies REG $\subseteq$ sCRCL, in other words, whether a CRCL-system R for a regular language L can be extended to a sCRCL-system for L.

For some languages $L \in$ REG, it is possible to obtain a sCRCL-system for L from the syntactic monoid Mon(L).

Theorem 7. *If $L \in$ REG and there exists a number n such that for each syntactic congruence class $C \in$ Mon(L)*

- *if C is finite, then $\forall w \in C : |w| < n$, and*
- *if C is infinite, then $\exists w_C \in C : |w_C| = n$,*

then L is strongly Church–Rosser congruential.

Proof. Assume L and n fulfill the conditions stated in the theorem, and fix a choice of w_C for each infinite class $C \in$ Mon(L). Define the string-rewriting system

$$R = \{(w \to w_C) \colon C \in \mathsf{Mon}(L), w \in A^{n+1} \cap C\}.$$

We show that R is a sCRCL-system for L.

Obviously, the system R is finite and length reducing, and $\leftrightarrow_R^*$ is a refinement of $\sim_L$.

Since each word of length $n+1$ appears in some infinite class $C \in$ Mon(L), the set of left hand sides of R is exactly A^{n+1}. Therefore, each $w \in A^{\geq n+1}$ is reducible, while each $w \in A^{\leq n}$ is not. It follows that IRR(R) = $A^{\leq n}$, and thus IRR(R) is finite.

We now prove the confluence of R, by showing that each $w \in A^*$ has exactly one irreducible descendant. If $|w| \leq n$, then it is irreducible. If $|w| \geq n+1$, then let $C \in$ Mon(L) be the syntactic congruence class of w.

In case $|w| = n+1$, the only R–reduction starting from w uses the rule $w \to w_C$, and thus w_C is the unique irreducible descendant. Assume $|w| > n+1$, and consider a R–reduction step $w = u\ell v \to urv = w'$ for some rule $(\ell \to r) \in R$. Since $(\ell \to r) \in R$ implies $\ell \sim_L r$, we also have $w = u\ell v \sim_L urv = w'$. Therefore w' is in the same class C. Since $|w'| + 1 = |w|$, the claim follows by induction. □

Corollary 8. *All finite languages are in sCRCL.*

Proof. If L is finite, let $n = 1 + \max\{|w| \colon w \in L\}$. Then Mon($L$) contains some finite classes, and exactly one infinite class C. This infinite class contains (at least) $A^{\geq n}$. For the choice of w_C, an arbitrary word of length n (and thus not in L) will do. □

Since sCRCL is closed under complement, this implies

Corollary 9. *All co–finite languages are in CRCL.* □

Finite and co–finite languages are at Level 1 of the Straubing-Thérien hierarchy, and indeed we will prove in Section 5 that each language from that level is in sCRCL.

4 CRCL and the Shuffle Operation

We recall that the *shuffle* $u \shuffle v$ of two words $u, v \in A^*$ is the set of words defined by $\varepsilon \shuffle u = u \shuffle \varepsilon = \{u\}$ and $u_1 \ldots u_m \shuffle v_1 \ldots v_n = u_1(u_2 \ldots u_m \shuffle v_1 \ldots v_n) \cup v_1(u_1 \ldots u_m \shuffle v_2 \ldots v_n)$. This operation is extended to languages L_1, L_2 by $L_1 \shuffle L_2 = \bigcup_{w_1 \in L_1, w_2 \in L_2} w_1 \shuffle w_2$.

Proposition 10. *If A and B are disjoint alphabets, and $L \subseteq A^*$ is a regular language, then $L \shuffle B^*$ is strongly Church–Rosser congruential.*

Proof. Let $L' = L \shuffle B^*$. The syntactic monoids $\mathsf{Mon}(L)$ and $\mathsf{Mon}(L')$ are isomorphic, because to each class $C \in \mathsf{Mon}(L)$, there corresponds a class $C' = C \shuffle B^* \in \mathsf{Mon}(L')$, and vice versa.

Now let $n = \max_{C \in \mathsf{Mon}(L)} \min_{w \in C} |w|$. (This is well-defined since each C is non-empty.) Then each class $C' \in \mathsf{Mon}(L')$ contains at least one word of length n, and so Theorem 7 can be applied. □

Example 11. For $A = \{a, b\}$, it is open whether $L = (A^3)^*$ is in CRCL, see Section 6. However, by Proposition 10 we know that $L' = L \shuffle c^*$ is in CRCL. A sCRCL-system for L' is $\{c^3 \to c^2, A^3 \to c^2, A \shuffle c^2 \to ac, A^2 \shuffle c \to aa\}$ (where we have abbreviated sets of rules with common right-hand sides). □

We can generalize this result as follows.

Proposition 12. *If A and B are alphabets with $B \setminus A \neq \emptyset$, and $L \subseteq A^*$ is a regular language, then $L \shuffle B^*$ is strongly Church–Rosser congruential.*

Proof. Let $B_1 = B \cap A$, and $B_2 = B \setminus A$. (Note that that B_2 is non-empty, and B_1 and B_2 are disjoint.) We have $L \shuffle B^* = L \shuffle (B_1 \cup B_2)^* = L \shuffle (B_1^* \shuffle B_2^*)$. The shuffle is associative, and so the claim follows from applying Proposition 10 to the language $L \shuffle B_1^* \subseteq (A \cup B_1)^*$ and the (now disjoint) alphabet B_2. □

5 Level 1 of the Straubing-Thérien Hierarchy Is in CRCL

The Straubing-Thérien hierarchy is a concatenation hierarchy that exhausts the star-free regular languages. It is defined inductively, starting from languages with trivial syntactic monoid, using the operations of *polynomial* and *boolean* closure.

Definition 13. *The* polynomial closure $\mathsf{Pol}(F)$ *of a family F of languages over A^* is the family of finite unions of languages of the form*

$$L_0 a_1 L_1 a_2 \ldots a_n L_n, \quad \text{where } a_i \in A \text{ and } L_i \in F.$$

Definition 14. *The* boolean closure $\mathsf{Bool}(F)$ *of a family F of languages over A^* is the family of finite boolean combinations (that is, unions, intersections, and complements w.r.t. A^*) of languages from F.*

Definition 15. *The* Straubing-Thérien hierarchy *is the sequence of families of languages* $\mathsf{Straub}_0 \subset \mathsf{Straub}_{1/2} \subset \mathsf{Straub}_1 \subset \ldots$ *given by*

$$\mathsf{Straub}_0 = \{\emptyset, A^*\}$$
$$\mathsf{Straub}_{n+1/2} = \mathsf{Pol}(\mathsf{Straub}_n), \quad \mathsf{Straub}_{n+1} = \mathsf{Bool}(\mathsf{Straub}_{n+1/2}).$$

The family Straub_k is called *level k* of the hierarchy. These families are in fact (positive) varieties of languages. The low levels of the hierarchy have additional characterizations.

Definition 16. *A language $L \subseteq A^*$ is a* shuffle ideal *iff $L \shuffle A^* \subseteq L$.*

Note that the language L' from Example 11 is *not* a shuffle ideal because the definition requires the shuffle with the complete alphabet.

Proposition 17. *[4] L is in $\mathsf{Straub}_{1/2}$ iff L is a shuffle ideal.* □

Example 18. The language $L = (abb \cup bba) \shuffle \{a,b\}^*$ is in $\mathsf{Straub}_{1/2}$. □

The languages from Straub_1 are boolean combinations of $\mathsf{Straub}_{1/2}$ languages. They are also called *piecewise testable* languages.

Example 19. $L' = a^+ba^+$ is an element of Straub_1.

Proof. $L' = (aba \shuffle A^*) \setminus (bb \shuffle A^*)$. □

Proposition 20. *[4] A language $L \subseteq A^*$ is in $\mathsf{Straub}_{3/2}$ iff L is a finite union of languages*
$$A_0^* a_1 A_1^* a_2 \ldots a_n A_n^*, \quad \text{where} \quad a_i \in A, A_i \subseteq A.$$
□

Lemma 21. *If $L \in \mathsf{Straub}_1$, then for each congruence class $C \in \mathsf{Mon}(L)$, it holds that $C \in \mathsf{Straub}_1$.*

Proof. (In fact this is true for any integer level of the hierarchy.) It is an immediate consequence of Eilenberg's variety theorem. Since L belongs to the variety Straub_1, its syntactic monoid $\mathsf{Mon}(L)$ belongs to the corresponding variety of J–trivial monoids. Since each C is recognized by $\mathsf{Mon}(L)$, it belongs, in turn, to the variety Straub_1 of languages that we started with. □

Example 22. See Examples 18 and 19. The congruence classes of L are
$$\{\varepsilon, a^+, b, ba^+, a^+b, bb^+, ba^+b, L', L\}.$$

Each one is in Straub_1. Note that $L \in \mathsf{Straub}_{1/2}$, but $L' \in \mathsf{Straub}_1 \setminus \mathsf{Straub}_{1/2}$. This is due to the half–integer levels not being boolean closed. □

Lemma 23. *If $L \in \mathsf{Straub}_{3/2}$, then there exists a number n such that*
$$L \subseteq A^{<n} \quad \text{or} \quad \forall n' \geq n : \emptyset \neq L \cap A^{n'}.$$

Proof. We use the presentation of L according to Proposition 20. If at least one of the A_i is nonempty, we can 'pump' some word w from L at the corresponding position, obtaining, for each $n' \geq n = |w|$, at least one word $w' \in L$ of length n'. In the case that all A_i are empty, the language L is finite, and we take n as $1 + \max\{|w| : w \in L\}$. □

Now we are able to state and prove the main result of this section:

Theorem 24. *Each language from Level 1 of the Straubing-Thérien hierarchy is strongly Church–Rosser congruential.*

Proof. Using the Lemmata 21 and 23, Theorem 7 can be applied by taking the maximum of the numbers n for the different congruence classes. □

Our method of proof does not seem to extend to higher levels of the hierarchy, because the 'pumping property' corresponding to Lemma 23 fails. On the other hand, there are regular languages known to be in CRCL that are not star-free, i.e. completely outside the Straubing-Thérien hierarchy.

It remains to be studied how families of CRCL languages behave with respect to Boolean and polynomial closure. Note that by Lemma 3, the family of regular languages of polynomial density is the polynomial closure of the family of "single loop" languages (of the form w^*, for $w \in A^*$).

We even know regular CRCL languages that are neither starfree nor polynomially dense. One such language is $(A_m^2)^*$, see the following section. Its syntactic equivalence classes $C_0 = A_m^{\text{even}}$ and $C_1 = A_m^{\text{odd}}$ obviously do not admit a selection of $w_i \in C_i$ with $|w_0| = |w_1|$. This already happens for $L = (a(b \cup c))^* \in \text{Straub}_2$. Its syntactic monoid consists of these classes

$$\{\varepsilon, A^*(a^2 \cup (b \cup c)^2)A^*, L \setminus \varepsilon, (b \cup c)L, La, (b \cup c)La\} \quad ,$$

where $A = \{a, b, c\}$. Here, $L \setminus \varepsilon$ and La avoid each other's word lengths.

6 A Family of CRCL Group Languages

In the previous section, we were looking at star-free languages. Their syntactic monoid is always *group–free*, that is, it does not contain a nontrivial sub-monoid that is a group.

Let us now turn to the opposite direction and consider some *group languages*. A language is called group language iff its syntactic monoid is a group. Especially, we look at some seemingly simple languages L whose syntactic monoid is a *cyclic* group.

For cycles of order two, we will show that the corresponding languages are in CRCL. The respective rewrite systems are by no means obvious. We found them by generalizing results of computer searches. The situation for longer cycles remains unsettled.

Example 25. Consider $L = (A_2^2)^* = \{w \in \{a, b\}^* : |w| \equiv 0 \bmod 2\}$. Define

$$R_2 = \{(aaa \to a), (aab \to b), (baa \to b), (bab \to b), (bbb \to b)\}.$$

R_2 is finite, length-reducing, and confluent. Its irreducible words are

$$\text{IRR}(R_2) = \{\varepsilon, a, b, aa, ab, ba, bb, aba, abb, bba, abba\}.$$

Further

$$\forall u, v \in \{a, b\}^* : u \leftrightarrow^*_{R_2} v \text{ implies } |u| \equiv |v| \bmod 2 \quad .$$

Thus $L = \bigcup \{[w]_{R_2} : w \in \text{IRR}(R_2) \wedge |w| \equiv 0 \bmod 2\}$. That is, L is a Church-Rosser congruential language, even $L \in $ sCRCL. □

This example will now be generalized. The dependency on the alphabet size is nontrivial.

Theorem 26. *For each $m > 0$, the language*

$$(A_m^2)^* = \{w \in A_m : |w| \equiv 0 \bmod 2\}$$

is Church-Rosser congruential.

Proof. Let $A_m = \{a_1, a_2, \ldots, a_m\}$. We define an ordering $a_1 < a_2 < \cdots < a_m$ among the symbols of A_m. Define a string rewriting system R by

$$R = \{(xyz \to \max(x, z)) : x, y, z \in A_m, y = \min(x, y, z)\}.$$

Claim. R is confluent.

Proof. R is terminating, so we show that R is locally confluent. There are two kinds of overlaps: those involving two symbols and those involving only one.

Case 1: Let $(x_1 y_1 z_1 \to m_1) \in R$ and $(x_2 y_2 z_2 \to m_2) \in R$ with $y_1 = x_2$ and $z_1 = y_2$. That is $x_1 y_1 z_1 z_2 = x_1 x_2 y_2 z_2$. By $y_1 = \min(x_1, y_1, z_1) = \min(x_1, y_1, y_2)$ and $y_2 = \min(x_2, y_2, z_2) = \min(y_1, y_2, z_2)$ it follows that $y_1 = y_2$.

Thus we have $x_1 y_1 y_1 z_2 \to m_1 z_2$ on the one hand and $x_1 y_1 y_1 z_2 \to x_1 m_2$. From the minimality of y_1 it follows that $\max(x_1, y_1, y_1) = x_1$ and that $\max(y_1, y_1, z_2) = z_2$. This implies $m_1 z_2 = x_1 z_2 = x_1 m_2$. Thus, in fact in the case of an overlap involving two symbols the critical pair is trivial.

Case 2: Let $(x_1 y_1 z_1 \to m_1) \in R$ and $(x_2 y_2 z_2 \to m_2) \in R$ with $z_1 = x_2$, that is, $x_1 y_1 z_1 y_2 z_2 = x_1 y_1 x_2 y_2 z_2$. Here we distinguish four sub-cases.

Case 2.1: If $x_1 = \max(x_1, y_1, z_1)$ and $x_2 = \max(x_2, y_2, z_2)$, then we have $x_1 y_1 z_1 y_2 z_2 \to x_1 y_2 z_2$ on the one hand and $x_1 y_1 z_1 y_2 z_2 = x_1 y_1 x_2 y_2 z_2 \to x_1 y_1 z_1 \to x_1$ on the other. As $x_2 = z_1$ it follows that $x_1 \geq x_2$ and thus $(x_1 y_2 z_2 \to x_1) \in R$. So, the critical pair is resolvable.

Case 2.2: If $x_1 = \max(x_1, y_1, z_1)$ and $z_2 = \max(x_2, y_2, z_2)$, then we have $x_1 y_1 z_1 y_2 z_2 \to x_1 y_2 z_2$ on the one hand and $x_1 y_1 z_1 y_2 z_2 = x_1 y_1 x_2 y_2 z_2 \to x_1 y_1 z_2$ on the other. If $x_1 \geq z_2$, then $(x_1 y_1 z_2 \to x_1) \in R$ and $(x_1 y_2 z_2 \to x_1) \in R$. If $x_1 \leq z_2$, then $(x_1 y_1 z_2 \to z_2) \in R$ and $(x_1 y_2 z_2 \to z_2) \in R$. In both cases the critical pair is resolvable.

Case 2.3: If $z_1 = \max(x_1, y_1, z_1)$ and $x_2 = \max(x_2, y_2, z_2)$ (note $x_2 = z_1$!), then we have $x_1 y_1 z_1 y_2 z_2 \to z_1 y_2 z_2 = x_2 y_2 z_2 \to x_2$ on the one hand and $x_1 y_1 z_1 y_2 z_2 = x_1 y_1 x_2 y_2 z_2 \to x_1 y_1 x_2 = x_1 y_1 z_1 \to z_1$ on the other. Thus, the critical pair is resolvable.

Case 2.4: If $z_1 = \max(x_1, y_1, z_1)$ and $z_2 = \max(x_2, y_2, z_2)$, then it follows similarly to Case 2.1 that this critical pair is resolvable.

Thus R is confluent.

Claim. $\leftrightarrow_R^*$ is a refinement of the syntactic congruence of $(A_m^2)^*$.

Proof. For $v, w \in A^*$ it follows from $v \to_R^* w$ that $|v| \equiv |w| \mod 2$.

Claim. The set $\mathrm{IRR}(R)$ is finite.

Proof. Define a language (compare with $\mathrm{IRR}(R_2)$ from Example 25)

$$\Lambda = \left\{ x_1 \ldots x_{m_1} y_{m_2} y_{m_2-1} \cdots y_1 \;\middle|\; \begin{array}{l} x_1, \ldots, x_{m_1}, y_1, \ldots, y_{m_2} \in A_m, \\ x_1 < \cdots < x_{m_1} \leq y_{m_2}, \\ \text{and } y_{m_2} > y_{m_2-1} > \cdots > y_1 \end{array} \right\}$$

More precisely, we show that $\mathrm{IRR}(R) \subseteq \Lambda$. Assume $v \notin \Lambda$. Thus v has a factor $u_1 u_2 u_3$, $u_1, u_2, u_3 \in A_m$, where $u_1 \geq u_2 \leq u_3$. Then $(u_1 u_2 u_3 \to \max(u_1, u_3)) \in R$, and thus v is reducible. So each irreducible word with respect to R is in Λ. As Λ is finite, so is $\mathrm{IRR}(R)$.

In all, this shows that R is a sCRCL-system for $(A_m^2)^*$, and therefore $(A_m^2)^* \in$ CRCL. □

Comparing this construction with the method presented in Section 3, it seems too simplicistic to assume that all $L \in$ REG admit a sCRCL-system R that maps each (long enough) word of a $\sim_L$ equivalence class C to a unique R–normal form w_C. Indeed the Thue congruence of the sCRCL-system for $(A_2^2)^*$ (see Example 25) generates 10 infinite congruence classes, while the syntactic congruence of $(A_2^2)^*$ generates only two.

7 Concluding Remarks

We have seen that apart from the regular languages of polynomial density also very different families of regular languages with exponential density are Church-Rosser congruential. So possible direction for further research include the following questions:

- Which higher Straubing-Thérien levels belong to CRCL?
- Is each star-free language in CRCL?
- Is every language $(A_m^k)^* = \{w \in A_m^* : |w| \equiv 0 \bmod k\}$ in CRCL?
- Is every group language in CRCL?

Acknowledgments. The construction used in Section 3 was suggested by Klaus Reinhardt. We thank Friedrich Otto for valuable discussions concerning languages of the form $(A_m^n)^*$ and their relation to CRCL, and Jean-Eric Pin for explanations on varieties.

References

1. Ronald V. Book and Friedrich Otto. *String-Rewriting Systems*. Texts and Monographs in Computer Science. Springer, New-York, 1993.
2. Robert McNaughton, Paliath Narendran, and Friedrich Otto. Church-Rosser Thue systems and formal languages. *Journal of the ACM*, 35:324–344, 1988.
3. Gundula Niemann. Regular Languages and Church-Rosser Congruential Languages. In Rudolf Freund and Alica Kelemenova, editors, *Proceedings of the International Workshop Grammar Systems 2000*, pages 359–370. Silesian University at Opava, Faculty of Philosophy and Science, Institute of Computer Science, 2000.
4. Jean-Eric Pin. Syntactic semigroups. In Rozenberg and Salomaa [5], pages 679–746.
5. G. Rozenberg and A. Salomaa, editors. *Handbook of Formal Languages*, volume 1. Springer, 1997.
6. A. Szilard, S. Yu, K. Zhang, and J. Shallit. Characterizing regular languages with polynomial densities. In *Proceedings of the 17th International Symposium on Mathematical Foundations of Computer Science*, number 629, pages 494–503, Berlin/New York, 1992. Springer.
7. Sheng Yu. Regular languages. In Rozenberg and Salomaa [5], pages 41–110.

On the Relationship between the McNaughton Families of Languages and the Chomsky Hierarchy

M. Beaudry[1], M. Holzer[2], G. Niemann[3], and F. Otto[3]

[1] Département de mathématique et informatique, Université de Sherbrooke
Sherbrooke, Québec, Canada J1K 2R1
beaudry@dmi.usherb.ca
[2] Institut für Informatik, Technische Universität München, Arcisstraße 21
80290 München, Germany[†]
holzer@informatik.tu-muenchen.de
[3] Fachbereich Mathematik/Informatik, Universität Kassel, 34109 Kassel, Germany
{niemann,otto}@theory.informatik.uni-kassel.de

Abstract. By generalizing the Church-Rosser languages the *McNaughton families of languages* are obtained. Here we concentrate on those families that are defined by monadic or special string-rewriting systems. We investigate the relationship of these families to each other and to the lower classes of the Chomsky hierarchy and present some closure and some non-closure properties for them. Moreover, we address some complexity issues for their membership problems.

1 Introduction

In [16] (see also [15]) the notion of *Church-Rosser languages* is introduced. Such a language is defined by using the set of ancestors of a certain symbol with respect to a finite, length-reducing, and confluent string-rewriting system. In this definition apart from the final symbol, other nonterminal symbols are also allowed, e.g., for marking the left and the right border of the input strings.

This concept can easily be generalized by admitting other classes of finite string-rewriting systems in the definition. This leads to the various *McNaughton families of languages* [1]. In fact, there are two kinds of families: those defined by finite (terminating) string-rewriting systems, and those defined by finite, terminating and confluent systems. The former can be shown to correspond to nondeterministic complexity and language classes and the latter to deterministic complexity and language classes.

Of particular interest are the lower members of this collection of families, as they do not seem to coincide with any well-known language classes. These are the McNaughton families that are defined by monadic or special string-rewriting systems (see Section 2 for the definitions). Since we consider confluent and non-confluent systems, we obtain four different families.

This paper is structured as follows. In Section 2 the basic definitions on string-rewriting systems are given, and the McNaughton families of languages are defined. Then we consider the relationships between well-known language classes and the McNaughton

[†] Most of the work presented here was performed, while M. Holzer was visiting the Département d'I.R.O., Université de Montréal, Montréal, Québec, Canada.

families of languages, concentrating on those families that are defined by monadic or special string-rewriting systems. In Section 4 we address the closure and non-closure properties of these families, and in Section 5 we investigate complexity issues for the fixed and the general membership problems for them. The paper closes with a short discussion of some open problems and possible directions for future research on the McNaughton families of languages.

2 Definitions and First Results

Let Σ denote a finite alphabet. Then Σ^* is the set of all strings over Σ including the empty string λ, and Σ^+ is the set of all non-empty strings over Σ. As usual $|w|$ will denote the length of the string w.

We assume that the reader is familiar with the basic concepts of formal language and automata theory. As our standard reference concerning this field we use the monograph by Hopcroft and Ullman [10]. In addition we need some background from the theory of string-rewriting systems, where [4] serves as our main reference.

A *string-rewriting system* on Σ is a set R of pairs of strings from Σ^*. Usually a pair $(\ell, r) \in R$ is written as $(\ell \to r)$, and it is called a *rewrite rule*. The *reduction relation* $\to_R^*$ induced by R is the reflexive, transitive closure of the relation $\to_R := \{ (u\ell v, urv) \mid (\ell \to r) \in R, u, v \in \Sigma^* \}$. For a string $u \in \Sigma^*$, if there exists a string v such that $u \to_R v$ holds, then u is called *reducible* mod R, v is a *direct descendant* of u, and u is a *direct ancestor* of v. If such a string v does not exist, then u is called *irreducible* mod R. By $\Delta_R^*(u)$ we denote the set of all descendants of u, that is, $\Delta_R^*(u) := \{ v \mid u \to_R^* v \}$, $\nabla_R^*(v) := \{ u \mid u \to_R^* v \}$ is the set of all ancestors of v, and IRR(R) denotes the set of all irreducible strings mod R. By $\leftrightarrow_R^*$ we denote the *Thue congruence* on Σ^* that is induced by R. It is the smallest equivalence relation on Σ^* containing the relation $\to_R$. For $w \in \Sigma^*$, $[w]_R$ denotes the *congruence class* of w mod $\leftrightarrow_R^*$.

Here we will be using certain restricted types of string-rewriting systems. A string-rewriting system R is called

- *terminating* if there is no infinite sequence of reduction steps,
- *confluent* if, for all $v, w \in \Sigma^*$, $\nabla_R^*(v) \cap \nabla_R^*(w) \neq \emptyset$ implies that $\Delta_R^*(v) \cap \Delta_R^*(w) \neq \emptyset$,
- *convergent* if it is both terminating and confluent.

If reduction sequences are seen as computations, then the termination property expresses the fact that each computation terminates after finitely many steps, and the confluence property expresses the fact that the resulting irreducible descendant of a given string is unique, if it exists. For a convergent system R, the set IRR(R) of irreducible strings is a complete set of unique representatives for the Thue congruence $\leftrightarrow_R^*$.

The following definition is a straightforward generalization of the definition of a Church-Rosser language (CRL) [15,16]. Let $\mathcal{C}$ be a class of string-rewriting systems. Then $\mathcal{C}$ yields a family of languages $\mathcal{L}(\mathcal{C})$ as follows. A language $L \subseteq \Sigma^*$ belongs to $\mathcal{L}(\mathcal{C})$, if there exist a finite alphabet Γ strictly containing Σ, a finite string-rewriting system $R \in \mathcal{C}$ on Γ, strings $t_1, t_2 \in (\Gamma \smallsetminus \Sigma)^* \cap$ IRR(R), and a letter $Y \in (\Gamma \smallsetminus \Sigma) \cap$ IRR(R) such that, for all $w \in \Sigma^*$, $w \in L$ holds if and only if $t_1 w t_2 \to_R^* Y$, that is, $t_1 w t_2 \in \nabla_R^*(Y)$.

Here the symbols of Σ are *terminals*, while those of $\Gamma \smallsetminus \Sigma$ can be seen as *nonterminals*. The language L is said to be *specified* by the four-tuple (R, t_1, t_2, Y). This fact will be denoted as $L = L(R, t_1, t_2, Y)$. The class $\mathcal{L}(\mathcal{C})$ is called the *McNaughton family of languages* that is specified by $\mathcal{C}$.

If $\mathcal{C}$ is the class of all finite string-rewriting systems, then the corresponding McNaughton family of languages McNL coincides with the class RE of recursively enumerable languages [1]. In fact, RE also coincides with the McNaughton family con-McNL, which is specified by the class of all finite confluent string-rewriting systems.

The McNaughton family non-in-McNL is obtained by using finite terminating string-rewriting systems R that are *non-increasing*, that is, $|\ell| \geq |r|$ holds for each rule $(\ell \to r) \in R$. By considering systems that are *length-reducing* we obtain the family lr-McNL. By requiring in addition that the string-rewriting systems used in the specifications are confluent, we obtain the families con-non-in-McNL and con-lr-McNL.

For these families we have the following results, where CSL (DCSL) denotes the class of (*deterministic*) *context-sensitive languages*, and GCSL is the class of *growing context-sensitive languages* (see, e.g., [5]).

Theorem 1. [1] non-in-McNL = CSL, con-non-in-McNL = DCSL, lr-McNL = GCSL, con-lr-McNL = CRL.

Finally we introduce those McNaughton families of languages that are our main concern here. A string-rewriting system R on Γ is called *monadic* if it is length-reducing and if $|r| \leq 1$ holds for each rule $(\ell \to r)$ of R, and it is called *special* if it is length-reducing, and the right-hand side of each rule is the empty string.

By sp-McNL, respectively mon-McNL, we denote the McNaughton family that is obtained by using finite string-rewriting systems that are special, respectively monadic. By requiring in addition that the string-rewriting systems are confluent, we obtain the families con-sp-McNL and con-mon-McNL.

3 Inclusion and Non-inclusion Results

Here we compare the special and monadic McNaughton families to each other and to the lower classes of the Chomsky hierarchy.

In [3] Theorem 2.2 it is shown that for a monadic string-rewriting system R, the set $\nabla_R^*(Y)$ is context-free. It follows that each language from the family mon-McNL is context-free. Conversely, it can be shown that the reverse of the derivation process of a context-free grammar in Chomsky normal form can be simulated by a finite monadic string-rewriting system. Hence, we have the following characterization, where CFL denotes the class of *context-free languages*.

Theorem 2. mon-McNL = CFL.

Obviously, con-mon-McNL $\subseteq$ mon-McNL holds. On the other hand, we have the following proper inclusions, where DCFL denotes the class of *deterministic context-free languages*.

Theorem 3. REG $\subsetneq$ con-mon-McNL $\subsetneq$ DCFL.

Proof. The first inclusion follows from the proof of [15] Theorem 2.1, where a deterministic finite-state acceptor is simulated by a string-rewriting system that is finite, monadic, and confluent. As each Dyck language is contained in con-mon-McNL, this inclusion is proper.

To prove the second inclusion, let R be a finite, confluent, and monadic string-rewriting system such that, for each $w \in \Sigma^*$, $w \in L$ iff $t_1 w t_2 \to_R^* Y$, that is, $t_1 \cdot L \cdot t_2 = [Y]_R \cap t_1 \cdot \Sigma^* \cdot t_2$. By [3] Theorem 3.9, $[Y]_R \in$ DCFL, and so $L \in$ DCFL.

Finally, let $L := \{ c a^n b^n \mid n \geq 1 \} \cup \{ d a^n b^{2n} \mid n \geq 1 \} \in$ DCFL. It is well-known that $L^R = \{ b^n a^n c \mid n \geq 1 \} \cup \{ b^{2n} a^n d \mid n \geq 1 \} \notin$ DCFL. It is easily seen from the definition that the class con-mon-McNL is closed under reversal. This means that $L \notin$ con-mon-McNL. □

The proof above uses the fact that DCFL is not closed under reversal. Hence, it arises the question of whether a language L is in con-mon-McNL, whenever L and L^R are both deterministic context-free. Next we will answer this question in the negative.

Let $A := \{ a^n b^n \mid n \geq 1 \}$ and $B := \{ a^m b^{2m} \mid m \geq 1 \}$, and let $L_p := A \cdot B = \{ a^n b^n a^m b^{2m} \mid n, m \geq 1 \}$. It is obvious that L_p as well as L_p^R are deterministic context-free. In contrast to this observation we have the following result [1].

Lemma 1. *The language L_p does not belong to the family* con-mon-McNL.

This fact yields the following consequence.

Corollary 1. con-mon-McNL $\subsetneq \{ L \mid L \in$ DCFL *and* $L^R \in$ DCFL $\}$.

It remains to determine the exact position of the families con-sp-McNL and sp-McNL within the Chomsky hierarchy.

Theorem 4. REG $\not\subset$ sp-McNL *and* con-sp-McNL $\not\subset$ REG.

Proof. Let $L_1 := ab^* ab^* a \in$ REG, and let R be a special string-rewriting system such that $t_1 ab^n ab^m a t_2 \to_R^* Y$ holds for all $n, m \geq 0$. Since R is a finite special system, it must contain a rule $(\ell \to \lambda)$ such that $\ell \in \{a, b\}^+$. This, however, implies that the language L specified by (R, t_1, t_2, Y) differs from L_1, that is, REG $\not\subset$ sp-McNL.

Conversely, let $R := \{ ab \to \lambda, ba \to \lambda \}$, $t_1 := \lambda$, and $t_2 := Y$. Then, for all $w \in \{a, b\}^*$, $t_1 w t_2 \to_R^* Y$ iff $w \to_R^* \lambda$ iff $|w|_a = |w|_b$. Hence, the language L defined by (R, t_1, t_2, Y) is the two-sided Dyck language D_1^*. As R is special and confluent, we see that con-sp-McNL $\not\subset$ REG holds. □

Hence, the inclusions con-sp-McNL $\subseteq$ con-mon-McNL and sp-McNL $\subseteq$ mon-McNL are proper.

Next we compare the family sp-McNL to con-mon-McNL and to DCFL. From Theorem 4 we already see that con-mon-McNL $\not\subset$ sp-McNL holds. However, also the converse inclusion does not hold. In fact we have the following result showing that DCFL and sp-McNL are incomparable under inclusion.

Theorem 5. sp-McNL $\not\subset$ DCFL.

Proof. Let $R := \{ ab \to \lambda, abb \to \lambda \}$, $t_1 := \lambda$, and $t_2 := Y$, where Y is a new symbol. By L we denote the language that is specified by (R, t_1, t_2, Y). Then $L \in$ sp-McNL. Now consider the subset $L \cap a^* b^*$. It is easily seen that $L \cap a^* b^* = \{ a^n b^m \mid 0 \leq n \leq m \leq 2n \}$. Since this set is not in DCFL, and since DCFL is closed under intersection with regular languages, it follows that L does not belong to DCFL. Thus, sp-McNL $\not\subset$ DCFL. □

We close this section by comparing the McNaughton families to still another class of languages. A language $L \subseteq \Sigma^*$ is called *Church-Rosser congruential* if there exist a finite, length-reducing, and confluent string-rewriting system R on Σ and finitely many strings $w_1, w_2, \ldots, w_n \in \mathrm{IRR}(R)$ such that $L = \bigcup_{i=1}^n [w_i]_R$. In [15] it is shown that the class CRCL of all Church-Rosser congruential languages is properly contained in the class CRL, and that it contains languages that are not context-free. Further, the language $L := \{\, a^n b^n \mid n \geq 1 \,\} \cup b^*$, which belongs to con-mon-McNL [1], is not congruential, and hence, CRCL is incomparable to con-mon-McNL. Here we want to compare CRCL to the families con-sp-McNL and sp-McNL. For that we will make use of the following characterization of con-sp-McNL.

Lemma 2. *Let $L \subseteq \Sigma^*$ be a member of the family* con-sp-McNL. *Then there exist an alphabet $\Gamma = \Sigma \cup \Theta \cup \{\cent, \$, Y\}$, a finite, special, and confluent string-rewriting system $R = R_1 \cup R_2$ on Γ, and irreducible strings $t_1, t_2 \in \Theta^*$ such that the following conditions are satisfied:*

(a) R_1 *is a confluent string-rewriting system on Σ,*
(b) $R_2 \subset (\cent t_1 \cdot \mathrm{IRR}(R_1) \cdot t_2 \$) \times \{\lambda\}$ *has no critical pairs, and*
(c) *for all $w \in \Sigma^*$, $w \in L$ if and only if there exists some $\hat{w} \in \mathrm{IRR}(R_1)$ such that $\cent t_1 w t_2 \$ \to_{R_1}^* \cent t_1 \hat{w} t_2 \$ \to_{R_2} \lambda$.*

Based on this characterization we obtain the following inclusion.

Theorem 6. con-sp-McNL $\subsetneq$ CRCL.

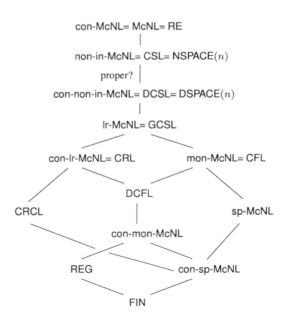

Fig. 1. The hierachy of McNaughton families of languages

On the other hand, for $R = \{ab \to \lambda, abb \to \lambda\}$ the language specified by (R, λ, Y, Y) is not Church-Rosser congruential [1], and so we obtain sp-McNL $\not\subset$ CRCL.

Thus, in summary we have the hierarchy of McNaughton families of languages depicted in Figure 1, where FIN denotes the class of all finite languages, and apart from the well-known LBA-problem all inclusions are known to be proper.

At this time it remains open whether or not sp-McNL is contained in CRL. Also it is a long-standing open problem whether or not each regular language belongs to CRCL. For a partial result showing that at least all regular languages of polynomial density are in CRCL see [17]. Further results on regular languages that are Church-Rosser congruential can be found in [18].

4 Some Closure and Non-closure Properties

As mentioned before it is immediate from the definition that each McNaughton family is closed under reversal. In addition we have the following closure property.

Theorem 7. *Each McNaughton family of languages is closed under left- and right-derivatives.*

Proof. Actually it suffices to consider the special case of a single letter. For this case the proof given in [15] for the class CRL also applies to con-mon-McNL and all the families above it. For the families sp-McNL and con-sp-McNL a proof can be found in [1]. □

It is known that the class CRL is neither closed under union nor under intersection [19]. For the families con-mon-McNL, con-sp-McNL and sp-McNL we have the following non-closure properties.

Theorem 8. (a) con-sp-McNL, sp-McNL, *and* con-mon-McNL *are neither closed under union, nor under intersection, nor under product.*
(b) con-sp-McNL *and* sp-McNL *are neither closed under complementation, nor under intersection with regular sets.*

The closure of the family con-mon-McNL with respect to complementation or intersection with regular sets remains open at this time.

We close this section by presenting an adaptation of results obtained by Herzog [8] for the context-free languages to the families con-mon-McNL and con-sp-McNL. In his paper he considers two families of context-free languages, which are also of interest to us.

For $k \geq 1$ and $i = 1, \ldots, k$, let $B_{k,i} := \{ b_1^n b_2^{i \cdot n} \mid n \geq 1 \}$, let

$$A_{k,i} := \{ a_1^{n_1} a_2^{n_2} a_3^{n_2} \cdots a_{2i-1}^{n_i} a_{2i}^{n_1} a_{2i+1}^{n_{i+1}} a_{2i+2}^{n_{i+1}} \cdots a_{2k-1}^{n_k} a_{2k}^{n_k} \mid n_1, \ldots, n_k \geq 1 \},$$

and let $B_k := \bigcup_{i=1}^{k} B_{k,i}$ and $A_k := \bigcup_{i=1}^{k} A_{k,i}$.

Herzog shows that the language A_k is inherently ambiguous of degree k, that the language B_ℓ is unambiguous and nondeterministic of degree ℓ, and that the union $A_k \cup B_\ell$ is inherently ambiguous of degree k and nondeterministic of degree ℓ for $\ell \geq k$.

As con-mon-McNL and con-sp-McNL are not closed under union, we are interested in the classes $\bigcup^k$con-mon-McNL and $\bigcup^k$ con-sp-McNL. Obviously, we have $\bigcup^k$ con-sp-McNL $\subseteq \bigcup^k$ con-mon-McNL $\subseteq \bigcup^k$ DCFL. For $k \geq 1$, A_k, B_k

can be shown to belong to $\bigcup^k$ con-mon-McNL. However, for all $k \geq 2$, $B_k \in \bigcup^k$ DCFL $\smallsetminus \bigcup^{k-1}$ DCFL, and $B_k \notin \bigcup^k$ con-sp-McNL.

The language $L := \{\, ca^n b^n \mid n \geq 1 \,\} \cup \{\, da^n b^{2n} \mid n \geq 1 \,\}$ is deterministic context-free, but it does not belong to con-mon-McNL. Since the alphabets $\{a, b, c, d\}$ and $\{b_1, b_2\}$ are disjoint, the product $C_k := B_k \cdot L$ belongs to the class $\bigcup^k$ DCFL. On the other hand, since $L \in \bigcup^2$ con-mon-McNL, it follows that $C_k \in \bigcup^{2k}$ con-mon-McNL. Here the bound $2k$ is sharp, as it can be shown that the language C_k does not belong to the class $\bigcup^\ell$ con-mon-McNL for any $\ell < 2k$ by using the pumping lemma for deterministic context-free languages (see [8] Lemma 5). Hence, we have the following results, where part (a) can be shown by using another family of example languages.

Theorem 9. (a) *The classes $\bigcup^k$ con-sp-McNL, $k \geq 1$, form a strict hierarchy.*
(b) *The classes $\bigcup^k$ con-mon-McNL, $k \geq 1$, form a strict hierarchy.*
(c) *For each $k \geq 1$, $\bigcup^k$ con-sp-McNL $\subsetneq \bigcup^k$ con-mon-McNL $\subsetneq \bigcup^k$ DCFL.*

5 On the Complexity of McNaughton Languages

For the following considerations we will need the complexity classes NC^1, L, LOG(DCFL), LOG(CFL) and P. We refer to [11], Section 5, for the definitions. It is known that $NC^1 \subseteq L \subseteq LOG(DCFL) \subseteq LOG(CFL) \subseteq P$.

We are interested in the complexity of the following decision problems for certain McNaughton families of languages. The *fixed membership problem* for a language $L \subseteq \Sigma^*$ is the problem of deciding, given a string $w \in \Sigma^*$ as input, whether or not w belongs to the language L. The *general membership problem* for a McNaughton family $\mathcal{L}$ is the problem of deciding, given a specification (R, t_1, t_2, Y) of a language $L \in \mathcal{L}$ and a string w as input, whether or not w belongs to the language L.

As CFL $=$ mon-McNL, the fixed membership problem for the family mon-McNL is LOG(CFL)-complete. The existence of a hardest context-free language implies that there even exists a fixed language in mon-McNL such that the membership problem for that language is LOG(CFL)-complete. Further, it can be shown that the general membership problem for mon-McNL is in LOG(CFL). Thus, in summary we have the following results.

Theorem 10. *The fixed and the general membership problems for the family mon-McNL are LOG(CFL)-complete with respect to many-one log-space reductions.*

This nicely contrasts the situation for context-free grammars, as the general membership problem for them is P-complete [12]. The hardest deterministic context-free language given in [20] is not contained in con-mon-McNL, as its reversal is not deterministic context-free. Therefore, we only obtain that LOG(DCFL) is an upper bound for the membership problems for the family con-mon-McNL.

In [15] DCFL $\subseteq$ CRL is proved by showing how to construct a confluent and length-reducing string-rewriting system that simulates the shift-reduce deterministic parsing of an LR(1) grammar. Motivated by this construction we have found a way to construct a confluent and monadic string-rewriting system R and irreducible strings t_1, t_2, Y from a linear LR(1) grammar G such that the membership problem for the language $L(G)$ is NC^1-reducible to the membership problem for the language specified by (R, t_1, t_2, Y) [1]. As there exists a linear LR(1) language for which the membership problem is L-complete [9], this yields the following result.

Theorem 11. *The fixed and the general membership problems for the family* con-mon-McNL *are* L-*hard with respect to* NC^1-*reductions, and they belong to* LOG(DCFL).

Finally we turn to the special McNaughton families of languages.

Theorem 12. *The fixed and the general membership problems for the family* con-sp-McNL *are* NC^1-*hard with respect to deterministic log-time reductions, and they belong to* LOG(DCFL).

Proof. The *Boolean formula evaluation problem* is the problem of determining the truth value of a given Boolean formula $\alpha \in \{0, 1, (,), \wedge, \vee, \sim\}^*$. According to [6] this problem is NC^1-complete under *deterministic log-time reductions*.

Let $\Sigma := \{0, 1, (,), \wedge, \vee, \sim\}$, $\Gamma := \Sigma \cup \{\#, Y\}$, $t_1 := \#$, $t_2 := Y$, and let R be the special string-rewriting system that consists of the following rules:

$$(0 \wedge 0)1 \to \lambda, \quad 0(1 \wedge 1) \to \lambda, \quad (0 \vee 0)1 \to \lambda, \quad 0(1 \vee 1) \to \lambda,$$
$$(0 \wedge 1)1 \to \lambda, \quad 0(\sim 0) \to \lambda, \quad 0(0 \vee 1) \to \lambda, \quad \#1 \to \lambda,$$
$$(1 \wedge 0)1 \to \lambda, \quad (\sim 1)1 \to \lambda, \quad 0(1 \vee 0) \to \lambda.$$

As R has no non-trivial overlaps, it is confluent. Hence, the language L specified by (R, t_1, t_2, Y) belongs to con-sp-McNL.

Let $\alpha \in \Sigma^*$ be a Boolean formula. We encode α by inserting 0 in front of every opening parenthesis and 1 after each closing parenthesis. Obviously, this transformation can be realized through a deterministic log-time reduction, and it is easily verified that a formula α is true if and only if its encoding belongs to L. □

As pointed out by Lohrey [13] the general membership problem for con-sp-McNL can even be shown to be L-hard under NC^1-reductions by a reduction from *direct forest accessibility* [7]. For the family sp-McNL we have the following result.

Theorem 13. *The fixed and the general membership problems for the family* sp-McNL *are* LOG(CFL)-*complete under deterministic log-time reductions.*

6 Conclusion

The Church-Rosser congruential languages can be seen as the *pure* variants of the Church-Rosser languages. Accordingly, this notion can be generalized similarly to yield pure McNaughton families of languages. Also instead of only distinguishing between confluent and non-confluent string-rewriting systems we could also consider various classes of *weakly confluent* systems [14], which would lead to weakly confluent McNaughton families of languages. We close with a list of open problems:

1. Is the family sp-McNL contained in CRL?
2. Is con-mon-McNL closed under complementation or under intersection with regular sets?
3. Which regular languages belong to the families con-sp-McNL and sp-McNL?
4. For the families mon-McNL and sp-McNL the fixed and the general membership problems have the same degree of complexity. Is this also true for the corresponding confluent families?
5. Are the membership problems for the families con-sp-McNL and con-mon-McNL even LOG(DCFL)-hard?

6. What is the exact relationship of con-mon-McNL to the class NTS of *nonterminal separated languages* [2]? As con-mon-McNL contains languages that are not congruential, it is not contained in NTS, but it is open whether or not NTS is a subclass of con-mon-McNL.

References

1. M. Beaudry, M. Holzer, G. Niemann, and F. Otto. McNaughton languages. *Mathematische Schriften Kassel* 26/00, Universität Kassel, November 2000. A revised version will appear in *Theoret. Comput. Sci.*
2. L. Boasson and G. Sénizergues. NTS languages are deterministic and congruential. *J. Comput. System Sci.*, 31:332–342, 1985.
3. R.V. Book, M. Jantzen, and C. Wrathall. Monadic Thue systems. *Theoret. Comput. Sci.*, 19:231–251, 1982.
4. R.V. Book and F. Otto. *String-Rewriting Systems*. Springer-Verlag, New York, 1993.
5. G. Buntrock and F. Otto. Growing context-sensitive languages and Church-Rosser languages. *Inform. and Comput.*, 141:1–36, 1998.
6. S. Buss. The Boolean formula value problem is in ALOGTIME. In *Proc. of 19th STOC*, pp. 123–131. ACM Press, 1987.
7. S.A. Cook and P. McKenzie. Problems complete for deterministic logarithmic space. *J. Algorithms*, 8:385–394, 1987.
8. C. Herzog. Pushdown automata with bounded nondeterminism and bounded ambiguity. *Theoret. Comput. Sci.*, 181:141–157, 1997.
9. M. Holzer and K.-J. Lange. On the complexities of linear LL(1) and LR(1) grammars. In Z. Ésik, editor, *Proceedings 9th FCT*, Lect. Notes Comput. Sci. 710, pp. 299–308. Springer-Verlag, Berlin, 1993.
10. J.E. Hopcroft and J.D. Ullman. *Introduction to Automata Theory, Languages, and Computation*. Addison-Wesley, Reading, M.A., 1979.
11. D.S. Johnson. A catalog of complexity classes. In J. van Leeuwen, editor, *Handbook of Theoretical Computer Science*, pp. 67–161. The MIT Press/Elsevier, Cambridge, MA/Amsterdam, 1990.
12. N.D. Jones and W.T. Laaser. Complete problems for deterministic polynomial time. *Theoret. Comput. Sci.*, 3:105–117, 1977.
13. M. Lohrey. Private communication, November 2000.
14. K. Madlener, P. Narendran, F. Otto, and L. Zhang. On weakly confluent monadic string-rewriting systems. *Theoret. Comput. Sci.*, 113:119–165, 1993.
15. R. McNaughton, P. Narendran, and F. Otto. Church-Rosser Thue systems and formal languages. *J. ACM*, 35:324–344, 1988.
16. P. Narendran. *Church-Rosser and related Thue systems*. PhD thesis, Rensselaer Polytechnic Institute, Troy, New York, 1984.
17. G. Niemann. Regular Languages and Church-Rosser congruential languages. In R. Freund and A. Kelemenová, editors, *Grammar Systems 2000, Proceedings*, pp. 359–370. Silesian University, Opava, 2000.
18. G. Niemann and J. Waldmann. Some regular languages that are Church-Rosser congruential. This volume.
19. G. Niemann and F. Otto. The Church-Rosser languages are the deterministic variants of the growing context-sensitive languages. In M. Nivat, editor, *Proceedings FoSSaCS'98*, Lect. Notes Comput. Sci. 1378, pp. 243–257. Springer-Verlag, Berlin, 1998.
20. H. Sudborough. On the tape complexity of deterministic context-free languages. *J. ACM*, 25:405–414, 1978.

Forbidden Factors and Fragment Assembly*

Filippo Mignosi, Antonio Restivo, and Marinella Sciortino

University of Palermo, Dipartimento di Matematica ed Applicazioni,
Via Archirafi 34, 90123 Palermo, Italy
{mignosi,restivo,mari}@math.unipa.it

Abstract. In this paper we approach the fragment assembly problem by using the notion of minimal forbidden factors introduced in previous paper. Denoting by $M(w)$ the set of minimal forbidden factors of a word w, we first focus on the evaluation of the size of elements in $M(w)$ and on designing of an algorithm to recover the word w from $M(w)$. Actually we prove that for a word w randomly generated by a memoryless source with identical symbol probabilities, the maximal length $m(w)$ of words in $M(w)$ is logarithmic and that the reconstruction algorithm runs in linear time. These results have an interesting application to the fragment assembly problem, i.e. reconstruct a word w from a given set I of substrings (fragments). Indeed under a suitable hypothesis on the set of fragments I, one can detect the elements of $M(w)$ by looking at the minimal forbidden factors of elements in I and then apply the reconstruction algorithm.

1 Introduction

Let w be a word over the alphabet A and let $L(w)$ denote the set of factors of w. A word v is a *minimal forbidden factor* of w if $v \notin L(w)$ and all proper factors of v belong to $L(w)$. Denote by $M(w)$ the set of minimal forbidden factors of the word w. The notion of minimal forbidden factors is a very basic one in combinatorics on words: some relevant information on the structure of a word w can be detected by looking at the set $M(w)$ (cf. [6], [8]). This leads to some important applications in Data Compression (cf. [7]) and Symbolic Dynamics (cf. [2]).

In this paper we focus on the evaluation of the size of the set $M(w)$ and on an algorithm for reconstructing the word w from $M(w)$. In both cases the results are based on a close relation, stated in [6], between the set $M(w)$ and the *factor automaton* of w, i.e. the minimal deterministic automaton recognizing $L(w)$. In particular, as to concern the size, we show that, in general, the size of words in $M(w)$ is "very small" with respect to the size of w: we prove that, for a word w randomly generated by a memoryless source with identical symbol probabilities, the maximal length $m(w)$ of words in $M(w)$ is $O(\log_d(|w|))$ where d is the cardinality of the alphabet. As to concern the reconstruction, we prove that there is a linear time algorithm to recover word w from the set $M(w)$ of its minimal forbidden factors.

These results lead to an interesting applications to the fragment assembly problem. The *fragment assembly problem* can be formulated as follows: given a set I of substrings

---

* Partially supported by MURST projects: *Modelli di calcolo innovativi: metodi sintattici e combinatori* and *Bioinformatica e Ricerca Genomica*.

(*fragments*) of a word w, i.e. $I \subseteq L(w)$, reconstruct w from I. It is obvious that, without any additive hypothesis on I and w, it is not in general possible to infer w from I, i.e. in general, given I, there exist several different words w compatible with I, i.e. such that $I \subseteq L(w)$. In this paper we introduce an hypothesis concerning the set of fragments I and the size of the set of minimal forbidden words $M(w)$. Such hypothesis allows us to reconstruct uniquely the word w from the set I in linear time. The basic idea is the following. From the results on the size of $M(w)$, it is very reasonable to suppose that, in several practical applications (as for instance in the case of DNA sequences) the set of fragments I is such that any factors of w of length $m(w)$ is "covered" by at least one element of I. Under such hypothesis, one can detect the minimal forbidden factors of the whole word w only by looking at its "fragments" in I. By using the reconstruction algorithm one can recover in linear time the word w from its fragments. Further details on the assembly algorithm described in this paper and a discussion on its adequacy to biological sequences can be found in [9], which also contains another linear time assembly algorithm based on a more classic approach. In [4], by using the notions of special and univalent factor, an assembly algorithm is proposed, which presents some analogies with that ones of the present paper.

2 Preliminaries

Let A be a finite alphabet and let A^* be the set of finite words over the alphabet A, the empty word ϵ included. Let $L \subseteq A^*$ be a *factorial language*, i.e. a language satisfying that $\forall u, v \in A^*$, $uv \in L \Rightarrow u, v \in L$. From an algebraic point of view we observe that the complement language $L^c = A^* \setminus L$ is a two-sided ideal of the free monoid A^*. Denote by $M(L)$ the base of this ideal, i.e. $L^c = A^* M(L) A^*$. The set $M(L)$ is called the set of *minimal forbidden words* for L. A word $v \in A^*$ is *forbidden* for the factorial language L if $v \notin L$, which is equivalent to say that v occurs in no word of L. In addition, v is *minimal* if it has no proper factor that is forbidden. From the minimality of its words follows that $M(L)$ is an *antifactorial language*, i.e. $\forall u, v \in M(L), u \neq v \Rightarrow u$ is not a factor of v.

Remark 1. One can note that the set $M(L)$ uniquely characterizes L, just because

$$L = A^* \setminus A^* M(L) A^* \qquad (1)$$

Conversely the following remark shows that L also uniquely characterizes $M(L)$.

Remark 2. A word $v = a_1 a_2 \cdots a_n$ belongs to $M(L)$ if and only if the two conditions hold:
- v is forbidden, (*i.e.* $v \notin L$),
- both $a_1 a_2 \cdots a_{n-1} \in L$ and $a_2 a_3 \cdots a_n \in L$ (the prefix and the suffix of v of length $n-1$ belong to L).

Hence we have that

$$M(L) = AL \cap LA \cap (A^* \setminus L). \qquad (2)$$

From the equalities 1 and 2 it follows that $M(L)$ uniquely characterizes L and L uniquely characterizes $M(L)$ respectively. Recall that a language $L \subset A^*$ is *rational* if it is recognized by a finite state automaton. From the equalities 1 and 2, one also derives that L and $M(L)$ are simultaneously rational, i.e. L is rational if and only if $M(L)$ is a rational language. By using the classical algorithms of automata theory, in the case L is rational, one can effectively construct $M(L)$ from L, and conversely.

In this paper we are interested to the language $L(w)$ of factors of a single word w and consequently on the set $M(L(w))$, which is here denoted simply by $M(w)$ and is called the set of *minimal forbidden factors* of the word w.

Example 1. Let us consider the word $w = acbcabcbc$ on the alphabet $\{a, b, c\}$. One has that
$$M(w) = \{aa, ba, bb, cc, aca, cac, cbcb, abca, bcbca\}.$$

It is obvious that $M(w)$ is a finite set uniquely characterizing the word w.

3 $M(w)$ and Factor Automaton

In this section we report some results of [6] showing that there is a close relation between the set $M(w)$ and the *factor automaton* of the word w, i.e. the minimal deterministic automaton recognizing $L(w)$.

Given the set $M(w)$, we define a finite automaton $\mathcal{A}(w) = (Q, A, i, T, F)$ where

- the set Q of states is $\{v \mid v$ is a prefix of a word in $M(w)\}$,
- A is the current alphabet,
- the initial state i is the empty word ϵ,
- the set T of terminal states is $Q \setminus M(w)$.

States of $\mathcal{A}(w)$ that are words of $M(w)$ are sink states. The set F of transitions is partitioned into the three (pairwise disjoint) sets F_1, F_2, and F_3 defined by:

- $F_1 = \{(u, a, ua) \mid ua \in Q, a \in A\}$ (forward edges or tree edges),
- $F_2 = \{(u, a, v) \mid u \in Q \setminus M(w), a \in A, ua \notin Q, v$ longest suffix of ua in $Q\}$ (backward edges),
- $F_3 = \{(u, a, u) \mid u \in M(w), a \in A\}$ (loops on sink states).

Theorem 1. *For any $w \in A^*$, the automaton obtained from $\mathcal{A}(w)$ by removing its sink states is the factor automaton of w (i.e. the minimal deterministic finite automaton $\mathcal{F}(w)$ accepting the language $L(w)$).*

Note that, as it is showed in [6], previous construction is possible even for an antifactorial language M and in this case it provides an automaton $\mathcal{A}(M)$ recognizing the corresponding language L. However, in the general case, $\mathcal{A}(M)$ is not minimal. In [6] the above definition of $\mathcal{A}(M)$ is turned into an algorithm, called L-AUTOMATON that builds the automaton from a finite antifactorial set of words. The input is the trie $\mathcal{T}$ that represents M. The procedure can be adapted to test whether $\mathcal{T}$ represents an antifactorial set, or even to generate the trie of the antifactorial language associated with a set of words. In [6] it is proved that this algorithm runs in time $O(|Q| \times |A|)$, where Q and A are respectively the set of states and the alphabet of the input trie where the transition functions are implemented by transition matrices.

Recall also that there are some interesting results (cf. [3], [5]) about the size of factor automaton of a word w. Indeed, by denoting with Q and E the set of states and edges respectively, it is proved that the size of $\mathcal{F}(w)$ is linear with the length of the word w. In particular, if $|w| \leq 2$ then $|Q| = |w| + 1$ and $|w| \leq |E| \leq 2|w| - 1$. If $|w| \geq 3$ then $|w| + 1 \leq |Q| \leq 2|w| - 2$ and $|w| \leq |E| \leq 3|w| - 4$. These bounds will be useful in next sections.

4 On the Size of $M(w)$

In this section we are interested in a valuation of the size of $M(w)$. Given a finite word w, we consider the following parameters:

$$c(w) = Card(M(w))$$

$$m(w) = \max\{|v|, v \in M(w)\}.$$

Example 1 (continued) Let $w = acbcabcbc$. We have $c(w) = 9$, $m(w) = 5$.

The first result of this section states in particular an upper bound of $c(w)$, which linearly depends on the length of the word w. Remark that the cardinality of $L(w)$, the set of factors of w, is $O(|w|^2)$. Denote by d the cardinality of the alphabet A and by $d(w)$ the number of the letters of A occurring in w, i.e. $d(w) = Card(alph(w))$, where $alph(w)$ denotes the set of letters occurring in w.

A complete proof of the following two theorems can be found in [8].

Theorem 2. *Let $w = a_1 \ldots a_n$ be a finite word over the alphabet A. The following inequalities hold:*

$$d \leq c(w) \leq (d(w) - 1)(|w| - 1) + d.$$

Moreover previous inequalities are sharp.

In the next theorem we derive lower and upper bounds on the parameter $m(w)$. We will consider $d > 1$, because if A has just one element, it is trivial that $m(w) = |w| + 1$. Recall that, for any real number α, $\lceil \alpha \rceil$ denotes the smallest integer greater than or equal to α.

Theorem 3. *Let w be a finite word over the alphabet A with at least two elements. The following inequalities hold:*

$$\lceil \log_d(|w| + 1) \rceil \leq m(w) \leq |w| + 1.$$

Furthermore the bounds are actually attained.

We now evaluate this parameter for a word w which is randomly generated by a memoryless source with identical symbol probabilities. For simplicity we consider sources over a binary alphabet. All the results presented here can easily be generalized to alphabets of more letters.

Remark 3. It is easy to prove that $k < m(w) - 1$ if and only if there exists a factor of the word w of length k having at least two occurrences.

Theorem 4. *Let w be a word over a binary alphabet which is randomly generated by a memoryless source with identical symbol probabilities. The probability that there exists a word v of length k that appears at least twice as factor of w is smaller than $(n-k)(k-1)2^{-k} + (n-2k+1)^2 2^{-k}$, where n is the length of w.*

For a proof of previous theorem cf. [9].
From Remark 3 and previous theorem we have the following

Corollary 1. *The probability that $m(w)$ is smaller than or equal to $3\log_2 n + 1$ is about 1 for large enough values of n.*

For a memoryless source where the symbols have fixed probabilities different than $1/2$, analogous computation can be done. From previous corollary easily follows that for a word w randomly generated by a memoryless source the parameter $m(w)$ approximates $O(\log_d(n))$ where n is the length of w and d is the cardinality of the alphabet.

5 Reconstruction Algorithm

As we showed in Section 2, $M(w)$ is a finite set uniquely characterizing the word w. As to concern the construction of $M(w)$ from the word w, we refer to (cf. [6]) the algorithm MF-TRIE that constructs the trie accepting the language $M(w)$ from the factor automaton of w.

The input of algorithm MF-TRIE is the factor automaton $\mathcal{F}(w)$ of word w. It includes the *suffix function* that is a failure function defined on the states of the automaton. This function is a by-product of efficient algorithms that build the factor automaton (cf. [5]). In [6] it is proved that the algorithm MF-TRIE runs in time $O(|w| \times |A|)$ on input $\mathcal{F}(w)$ if transition functions are implemented by transition matrices. The result is a consequence of the linear size of $\mathcal{F}(w)$.

The construction of factor automaton $\mathcal{F}(w)$ from the word w is also known (cf. [3], [5]). In [5] the algorithm FACTORAUTOMATON is given and its construction can be implemented to work on the input word w in time $O(|w| \times \log |A|)$ within $O(|w|)$ space if one uses adjacency lists.

Hence it is possible to construct the set $M(w)$ from the word w over a fixed alphabet in linear time with the length of w.

In this section we are interested on the converse of previous transformation, that is reconstructing the word w from the set $M(w)$. We show that, by using close relation between $M(w)$ and factor automaton of w (cf. Section 3) and its linear size, it is possible to design a linear algorithm solving such problem. This algorithm involves, besides factor automaton, another known construction that is topological sort of a directed acyclic graph (cf.[1]). Our algorithm is also described in [8]. The main procedures used in the algorithm are L-AUTOMATON (cf. [6] and see also Section 3), BUILDWORD and TOPOLOGICALSORT (cf. [1]). We will give a description of the algorithm, dwelling upon the procedure BUILDWORD.

Procedure L-AUTOMATON takes as input the trie of the set $M(w)$ and returns a complete deterministic automaton accepting $L(w)$. Recall that from this automaton it is possible, by removing its sink states, to obtain the minimal deterministic automaton $\mathcal{F}(w)$ (see Theorem 1).

Procedure BUILDWORD, that is related to the search for the longest path in a directed acyclic graph, works as follows: it first call Procedure TOPOLOGICALSORT that produces a linear ordering of all vertices of the transition graph $\mathcal{G}(w)$ of the factor automaton $\mathcal{F}(w)$ by using a depth first search procedure. If $\mathcal{G}(w)$ contains an edge (q,p), then q appears before p in the ordering. Recall that the transition graph of a factor automaton is a directed acyclic graph (if the graph is not acyclic, then no linear ordering is possible).

Then in Procedure BUILDWORD we create the *precedence-lists* B of the factor automaton $\mathcal{F}(w) = (Q, A, q_0, T, \delta)$. If n is the number of the states of $\mathcal{F}(w)$ then B consists of an array of n lists, one for each state. For each state q the *precedence-list* $B(q)$ contains (pointers to) all states s such that $\delta(s, a) = q$ for some $a \in A$:

$$B(q) = \{s| \exists a \in A \text{ such that } \delta(s, a) = q\}.$$

Moreover we define a function $\pi : Q \mapsto \mathbb{N}$ such that, for each state q, $\pi(q)$ is the length of the longest path from q_0 to q in the transition graph $\mathcal{G}(w)$ of the factor automaton $\mathcal{F}(w)$. A recursive definition of the function π is the following:

$$\pi(q_0) = 0$$
$$\pi(q) = \max\{\pi(s)|s \in B(q)\} + 1$$

where q_0 is the initial state. Remark that, if $s \in B(q)$ then $s < q$ in the topological sort.

We also remark that the factor automaton $\mathcal{F}(w)$ satisfies the property that if x and y are respectively the label of two paths from state q to the state p, then $|x| \neq |y|$. From this property it follows that, for any $q \in Q \setminus \{q_0\}$, there exists a unique state $s_q \in B(q)$ such that $\pi(s_q) = \max\{\pi(s)|s \in B(q)\}$. Moreover, according with the previous property, we can define the following partial function $l : Q \times Q \mapsto A$ such that

$$l(p, q) = a \quad \text{if} \quad \delta(p, a) = q.$$

Finally, if $q = q_{n-1}$ is the last state in the linear ordering produced by Procedure TOPOLOGICALSORT, the word is built by taking the letter $l(s_q, q)$, by setting q equal to s_q and updating it by concatenating on the left the letter $l(s_q, q)$. This cycle has to be carried out while q is different from the initial state.

```
BUILDWORD (factor automaton F(w) = (Q, A, i, T, δ))
1.  {q_0, q_1, ..., q_{n-1}} ← TOPOLOGICALSORT(G(w));
2.  for each state q_i, 0 ≤ i ≤ n − 1, create the precedence-list B(q_i);
3.  π(q_0) ← 0;
       for each state q_i, 1 ≤ i ≤ n − 1
           Find the maximum m_i of the set {π(s), s ∈ B(q_i)};
           return the unique state s_{q_i} such that π(s_{q_i}) = m_i;
           π(q_i) ← m_i + 1;
4.  w ← ε;
5.  q ← q_{n-1};
6.  while q ≠ q_0 do
           a ← l(s_q, q);
           w ← aw;
           q ← s_q;
7.  return w;
```

The following proposition, by using the fact that $\sum_{i=0}^{n-1} |B(q_i)| = |E|$, establishes an upper bound on the time required for Procedure BUILDWORD.

Proposition 1. *The execution time for Procedure* BUILDWORD *is* $O(|Q| + |E|)$, *where Q and E are respectively the set of the states and the set of the edges of factor automaton $\mathcal{F}(w)$ of a word w over a fixed alphabet A.*

```
w-RECONSTRUCTION (Trie T(w) representing the set M(w))
1.  A(w) ← L-AUTOMATON(T(w));
2.  let F(w) be the automaton obtained by removing sink states of A(w);
3.  w ← BUILDWORD(F(w));
4.  return w;
```

From Proposition 1, from linear size of factor automaton and since L-AUTOMATON procedure runs in time $O(|Q| \times |A|)$, we can conclude that:

Proposition 2. *Given the set $M(w)$ of minimal forbidden factors of a word w over a fixed alphabet A, it is possible to reconstruct w in linear time with the size of the trie representing the set $M(w)$.*

6 Applications to Fragment Assembly

The *fragment assembly problem* can be formulated as follows: given a set I of substrings (*fragments*) of a word w, i.e. $I \subseteq L(w)$, reconstruct w from I. It is obvious that, without any additive hypothesis on I and w, it is not in general possible to infer w from I, i.e. in general, given I, there exist several different words w compatible with I, i.e. such that $I \subseteq L(w)$. In this paper we introduce an hypothesis concerning the set of fragments I and the size of the set of minimal forbidden words $M(w)$. Such hypothesis allows us to reconstruct uniquely the word w from the set I. The basic idea of the algorithm is the following. In Section 5 we showed that, given the set $M(w)$, it is possible to reconstruct in linear time the word w. We can obtain this set from the set of minimal forbidden factors of another word w_1 obtained by concatenating all fragments in I. Indeed, if the set of fragments is such that any factor of w of length $m(w)$ is "covered" by at least one element of I, we show that one can obtain all minimal forbidden factors of original word w from those of w_1. Actually the estimate of the parameter $m(w)$ in previous section shows that the minimal forbidden factors are "short enough" with respect to the length of w. So the previous hypothesis on fragments is appropriate in some practical applications. We broach now these arguments more rigorously. For the complete proofs of the results of this section see [9].

Definition 1. *A set of substrings I of a word w is a k-cover for w if every substring of length k of w is a substring of at least one string in I. The covering index of I, denoted $C(I)$, is the largest value of k such that I is a k-cover of w.*

Clearly, I is a k-cover for w for all $k \leq C(I)$. We can now state the following theorem that will be proved in the sequel. Recall that we use the notation $||I||$ to denote the sum of the lengths of all words in I.

Theorem 5. *Let w be a word over a fixed alphabet A and let I be a set of substrings of w such that*

$$m(w) \leq C(I).$$

Then the word w can be uniquely reconstructed by the set I and this reconstruction can be done in linear time $O(||I||)$.

We note that in previous theorem we do not need to know the length of the string w to recover it. The "linear time" has to be considered under the usual standard assumption that we can store, add and compare integers in constant time. Note also that the condition $m(w) \leq C(I)$ in previous theorem is tight in the sense that, if there exists a substring v that occurs in w at least twice and such that v never appears as proper factor of any element of I, then the reconstruction can be ambiguous, i.e. there could exist several words w compatible with I.

In this section we present an algorithm, called ASSEMBLY, that, under the hypothesis of Theorem 5, solves the fragment assembly problem in linear time.

In the first step the following easy CONCAT algorithm is used. Its input is the set of fragments I and a symbol \$ that is not in the alphabet A of all fragments in I. The output is a word w_1 over the alphabet $A \cup \{\$\}$ that is the concatenation of all strings in I interspersed with the symbol \$, i.e. between two consecutive strings there is one dollar. The set I is structured as a simple stack. The operation of concatenation between words is denoted by a simple dot "." and the empty word is denoted by ε.

CONCAT (set I, symbol \$)
1. $w_1 \leftarrow \varepsilon$;
2.   **while** $I \neq \emptyset$
3.   extract v from I;
4. $w_1 \leftarrow w_1.v.\$$;
5.   **return** (w_1);

The second goal of ASSEMBLY algorithm consists in the construction of the trie of minimal forbidden factors of w_1 having length smaller than or equal to $m(w)$ and not containing the symbol $. This is done by FACTORAUTOMATON algorithm and CREATE_TRIE algorithm that is a light variation of the MF-TRIE algorithm described in [6], where the trie of all minimal forbidden factors of a word was obtained by its factor automaton. Here the correctness of CREATE_TRIE algorithm is a simple consequence of the correctness of MF-TRIE algorithm, together with the fact, stated in next Proposition 3, that the set of minimal forbidden factors of original string w is exactly the set of words that are found by CREATE_TRIE algorithm. Note that the hypothesis of next proposition is essential to its proof.

Proposition 3. *Let w be a word over a fixed alphabet A and let I be a set of substrings of w such that*

$$C(I) \geq m(w).$$

Then the set of minimal forbidden factors of the word w is exactly the set of all the minimal forbidden factors of w_1 that do not contain the symbol $ and that have length smaller than or equal to $m(w)$, i.e.

$$M(w) = M(w_1) \cap A^{\leq m(w)}.$$

CREATE_TRIE (factor automaton $\mathcal{F}(w_1) = (Q, A \cup \{\$\}, i, T, \delta)$, suffix function h)
1. **for** each state $p \in Q$ in breadth-first search from i **and** each $a \in A$
2.   **if** $\delta(p, a)$ undefined **and**$(p = i$ **or** $\delta(h(p), a)$ defined$)$
3. $\delta'(p, a) \leftarrow$ new sink;
4.   **else if** $\delta(p, a) = q$ **and** q is distant from i more than p
5. $\delta'(p, a) \leftarrow q$;
6. in a depth-first search respect to δ' prune all branches of trie $\mathcal{T}(w)$ not ending in a state that is sink and has depth smaller than or equal to $m(w)$;
7. **return** $\mathcal{T}(w) = (Q', A, i, \{sinks\}, \delta')$;

Recall that the suffix function h is defined as follows. Let $u \in (A \cup \{\$\})^+$ and $p = \delta(i, u)$. Then $h(p) = \delta(i, u')$ where u' is the longest suffix of u for which $\delta(i, u) \neq \delta(i, u')$.

Note that, in Line 6, not all the states of $\mathcal{F}(w_1)$ are reachable starting from the root i, because δ' does not represent edges labelled by $. All states and edges that are not reachable from i are implicitly pruned in $\mathcal{T}(w)$. The implicit and explicit pruning operations in line 6 obviously change the set of states, the set of sinks and the function δ'.

Remark 4. It is easy to see that the elements of $M(w_1) \cap A^*$ that don't belong to $M(w)$ (i.e. having length greater than $m(w)$) are of the form avb, where $av\$$ and $\$vb$ are factors of the word w_1. By using this remark it is possible to modify the CREATE_TRIE algorithm in order to obtain the elements of $M(w)$, without the explicit knowledge of the value of $m(w)$.

The final step of algorithm ASSEMBLY consists in recovering the word w from $\mathcal{T}(w)$. This is done by w-RECONSTRUCTION algorithm showed in Section 5 (cf. also [8]).

The overall ASSEMBLY algorithm is thus

> ASSEMBLY (set of fragments I)
> 1. $w_1 \leftarrow$ CONCAT (set $I, \$$);
> 2. $\mathcal{F}(w_1) = (Q, A \cup \{\$\}, i, T, \delta) \leftarrow$ FACTORAUTOMATON (w_1);
> 3. $\mathcal{T}(w) = (Q', A, i, \{sinks\}, \delta') \leftarrow$ CREATE_TRIE $(\mathcal{F}(w_1), h)$;
> 4. $w \leftarrow w$-RECONSTRUCTION $(\mathcal{T}(w))$;
> 5. **return** w;

From Proposition 3 and from linear time complexity of all procedures used in previous algorithm the proof of Theorem 5 follows.

References

1. Aho, A. V., Hopcroft, J. E., Ullman, J. D.: Data Structures and Algorithms. Addison Wesley, Reading, Mass, 1983
2. Béal, M.-P., Mignosi, F., Restivo, A., Sciortino, M.: Forbidden Words in Symbolic Dynamics. Advances in Appl. Math. **25** (2000) 163–193
3. Blumer, A., Blumer, J., Ehrenfeucht, A., Haussler, D., Chen, M.T., Seiferas, J.: The smallest automaton recognizing the subwords of a text. Theoret. Comput. Sci. **40** (1985) 31–55
4. Carpi, A., de Luca, A., Varricchio, S.: Words, univalent factors, and boxes. Report, Università di Roma "La Sapienza", Dipartimento di Matematica "Guido Castelnuovo", 2000.
5. Crochemore, M., Hancart, C.: Automata for matching patterns. Handbook of Formal Languages, volume 2, chapter 9, 399–462, G.Rozenberg, A.Salomaa (Eds.), Springer Verlag, 1997
6. Crochemore, M., Mignosi, F., Restivo, A.: Automata and forbidden words. Inf. Proc. Lett. **67** (1998) 111-117
7. Crochemore, M., Mignosi, F., Restivo, A., Salemi, S.: Data Compression using antidictionaries. Proceedings of the IEEE, Special Issue on Lossless Data Compression (J. A. Storer Ed.) **88**:11 (2000) 1756-1768
8. Mignosi, F., Restivo, A., Sciortino, M.: Words and Forbidden Factors. Theoret. Comput. Sci. **273**(1-2) (2001) 99–117
9. Mignosi, F., Restivo, A., Sciortino, M., Storer, J.: On Sequence Assembly. Technical report Brandeis cs-00-210

Parallel Communicating Grammar Systems with Incomplete Information Communication*

Erzsébet Csuhaj-Varjú and György Vaszil

Computer and Automation Research Institute
Hungarian Academy of Sciences
Kende utca 13-17, 1111 Budapest, Hungary
{csuhaj,vaszil}@sztaki.hu

Abstract. We examine the generative power of parallel communicating (PC) grammar systems with context-free or $E0L$ components communicating incomplete information, that is, only subwords of their sentential forms. We prove that these systems in most cases, even with $E0L$ components, generate all recursively enumerable languages.

1 Introduction

Parallel communicating (PC) grammar systems are grammatical models of parallel and distributed computation ([8]). In these systems several grammars generate their own sentential forms in a synchronized manner and their work is organized in a communicating system to generate a single language. During the work of the system each grammar executes one rewriting step in each time unit and communication is realized by so-called query symbols, one different symbol referring to each component grammar of the system. When a component introduces a query symbol in its sentential form, the rewriting process stops and one or more communication steps are performed by substituting all occurrences of the query symbols with the current sentential forms of the queried component grammars, supposing that the requested string is query free. When no more query symbol is present in any of the sentential forms, the rewriting process starts again. In so-called returning systems, after communicating its current sentential form the component returns to its start symbol and begins to generate a new string. In non-returning systems the components continue the rewriting of their current sentential forms. The language defined by the system is the set of terminal words obtained as sentential forms of a dedicated component grammar, the master, during the performed rewriting steps and communication.

PC grammar systems have been investigated in details during the years. For a summary of results the reader is referred to [1] and [3] and for an on-line bibliography see http://www.sztaki.hu/mms/bib.html.

One central question concerning PC grammar systems is, how powerful the different classes of these systems are, and whether this power changes or not if we modify some of the basic parameters of the contstruct.

---

* Research supported in part by the Hungarian Scientific Research Fund "OTKA" grant no. T 029615.

In [2] and [6] it was shown that PC grammar systems with context-free components are as powerful as Turing machines. In [7] a variant with a modified way of communication was introduced and examined: it was proved that any recursively enumerable language can be obtained as the homomorphic image of the intersection of a regular language and the language of a PC grammar system where the components communicate only non-empty prefixes of the requested sentential forms.

Continuing this line of investigations, in this paper we deal with PC grammar systems with incomplete information communication. We prove that the power - the computational completeness - of returning PC grammar systems with context-free components does not change if the component grammars communicate only a (non-empty) subword, a prefix, or a suffix of the requested string. Furthermore, we show that the computational completeness is also preserved in the case of non-returning context-free PC grammar systems with prefix or suffix mode of communication. In contrast to systems with context-free components, the generative power of systems with $E0L$ components is not known in the generic case. We prove that using the prefix or suffix mode of communication, returning PC $E0L$ systems determine the class of recursively enumerable languages. Furthermore, in all the above cases, to generate a recursively enumerable language, a PC grammar system or a PC $E0L$ system with a reasonably small number of components can be chosen.

The question how powerful are non-returning context-free PC grammar systems with subword communication is still open and this is the case of determining the power of returning PC $E0L$ systems with subword and non-returning PC $E0L$ systems with any of the above incomplete information communication modes.

2 Preliminaries

The reader is assumed to be familiar with formal language theory; for further information we refer to [5].

The set of nonempty words over an alphabet V is denoted by V^+, if the empty word, ε, is included, then we use notation V^*. A set of words $L \subseteq V^*$ is called a language over V, for $w \in L$ we denote the length of w by $lng(w)$, and the number of occurences of symbols from a set X in w by $|w|_X$. The class of recursively enumerable languages is denoted by $\mathcal{L}(RE)$, CF denotes the class of context-free grammars and $E0L$ refers to the class of $E0L$ systems. Now we recall the notion of a parallel communicating grammar system and that of a parallel communicating $E0L$ system. For further details consult [3].

A *parallel communicating grammar system* with n components (a PC grammar system in short) is an $(n+3)$-tuple $\Gamma = (N, K, T, (P_1, \omega_1), \ldots, (P_n, \omega_n))$, $n \geq 1$, where the set N is the *nonterminal alphabet*, T is the *terminal alphabet*, and $K = \{Q_1, Q_2, \ldots, Q_n\}$ is an alphabet of *query symbols*, with N, T, and K being pairwise disjoint. (P_i, ω_i), $1 \leq i \leq n$, is called a *component* of Γ. It refers to a usual Chomsky grammar or an $E0L$ system with nonterminal alphabet $N \cup K$, terminal alphabet T, set of rewriting rules P_i, and *axiom* $\omega_i \in (N \cup T)^*$, with $\omega_i = S_i \in N$ in the case of Chomsky grammars. (P_1, ω_1) is said to be the *master* of Γ. An n-tuple $(x_1, \ldots, x_n)$, where $x_i \in (N \cup T \cup K)^*$, $1 \leq i \leq n$, is called a *configuration* of Γ. $(\omega_1, \ldots, \omega_n)$ is said to be the *initial configuration*.

For PC grammar systems, the notion of a direct derivation step with *prefix* communication was introduced in [7]. We extend the notion to *subword* communication (*suffix* communication), both for PC grammar systems and PC $E0L$ systems.

Definition 1. *Let Γ be a PC grammar system (a PC $E0L$ system) as above and let $(x_1, \ldots, x_n)$, $(y_1, \ldots, y_n)$ be two configurations of Γ. We say that $(x_1, \ldots, x_n)$ directly derives $(y_1, \ldots, y_n)$ with subword communication, denoted by $(x_1, \ldots, x_n) \Rightarrow_{sub} (y_1, \ldots, y_n)$, if one of the following two cases holds:*

1. For each i, $1 \leq i \leq n$, $x_i \in (N \cup T)^*$. Then, y_i is obtained from x_i by a derivation step according to the ith component, that is, $x_i \Rightarrow_{P_i} y_i$, $1 \leq i \leq n$. In case of Chomsky grammars $x_i = y_i$ for $x_i \in T^*$, $1 \leq i \leq n$.

2. There is some x_i, $1 \leq i \leq n$, with $|x_i|_K \neq 0$. Then for each x_i, $1 \leq i \leq n$, with $|x_i|_K \neq 0$ we write $x_i = z_1 Q_{i_1} z_2 Q_{i_2} \ldots z_t Q_{i_t} z_{t+1}$, where $z_j \in (N \cup T)^*$, $1 \leq j \leq t+1$, and $Q_{i_l} \in K$, $1 \leq l \leq t$, and for each x_j, $1 \leq j \leq n$, with $|x_j|_K = 0$ we write $x_j = \alpha_j x'_j \beta_j$, where $\alpha_j \beta_j \in (N \cup T)^*$ and $x'_j = \varepsilon$ if and only if $x_j = \varepsilon$.

Then y_i is obtained from x_i, $1 \leq i \leq n$, as follows: If $|x_i|_K \neq 0$, where $x_i = z_1 Q_{i_1} z_2 Q_{i_2} \ldots z_t Q_{i_t} z_{t+1}$ and $|x_{i_l}|_K = 0$ for each i_l, $1 \leq l \leq t$, then $y_i = z_1 x'_{i_1} z_2 x'_{i_2} \ldots z_t x'_{i_t} z_{t+1}$, where x'_{i_l} is the subword of x_{i_l} defined above. In returning systems, $y_{i_l} = \alpha_{i_l} \beta_{i_l}$ for $\alpha_{i_l} \beta_{i_l} \neq \varepsilon$ and $y_{i_l} = \omega_{i_l}$ for $\alpha_{i_l} = \beta_{i_l} = \varepsilon$. In non-returning systems, $y_{i_l} = x_{i_l}$. For all j, $1 \leq j \leq n$, for which y_j is not specified above, $y_j = x_j$.

We say that $(x_1, \ldots, x_n)$ directly derives $(y_1, \ldots, y_n)$ with suffix communication (prefix communication), denoted by $\Rightarrow_{suf}$ ($\Rightarrow_{pref}$), if in the above definition $x_{i_l} = \alpha_{i_l} x'_{i_l}$ ($x_{i_l} = x'_{i_l} \beta_{i_l}$) is required. Let $\Rightarrow^*_X$ denote the reflexive and transitive closure of $\Rightarrow_X$, $X \in \{pref, suf, sub\}$.

The language generated by a parallel communicating grammar system Γ, as above, with incomplete information communication is $L_X(\Gamma) = \{\alpha_1 \in T^* \mid (\omega_1, \ldots, \omega_n) \Rightarrow^*_X (\alpha_1, \ldots, \alpha_n)\}$, where $X \in \{pref, suf, sub\}$ and (P_1, ω_1) is the master of Γ.

Let the class of languages generated by returning or non-returning PC grammar systems with incomplete information communication having n components of type $Y \in \{E0L, CF\}$ be denoted by $\mathcal{L}_X(PC_n Y)$ and $\mathcal{L}_X(NPC_n Y)$, $X \in \{pref, suf, sub\}$, respectively.

To prove our results, we simulate the generation of languages of so-called Extended Post Correspondences in PC grammar systems (PC $E0L$ systems) with subword (prefix, suffix) communication.

Let $T = \{a_1, \ldots, a_n\}$, $1 \leq n$, be an alphabet. An *Extended Post Correspondence* (an EPC in short) is a pair $P = (\{(u_1, v_1), \ldots, (u_m, v_m)\}, (z_{a_1}, \ldots, z_{a_n}))$, where $u_i, v_i, z_{a_j} \in \{0,1\}^*$, $1 \leq i \leq m$, $1 \leq j \leq n$. The *language represented by* P is $L(P) = \{x_1 \ldots x_r \in T^* \mid \text{there are } i_1, \ldots i_s \in \{1, \ldots, m\}, s \geq 1, \text{ such that } v_{i_1} \ldots v_{i_s} = u_{i_1} \ldots u_{i_s} z_{x_1} \ldots z_{x_r}\}$.

It is known (see [4]) that for each recursively enumerable language L there exists an EPC P such that $L(P) = L$. Notice that the statement remains true if words u_i, v_i, z_{a_j},

$1 \leq i \leq m$, $1 \leq j \leq n$ are defined over alphabet $\{1,2\}$. We shall use this modified version of the EPC in the sequel to be able to express the equality of the strings on the two sides of the equation by the equality of two integers represented in base three notation. For a string $u \in \{1,2\}^*$, we denote by $val(u)$ the value of the integer represented by u in base three notation.

3 Results

In this section we show how the class of recursively enumerable languages can be described by PC grammar systems with components communicating incomplete information.

Theorem 1. $\mathcal{L}_X(PC_{13}CF) = \mathcal{L}(RE)$, where $X \in \{sub, pref, suf\}$.

Proof sketch. Let L be a recursively enumerable language over an alphabet $T = \{a_1, \ldots, a_n\}$, and let P be a modified EPC as above with $L = L(P)$. We first construct a context-free returning PC grammar system Γ which generates $L(P)$ with subword communication, based on the following considerations. For any word $w = x_1 \ldots x_s$, $x_i \in T$, $1 \leq i \leq s$, or $w = \varepsilon$ we can decide whether or not $w \in L(P)$ as follows: we build two strings u and v of the form $u = u_{i_1} \ldots u_{i_r} z_{x_1} \ldots z_{x_s}$ and $v = v_{i_1} \ldots v_{i_r}$, $r \geq 1$, and check whether or not $u = v$. The constructed PC grammar system will simulate this procedure. However, instead of building u and v, Γ will build two strings $\alpha = A^t$ and $\beta = A^p$, where A is a symbol and t and p are the integer values of $\bar{u} = 1u$ and $\bar{v} = 1v$ according to the base three notation, respectively, and then Γ will decide whether or not $t = p$ holds. The reader can easily notice that the equality of t and p holds if and only if the represented strings u and v are equal.

Let

$$\Gamma = (N, K, T, (P^M, S), (P^{M1}, S), (P^{sel}, S), (P^{ini}, S), (P^{sel1}, \bar{E}),$$
$$(P^{st1}, S), (P^{st2}, S), (P^{exe1}, S), (P^{exe2}, S),$$
$$(P^{st3}, S_3), (P^{st4}, S_4), (P^{ch1}, S), (P^{ch2}, S)),$$

where (P^M, S) is the master, the terminal alphabet T is as above, the query symbols of K are denoted by the symbol Q with the appropriate superscript, and all other symbols belong to the nonterminals of N. In the following we will speak about the components of Γ without mentioning their startsymbols. We define the components as follows. Let

$$P^M = \{S \to S^+, S^+ \to Q^{sel}, E \to E', E' \to Q^{sel1}\}$$
$$\cup \{(i) \to (i)', (i)' \to (i)'', (i)'' \to Q^{sel} \mid 1 \leq i \leq m\}$$
$$\cup \{[j] \to [j]', [j]' \to a_j[j]'', [j]'' \to Q^{sel} \mid 1 \leq j \leq n\}$$
$$\cup \{F \to Q^{st3}Q^{st4}Q^{ch1}Q^{ch2}, C \to \varepsilon, \bar{E} \to \bar{E}', \bar{E}' \to Q^{sel1}\},$$

$$P^{M1} = \{S \to S^+, S^+ \to Q^{sel}, E \to E', E' \to Q^{sel1}\}$$
$$\cup \{(i) \to (i)', (i)' \to (i)'', (i)'' \to Q^{sel} \mid 1 \leq i \leq m\}$$
$$\cup \{[j] \to [j]', [j]' \to [j]'', [j]'' \to Q^{sel} \mid 1 \leq j \leq n\}$$
$$\cup \{F \to F_1, F_1 \to F_2, F_2 \to F_3, \bar{E} \to \bar{E}', \bar{E}' \to Q^{sel1}\}.$$

P^M builds the terminal word $w = x_1 \ldots x_s$ (or $w = \varepsilon$), and finishes the checking of the equality of the strings representing u and v constructed for w. Terminal words can only be obtained at P^M if the equality holds. In this work, the master is assisted by component P^{M1}. Let

$$P^{sel} = \{S \to S', S \to Q^{ini}, E' \to E'', E'' \to E\}$$
$$\cup \{S' \to (i_1), (i_1)' \to (i_1)'', (i_1)'' \to (i_2), (i_1)'' \to [j_1],$$
$$[j_1] \to [j_1]', [j_1]' \to [j_1]'', [j_1]'' \to [j_2],$$
$$(i_1)'' \to E, [j_1]'' \to E \mid 1 \leq i_1, i_2 \leq m, 1 \leq j_1, j_2 \leq n\}.$$

This component selects a pair (u_i, v_i) or (z_{a_j}, a_j) of P, appending of which is simulated when the representation of the two strings u and v, above, are built, or P^{sel} indicates that simulation of the checking phase of their equality has started. The following two components assist the scheduling: P^{ini} stores the information concerning the choice made by P^{sel}, and P^{sel1} controls the checking procedure of the equality of u and v. Thus, let

$$P^{ini} = \{S \to S^+, S^+ \to Q^{sel}, E \to E'\}$$
$$\cup \{(i) \to (i)', [j] \to [j]' \mid 1 \leq i \leq m, 1 \leq j \leq n\},$$

$$P^{sel1} = \{\bar{E} \to \bar{E}, \bar{E} \to F\}.$$

The following four components build the base three representations of the string pair u, v constructed for $w = x_1 \ldots x_s$ (or $w = \varepsilon$). P^{st1} and P^{st2} store the base three value representations of the strings that had already been built, while P^{exe1} and P^{exe2} simulate the effect of appending a new pair of P to these string pairs. For $Y = 1, 2$ let

$$P^{stY} = \{S \to Q^{exeY}D, D \to D', D' \to \varepsilon, E \to E\},$$

$$P^{exe1} = \{S \to Q^{sel}, S \to A, D \to \varepsilon, E \to E\}$$
$$\cup \{(i) \to (Q^{st1})^{3^{lng(u_i)}} A^{val(u_i)} D \mid 1 \leq i \leq m\}$$
$$\cup \{[j] \to (Q^{st1})^{3^{lng(z_{a_j})}} A^{val(z_{a_j})} D \mid 1 \leq j \leq n\},$$

$$P^{exe2} = \{S \to Q^{sel}, S \to A, D \to \varepsilon, E \to E\}$$
$$\cup \{(i) \to (Q^{st2})^{3^{lng(v_i)}} A^{val(v_i)} D \mid 1 \leq i \leq m\}$$
$$\cup \{[j] \to Q^{st2}D \mid 1 \leq j \leq n\}.$$

The following four components decide whether or not $u = v$ holds. After being generated, the representations of u and v are forwarded to P^{st3} and P^{st4}, respectively. Then, in parallel, P^{ch1} and P^{ch2} eliminate the symbols in these sentential forms by repeatedly requesting subwords of these strings. The generation of $x_1 \ldots x_s$ correctly terminates at the master if and only if in each communication the communicated subword is a single symbol B and the same number of communications is needed to eliminate all symbols in both sentential forms. For $Y = 3, 4$ and $Z = 1, 2$ let

$$P^{stY} = \{S_Y \to S^+, S^+ \to Q^{sel}, S^+ \to C, E \to Q^{st(Y-2)}B, A \to A',$$
$$A' \to B\} \cup \{(i) \to (i)', (i)' \to (i)'', (i)'' \to Q^{sel} \mid 1 \leq i \leq m\}$$
$$\cup \{[j] \to [j]', [j]' \to [j]'', [j]'' \to Q^{sel} \mid 1 \leq j \leq n\},$$

$$P^{chZ} = \{S \to \bar{S}, \bar{S} \to Q^{sel}, E \to Q^{st(Z+2)}, \bar{E} \to Q^{st(Z+2)}, B \to Q^{sel1},$$
$$F \to \varepsilon\} \cup \{(i) \to (i)', (i)' \to (i)'', (i)'' \to Q^{sel} \mid 1 \leq i \leq m\}$$
$$\cup \{[j] \to [j]', [j]' \to [j]'', [j]'' \to Q^{sel} \mid 1 \leq j \leq n\}.$$

Now we briefly explain how Γ functions. Starting from the initial configuration, Γ in a few steps enters a configuration indicating that appending of the ith pair of P to two strings u' and v' both represented by A will be simulated, $((i), (i), S, (i), \bar{E}, AD', AD', (i), (i), (i), (i), (i), (i))$. Then we obtain the configuration $((i)'', (i)'', (i)'', S^+, \bar{E}, A^k D, A^l D, S, S, (i)'', (i)'', (i)'', (i)'')$, with $k = val(1u_i)$ and $l = val(1v_i)$, which means that the simulation of adding the ith pair of P to u' and v' has been performed and Γ is ready to perform the next simulation step. Then $(Q^{sel}, Q^{sel}, X, Q^{sel}, \bar{E}, A^k D, A^l D, Q^{sel}, Q^{sel}, Q^{sel}, Q^{sel}, Q^{sel}, Q^{sel})$ is the next possible configuration, where X denotes either (i), $1 \leq i \leq m$, or $[j]$, $1 \leq j \leq n$, or E. If $X = (i)$ or $X = [j]$, then the above procedure is repeated to simulate the appending of the chosen pair of P to the representing strings. If the chosen nonterminal is $[j]$ for some j, $1 \leq j \leq n$, then the letter a_j is appended to the sentential form of P^M. The correct order of appending the pairs is guaranteed by the definition of P^{sel}. Although Γ uses subword communication, in this phase only complete sentential forms can be communicated, otherwise the derivation would be blocked at the next step with abnormal termination. Moreover, the choice of other configurations during the above simulation procedure would lead to blocking situations without obtaining a terminal word.

If letter E is chosen by P^{sel}, then checking of the equality of t and p will start, where $A^t D'$ and $A^p D'$ are obtained at P^{st1} and P^{st2}. After a few steps, A^t and A^p are forwarded to P^{st3} and P^{st4}, and then a subword with an occurrence of B of the sentential form $A^t B$ and $A^p B$ found at P^{st3} and P^{st4}, respectively, is sent to P^{ch1} and P^{ch2}. In the next steps P^{st3}, P^{st4} and P^{ch1}, P^{ch2} eliminate symbols A from the sentential forms of P^{st3} and P^{st4}. Components P^{st3} and P^{st4} repeatedly rewrite one occurrence of A to B that is forwarded to P^{ch1} and P^{ch2}. This procedure is controlled by component P^{sel1}. The other components in this phase work without interfering with this part of the system. Let symbol F, indicating that the derivation must end in a few steps, be introduced by P^{sel1} when the sentential forms of P^{st3} and P^{st4} are equal to B. Then, after a few steps, the sentential forms of these components and of P^{ch1} and P^{ch2} are forwarded to the master component which has to finish the derivation after two steps, with the assistance of P^{M1}. P^M can produce a terminal word if and only if the obtained sentential forms of P^{st3}, P^{st4}, P^{ch1}, and P^{ch2}, is C, C, ε, and ε, respectively. This is possible if and only if P^{st3} and P^{st4}, when eliminating As, send at each communication step exactly one symbol B to P^{ch1} and P^{ch2}, and their strings reduce to a single B at the same derivation step, that is, $t = p$ holds. From this explanation we can see that any terminal word of Γ satisfies EPC P, and for any $w \in L(P)$ a terminating derivation in Γ can be easily determined. The reverse inclusion, $\mathcal{L}_{sub}(PC_{13}CF) \subseteq \mathcal{L}(RE)$, can be proved by standard simulation techniques. By the above construction, the result follows also for systems with suffix communication. The proof for the case of prefix communication can be obtained with the obvious modifications of the rule sets P^{st3} and P^{st4}. □

Theorem 2. $\mathcal{L}_X(PC_{10}E0L) = \mathcal{L}(RE)$, $X \in \{pref, suf\}$.

Proof sketch. The proof is based on a construction similar to that of the proof of Theorem 1., therefore we only indicate the differences. The PC E0L system Γ designed for simulating the generation of a language L defined by a modified EPC P is as follows:

$$\Gamma = (N, K, T, (P^M, S), (P^{M1}, S), (P^{sel}, S), (P^{ini}, S), (P^{ini1}, S), (P^{sel1}, S),$$
$$(P^{exe1}, S), (P^{exe2}, S), (P^{ch1}, S), (P^{ch2}, S)).$$

As above, we mention the components only by their production sets; if for a symbol no explicit rule is given, then we assume the presence of an identical rule for terminals and a self-query rule for nonterminals. The generation of strings representing u and v, above, is realized by only two components, P^{exe1} and P^{exe2}, because the multiplication of the number of nonterminals can be achieved by parallel rewriting. These are

$$P^{exe1} = \{S \to AD, A \to Q^{sel}, D \to D', D' \to Q^{ini1}, D'' \to D',$$
$$A' \to A'', A'' \to Q^{sel}\} \cup$$
$$\{(i) \to (A')^{3^{lng(u_i)}}, (i)' \to (A'')^{val(u_i)} D'', [j] \to (A')^{3^{lng(z_{a_j})}},$$
$$[j]' \to (A'')^{val(z_{a_j})} D'' \mid 1 \le i \le m, 1 \le j \le n\} \cup$$
$$\{E \to BC, B \to B, C \to C, E' \to E'', E'' \to E''\},$$

$$P^{exe2} = \{S \to AD, A \to Q^{sel}, D \to D', D' \to Q^{ini1}, D'' \to D',$$
$$A' \to A'', A'' \to Q^{sel}\} \cup$$
$$\{(i) \to (A')^{3^{lng(v_i)}}, (i)' \to (A'')^{val(v_i)} D'', [j] \to A',$$
$$[j]' \to D'' \mid 1 \le i \le m, 1 \le j \le n\} \cup$$
$$\{E \to BC, B \to B, C \to C, E' \to E'', E'' \to E''\}.$$

The checking of the equality of the strings representing $1u$ and $1v$ is done with two components, P^{ch1} and P^{ch2}, by modifying the corresponding production sets of the construction of the proof of Theorem 1. as follows: we replace productions in $\{E \to Q^{stX}, A \to Q^{sel1}, F \to \varepsilon \mid X = 3, 4\}$ with $E \to E', E' \to E', E' \to Q^{exe1} Q^{exe2} E'$, and add rule $B \to \varepsilon$ to P^{ch1} and rule $C \to \varepsilon$ to P^{ch2}. The correctness of the simulation is checked by the master grammar obtained from the master of the construction of the previous proof if we replace productions $F \to Q^{st3} Q^{st4} Q^{ch1} Q^{ch2}, C \to \varepsilon$ with productions $F \to Q^{exe1} Q^{exe2}, E'' \to \varepsilon$. The rules of the additional component P^{ini1} are the rules of P^{ini} and production $E' \to E'$.

To obtain a system which generates L with suffix communication, it is sufficient to replace the right hand sides of the rules of components other than P^M with their mirror image. □

Theorem 3. $\mathcal{L}_X(NPC_{18}CF) = \mathcal{L}(RE), X \in \{pref, suf\}$.

Proof sketch. Let $T = \{a_1, \ldots a_n\}$ be an alphabet and $L \in \mathcal{L}(RE), L \subseteq T^*$ be a language with $L = L(P)$ for an EPC P as above. Here we define the strings of P over alphabet $\{0, 1\}$ instead of $\{1, 2\}$, as in the previous proofs, because we will check the equality of strings, not integer values. We construct a non-returning PC grammar system Γ with context-free components which generates L with prefix communication.

Now, let

$$\Gamma = (N, K, T, (P^M, S), (P^{s1}, S), (P^v, S), (P^u, S), (P^{s2}, S),\\
(P^{rv}, S), (P^{va}, S), (P^{ru}, S), (P^{ua}, S), (P^{0v}, S), (P^{0u}, S),\\
(P^{1v}, S), (P^{1u}, S), (P^{01v}, S), (P^{01u}, S), (P^{s3}, S), (P^{cu}, S), (P^{cv}, S)),$$

where (P^M, S) is the master, the terminal alphabet T is as above, the query symbols of K are denoted by the symbol Q with the appropriate superscript, and all other symbols belong to the set of nonterminals of N.

In the following, we denote components of the system by their rule sets only. The rules of the components are as follows. The first group of components P^M, P^{s1}, P^v, P^u generates strings $u = u_{i_1} \ldots u_{i_r} z_{x_1} \ldots z_{x_s}$ and $v = v_{i_1} \ldots v_{i_r}, r \geq 1, s \geq 0$, according to EPC P. The sentential forms of P^v and P^u represent v and u, the sentential form of P^M represents the corresponding terminal word $x_1 \ldots x_s$. The representation of u and v, respectively, is not based on their value in base three notation, a sequence of symbols over the alphabet $\{0, 1\}$ is represented by the same sequence of nonterminals $0, 1 \in N$. Let

$$P^M = \{S \to Q^{s1}, (i) \to Q^{s1}, [j] \to a_j Q^{s1} \mid 1 \leq i \leq m, 1 \leq j \leq n\}\\
\cup \{F \to Q^{s2}, 0 \to 0', 0' \to Q^{s2}, 1 \to 1', 1' \to Q^{s2},\\
E \to E_1, E_1 \to Q^{cu} Q^{cv} Q^{0v} Q^{0u} Q^{1v} Q^{1u} Q^{01v} Q^{01u}\}.$$

This is the master component. Its sentential form contains the terminal word corresponding to the equation, which will be generated by the system if the strings u and v are equal. Let

$$P^{s1} = \{S \to (i), (i_1) \to (i_2), (i) \to [j], [j_1] \to [j_2], (i) \to F,\\
[j] \to F, F \to F \mid 1 \leq i, i_1, i_2 \leq m, 1 \leq j, j_1, j_2 \leq n\}.$$

This component selects the pair of strings (u_i, v_i) or the string z_{a_j} to be appended to the representation of u and v or indicates that the equality of these strings will be checked. Let

$$P^v = \{S \to Q^{s1}, (i) \to v_i Q^{s1}, [j] \to Q^{s1} \mid 1 \leq i \leq m, 1 \leq j \leq n\}\\
\cup \{F \to C, C \to C\},$$

$$P^u = \{S \to Q^{s1}, (i) \to u_i Q^{s1}, [j] \to z_{a_j} Q^{s1} \mid 1 \leq i \leq m,\\
1 \leq j \leq n\} \cup \{F \to C, C \to C\}.$$

These components generate the representations of u and v by appending u_i, v_i, or z_{a_j} to the representing strings. When their work is finished, the system checks whether $u = v$ holds or not. If the equality holds, the corresponding terminal word is generated by P^M. First P^{s2} guesses the leftmost symbol of u and v. Let

$$P^{s2} = \{S \to Q^{s1}, (i) \to Q^{s1}, [j] \to Q^{s1} \mid 1 \leq i \leq m, 1 \leq j \leq n\}\\
\cup \{F \to 0, F \to 1, 0 \to 0, 0 \to 1, 1 \to 0,\\
1 \to 1, 0 \to E, 1 \to E, E \to E\}.$$

The next two components P^{rv} and P^{ru} receive the strings representing u and v with the assistance of P^{va} and P^{ua}, and repeatedly delete their leftmost symbols. The equality

$u = v$ holds if and only if the deleted symbols are always the same and the deletion process ends in the same step in both components. For $X \in \{u, v\}$, let

$$P^{rX} = \{S \to Q^{s1}, (i) \to Q^{s1}, [j] \to Q^{s1} \mid 1 \le i \le m,\ 1 \le j \le n\}$$
$$\cup\ \{F \to Q^X, 0 \to \bar{0}, \bar{0} \to \varepsilon, 1 \to \bar{1}, \bar{1} \to \varepsilon, C \to C_1, C_1 \to C_2\},$$

$$P^{Xa} = \{S \to Q^{s1}, (i) \to Q^{s1}, [j] \to Q^{s1} \mid 1 \le i \le m,\ 1 \le j \le n\}$$
$$\cup\ \{F \to Q^X, C \to C\}.$$

The next four components make sure that the deleted symbols were really the leftmost and that they were the same as guessed by component P^{s2}. For $X \in \{u, v\}$, let

$$P^{0X} = \{S \to Q^{s1}, (i) \to Q^{s1}, [j] \to Q^{s1} \mid 1 \le i \le m,\ 1 \le j \le n\}$$
$$\cup\ \{F \to Q^{s2}, 0 \to Q^{rX}, \bar{0} \to Q^{s2}, 1 \to 1', 1' \to Q^{s2},$$
$$E \to \varepsilon\},$$

$$P^{1X} = \{S \to Q^{s1}, (i) \to Q^{s1}, [j] \to Q^{s1} \mid 1 \le i \le m,\ 1 \le j \le n\}$$
$$\cup\ \{F \to Q^{s2}, 0 \to 0', 0' \to Q^{s2}, 1 \to Q^{rX}, \bar{1} \to Q^{s2},$$
$$E \to \varepsilon\}.$$

Components P^{01u} and P^{01v} make sure that the symbols selected for deletion were really deleted. They are assisted by P^{s3}. For $X \in \{u, v\}$, let

$$P^{s3} = \{S \to Q^{s1}, (i) \to Q^{s1}, [j] \to Q^{s1} \mid 1 \le i \le m,\ 1 \le j \le n\}$$
$$\cup\ \{F \to \bar{F}, \bar{F} \to F^2, F^2 \to \bar{F}\}.$$

$$P^{01X} = \{S \to Q^{s1}, (i) \to Q^{s1}, [j] \to Q^{s1} \mid 1 \le i \le m,\ 1 \le j \le n\}$$
$$\cup\ \{F \to F^1, F^1 \to F^2, F^2 \to Q^{rX}, 0 \to Q^{s3}, 1 \to Q^{s3},$$
$$C \to C_1, C_1 \to \varepsilon\},$$

Finally, the next two components make sure that the same number of steps is necessary for deleting the symbols of the strings representing u and v. Let for $X \in \{u, v\}$

$$P^{cX} = \{S \to Q^{s1}, (i) \to Q^{s1}, [j] \to Q^{s1} \mid 1 \le i \le m,\ 1 \le j \le n\}$$
$$\cup\ \{F \to Q^{s2}, 0 \to Q^{s2}, 1 \to Q^{s2}, E \to Q^{rX}, C_1 \to \varepsilon\}.$$

The system works in the following way. Starting from the initial configuration, after a rewriting and a communication step, all components receive the nonterminal selected by P^{s1}, say (i). This indicates, that the building of the representations of u and v starts with the pair (u_i, v_i). After the next rewriting step the sentential forms of P^v and P^u are $v_i Q^{s1}$ and $u_i Q^{s1}$, respectively. Component P^{s1} has already chosen the index of the next pair, say (j), and the other components have Q^{s1} as sentential forms. In the rewriting step after the communication, elements of the pair (u_j, v_j) will be added to the representation. When P^{s1} introduces a symbol $[i]$, $1 \le i \le n$, then the elements of the pair (z_{a_i}, a_i) will be added to the sentential forms of P^u and P^M, respectively, generating the terminal string $x_1 \ldots x_s$ at the master component.

If P^{s1} selects the nonterminal F, the representations of u and v are complete. Now, their equality must be checked. After one rewriting step, components P^v and P^u have

vC and uC as their sentential forms which are communicated to P^{rv} and P^{ru}, the communication is assisted by P^{va}, P^{ua}, and P^{s2} introduces the nonterminal 0 or 1, guessing the leftmost symbol of u and v. Let us assume, that it has selected 0, in which case components P^{0v} and P^{0u} start to work. After a rewriting step, they query components P^{rv} and P^{ru}, while these rewrite one of their symbols to $\bar{0}$ or $\bar{1}$. If the prefixes received by P^{0v} and P^{0u} do not contain $\bar{0}$, the system is blocked. If they contain any other symbol, then these other symbols can never be rewritten. This means, that if the sentential forms of P^{0v} and P^{0u} do not contain any non-barred symbols and the system is not blocked, than the leftmost symbols of both u and v were 0.

Now, the barred nonterminals must be erased from the sentential forms of P^{rv} and P^{ru}. This is checked by components P^{01v} and P^{01u} with the assistance of P^{s3} when after the next rewriting step, they receive prefixes of the sentential forms of P^{rv} and P^{ru}. If the barred letters, which were the leftmost symbols of these strings, are not erased, then the system is either blocked, or at least one of P^{01v} or P^{01u} contains a barred symbol in its sentential form.

The work of the system can continue in similar cycles with the simultaneous deletion of symbols of u and v, until P^{s2} introduces the nonterminal E, which indicates that the checking phase will be finished in a few steps. In these steps, P^{cu} and P^{cv} make sure that the symbols of strings representing u and v were deleted in the same number of steps and P^M, by introducing a query, requests sentential forms of P^{cu}, P^{cv}, P^{0v}, P^{0u}, P^{1v}, P^{1u}, P^{01v}, and P^{01u}. At this point, if all these sentential forms are empty, then the equation $v = u$ holds, that is, $w = x_{i_1} \ldots x_{i_s} \in L$ and the sentential form of P^M will be equal to w. If $v = u$ does not hold, then one or more of the strings above must be non-empty which would lead to the blocking of the system without generating a terminal word. Thus, we proved the inclusion. The reverse inclusion, $\mathcal{L}_{pref}(NPC_{18}CF) \subseteq \mathcal{L}(RE)$, can be proved by standard simulation techniques. To obtain a system generating L with suffix communication, it is sufficient to replace the right hand sides of the rules from the rule sets other than P^M with their mirror image. □

References

1. E. Csuhaj-Varjú, J. Dassow, J. Kelemen, Gh. Păun, *Grammar Systems. A Grammatical Approach to Distribution and Cooperation,* Gordon and Breach, Yverdon, 1994.
2. E. Csuhaj-Varjú, Gy. Vaszil, On the computational completeness of context-free parallel communicating grammar systems. *Theoretical Computer Science,* 215 (1999), 349-358.
3. J. Dassow, Gh. Păun, G. Rozenberg, Grammar Systems. *Handbook of Formal Languages,* Volume II, Chapter 4, ed. by G. Rozenberg and A. Salomaa, Springer, Berlin, 1997, 155-213.
4. V. Geffert, Context-free-like forms for phrase structure grammars. In: Proc. MFCS'88, LNCS 324, Springer-Verlag, Berlin, 1988, 309-317.
5. *Handbook of Formal Languages,* ed. by G. Rozenberg and A. Salomaa, Springer, Berlin, 1997.
6. N. Mandache, On the computational power of context-free PC grammar systems. *Theoretical Computer Science,* 237 (2000), 135-148.
7. Gh. Păun, Parallel communicating grammar systems: Recent results, open problems. *Acta Cybernetica,* 12 (1996), 381-395.
8. Gh. Păun, L. Sântean, Parallel communicating grammar systems: the regular case, *Ann. Univ. Bucharest, Ser. Matem.-Inform.* 38, 2 (1989), 55-63.

Eliminating Communication by Parallel Rewriting[*]

Branislav Rovan and Marián Slašt'an

Department of Computer Science,
Comenius University
Mlynská Dolina
842 48 Bratislava
Slovakia
{rovan, slastan}@dcs.fmph.uniba.sk

Abstract. We shall show that simple communication can be substituted by parallel rewriting and nondeterminism without a time penalty. This is no longer true for more complex communication. In particular, we shall show that time preserving simulation of regular PCGS by g-systems is possible whereas time preserving simulation of context-free PCGS is impossible.

1 Introduction

Parallelism has been studied in various settings for many years. First grammar like models were introduced in the late sixties (L-Systems). The early models studied various forms of parallel application of rewriting rules. Several general frameworks for studying this type of parallelism were introduced. The one we shall use in this paper is that of generative systems (g-systems) introduced in [2]. In 1989 a model of parallelism enabling to study communication, parallel communicating grammar systems (PCGS), was introduced in [5]. Various aspects of communication were studied, e.g. the number of communicating components, the communication structure, etc. An extensive bibliography of PCGS-systems can be found in [6]. In the present paper we shall investigate the possibility of eliminating communication by using parallel rewriting, i.e., the possibility to simulate PCGS by g-systems. We shall be interested in time preserving simulations.

We shall show that simple communication (regular PCGS) can be substituted by the massive nondeterminism of g-systems. We shall also show that for more complex communication this is no longer possible. In particular we shall show that time preserving simulation of context-free PCGS by g-systems is not possible. Due to the space constraint we cannot present details of the proofs in this paper. The full version of this paper will be published elsewhere.

In what follows we shall use standard notions and notation of formal languages theory. We refer to [1] for a formal definition of PCGS and the notation used. We only give semi-formal definitions here.

A parallel communicating grammar system of degree r, $r \geq 1$ is an $r + 3$-tuple $\Gamma = (N, Q, T, G_1, \ldots, G_r)$, where N, T, Q are mutually disjoint alphabets of nonterminal, terminal and query symbols resp., where $Q = \{Q_1, \ldots, Q_r\}$, and $G_i = (N \cup Q, T, P_i, \sigma_i), 1 \leq i \leq r$, are usual Chomsky grammars.

---
[*] This research was supported in part by the grant VEGA 1/7155/20.

The r grammars of the PCGS Γ cooperate in producing a word of the language $L(\Gamma)$ defined by Γ as follows. Γ operates on an r-sentential form $(X_1, \ldots, X_r)$ and each G_i operates on its own sentential form X_i, initially σ_i. In one *rewriting step* of Γ, each component performs the usual (Chomsky) type rewriting step on its own sentential form and terminal forms are just copied (note that the grammars are synchronized). Such rewriting steps are performed until a symbol from Q appears in some of the sentential forms. At this point a communication step of Γ is performed, if possible (the derivation is blocked otherwise). Each grammar of Γ is either receiving, providing, or inactive. The sentential form of a receiving grammar contains at least one query symbol and for each Q_i occurring in its sentential form the sentential form of G_i is in $(N \cup T)^*$. The grammar G_i is providing if its sentential form is in $(N \cup T)^*$ and Q_i occurs in some sentential form. G_i is inactive if it is neither receiving, nor providing. In one *communication step* all Q_i in the sentential forms of receiving grammars are replaced by the sentential form of the corresponding providing grammar G_i, the sentential forms of the providing grammars are reset to their initial symbols σ_i and the sentential forms of the inactive grammars remain unchanged[1].

The derivation stops when G_1 derives a terminal string. $L(\Gamma)$ is the set of terminal strings derived by G_1.

A PCGS is said to be regular, linear, context-free, etc., when the rules of its components are of these types.

In what follows we shall briefly define the notions concerning generative systems. See [2] or [4] for more details and motivation. In essence a g-system is a "grammar" in which the derivation step $\Rightarrow$ is defined via a finite state transduction.

A *one-input finite state transducer with accepting states* (we shall consider one-input transducers only, and call them briefly a-transducer) is a 6-tuple $M = (K, X, Y, H, q_0, F)$, where K is a finite set of states, X and Y are finite alphabets (input and output resp.), q_0 in K is the initial state, $F \subseteq K$ is the set of accepting (final) states, and H is a finite subset of $K \times X \times Y^* \times K$.

By a *computation* of such an a-transducer a word $h_1 \ldots h_n$ in H^* is understood such that (i) $pr_1(h_1) = q_0$, (ii) $pr_4(h_n)$ is in F, and (iii) $pr_1(h_{i+1}) = pr_4(h_i)$ for $1 \leq i \leq n-1$, where pr_i are homomorphisms on H^* defined by $pr_i((x_1, x_2, x_3, x_4)) = x_i$ for $i = 1, 2, 3, 4$. The set of all computations of M is denoted by C_M.

An *a-transducer mapping* is then defined for each language $L \subseteq X^*$ by $M(L) = pr_3(pr_2^{-1}(L) \cap C_M)$. For a word w let $M(w) = M(\{w\})$.

A *generative system* (g-system) is a 4-tuple $G = (N, T, M, S)$, where N and T are finite alphabets of nonterminal and terminal symbols respectively (not necessarily disjoint), S in N is the initial nonterminal symbol, and M is an a-transducer mapping, with $M(w) = \emptyset$ for each w in $(T - N)^*$.

The *language generated by a g-system* $G = (N, T, M, S)$ is the language $L(G) = \{w \in T^* \mid S \Rightarrow_G^* w\}$, where $\Rightarrow_G^*$ is the transitive and reflexive closure of the *derivation step* $\Rightarrow_G$ defined by $u \Rightarrow_G v$ iff v is in $M(u)$.

In what follows we shall be interested in time preserving simulations. The standard notion of time complexity based on the number of derivation steps is used both for PCGS and g-systems.

---
[1] This variant of PCGS is called returning. We shall not consider non-returning PCGS here.

2 Time Preserving Simulation of Regular PCGS

A straightforward simulation of a PCGS Γ, by a g-system G (e.g. via simple encoding of an r-sentential form of Γ in a sentential form of G), clearly cannot be time preserving. [A communication step of Γ may require copying of large parts of the sentential form of G. In a straightforward implementation this requires a number of derivation steps proportional to the length of the communicated string.] We shall show that in the case of regular PCGS a time preserving simulation by g-systems is possible. Thus for regular PCGS the communication can be substituted by nondeterminism (and parallel rewriting) without a time penalty.

The simulation is based on the following two simple observations. Given a derivation of $u \in L(\Gamma)$ in Γ it is easy to see that:

1. Each letter of u was generated by a particular grammar G_i at a particular time (rewriting step) t.
2. A letter generated by a particular grammar G_i in a rewriting step at time t may occur at several places in u (due to simultaneous querying, as illustrated by the PCGS in the example below).

G will achieve a time efficient simulation by generating *in parallel*, in the derivation step t of G, all occurrences of letters in u generated by each of G_i in the derivation step t of Γ. (Note that the "total synchrony" of the derivations in Γ and G is an oversimplification used to elucidate the main idea.) The positions in u where these letters are to be generated by G are nondeterministically guessed (checking the correctness of these guesses is performed during the simulation of the communication step of Γ).

A detailed construction of G appears in the Master Thesis of the second author [7] and will be published in a full version of this paper. We shall describe the simulation informally and subsequently illustrate the construction on a particular example.

After the initialization phase, G shall proceed by performing two types of "macro-steps" - componentwise derivation and communication.

During the componentwise derivation macrostep G selects a rule for each grammar G_i and consequently applies this rule in each G_i-segment of its sentential form (for each i).

In the communication macrostep G checks whether each query symbol Q_i is followed by the requested G_i-segment, and vice versa, each G_i-segment is preceded by Q_i. Since we have regular components there is at most one nonterminal in each G_i-segment. The last step is guessing the next target positions for the communicated components (after the communication, the communicated components are reset to the initial nonterminal σ_i).

Example 1. Let Γ be a parallel communicating grammar system with regular components, such that $\Gamma = (\{\sigma_1, \sigma_2, \sigma_3, \omega_1, \omega_1', \omega_2, \sigma_1'\}, \{Q_1, Q_2, Q_3\}, \{a, b\}, G_1, G_2, G_3)$, where the rules in the grammars G_1, G_2, G_3 are as follows:

G_1	G_2	G_3
$\sigma_1 \to a\sigma_1 \mid b\sigma_1$	$\sigma_2 \to \sigma_2$	$\sigma_3 \to \sigma_3$
$\sigma_1 \to Q_2$	$\sigma_1 \to \omega_1$	$\sigma_1 \to \omega_1'$
$\omega_1 \to Q_3$	$\sigma_1' \to \omega_1$	$\sigma_1' \to \omega_1'$
$\omega_2 \to \sigma_1'$	$\sigma_2 \to Q_1$	$\omega_1' \to \omega_2$
$\sigma_1' \to \epsilon$		$\sigma_3 \to Q_1$

It holds that $L(\Gamma) = \{w^{2^n} \mid w \in \{a,b\}^*, n \geq 1\}$. We leave a formal proof of this assertion to the reader.

Informally, Γ operates in two main phases : *the generation phase* and *the communication phase*.

Generation phase

In this phase the grammar G_1 generates a word $w\sigma_1$, where $w \in \{a,b\}^*$ (using the rules $\sigma_1 \to a\sigma_1 \mid b\sigma_1$). The grammars G_2, G_3 keep copying σ_2, σ_3 respectively. Here is an example of a derivation in Γ during the generation phase

$$(\sigma_1, \sigma_2, \sigma_3) \Rightarrow (a\sigma_1, \sigma_2, \sigma_3) \Rightarrow \ldots \Rightarrow (ababba\sigma_1, \sigma_2, \sigma_3) \Rightarrow \ldots$$

Communication phase

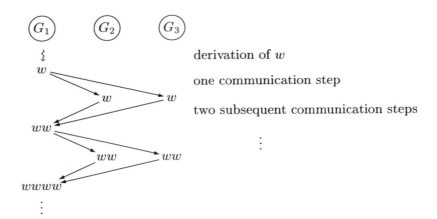

Fig. 1. Sentential form flow among G_1, G_2, G_3

The phase starts, when G_2 and G_3 introduce the Q_1 symbol (note that they have to do it at the same time, otherwise the derivation would halt), i.e. the sentential form of G_1 is concurrently copied to G_2 and G_3. Afterwards, G_1 acquires the sentential forms of G_2 and G_3 in two subsequent communication steps.

Thus, after few steps the grammar G_1 contains two copies of its previous sentential form. Figure 1 illustrates the flow of sentential forms among G_1, G_2 and G_3.

Hence, the derivation proceeds as follows

$$\ldots \Rightarrow (w\sigma_1, Q_1, Q_1) \Rightarrow (\sigma_1, w\sigma_1, w\sigma_1) \Rightarrow^6$$
$$\Rightarrow^6 (w^2 w_2, \sigma_2, \sigma_3) \Rightarrow^6 (w^4 w_2, \sigma_2, \sigma_3) \Rightarrow^6 \ldots \Rightarrow^6 (w^{2^n} w_2, \sigma_2, \sigma_3) \Rightarrow \ldots$$

Simulation by the g-system

For the purpose of this example we shall consider a "simplified" version of the g-system G simulating Γ, the detailed construction needed for the proof of Theorem 1 is more involved.

Consider the following terminal derivation in Γ

$$(\sigma_1, \sigma_2, \sigma_3) \Rightarrow (a\sigma_1, \sigma_2, \sigma_3) \Rightarrow (ab\sigma_1, Q_1, Q_1) \Rightarrow$$
$$\Rightarrow (\sigma_1, ab\sigma_1, ab\sigma_1) \Rightarrow (Q_2, abw_1, abw_1') \Rightarrow (abw_1, \sigma_2, abw_1') \Rightarrow$$
$$\Rightarrow (abQ_3, \sigma_2, abw_2) \Rightarrow (ababw_2, \sigma_2, \sigma_3) \Rightarrow (ababo_1', \sigma_2, \sigma_3) \Rightarrow$$
$$\Rightarrow (abab, \sigma_2, \sigma_3)$$

At the beginning of the simulation G generates the "space" required for subsequent guesses (because the a-transducer cannot rewrite the empty symbol)

$$S \Rightarrow^* ♮♮♮♮♮♮ \Rightarrow$$

In the next step, G guesses the positions of all grammars. In our case there are exactly two occurrences of ab in the result, i.e. G has to guess the two positions for G_1 and one for G_2 and G_3. Note that to maintain the consistency, the positions of G_2 and G_3 must be followed by G_1, due to the subsequent communication. The guessing

$$\Rightarrow ♮\$_2 \sigma_2 \$_1 \sigma_1 \$_3 \sigma_3 \$_1 \sigma_1 ♮♮ \Rightarrow$$

is followed by the application of the respective derivation rules in each component until the next communication step

$$\Rightarrow ♮\$_2 \sigma_2 \$_1 a\sigma_1 \$_3 \sigma_3 \$_1 a\sigma_1 ♮♮ \Rightarrow ♮\$_2 Q_1 \$_1 ab\sigma_1 \$_3 Q_1 \$_1 ab\sigma_1 ♮♮ \Rightarrow$$

The simulation of a communication step is very simple, because we do not have to copy strings, we just need to remove the auxiliary symbols "$Q_1\$_1$" and the two sentential forms (the requesting and the requested) are merged, i.e.

$$\Rightarrow ♮\$_2 ab\sigma_1 \$_3 ab\sigma_1 ♮♮ \Rightarrow$$

Afterwards we have to guess the new positions of communicated grammars (G_1)

$$\Rightarrow \$_1 \sigma_1 \$_2 ab\sigma_1 \$_3 ab\sigma_1 ♮♮ \Rightarrow$$

and repeat the previous steps until the end

$$\$_1 Q_2 \$_2 abw_1 \$_3 abw_1' ♮♮ \Rightarrow^2 \$_1 abw_1 \$_3 abw_1' \$_2 \sigma_2 ♮ \Rightarrow \$_1 abQ_3 \$_3 abw_2 \$_2 \sigma_2 ♮ \Rightarrow^2$$
$$\$_1 ababw_2 \$_2 \sigma_2 \$_3 \sigma_3 \Rightarrow \$_1 ababo_1' \$_2 \sigma_2 \$_3 \sigma_3 \Rightarrow \$_1 abab \$_2 \sigma_2 \$_3 \sigma_3 \Rightarrow abab$$

We have thus illustrated the main ideas of the proof of the following theorem.

Theorem 1. *For every PCGS Γ with regular components, there is a g-system G, such that $L(G) = L(\Gamma)$. Moreover, there exists a constant c such that for each derivation of w of length n in Γ there exists a derivation of w in G of length at most $c \cdot n$.*

Corollary 1. *REG-PCGS-TIME($f(n)$) $\subseteq$ GTIME($f(n)$).*

3 The Context-Free PCGS Case

We shall now show that time preserving simulation of context-free PCGS by g-systems is not possible. First, let us note, that the approach we used for simulating regular PCGS does not work in the context-free case.

There are some technical difficulties, such as the fact that query symbol does not necessarily have to be at the end of a sentential form. But this is not a major problem. Let us consider a sentential form in which the simulating g-system is supposed to produce the sentential form of G_i at two places. Let it be of the form

$$\ldots \$_i \alpha \beta \alpha \beta \#_i \ldots \$_i \alpha \beta \alpha \beta \#_i \ldots .$$

To simulate a derivation step in the i-th component (e.g. using the rule $\alpha \to \gamma \beta$), we have to assure not just the use of the same rewriting rule, but also that it is applied to the same occurrence of the nonterminal α in both copies of the sentential form of G_i. In our example there are two occurrences α, but in general an arbitrary number of occurrences of a particular nonterminal in the sentential form is possible. Since our g-system has only finite memory, it cannot guarantee the same occurrence of α in each copy of the sentential form of G_i will be rewritten.

Having demonstrated the difficulty in attempting to use the same construction that worked for regular PCGS we shall now prove that time preserving simulation of context-free PCGS by g-systems is not possible. We shall proceed by presenting a language L that can be generated by context-free PCGS in time $O(\log n)$ while each g-system generating L needs time $\Omega(\log n \cdot \log \log n)$.

Clearly no infinite language is in CF-PCGS-TIME($f(n)$) for $f(n) = o(\log n)$. Choosing L in CF-PCGS-TIME($\log n$) thus guarantees that a PCGS generating L in $O(\log n)$ time is the best PCGS for L. Since g-systems also cannot generate words in an infinite language faster than in $O(\log n)$ steps we know that no g-system can generate L faster than this PCGS.

For reasons that will become apparent in the proof of the lower bound we shall choose L to be a suitable subset of $\{ww | w \in \{0,1\}^*\}$. The more words of any given length we shall have in L, the better lower bound for g-system we shall be able to prove.

Let us first concentrate on the problem of generating many words of the form ww in time $\log n$ by a context-free PCGS. First observe that the number of nonterminals occurring in a component of an r-sentential form that contributes to the generated word of length n cannot exceed $\log n$. This is due to the fact that in each derivation step at most one nonterminal in each component can be rewritten to a terminal word and the length of the derivation is $\log n$.

Using the idea from the regular PCGS from the Example 1 in the previous section it is possible to generate words of length n in $\log n$ steps. However, these words will consist of a large number of identical "short" substrings.

Since the number of terminal derivations of length $\log n$ for a given PCGS Γ is $O((\log n)^{\log n})$ we can generate at most $O((\log n)^{\log n})$ words of length n in $\log n$ step. In general, different derivations may generate the same word. We shall now present a context-free PCGS Γ generating $[\log n]!$ words of length $2n$ in $\{ww|w \in \{0,1\}^*\}$. These words will be of the form $u1^t u1^t$, where $[\log n]!$ different words u will be generated using $\log n$ nonterminals in the sentential forms and 1^t is a padding of appropriate length.

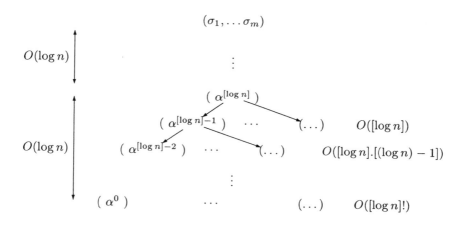

Fig. 2. The tree of al derivations of a PCGS for the "factorial" language working in $O(\log n)$ time.

Figure 2 shows a tree of all possible derivations of the desired PCGS. The PCGS works in the following way. First, it generates a nonterminal string $\alpha^{[\log n]}$ in $O(\log n)$ steps. This string is rewritten in the subsequent $O(\log n)$ steps to a terminal string so that each α is rewritten by a different terminal string (e.g. by $01^i 0$ in the i-th step). Note that these strings have to be different in order to ensure we obtain $O([\log n]!)$ different resulting words.

Now we can proceed to a formal description of this PCGS. Let Γ be a PCGS system with context-free components defined as follows.

$$\Gamma = (\{\sigma_1, \sigma_1', \sigma_1'', \sigma_2, \sigma_2', \sigma_3, \sigma_4, \sigma_5, \sigma_5', \sigma_5'', \sigma_5''', \sigma_6, \sigma_7, \sigma_8, \alpha, \beta, \beta', \gamma, \gamma', \omega\},$$
$$\{Q_1, Q_2, Q_3, Q_4, Q_5, Q_6, Q_7, Q_8\}, \{0,1\}, G_1, G_2, G_3, G_4, G_5, G_6, G_7, G_8)$$

where

$P_1 = \{\sigma_1 \to \sigma_1',$
$\quad \sigma_1' \to \sigma_1'',$
$\quad \sigma_1'' \to \sigma_1'',$
$\quad \sigma_1'' \to Q_2 Q_7 Q_2 Q_7,$
$\quad \sigma_5' \to 0\}$

$P_2 = \{\sigma_2 \to \sigma_2',$
$\quad \sigma_2' \to \sigma_2'\alpha,$
$\quad \sigma_2' \to Q_6,$
$\quad \alpha \to Q_5 Q_3,$
$\quad \beta \to 0\}$

$P_3 = \{\sigma_3 \to Q_4,$
$\quad \gamma \to \beta,$
$\quad \gamma' \to \beta'\}$

$P_4 = \{\sigma_4 \to \gamma',$
$\quad \sigma_4 \to 1 Q_3,$
$\quad \beta \to \gamma,$
$\quad \beta' \to \gamma\}$

$P_5 = \{\sigma_5 \to \sigma_5''',$
$\quad \sigma_5''' \to \sigma_5',$
$\quad \sigma_5'' \to \sigma_5'',$
$\quad \sigma_5'' \to \sigma_5''\}$

$P_6 = \{\sigma_6 \to \sigma_6,$
$\quad \sigma_6 \to Q_5,$
$\quad \sigma_5'' \to \sigma_5'\}$

$P_7 = \{\sigma_7 \to 1,$
$\quad \sigma_7 \to Q_8,$
$\quad \omega \to 1\}$

$P_8 = \{\sigma_8 \to Q_7 Q_7 \gamma,$
$\quad \gamma \to \omega\}$

Using a complex case analysis it is possible to show that the following holds.

Lemma 1. $L(\Gamma) = \{ww \mid w = 001^{\Phi(n)}0\ldots 01^{\Phi(3n-1)}01^{2^{(3n-1)+2}-1}, n \geq 2$ and Φ is a permutation of $n, \ldots, 3n-1\}$

Furthermore, based on the Γ above it is not difficult to see that the following two lemmas hold.

Lemma 2. $L(\Gamma) \in$ CF-PCGS-TIME$(O(\log n))$.

Lemma 3. Let $L(\Gamma)[n]$ be the set $L(\Gamma) \cap \{0,1\}^n = \{w \mid w \in L(\Gamma)$ and $|w| = n\}$. Then $|L(\Gamma)[n]| = \Omega(\lceil \log n^{\frac{1}{3}} \rceil !)$.

We shall now show that no g-system can generate $L(\Gamma)$ in $O(\log n)$ time. In proving the lower bound we shall exploit the lower bound result on nondeterministic communication complexity from [3] and the characterization of time for g-systems by one way nondeterministic space from [4].

Theorem 2 ([4]). *For every* $f(n) = \Omega(\log n)$ *it holds that*
$$\text{GTIME}(f(n)) = 1\text{NSPACE}(f(n))$$

Thus, in order to prove that $L \notin$ GTIME$(\log n)$ it suffices to show that $L \notin$ 1NSPACE$(\log n)$. To prove a lower bound for one-way nondeterministic space we use a result from [3] which gives a lower bound on 1NSPACE based on the nondeterministic communication complexity of the language.

First we need some notions from the communication complexity theory. Intuitively, the communication complexity measures the number of bits, that two unrestricted Turing machines need to exchange to accept a binary input (or to compute a Boolean function).

Initially each of the machines receives half of the input bits. We consider the worst case scenario, i.e., the most unfavorable splitting of the input for the two machines.

As shown in [3] the nondeterministic communication complexity of L given a particular distribution of input bits to the two machines, can be computed by using monochromatic matrix covers of a Boolean matrix representation of language L. We shall make this more precise now.

Let L be a language over the alphabet $\{0, 1\}$. Then, for any $n \in \mathbb{N}$, we define $h_{L,n}$ to be a Boolean function of n variables such that $h_{L,n}(\alpha) = 1$ iff $\alpha \in L[n] = L \cap \{0,1\}^n$.

Let f be a Boolean function of n variables in $X = \{x_1, \ldots, x_n\}$. Let $X_1, X_2 \subseteq X$, $X_1 \cup X_2 = X$, $X_1 \cap X_2 = \emptyset$, and let $\Pi(x) = b$ for every $x \in X_b$ ($b = 1, 2$). Π is called a *partition of* $\mathbf{X}$ (sometimes we write $\Pi = (X_1, X_2)$). Let $|X_1| = r, |X_2| = s$. Then the Boolean matrix $M(f, \Pi) = [a_{ij}]_{i=1,\ldots,2^r, j=1,\ldots,2^s}$, where $a_{ij} = f(\Pi^{-1}(\alpha_i, \beta_j))$ for the lexicographically i-th α_i in $\{0,1\}^r$ and the lexicographically j-th β_j in $\{0,1\}^s$ is the *matrix representation of f according to Π*.

Let $\Pi = (X_1, X_2)$ be a partition of $X = \{x_1, \ldots, x_n\}$. We say that Π is *balanced* if $||X_1| - |X_2|| \leq 1$.

Let $M = [a_{ij}]$ be a Boolean matrix of a size $n \times m$ for some $n, m \in \mathbb{N} - \{0\}$. Let $S_1 = \{i_1, i_2, \ldots, i_k\} \subseteq \{1, \ldots, n\}$, and $S_2 = \{j_1, j_2, \ldots, j_l\} \subseteq \{1, \ldots, m\}$. Let $M[S_1, S_2]$ denotes the $k \times l$ submatrix of M consisting exactly of the elements $[b_{rs}]_{r=1,\ldots,k, s=1,\ldots,l}$ for $b_{rs} = a_{i_r j_s}$.

Let M be a Boolean matrix, and let $M[S_1, R_1], \ldots, M[S_k, R_k]$ be some 1-monochromatic submatrices of M (not necessarily pairwise disjoint). We say $M[S_1, R_1], \ldots, M[S_k, R_k]$ *cover all ones of* M if each 1-element of M is also an element of some of the matrices $M[S_1, R_1], \ldots, M[S_k, R_k]$. Let $Cov(M)$ be the least natural number t such that all 1's of M can be covered by t 1-monochromatic submatrices.

Theorem 3 ([3]). *Let f be a Boolean function with a set of input variables X, and let Π be a balanced partition of X. Then the nondeterministic communication complexity of computing f given Π is $ncc(f, \Pi) = \lceil \log Cov(M(f, \Pi)) \rceil$.*

The reason for considering $L \subseteq \{ww | w \in \{0,1\}^*\}$ becomes apparent now. Choosing the balanced partition $\overline{\Pi}_n$ splitting the input to the first and second halves implies that all ones in the matrix representation of $L[n]$ appear at the diagonal of M. Thus, $Cov(M)$ is equal to the number of these 1's, i.e., the cardinality of $L[n]$. Using Lemma 3 we thus obtain:

Lemma 4. *Consider the $PCGS$ system Γ above. Then, for any $n \in \mathbb{N}$*

$$ncc(h_{L(\Gamma),n}, \overline{\Pi}_n) = \Omega(\log n \cdot \log \log n)$$

We are now ready to use the lower bound result form [3].

Theorem 4 ([3]). *For every $L \subseteq \{0,1\}^*$, $n \in \mathbb{N}$ and every on-line nondeterministic sequential machine M accepting L, it holds that*
$$ncc(h_{L,n}, \overline{\Pi}_n) \leq SPACE_M(n) + 1.$$

Combining this result, Lemma 4, Theorem 2 and Lemma 2 we obtain the main result of this section.

Theorem 5. *There is a language L, such that $L \in$ CF-PCGS-TIME$(\log n)$ and $L \notin$ GTIME$(\log n)$*

References

1. E. Csuhaj-Varjú, J. Dassow, J. Kelemen, Gh. Păun, *Grammar Systems: A Grammatical Approach to Distribution and Cooperation*, Gordon and Breach Science Publishers Ltd., London, 1994
2. B. Rovan, *A framework for studying grammars*, in Lecture notes in Computer Science 118, Springer Verlag, 1981, pp. 473-482
3. J. Hromkovič, *Communication Complexity and Parallel Computing*, Springer Verlag, 1997
4. P. Gvozdjak, B. Rovan, *Time-Bounded Parallel Rewriting*, in Development in Language Theory II (J. Dassow, G. Rozenberg, A. Salomaa eds.), World Scientific 1996, pp. 79 - 87.
5. Gh. Păun, L. Santean, Parallel communicating grammar systems: the regular case. *An. Univ. Buc., Ser. Mat.-Inf.* 38(1989), pp. 55-63.
6. E. Csuhaj-Varjú, Gy. Vaszil, An annotated bibliography of Grammar Systems. http://www.sztaki.hu/mms/bib.html
7. M. Slašťan, Parallel Communicating Grammar Systems, Master Thesis, Dept. of Computer Science Comenius University, 2000
   (http://www.dcs.fmph.uniba.sk/~rovan/thesis.ps)

String Rewriting Sequential P-Systems and Regulated Rewriting*

Petr Sosík[1] and Rudolf Freund[2]

[1] Institute of Computer Science
Silesian University
Opava, Czech Republic
petr.sosik@fpf.slu.cz

[2] Department of Computer Science
Technische Universität Wien
Wien, Austria
rudi@emcc.at

Abstract. We investigate the computational power of generalized P-systems of specific types in comparison with the computational power of certain control mechanisms for string rewriting grammars. An important restriction dwells in using sets of operators instead of multisets within sequential P-systems; this restriction is shown to be substantial, i.e., sequential P-systems of specific type using multisets are more powerful than the corresponding sequential P-systems using only sets.

1 Introduction

P-systems were introduced by Gheorghe Păun in [4] and [5] as membrane structures consisting of membranes hierarchically embedded in the outermost *skin* membrane. Every membrane encloses a *region* possibly containing other membranes; the part delimited by the membrane labelled by k and its inner membranes is called *compartment k*. A region delimited by a membrane not only may enclose other membranes but also specific objects and operators (used in the multiset sense), as well as evolution rules, which in *generalized P-systems (GP-systems)* as introduced in [3] are evolution rules for the operators. Besides operators working on (multisets of) objects, in GP-systems we also use ground operators as well as transfer operators allowing to transfer objects or operators (or even rules) either to the outer compartment or to an inner compartment delimited by a membrane of specific kind with also checking for some permitting and/or forbidding conditions on the objects to be transferred. In P-systems as introduced by Gheorghe Păun in [4] and [5], all objects are affected in parallel by the rules, whereas in GP-systems only one evolution rule is carried out in each step.

In the following section we give the defintions of grammars with specific control mechanisms (graph-controlled grammars, random context grammars), and then we describe the model of GP-systems as introduced in [3]. In the fourth section we show that

---

* Research supported by the Czech-Austrian Aktion projekt No. 28p9 and by the Grant Agency of Czech Republic, grant No. 201/98/P047

GP-systems of specific type have the same computational power as random context grammars without appearance checking. In the fifth section we compare the computational power of graph-controlled grammars and GP-systems using sets instead of multisets of objects and operators; finally, we elaborate an example of a language that can be generated by a GP-system using multisets, but not by a GP-system using only sets.

2 Definitions

For an alphabet V, by V^* we denote the free monoid generated by V under the operation of concatenation; the *empty string* is denoted by λ, and $V^* \setminus \{\lambda\}$ is denoted by V^+. Any subset of V^+ is called a *λ-free (string) language*. We say that two languages L_1, L_2 are equal if $L_1 \setminus \{\lambda\} = L_2 \setminus \{\lambda\}$. For more notions as well as basic results from the theory of formal languages, the reader is referred to [1].

In order to prove our results in a general setting, we use the following general notion of a grammar:

A *grammar* is a quadruple $G = (B, B_T, P, A)$, where B and B_T are sets of *objects* and *terminal objects*, respectively, with $B_T \subseteq B$, P is a finite set of *productions*, and $A \in B$ is the axiom. A production p in P in general is a partial recursive relation $\subseteq B \times B$, where we also demand that the domain of p is recursive (i.e., given $w \in B$ it is decidable if there exists some $v \in B$ with $(w,v) \in p$) and, moreover, that the range for every w is finite, i.e., for any $w \in B$, $card(\{v \in B \mid (w,v) \in p\}) < \infty$. The productions in P induce a derivation relation $\Longrightarrow_G$ on the objects in B; $\Longrightarrow_G^*$ denotes the reflexive and transitive closure of $\Longrightarrow_G$. The *language generated by G* is $L(G) = \{w \in B_T \mid A \Longrightarrow_G^* w\}$.

For example, a usual string grammar (V_N, V_T, P, S) in this general notion is written as $((V_N \cup V_T)^*, V_T^*, P, S)$. Such a string grammar is called to be of type RE, CF, and REG, if and only if all productions in P are arbitrary, context-free, and regular productions, respectively.

A *graph-controlled grammar (with appearance checking)* is a construct $G_C = (B, B_T, (R, L_{in}, L_{fin}), A)$; B and B_T are sets of *objects* and *terminal objects*, respectively, with $B_T \subseteq B$, $A \in B$ is the axiom; R is a finite set of rules r of the form $(l(r) : p(l(r)), \sigma(l(r)), \varphi(l(r)))$, where $l(r) \in Lab(G_C)$, $Lab(G_C)$ being a set of labels associated (in a one-to-one manner) to the rules r in R, $p(l(r))$ is a production over B, $\sigma(l(r)) \subseteq Lab(G_C)$ is the *success field* of the rule r, and $\varphi(l(r)) \subseteq Lab(G_C)$ is the *failure field* of the rule r; $L_{in} \subseteq Lab(G_C)$ is the set of initial labels, and $L_{fin} \subseteq Lab(G_C)$ is the set of final labels. For $r = (l(r) : p(l(r)), \sigma(l(r)), \varphi(l(r)))$ and $v, w \in B$ we define $(v, l(r)) \Longrightarrow_{G_C} (w, k)$ if and only if

- **either** $p(l(r))$ is applicable to v, the result of the application of the production $p(l(r))$ to v is w, and $k \in \sigma(l(r))$,
- **or** $p(l(r))$ is not applicable to v, $w = v$, and $k \in \varphi(l(r))$.

The language generated by G_C is

$$L(G_C) = \{w \in B_T \mid (w_0, l_0) \Longrightarrow_{G_C} (w_1, l_1) \Longrightarrow_{G_C} \ldots (w_k, l_k), k \geq 1, \\ w_j \in B \text{ and } l_j \in Lab(G_C) \text{ for } 0 \leq j \leq k, \\ w_0 = A, w_k = w, l_0 \in L_{in}, l_k \in L_{fin}\}.$$

If the failure fields $\varphi(l(r))$ are empty for all $r \in R$, then G_C is called a *graph-controlled grammar without appearance checking*. G_C is said to be of type X if the corresponding underlying grammar $G = (B, B_T, P, A)$, where $P = \{p(q) \mid q \in Lab(G_C)\}$, is of type X.

In the following we shall restrict ourselves to the case $B_T = V_T^*$ (where V_T is a set of terminal symbols) and $B = (V_T \cup V_N)^*$ (where V_N is a set of non-terminal symbols, $V_N \cap V_T = \emptyset$) in all types of grammars defined in this paper.

A *random context grammar (with appearance checking)* is a construct $G_{RC} = ((V_T \cup V_N)^*, V_T^*, P, A)$; V_T is a set of terminal symbols, V_N is a set of non-terminal symbols, $V_N \cap V_T = \emptyset$, $A \in (V_T \cup V_N)^*$ is the axiom; P is a finite set of rules r of the form $(\alpha \to \beta, Q, R)$, where $\alpha \to \beta$ is a rewriting rule over $V = V_N \cup V_T$, and Q, R are finite subsets of V. For $x, y \in V^*$ we write $x \Longrightarrow_G y$ if and only if there is a rule $(\alpha \to \beta, Q, R)$ in P such that $x = x'\alpha x''$, $y = x'\beta x''$ for some $x', x'' \in V^*$, all non-terminal symbols of Q appear in x and no non-terminal symbol of R appears in x. If every set R in the rules $(\alpha \to \beta, Q, R) \in P$ is empty, then G_{RC} is called a *random context grammar without appearance checking (without ac)*. G_{RC} is said to be of type X if the corresponding underlying grammar $G = ((V_T \cup V_N)^*, V_T^*, P', A)$, where $P' = \{\alpha \to \beta \mid (\alpha \to \beta, Q, R) \in P\}$, is of type X.

3 Generalized P-Systems (GP-Systems)

A *generalized P-system (GP-system) of type X* is a construct G_P of the form $G_P = (B, B_T, P, A, \mu, I, O, R, f)$ where

- (B, B_T, P, A) is a grammar of type X;
- μ is a membrane structure (with the membranes labelled by natural numbers $0, ..., n$);
- $I = (I_0, ..., I_n)$, where I_k is the initial contents of compartment k containing a finite multiset of objects from B as well as a finite multiset of operators from O and of rules from R;
- O is a finite set of operators (described in detail below);
- R is a finite set of evolution rules of the form

$$(op_1, ..., op_l; op'_1, ..., op'_m)$$

with $l \geq 1$ and $m \geq 0$, where $op_1, ..., op_l, op'_1, ..., op'_m$ are operators from O;
- $f \in \{0, 1, ..., n\}$ is the label of the final compartment.

The main power of GP-systems lies in the operators, which can be of the following types:

- $P \subseteq O$, i.e., the productions working on the objects from B are operators;
- $O_0 \subseteq O$, where O_0 is a finite set of special symbols, which are called *ground operators*;
- $Tr_{in} \subseteq O$, where Tr_{in} is a finite set of transfer operators on objects from B of the form $(\tau_{in,k}, E, F)$, $E \subseteq P$, $F \subseteq P$; the operator $(\tau_{in,k}, E, F)$ transfers an object w from B being in compartment m into compartment k provided

1. region m directly contains membrane k,
2. every production from E could be applied to w (hence, E is also called the permitting transfer condition),
3. no production from F can be applied to w (hence, F is also called the forbidding transfer condition);

- $Tr_{out} \subseteq O$, where Tr_{out} is a finite set of transfer operators on objects from B of the form (τ_{out}, E, F), $E \subseteq P, F \subseteq P$; the operator (τ_{out}, E, F) transfers an object w from B being in compartment m into compartment k provided
  1. region k directly contains membrane m,
  2. every production from E could be applied to w,
  3. no production from F can be applied to w;
- $Tr'_{in} \subseteq O$, where Tr'_{in} is a finite set of transfer operators working on operators from P, O_0, Tr_{in}, and Tr_{out} or even on rules from R; a transfer operator $\tau'_{in,k}$ moves such an element in compartment m into compartment k provided region m directly contains membrane k;
- $Tr'_{out} \subseteq O$, where Tr'_{out} is a finite set of transfer operators working on operators from P, O_0, Tr_{in}, and Tr_{out} or even on rules from R; a transfer operator τ'_{out} transfers such an element in compartment m into the surrounding compartment.

In sum, O is the disjoint union of P, O_0, and Tr, where Tr itself is the (disjoint) union of the sets of transfer operators Tr_{in}, Tr_{out}, Tr'_{in}, and Tr'_{out}. In the following we shall assume that the transfer operators in Tr'_{in}, and Tr'_{out} do not work on rules from R; hence, the distribution of the evolution rules is static and given by I. If in all transfer operators in Tr_{in} and Tr_{out} the permitting/forbidding sets are empty, then G_P is called a GP-system without permitting/forbidding transfer checking.

A *computation* in G_P starts with the initial configuration with I_k being the contents of compartment k. A transition from one configuration to another one is performed by evaluating one evolution rule $(op_1, ..., op_l; op'_1, ..., op'_m)$ in some compartment k, which means that the operators $op_1, ..., op_l$, are applied to suitable (non-deterministically chosen) elements in compartment k in the multiset sense, and we also obtain the operators $op'_1, ..., op'_m$ (in the multiset sense) in compartment k.

The language generated by G_P is the set of all terminal objects $w \in B_T$ obtained in the terminal compartment f at any time during a computation in G_P.

4 GP-Systems and Random Context Grammars

In this section we show that a certain class of GP-systems can be characterized by the class of random context grammars without ac.

Theorem 1 *Any random context grammar $G = (V^*, V_T^*, P, S)$ without ac of type X, $X \in \{CF, REG\}$, can be simulated by a GP-system with the same type of productions, without forbidding transfer checking and without transfer of operators, with the structure $[_0[_1]_1[_2]_2 \cdots [_n]_n]_0$, where $n = \mathrm{card}(P)$, and with evolution rules of the form $(r_i; r_i)$ only (and with only one object from $V \setminus V_T$ in I).*

Proof. Let $G_P = (V^*, V_T^*, P', S, [_0[_1]_1[_2]_2 \cdots [_n]_n]_0, I, O, R, 0)$, $V_N = V \setminus V_T$, and let $(\alpha_i \to \beta_i, Q_i, \emptyset)$ be the i-th production in P, $1 \leq i \leq n$. Then we define

$P' = \{\alpha_i \to \beta_i \mid 1 \leq i \leq n\}$,
$P'_i = \{B \to b \mid B \in Q_i\}$, $1 \leq i \leq n$ (for some fixed $b \in V_T$),
$O = \{(\tau_{in,i}, P'_i, \emptyset), \alpha_i \to \beta_i \mid 1 \leq i \leq n\} \cup \{(\tau_{out}, \emptyset, \emptyset)\}$,
$I_0 = \{S\} \cup \{(\tau_{in,i}, P'_i, \emptyset), ((\tau_{in,i}, P'_i, \emptyset); (\tau_{in,i}, P'_i, \emptyset)) \mid 1 \leq i \leq n\}$,
$I_i = \{\alpha_i \to \beta_i, (\alpha_i \to \beta_i; \alpha_i \to \beta_i)\} \cup$
$\quad \{(\tau_{out}, \emptyset, \emptyset), ((\tau_{out}, \emptyset, \emptyset); (\tau_{out}, \emptyset, \emptyset))\}$, $1 \leq i \leq n$,
$R = \{(p;p) \mid p \in O\}$.

The feature of checking for the appearance of special non-terminal symbols by the sets Q_i in the random context grammar G now is captured by the corresponding checks in the transfer operators of the GP-system G_P. For each production in P we use a seperate membrane; in the region enclosed by this membrane we can use the corresponding production. We have to point out that this production possibly can be applied even more than once, but the application of the production only consumes one specific non-terminal symbol, but does not change the occurrences of the other non-terminal symbols in the underlying string object, i.e., the applicability condition is not violated by applying the production more than once. Hence, we conclude $L(G) = L(G_P)$. □

Theorem 2 *Any GP-system $G_P = (V^*, V_T^*, P, A, \mu, I, O, R, f)$ of type X, $X \in \{CF, REG\}$, without forbidding transfer checking, without transfer of operators, with evolution rules of the form $(r_i; r_i)$, and with the initial objects in I only being from $V \setminus V_T$ can be simulated by a random context grammar of the same type without ac.*

Proof. Let G_P have $n+1$ compartments. Without loss of generality we may assume that an operator m is in I_i if and only if the rule $(m; m)$ is in I_i, $0 \leq i \leq n$ (otherwise the operator and the rule would not be applicable). We construct a random context grammar $G = ((V_T' \cup V_N')^*, (V_T')^*, P', S)$, simulating G_P. Notice that in G_P the evolution of each string is fully independent from the evolution of other strings. Hence G can non-deterministically choose an arbitrary string in I and then simulate its evolution.

(i) Let P have only regular productions. Then we take

$$V_N = V \setminus V_T, \quad V_N' = \{A_i \mid A \in V_N, 0 \leq i \leq n\} \cup \{S\}, \quad V_T' = V_T,$$

with new symbols A_i, S not in V_N, and P' contains the following two types of productions:

- $(S \to aD_j, \emptyset, \emptyset)$ for each production $C \to aD$ in I_j and an object $C \in I_i$, $0 \leq i, j \leq n$, such that either $i = j$ or else there is a transfer operator $(\tau_{in,j}, Q, \emptyset)$ or a transfer operator $(\tau_{out}, Q, \emptyset)$ in compartment i transferring the object C from compartment i to compartment j (where the application of the production $C \to aD$ is simulated) with $Q \subseteq \{C \to \beta \mid C \to \beta \in P\}$;

- $(C_i \to aD_j, \emptyset, \emptyset)$ for each production $C \to aD$ in I_j, $0 \le i,j \le n$, such that either $i = j$ or else there is a transfer operator $(\tau_{in,j}, Q, \emptyset)$ or a transfer operator $(\tau_{out}, Q, \emptyset)$ in compartment i transferring a string containing C from compartment i to compartment j with $Q \subseteq \{C \to \beta \mid C \to \beta \in P\}$;
- $(S \to a, \emptyset, \emptyset)$ for each production $C \to a$ in I_j and an object $C \in I_i$, $0 \le i,j \le n$, such that either $i = j$ or else there is a transfer operator $(\tau_{in,j}, Q, \emptyset)$ or a transfer operator $(\tau_{out}, Q, \emptyset)$ in compartment i transferring the object C from compartment i to compartment j (where the application of the production $C \to a$ is simulated) with $Q \subseteq \{C \to \beta \mid C \to \beta \in P\}$; moreover, there have to be transfer operators $(\tau'_1, \emptyset, \emptyset), \ldots, (\tau'_\ell, \emptyset, \emptyset)$, $\ell \ge 0$, in proper compartments of I, transferring subsequently a terminal object from compartment j to compartment f. Notice that a terminal string can satisfy only the empty permitting transfer condition.
- $(C_i \to a, \emptyset, \emptyset)$ for each production $C \to a$ in I_j, $0 \le i,j \le n$, such that either $i = j$ or else there is a transfer operator $(\tau_{in,j}, Q, \emptyset)$ or a transfer operator $(\tau_{out}, Q, \emptyset)$ in compartment i transferring the object C from compartment i to compartment j with $Q \subseteq \{C \to \beta \mid C \to \beta \in P\}$; moreover, there have to be transfer operators $(\tau'_1, \emptyset, \emptyset), \ldots, (\tau'_\ell, \emptyset, \emptyset)$, $\ell \ge 0$, in proper compartments of I, transferring subsequently a terminal object from compartment j to compartment f.

It follows from the description given above that $L(G) = L(G_P)$, and because G has only empty condition sets, it obviously is a random context grammar.

(ii) Let P also contain context-free productions. Then we take

$$V'_N = V_N \cup \{K_i \mid 0 \le i \le n\} \cup \{S\}, \quad V'_T = V_T \cup \{c\},$$

with new symbols K_i, S and c not in $V_N \cup V_T$, and P' contains the following types of productions:

- $(S \to XK_i, \emptyset, \emptyset)$ for each non-terminal symbol $X \in I_i$, $0 \le i \le n$;
- $(\alpha \to \beta, \{K_i\}, \emptyset)$ for each production $\alpha \to \beta$ in I_i, $0 \le i \le n$;
- $(K_i \to K_j, H, \emptyset)$ for each operator $(\tau_{in,j}, E, \emptyset)$ in I_i, $0 \le i,j \le n$, such that compartment i directly contains membrane j and $H = \{X \mid X \to \beta \in E\}$;
- $(K_i \to K_j, H, \emptyset)$ for each operator $(\tau_{out}, E, \emptyset)$ in I_i, $0 \le i,j \le n$, such that membrane i is directly contained in compartment j and $H = \{X \mid X \to \beta \in E\}$;
- $(K_f \to c, \emptyset, \emptyset)$.

From the description given above, we immediately obtain that G is a random context grammar without ac with $L(G) = L(G_P) \cdot \{c\}$. The family of languages generated by (λ-free) context-free random context grammars is closed under quotient with $\{c\}$, as shown in [1] (Corollary p. 47), which observation concludes the proof. $\square$

Corollary 3 *The class of GP-systems of type X, $X \in \{CF, REG\}$, without forbidding transfer checking, without transfer of operators, with evolution rules of the form $(r_i; r_i)$, and with the initial objects only being non-terminal symbols and the class of random context grammars of the same type without ac are computationally equivalent.*

As can be seen from the proofs elaborated above, the use of multisets in the GP-systems of such a restricted form as we used here is not important. The main power lies in the (transfer through the) membranes of the GP-systems, which act like filters. Hence, the family of languages generated by random context grammars without ac can also be characterized by GP-systems of the specific types described above, where in addition we do not take care of the number of objects and operators in the compartments. In fact, as we have no transfer of operators, what counts is only the existence or non-existence of an object in a compartment; it even makes no difference whether we assume an object to vanish from a compartment i when we transfer it to another one, or else we allow the object to remain in compartment i, too.

5 GP Systems and Graph-Controlled Grammars

In [3] it was shown that any graph-controlled grammar of arbitrary type can be simulated by a GP-system of the same type with the simple membrane structure $[_0[_1]_1[_2]_2]_0$. The following result improves [3], Theorem 1.

Theorem 4 *Any graph-controlled grammar of arbitrary type can be simulated by a GP-system of the same type with simple membrane structure $[_0[_1]_1[_2]_2]_0$, without permitting transfer conditions and with evolution rules of the (context-free) form $(op_1;)$, $(op_1; op_2)$ and $(op_1; op_2, op_3)$.*

Proof. Let $G_C = (B, B_T, (R, L_{in}, L_{fin}), A)$ be a graph-controlled grammar of type X and let $G = (B, B_T, P, A)$ be the corresponding underlying grammar of type X with $P = \{p(q) \mid q \in Lab(G_C)\}$.

Let $G_P = (B, B_T, P, A, [_0[_1]_1[_2]_2]_0, I, O, R', 1)$ be a GP-system, where O contains P, the set of ground operators $Lab(G_C)$, a special ground operator q_0, $q_0 \notin Lab(G_C)$, the primed set $\{q' \mid q \in Lab(G_C)\}$, and the transfer operators described below. Let I_0 contain the axiom A, the ground operator q_0 and the following rules:

1. for the initial configuration: $(q_0; q_{in})$ for each $q_{in} \in L_{in}$;
2. for the success case: $(q; p(q), r')$ for each $q \in Lab(G_C)$, $r \in \sigma(q)$, and $(p(q); \tau_{in,1})$;
3. for the failure case: $(q; (\tau_{in,2}, \emptyset, p(q)), s')$ for each $q \in Lab(G_C)$, $s \in \varphi(q)$, and $((\tau_{in,2}, \emptyset, \{p(q)\}); \tau_{in,1})$;
4. $(\tau_{in,1};)$ for both the success and failure cases mentioned above;
5. for terminal objects: $(q_{fin}; (\tau_{in,1}, \emptyset, \emptyset))$ for each $q_{fin} \in L_{fin}$, and $((\tau_{in,1}, \emptyset, \emptyset); (\tau_{in,1}, \emptyset, \emptyset))$.

Moreover, let

$$I_1 = \{\tau_{out}, (\tau_{out}; \tau_{out})\} \cup \{(q'; q) \mid q \in Lab(G_C)\},$$
$$I_2 = \{(\tau_{out}, \emptyset, \emptyset), ((\tau_{out}, \emptyset, \emptyset); (\tau_{out}, \emptyset, \emptyset))\}.$$

The contents of R follows from the description of I.

Any derivation step $(v,l) \implies (w,k)$ of G_C is simulated by G_P in the following way: At the beginning, the object v and the operator l are present in compartment 0. Then either the rules in 2 or 3 are chosen non-deterministically and applied (notice that these rules cannot be mixed). In the success case, v is changed to w via the production $p(l)$, and also k' is generated. Then k' is transferred into compartment 1, from where it returns as k. In the failure case, v is transferred into compartment 2 in such a way that the non-applicability of $p(l)$ is checked, from where it immediately returns back. Finally, evaluating $(\tau_{in,1};)$ transfers k' into compartment 1, where it is replaced by k and transferred back. At the end, a terminal object is transferred into the final compartment 1 via the rules in 5. Hence, we conclude $L(G_C) = L(G_P)$. □

In the case of non-ac, we need neither transfer conditions nor compartment 2. The following corollary shows that in the string rewriting case the infinite hierarchy in [3], Theorem 2 collapses (in [3], for each node of the control graph, a seperate compartment was needed).

Corollary 5 *Any graph-controlled grammar of an arbitrary type without ac can be simulated by a GP-system of the same type without transfer conditions, with the structure $[_0[_1]_1]_0$ and with rules of the form $(op_1;)$, $(op_1;op_2)$ and $(op_1;op_2,op_3)$.*

As can be seen from the proofs given above, the number of occurrences of objects and operators in a compartment does not count. For each compartment k and each element $x \in B \cup O \cup R$, it is only important whether x is or is not present in compartment k. Hence, in the rest of this section we consider GP-systems, where the number of occurrences of objects and operators in a compartment does not count, and we call such systems *non-counting GP-systems*. Of course, the same element x can be present in more compartments simultaneously, but we do not distinguish whether there is one or more occurrences of x in a particular compartment. If x is transferred into a compartment where x is already present or produced in a further copy by a production, then the contents of the compartment does not change. If $x \in B$ is consumed by the application of a production or $x \in O$ is consumed by the application of an evolution rule or a transfer operator, then x is no more present in the compartment, until it again appears due to the application of a production or a transfer operator.

Because lack of space, for a proof of the following important results we have to refer the interested reader to a longer version of this paper (see [6]).

Theorem 6 *Any non-counting GP-system with forbidding transfer conditions but without permitting transfer conditions and with the single object A in I can be simulated by a graph-controlled grammar of the same type with ac.*

Corollary 7 *The class of non-counting GP-systems of arbitrary type with forbidding transfer conditions but without permitting transfer conditions and with the single object A in I, and the class of graph-controlled grammars of the same type with ac are computationally equivalent.*

Corollary 8 *Any non-counting GP-system G_P of arbitrary type with forbidding transfer conditions but without permitting transfer conditions and with the single object A in I can be simulated by a non-counting GP system G'_P without permitting transfer conditions with the simple membrane structure $[_0[_1]_1[_2]_2]_0$ and with evolution rules of the form $(op_1;)$, $(op_1; op_2)$ and $(op_1; op_2, op_3)$.*

The following theorem shows that GP-systems (with multisets of operators) of a certain type are strictly more powerful than the corresponding non-counting GP-systems, and hence the equivalence result from Corollary 7 cannot be extended to the multiset case.

Theorem 9 *There is a GP-system with regular productions, with forbidding transfer conditions but without permitting transfer conditions, with a single axiom in I (and with multisets of operators), which cannot be simulated by a non-counting GP-system with the same limitations.*

Proof. Consider the following GP system with multisets of operators (multisets are enclosed in brackets $[,]$); the contents of O and R is determined by the (multi)sets described below:

$G_P = (\{A, B, a, b\}, \{a, b\}, P, AB, [_0[_1]_1[_2]_2]_0, I, O, R, 2)$, where

$P = \{A \to aA, B \to bB, A \to a, B \to b\}$,

$O_0 = \{q_{A0}, q_{A1}, q_{B0}, q_{B1}\}$,

$I_0 = [AB, q_{A0}, q_{B0}, \tau_{in,1}, (q_{B0}; B \to bB), (B \to bB; q_{B1}),$
$(\tau_{in,1}; \tau_{in,1}), (\tau_{in,2}, \emptyset, \emptyset),$
$(q_{A0}; (\tau_{in,1}, \emptyset, \emptyset)), ((\tau_{in,1}, \emptyset, \emptyset); q_{A1}), ((\tau_{in,2}, \emptyset, \emptyset);)]$,

$I_1 = [\tau_{out}, (q_{B1}; B \to bB), (B \to bB; q_{B0}, q_{B0}), (\tau_{out}; \tau_{out}),$
$(q_{A1}; A \to aA), (A \to aA; (\tau_{out}, \emptyset, \emptyset)), ((\tau_{out}, \emptyset, \emptyset); q_{A0})]$,

$I_2 = [A \to a, B \to b, (A \to a;), (B \to b;)]$.

The language generated by G_P is $\{a^n b^m \mid m \leq 2^n, n \geq 1\}$.

On the other hand, this language cannot be generated by a non-counting GP-system with the additional limitations stated above: Such a system is computationally equivalent to a graph-controlled grammar with appearance checking and with regular productions, see Corollary 7. This graph-controlled grammar is further computationally equivalent to a matrix grammar with appearance checking, regular productions and with an arbitrary string as an axiom; such a matrix grammar is a special case of a contex-free matrix grammar of finite index, which can generate only semilinear languages, see Theorem 3.1.2 and 3.1.5 in [1]. The language $L(G_P)$ is clearly non-semilinear, which concludes the proof. □

References

1. Dassow, J., Păun, Gh.: Regulated Rewriting in Formal Language Theory. Springer-Verlag, Berlin Heidelberg New York (1989)

2. Dassow, J., Păun, Gh.: On the power of membrane computing. Journal of Universal Computer Science **5**, 2 (1999) 33–49 (http://www.iicm.edu/jucs)
3. Freund, R.: Generalized P-systems. Fundamentals of Computation Theory, FCT'99, Iasi, 1999. In: Ciobanu, G., Păun, Gh. (eds.): Lecture Notes in Computer Science, Vol. 1684. Springer-Verlag, Berlin Heidelberg New York (1999) 281–292
4. Păun, Gh.: Computing with Membranes. Journal of Computer and System Sciences **61**, 1 (2000) 108–143
5. Păun, Gh.: Computing with Membranes: An Introduction. Bulletin EATCS **67** (Febr. 1999) 139–152
6. Sosík, P., Freund, R.: String Rewriting Sequential P-systems and Regulated Rewriting. Techn. Report, TU Wien (2001).

Author Index

Ananichev, D.S. 166

Beaudry, M. 340

Cachat, T. 145
Calude, C.S. 1
Calude, E. 1
Câmpeanu, C. 186
Choffrut, C. 15
Csuhaj-Varjú, E. 359
Culik, K. 175

Domaratzki, M. 319

Eiter, T. 37
Engelfriet, J. 228
Ésik, Z. 21, 217

Fernau, H. 281
Freund, R. 379

Glaßer, C. 251
Gottlob, G. 37

Halava, V. 304
Harju, T. 57, 304
Holzer, M. 340
Hoogeboom, H.J. 293
Hromkovič, J. 262

Ito, M. 69

Karhumäki, J. 175
Kari, J. 175
Kudlek, M. 311
Kunimoch, Y. 69
Kuske, D. 206

Lischke, G. 272

Maneth, S. 228
Mignosi, F. 349

Németh, Z.L. 217
Niemann, G. 197, 330, 340

Otto, F. 340

Prodinger, H. 81

Razborov, A.A. 100
Restivo, A. 117, 349
Rogozhin, Y. 311
Rovan, B. 369

Salemi, S. 117
Salomaa, K. 186
Schmitz, H. 251
Schnitger, G. 262
Schwentick, T. 37, 239
Sciortino, M. 349
Shallit, J. 319
Slašt'an, M. 369
Sosík, P. 379
Staiger, L. 155
Stiebe, R. 281

Thérien, D. 239
Thomas, W. 130

Vágvölgyi, S. 186
Vaszil, G. 359
Volkov, M.V. 166
Vollmer, H. 239

Waldmann, J. 330
Woinowski, J.R. 197

Yu, S. 319

Lecture Notes in Computer Science

For information about Vols. 1–2235
please contact your bookseller or Springer-Verlag

Vol. 2236: K. Drira, A. Martelli, T. Villemur (Eds.), Co-operative Environments for Distributed Systems Engineering. IX, 281 pages. 2001.

Vol. 2237: P. Codognet (Ed.), Logic Programming. Proceedings, 2001. XI, 365 pages. 2001.

Vol. 2238: G.D. Hager, H.I. Christensen, H. Bunke, R. Klein (Eds.), Sensor-Based Intelligent Robots. Proceedings, 2000. VIII, 375 pages. 2002.

Vol. 2239: T. Walsh (Ed.), Principles and Practice of Constraint Programming – CP 2001. Proceedings, 2001. XIV, 788 pages. 2001.

Vol. 2240: G.P. Picco (Ed.), Mobile Agents. Proceedings, 2001. XIII, 277 pages. 2001.

Vol. 2241: M. Jünger, D. Naddef (Eds.), Computational Combinatorial Optimization. IX, 305 pages. 2001.

Vol. 2242: C.A. Lee (Ed.), Grid Computing – GRID 2001. Proceedings, 2001. XII, 185 pages. 2001.

Vol. 2243: G. Bertrand, A. Imiya, R. Klette (Eds.), Digital and Image Geometry. VII, 455 pages. 2001.

Vol. 2244: D. Bjørner, M. Broy, A.V. Zamulin (Eds.), Perspectives of System Informatics. Proceedings, 2001. XIII, 548 pages. 2001.

Vol. 2245: R. Hariharan, M. Mukund, V. Vinay (Eds.), FST TCS 2001: Foundations of Software Technology and Theoretical Computer Science. Proceedings, 2001. XI, 347 pages. 2001.

Vol. 2246: R. Falcone, M. Singh, Y.-H. Tan (Eds.), Trust in Cyber-societies. VIII, 195 pages. 2001. (Subseries LNAI).

Vol. 2247: C. P. Rangan, C. Ding (Eds.), Progress in Cryptology – INDOCRYPT 2001. Proceedings, 2001. XIII, 351 pages. 2001.

Vol. 2248: C. Boyd (Ed.), Advances in Cryptology – ASIACRYPT 2001. Proceedings, 2001. XI, 603 pages. 2001.

Vol. 2249: K. Nagi, Transactional Agents. XVI, 205 pages. 2001.

Vol. 2250: R. Nieuwenhuis, A. Voronkov (Eds.), Logic for Programming, Artificial Intelligence, and Reasoning. Proceedings, 2001. XV, 738 pages. 2001. (Subseries LNAI).

Vol. 2251: Y.Y. Tang, V. Wickerhauser, P.C. Yuen, C.Li (Eds.), Wavelet Analysis and Its Applications. Proceedings, 2001. XIII, 450 pages. 2001.

Vol. 2252: J. Liu, P.C. Yuen, C. Li, J. Ng, T. Ishida (Eds.), Active Media Technology. Proceedings, 2001. XII, 402 pages. 2001.

Vol. 2253: T. Terano, T. Nishida, A. Namatame, S. Tsumoto, Y. Ohsawa, T. Washio (Eds.), New Frontiers in Artificial Intelligence. Proceedings, 2001. XXVII, 553 pages. 2001. (Subseries LNAI).

Vol. 2254: M.R. Little, L. Nigay (Eds.), Engineering for Human-Computer Interaction. Proceedings, 2001. XI, 359 pages. 2001.

Vol. 2255: J. Dean, A. Gravel (Eds.), COTS-Based Software Systems. Proceedings, 2002. XIV, 257 pages. 2002.

Vol. 2256: M. Stumptner, D. Corbett, M. Brooks (Eds.), AI 2001: Advances in Artificial Intelligence. Proceedings, 2001. XII, 666 pages. 2001. (Subseries LNAI).

Vol. 2257: S. Krishnamurthi, C.R. Ramakrishnan (Eds.), Practical Aspects of Declarative Languages. Proceedings, 2002. VIII, 351 pages. 2002.

Vol. 2258: P. Brazdil, A. Jorge (Eds.), Progress in Artificial Intelligence. Proceedings, 2001. XII, 418 pages. 2001. (Subseries LNAI).

Vol. 2259: S. Vaudenay, A.M. Youssef (Eds.), Selected Areas in Cryptography. Proceedings, 2001. XI, 359 pages. 2001.

Vol. 2260: B. Honary (Ed.), Cryptography and Coding. Proceedings, 2001. IX, 416 pages. 2001.

Vol. 2261: F. Naumann, Quality-Driven Query Answering for Integrated Information Systems. X, 166 pages. 2002.

Vol. 2262: P. Müller, Modular Specification and Verification of Object-Oriented Programs. XIV, 292 pages. 2002.

Vol. 2263: T. Clark, J. Warmer (Eds.), Object Modeling with the OCL. VIII, 281 pages. 2002.

Vol. 2264: K. Steinhöfel (Ed.), Stochastic Algorithms: Foundations and Applications. Proceedings, 2001. VIII, 203 pages. 2001.

Vol. 2265: P. Mutzel, M. Jünger, S. Leipert (Eds.), Graph Drawing. Proceedings, 2001. XV, 524 pages. 2002.

Vol. 2266: S. Reich, M.T. Tzagarakis, P.M.E. De Bra (Eds.), Hypermedia: Openness, Structural Awareness, and Adaptivity. Proceedings, 2001. X, 335 pages. 2002.

Vol. 2267: M. Cerioli, G. Reggio (Eds.), Recent Trends in Algebraic Development Techniques. Proceedings, 2001. X, 345 pages. 2001.

Vol. 2268: E.F. Deprettere, J. Teich, S. Vassiliadis (Eds.), Embedded Processor Design Challenges. VIII, 327 pages. 2002.

Vol. 2269: S. Diehl (Ed.), Software Visualization. Proceedings, 2001. VIII, 405 pages. 2002.

Vol. 2270: M. Pflanz, On-line Error Detection and Fast Recover Techniques for Dependable Embedded Processors. XII, 126 pages. 2002.

Vol. 2271: B. Preneel (Ed.), Topics in Cryptology – CT-RSA 2002. Proceedings, 2002. X, 311 pages. 2002.

Vol. 2272: D. Bert, J.P. Bowen, M.C. Henson, K. Robinson (Eds.), ZB 2002: Formal Specification and Development in Z and B. Proceedings, 2002. XII, 535 pages. 2002.

Vol. 2273: A.R. Coden, E.W. Brown, S. Srinivasan (Eds.), Information Retrieval Techniques for Speech Applications. XI, 109 pages. 2002.

Vol. 2274: D. Naccache, P. Paillier (Eds.), Public Key Cryptography. Proceedings, 2002. XI, 385 pages. 2002.

Vol. 2275: N.R. Pal, M. Sugeno (Eds.), Advances in Soft Computing – AFSS 2002. Proceedings, 2002. XVI, 536 pages. 2002. (Subseries LNAI).

Vol. 2276: A. Gelbukh (Ed.), Computational Linguistics and Intelligent Text Processing. Proceedings, 2002. XIII, 444 pages. 2002.

Vol. 2277: P. Callaghan, Z. Luo, J. McKinna, R. Pollack (Eds.), Types for Proofs and Programs. Proceedings, 2000. VIII, 243 pages. 2002.

Vol. 2278: J.A. Foster, E. Lutton, J. Miller, C. Ryan, A.G.B. Tettamanzi (Eds.), Genetic Programming. Proceedings, 2002. XI, 337 pages. 2002.

Vol. 2279: S. Cagnoni, J. Gottlieb, E. Hart, M. Middendorf, G.R. Raidl (Eds.), Applications of Evolutionary Computing. Proceedings, 2002. XIII, 344 pages. 2002.

Vol. 2280: J.P. Katoen, P. Stevens (Eds.), Tools and Algorithms for the Construction and Analysis of Systems. Proceedings, 2002. XIII, 482 pages. 2002.

Vol. 2281: S. Arikawa, A. Shinohara (Eds.), Progress in Discovery Science. XIV, 684 pages. 2002. (Subseries LNAI).

Vol. 2282: D. Ursino, Extraction and Exploitation of Intensional Knowledge from Heterogeneous Information Sources. XXVI, 289 pages. 2002.

Vol. 2283: T. Nipkow, L.C. Paulson, M. Wenzel, Isabelle/HOL. XIII, 218 pages. 2002.

Vol. 2284: T. Eiter, K.-D. Schewe (Eds.), Foundations of Information and Knowledge Systems. Proceedings, 2002. X, 289 pages. 2002.

Vol. 2285: H. Alt, A. Ferreira (Eds.), STACS 2002. Proceedings, 2002. XIV, 660 pages. 2002.

Vol. 2286: S. Rajsbaum (Ed.), LATIN 2002: Theoretical Informatics. Proceedings, 2002. XIII, 630 pages. 2002.

Vol. 2287: C.S. Jensen, K.G. Jeffery, J. Pokorny, Saltenis, E. Bertino, K. Böhm, M. Jarke (Eds.), Advances in Database Technology – EDBT 2002. Proceedings, 2002. XVI, 776 pages. 2002.

Vol. 2288: K. Kim (Ed.), Information Security and Cryptology – ICISC 2001. Proceedings, 2001. XIII, 457 pages. 2002.

Vol. 2289: C.J. Tomlin, M.R. Greenstreet (Eds.), Hybrid Systems: Computation and Control. Proceedings, 2002. XIII, 480 pages. 2002.

Vol. 2291: F. Crestani, M. Girolami, C.J. van Rijsbergen (Eds.), Advances in Information Retrieval. Proceedings, 2002. XIII, 363 pages. 2002.

Vol. 2292: G.B. Khosrovshahi, A. Shokoufandeh, A. Shokrollahi (Eds.), Theoretical Aspects of Computer Science. IX, 221 pages. 2002.

Vol. 2293: J. Renz, Qualitative Spatial Reasoning with Topological Information. XVI, 207 pages. 2002. (Subseries LNAI).

Vol. 2295: W. Kuich, G. Rozenberg, A. Salomaa (Eds.), Developments in Language Theory. Proceedings, 2001. IX, 389 pages. 2002.

Vol. 2296: B. Dunin-Kęplicz, E. Nawarecki (Eds.), From Theory to Practice in Multi-Agent Systems. Proceedings, 2001. IX, 341 pages. 2002. (Subseries LNAI).

Vol. 2299: H. Schmeck, T. Ungerer, L. Wolf (Eds.), Trends in Network and Pervasive Computing – ARCS 2002. Proceedings, 2002. XIV, 287 pages. 2002.

Vol. 2300: W. Brauer, H. Ehrig, J. Karhumäki, A. Salomaa (Eds.), Formal and Natural Computing. XXXVI, 431 pages. 2002.

Vol. 2301: A. Braquelaire, J.-O. Lachaud, A. Vialard (Eds.), Discrete Geometry for Computer Imagery. Proceedings, 2002. XI, 439 pages. 2002.

Vol. 2302: C. Schulte, Programming Constraint Services. XII, 176 pages. 2002. (Subseries LNAI).

Vol. 2303: M. Nielsen, U. Engberg (Eds.), Foundations of Software Science and Computation Structures. Proceedings, 2002. XIII, 435 pages. 2002.

Vol. 2304: R.N. Horspool (Ed.), Compiler Construction. Proceedings, 2002. XI, 343 pages. 2002.

Vol. 2305: D. Le Métayer (Ed.), Programming Languages and Systems. Proceedings, 2002. XII, 331 pages. 2002.

Vol. 2306: R.-D. Kutsche, H. Weber (Eds.), Fundamental Approaches to Software Engineering. Proceedings, 2002. XIII, 341 pages. 2002.

Vol. 2307: C. Zhang, S. Zhang, Association Rule Mining. XII, 238 pages. 2002. (Subseries LNAI).

Vol. 2308: I.P. Vlahavas, C.D. Spyropoulos (Eds.), Methods and Applications of Artificial Intelligence. Proceedings, 2002. XIV, 514 pages. 2002. (Subseries LNAI).

Vol. 2309: A. Armando (Ed.), Frontiers of Combining Systems. Proceedings, 2002. VIII, 255 pages. 2002. (Subseries LNAI).

Vol. 2310: P. Collet, C. Fonlupt, J.-K. Hao, E. Lutton, M. Schoenauer (Eds.), Artificial Evolution. Proceedings, 2001. XI, 375 pages. 2002.

Vol. 2311: D. Bustard, W. Liu, R. Sterritt (Eds.), SoftWare 2002: Computing in an Imperfect World. Proceedings, 2002. XI, 359 pages. 2002.

Vol. 2313: C.A. Coello Coello, A. de Albornoz, L.E. Sucar, O.Cairó Battistutti (Eds.), MICAI 2002: Advances in Artificial Intelligence. Proceedings, 2002. XIII, 548 pages. 2002. (Subseries LNAI).

Vol. 2314: S.-K. Chang, Z. Chen, S.-Y. Lee (Eds.), Recent Advances in Visual Information Systems. Proceedings, 2002. XI, 323 pages. 2002.

Vol. 2315: F. Arhab, C. Talcott (Eds.), Coordination Models and Languages. Proceedings, 2002. XI, 406 pages. 2002.

Vol. 2318: D. Bošnački, S. Leue (Eds.), Model Checking Software. Proceedings, 2002. X, 259 pages. 2002.

Vol. 2319: C. Gacek (Ed.), Software Reuse: Methods, Techniques, and Tools. Proceedings, 2002. XI, 353 pages. 2002.

Vol. 2322: V. Mařík, O. Štěpánková, H. Krautwurmová, M. Luck (Eds.), Multi-Agent Systems and Applications II. Proceedings, 2001. XII, 377 pages. 2002. (Subseries LNAI).